TELECOMMUNICATIONS PRIMER

DATA, VOICE AND VIDEO COMMUNICATIONS

SECOND EDITION

ISBN 0-13-022155-4

9 780130 221551

90000

Feher/Prentice Hall Digital and Wireless Communication Series

Carne, E. Bryan. Telecommunications Primer: Data, Voice and Video Communications

Feher, Kamilo. Wireless Digital Communications: Modulation and Spread
Spectrum Applications

Pelton, N. Joseph. Wireless Satellite Telecommunications: The Technology,
the Market & the Regulations

Other Books by Dr. Kamilo Feher

Advanced Digital Communications: Systems and Signal Processing Techniques

Telecommunications Measurements, Analysis and Instrumentation

Digital Communications: Satellite/Earth Station Engineering

Digital Communications: Microwave Applications

Available from CRESTONE Engineering Books, c/o G. Breed, 5910 S. University
Blvd., Bldg. C-18 #360, Littleton, CO 80121, Tel. 303-770-4709,
Fax 303-721-1021, or from DIGCOM, Inc., Dr. Feher and Associates,
44685 Country Club Drive, El Macero, CA 95618, Tel. 916-753-0738,
Fax 916-753-1788.

TELECOMMUNICATIONS PRIMER

DATA, VOICE AND VIDEO COMMUNICATIONS

SECOND EDITION

E. BRYAN CARNE

Prentice Hall PTR
Upper Saddle River, New Jersey 07458
http://www.phptr.com

Library of Congress Cataloging-in-Publication Data

Carne, E. Bryan
    Telecommunications primer: data, voice and video communications
/E. Bryan Carne. — 2nd ed.
        p.   cm. — (Feher/Prentice Hall digital and wireless
communication series)
    Includes bibliographical references and index.
    ISBN 0-13-022155-4
    1. Telecommunication. I. Title. II. Series.
TK5105.C29863 1999
621.382—dc21

                                                                99-20391
                                                                CIP

Editorial/Production Supervision: *Kerry Reardon*
Editor in Chief: *Bernard Goodwin*
Buyer: *Alan Fischer*
Cover Design: *Talar Agasyan*
Cover Design Direction: *Jerry Votta*
Marketing Manager: *Bryan Gambrel*
Editorial Assistant: *Diane Spina*

© 1999 Prentice Hall PTR
Prentice-Hall, Inc.
Upper Saddle River, NJ 07458

Prentice Hall books are widely used by corporations and government agencies for training, marketing, and resale.

The publisher offers discounts on this book when ordered in bulk quantities. For more information, call the Corporate Sales Department at 800-382-3419; FAX: 201-236-714, email corpsales@prenhall.com or write Corporate Sales Department, Prentice Hall PTR, One Lake Street, Upper Saddle River, NJ 07458

Printed in the United States of America

10   9   8   7   6   5   4   3   2

ISBN 0-13-022155-4

Prentice-Hall International (UK) Limited, *London*
Prentice-Hall of Australia Pty. Limited, *Sydney*
Prentice-Hall Canada Inc., *Toronto*
Prentice-Hall Hispanoamericana, S.A., *Mexico*
Prentice-Hall of India Private Limited, *New Delhi*
Prentice-Hall of Japan, Inc., *Tokyo*
Prentice-Hall (Singapore) Pte. Ltd., *Singapore*
Editora Prentice-Hall do Brasil, Ltda., *Rio de Janeiro*

To JOAN

CONTENTS

SIGNALS

BUILDING BLOCKS

7 DATA COMMUNICATION   **289**

NETWORKS

PREFACE

This, the second edition of *Telecommunications Primer*, has the same objective as the first—to be a sourcebook on telecommunication facilities and protocols for knowledge workers. Published in 1995, the first edition is fast becoming out-of-date in many areas. In the past three or four years, major advances have been made in technology, concepts, and operations. Among them are

- **Growth of Internet:** from a network connecting some 50,000 networks containing 3 million hosts in 1994, Internet has grown rapidly. In 1998, it was estimated to connect some 2 million networks containing 40 million hosts. The growth of Internet has stimulated governments to plan national and global information infrastructures (NIIs and GIIs). Also, commercial users have invented intranets and extranets to share private information among employees, customers, and suppliers.

- **Scalability and Network Management:** the rapid growth of data traffic has caused all network providers to review their options for converting operations from a centralized format that is growth-limited, to a decentralized format that can grow indefinitely. As part of this strategy, they are establishing distributed management systems with automated agents.

- **Deployment of Digital Cellular Radio Systems:** North American time-division and code division multiple access systems (TDMA and CDMA) have been deployed that are compatible with earlier FM-AMPS. In Europe, Global System for Mobile Communications (GSM) has been expanded and updated. In Europe, Japan, and the United States, personal communication systems are being deployed in the 1.9-GHz personal communication services (PCS) band.

- **Medium- and Low-Earth Orbit Satellite Constellations (MEO and LEO):** several ambitious multisatellite systems are being deployed to provide worldwide communication services to mobile customers.

- **Optical Fiber Cables Encircle the Globe:** individual undersea cable systems that span the Atlantic, Indian, and Pacific basins have been interconnected. In addition, undersea cable systems are being deployed around Africa and South America. They all employ wavelength-division multiplexing. As a related matter, most fiber-optic network operators experienced *fiber exhaust* in 1977 and have adopted wave length division multiplexers (WDM) to increase capacity without laying new fibers.

- **Asynchronous Transfer Mode:** as the need for high-speed switching has become real, ATM has overcome early problems associated with flow control (and other operations) and is the heir apparent to the next generation switching machine.

- **Digital Video:** MPEG standards have created bandwidth efficient digital video and television signals. It has been deployed in direct broadcasting satellites and in terrestrial, high-definition television service.

- **1996 Telecommunications Act:** among other things, the Act opens local service to competition, allows local exchange companies to enter the long-distance market, permits cable companies to offer telephone services, and permits telephone companies to offer television services.

Without doubt, on the eve of the second millennium, the first edition does not describe all of the technology of importance to the practicing knowledge engineer. To cover the new environment, new material has been added, existing material has been updated to reflect the developments of the last three to four years, and the text has been reorganized. Principally, the changes and additions are

INTRODUCTION

- Expansion of this section to two chapters to bring the soft topics together in one place

SIGNALS

- Additional discussion of several topics including: noise and the creation of errors, 2B1Q pulse formats, scrambling and unscrambling, Shannon-Fano coding, MPEG video coding and digital television, spread spectrum modulation, etc.

BUILDING BLOCKS

- Addition of new chapter describing the characteristics of common bearers—wire cables, optical fibers (including optical amplifiers and wavelength-division multiplexing), cellular radios, and communication satellites

- Addition of new chapter focusing on multiplexing, digital subscriber lines, and hybrid voice, data, and video connections

- Expanded discussion of open systems architecture and network management procedures of the International Standards Organization (ISO)

- Expanded discussion of the relationship among packet switching, frame relay, and cell relay (SMDS, ATM)

- Additional discussion of modems, particularly higher-speed modems and cable modems
- Expanded discussion of automatic-repeat-request (ARQ) error correction, forward error correction coding and throughput

NETWORKS

- Addition of new chapter on transfer modes to emphasize the interdependence of switching/routing and multiplexing, including expanded discussion of relation among X.25, frame relay, SMDS, and ATM networks
- Inclusion of traffic engineering in discussion of local and long-distance networks
- Discussion of the use of ATM for high-speed networks, television distribution, and local-area networks (LANs)
- Description of GEO, MEO, and LEO satellite systems for mobile communications
- Additional discussion of NA-AMPS and GSM cellular radio systems
- New description of NA-TDMA and NA-CDMA cellular radio systems
- Additional description of PCS systems
- New chapter on LANs, MANs, and Internet; includes description of the operation of routers and gateways, intranets and extranets, and new initiatives such as Internet2 and vBNS
- Addition of discussion of national and global information infrastructure initiatives
- Inclusion of SNA in enterprise network discussion, including subarea and APPN operation
- Discussion of network management systems including SNMP, and management of non-OSI devices.

The result is a 40% larger book, with over 100 new diagrams. In addition, I have included review questions. Because my intent is to remind you of the information contained in each Section, they follow the sequence of the text and should present no difficulty for the reader.

For an author, the publication of a new edition is exciting. In my case it is tempered by the realization that it will all have to be done again in three or four years as the tempo of change continues to increase. I do hope you find the second edition of *Telecommunications Primer* a worthy addition to your technical library.

E. Bryan Carne
Peterborough, NH.

1

CONTEMPORARY COMMUNICATION

Contemporary telecommunications facilities augment human capabilities. In doing so, they have affected the scale and rhythm of our lives and become essential to the modern, global economy, on which we all depend.

1.1 EXTENDING PERSONAL CAPABILITIES

As human beings, we use technical means to augment and extend the natural capabilities of speaking, drawing, writing, and counting to provide a rich diversity of ways of communicating. The result is an impressive array of intercommunication and mass communication facilities. In this section, I describe the development, characteristics, and applications of telephone, facsimile, television, and electronic mail. They are the paramount telecommunications capabilities employed among persons.

1.1.1 COMMUNICATION, TELECOMMUNICATION, AND INFORMATION

In all studies, questions arise as to the meaning of words. Let's begin with some basic terms. What is *communication*? How does it differ from *telecommunication*? In addition, how do they differ from *telecommunications*? What is *information*? In this book, they are used to mean:

- **Communication**: the activity associated with distributing or exchanging information. Derived from the Latin *communicare*, meaning to make common, to share, to impart, or to transmit, communication may be
  - *One way* in the sense of an announcement
  - *Two ways*. The exchange may be *interactive* as in a conversation between persons in which information is transmitted one way at a time, or *simultaneous*, as in an exchange of data between machines in two directions at once.
- **Telecommunication**: the action of communicating at a distance. Derived from the Greek *τηλε*, i.e., *tele*, meaning far-off, and *communication*. In the broadest sense, it includes several ways of communicating (e.g., letters, newspapers, telephone, etc.); however, it is customary to associate it only with electronic communication (e.g., by telephone, data communication [including telegraph], radio, and television).

Communication and telecommunication may be between persons, persons and machines, and between machines. Note that both terms imply the act of distributing or exchanging information, but telecommunication also implies that the activity is undertaken between entities separated by a distance.

In a more general sense, the following terms refer to the ways in which telecommunication is achieved:

- **Intercommunication**: telecommunication in which information flows between two sites
  - when computers and data processors are involved, intercommunication takes place on demand, between units that are authorized to communicate
  - when persons are involved, intercommunication takes place by agreement at a mutually convenient time and is controlled by the participants—as their conversation develops, they control the format and content of the exchange.

So as not to interfere with others, in intercommunication each pair of users requires a separate, exclusive connection. This sets a limit to the number of parties that can be served simultaneously by a particular complement of equipment (a facility or network). Consequently, at times of peak use, some of those who wish to communicate may not be able to do so immediately. Intercommunication services can become congested during periods of peak demand so that some potential users are refused service.

- **Mass communication**: telecommunication in which information flows from a single (transmitting) site to a large number of (receiving) sites simultaneously, without response from the receivers

– among persons, mass communication is associated principally with the delivery of entertainment and information services; the originators control the contents of the messages and the times at which transmissions take place

– among computers and data processors, mass communication may be employed to update a database or data-file that is possessed by many secondary stations.

Mass communication services do not suffer capacity limitations in the sense of a limit on the number of sites that can receive a service. However, the number of originators who can operate simultaneously is limited. For instance, in radio broadcasting, only a certain number of frequency assignments are possible. In cable television, the characteristics of the cable used to distribute the signals set a limit to the number of channels that are available.

In contrast to telecommunication, telecommunications is not an activity but the technology that supports the activity.

- **Telecommunications**: the technology of communication at a distance. An *enabling* technology, it makes it possible for information that is created *anywhere* to be used *everywhere* without delay.

Telecommunications may be used as an adjective to describe hardware and services. Two examples of its use are

- **Telecommunications facility**: the combination of equipment, services, and associated support persons (if any) that implements a specific capability for communicating at a distance

- **Telecommunications network**: an array of facilities that provide custom routing and services for a large number of users so that they may distribute or exchange information simultaneously.

Finally, what is meant by information? It depends on the context; thus

- **Information**: that which is distributed or exchanged by communication
  - when associated with computers and data processors, it may be defined as *organized or processed data*—i.e., the output that results from processing data according to a given algorithm. Data, per se, are not information; information is created by organizing them.
  - when associated with persons, it may be defined as the *substance of messages*—i.e., that which is known about, or may be inferred from, particular facts or circumstances
  - when associated with the exchange of symbols, Claude Shannon (a 20th-century American mathematician) defined it in terms of the probability of the symbol (message) being sent.

The first implies a quantitative result—something that has resulted from processing data. The second implies a qualitative result—the intellectual product of communication. The third is the easiest with which to deal—it is the basis for information theory. We return to it later in Section 3.3.4(1).

1.1.2 TELEPHONES

At some time in prehistory, persons began to make specific noises that were commonly understood to represent feelings, things, actions, and, eventually, ideas. When married to 19th- and 20th-century technologies, speaking over great distances became possible.

In 1876, at the Centennial Exhibition in New York City, Alexander Graham Bell demonstrated the transmission of voice signals over a wire. This feat stimulated persons to purchase telephones for use over private circuits and led to the formation of local telephone companies under Bell licenses. Soon, as they competed with telegraph companies to serve public demand, the major thoroughfares were festooned with strands of wire attached to taller and taller poles. In 1900, there were approximately 1.6 million telephones in service in the United States. By 1913, some 20,000 telephone companies provided connections within local communities. Long-distance messages continued to be carried by telegraph. Today, there are more than 500 million telephones worldwide—and the vast majority of telephone subscribers have ready access to one another.

Between the world wars, the development of vacuum tube amplifiers, and other devices, allowed customers to talk to one another over greater and greater distances until transcontinental conversations were possible. In addition, electromechanical switches replaced an increasing number of manual switchboards, making the dial telephone a household commodity. In the late 1940s and early 1950s, reliable long-distance routes that used microwave radio relay equipment were installed across the United States. In 1956, the first voice cable to span the Atlantic Ocean was placed in service. In the 1960s, computer-controlled switches were introduced, and nationwide direct-distance-dialing (DDD) became a reality.

In 1970, *Viewdata,* an information retrieval system with the generic name *wired-Videotex,* was introduced in Great Britain, and a similar system was deployed in France. They used the telephone for access to a database and a television receiver for display of the data retrieved. In addition, in the 1970s, geostationary satellites were deployed, providing continental and intercontinental voice, video, and data links. In 1977, optical fibers were first used for interexchange connections, and toward the end of the decade, all-digital telephone switches were being installed. At the same time, in Europe, cellular radio systems were sanctioned and mobile telephone systems were deployed rapidly throughout the developed and developing countries of the world. In the United States, competitive cellular services were introduced.

In the 1970s, those who would provide alternative intercity transport facilities challenged the telecommunication establishment of the United States for the opportunity to serve the data needs of businesses. In 1984, after several years of protracted hearings before the Federal Communications Commission, and an antitrust suit brought against AT&T by the Department of Justice, some of the monopoly powers of AT&T were abolished. Under the settlement (called MFJ, Modified Final Judgment)

- The Bell System was divided into:
  - seven autonomous holding companies (Regional Bell Operating Companies, RBOCs)
  - a long-distance carrier with manufacturing capabilities (AT&T and Western Electric).
- A competitive long-distance telecommunications industry was established
  - independent carriers (such as MCI and Sprint) were confirmed as providers of long-distance services
  - the creation of new carriers was encouraged.

In the 1980s, the telephone carriers of the world agreed on standards for digital telephone networks (Integrated Services Digital Networks, ISDNs). In the 1990s, many of them have implemented a level of ISDNs and broadband ISDNs. To facilitate the development of multimedia services, a new sort of digital switching system called asynchronous transfer mode (ATM) has been standardized. In the United States, the carriers implemented intelligent networks (INs). Based on signaling system 7 (SS7, an integral part of ISDN), intelligent networks are market-driven vehicles for the speedy provision of whatever communication services users require (such as call forwarding, call waiting, caller ID, etc.).

In the 1990s

- Digital switches and high-speed optical fiber transmission facilities continue to replace existing analog facilities
- Personal (portable) communications systems (PCSs) are being developed to provide universal voice, video, and data (multimedia) services
- Low- and medium-earth orbiting satellite systems are being deployed to facilitate a new level of worldwide communication
- Undersea, optical fiber cable systems have been laid that encircle the earth with wideband channels
- Congress passed the 1996 Telecommunications Act. Under certain conditions, it provides for competition in all phases of telecommunication.

1.1.3 FACSIMILE MACHINES

Long before men and women could write, they created colorful shapes on two-dimensional surfaces to tell stories of everyday life. Married to 20th-century technologies, drawing has given rise to combinations of physical media, bearers, and transducers that transport images.

In the 1860s, telephotograph machines scanned relief photographs line by line to create an electrical signal that is proportional to the picture features. At the receiving station, a V-shaped cutting stylus produced lines of variable widths to create a facsimile of the source picture. Improved through the years, today's facsimile machines produce high-quality images of printed pages, as well as pictures. Digital versions transmit mixed text and graphical information with a resolution of several hundred dots per inch in 10 seconds per page, or less.

1.1.4 TELEVISION

Based on experiments conducted in the late 1920s and in the 1930s, black-and-white television was introduced to the public in the late 1930s. It brought visual messages to the home, capturing the attention of every member of the household with the sights of far-off events and with entertainment designed to attract audiences for messages from those paying for the service. In the 1960s, this window was filled with living color, and in the 1970s it was extended in real-time across the oceans, and around the world, by satellites. In addition, in the 1970s, data were added to the television signal to provide an electronic magazine. With the generic name of *broadcast-Videotex*, the service is generally known today as *Teletext*.

Techniques for recording and reproducing television signals became available in the 1970s. Quickly, wideband tape recorders permeated the television industry. With perseverance, they were developed into affordable consumer units that are used to provide alternative programming at the convenience of the viewer. Through the use of videocassette recorders (VCRs), television programs are recorded on videotapes for replaying at a later time, and movies that have been converted to videotape are played instead of current programming. In addition, digital videodiscs (DVDs) are available. They reproduce video sequences stored on metal or plastic disks.

Television is broadcast at frequencies that require the transmitter to have a clear line of sight to the receiver. For this reason, television transmitting antennas are placed on the top of very tall towers, or on very high buildings. However, if the receiving antenna's view of the transmitter's antenna is obstructed by hills, or by steel-frame buildings, the signal will be attenuated and may be lowered below the level required for satisfactory signal reception. In the late 1940s, to overcome this limitation, residents of a hilly section of Pennsylvania constructed an antenna on the top of a hill from which the transmitter's antenna was visible. They distributed the received signal to their homes by coaxial cable. In this manner, cable television (CATV) was born.

Throughout the 1950s, CATV was regarded as a technique for making network signals available to rural communities. However, in the late 1960s, 1970s, and 1980s, CATV was expanded to carry many more signals, including some produced specifically for cable systems, and was introduced into suburban and urban areas. Today, a majority of the U.S. population has access to cable service; modern single cable systems deliver over 50 channels to the subscriber, and two cable systems deliver over 100 channels. A few installations include a return channel so that customers can respond to the head-end (interactive CATV). Competing with cable, in the mid-1990s, geostationary satellites began to broadcast several hundred digitally encoded channels to micro dishes (approximately 18 inches in diameter) mounted on the subscriber's premises.

The broadcast of high-definition, digital television services began in 1998, and the FCC has directed that this format replace existing low-definition analog services some time in the early 21st century. Terrestrial broadcasters have been allocated additional spectrum space that they may fill with digital signals similar to those employed by satellite service providers. Encouraged by the 1996 Telecommunications Act, cable companies are beginning to provide telephone and high-speed data services to residential subscribers.

In the late-1960s, on a trial basis, some Bell System companies offered videotelephone service in selected locations. Providing a black-and-white, limited resolution, television picture, and telephone-quality voice, the service required special transmission and switching facilities. For many reasons, including lack of public interest, service was withdrawn.

At about the same time, geostationary satellite systems created an interest in videoconferencing. Usually limited to two studios, the service provided participants in separated locations with a standard color television picture of the parties at the remote end. In the early 1970s, several companies with nationwide facilities built private satellite-based teleconferencing systems for internal use. Toward the end of the 1970s, networks of public studios were created so that persons without access to a private studio could still participate in videoconferences. The service met with limited success.

In 1991, AT&T introduced a new version of videotelephone service. Using a telephone that incorporates a small screen on which still color pictures of the remote party are displayed, the service requires a single telephone channel. To give a semblance of motion, the individual frames are replaced every few seconds by a more recent one.

In the 1990s, the availability of powerful digital signal processors and the development and standardization of video coding and compression techniques have led to satisfactory (quasi-full motion) videoconferencing over digital telephone channels. Employing portable equipment, the service can be accessed in the office or at home. In addition, videoconferencing (of limited quality) is now possible over Internet.

1.1.5 ELECTRONIC MAIL

Most letters are private communications between individuals. On a personal level, they provide status reports between relatives and friends, informing, counseling and seeking help. In business, letters carry information that makes it possible for organizational entities to operate in the larger world of the corporation. Other letters involve exchanges between citizens and government; still others contain commercial messages; and some letters are written for publication.

In the last half of the 20th century, electronic technologies have made it possible to prepare and transport letters using electrical means. Delivered to an electronic mailbox, messages are called up on a video screen at the request of the receiver. Electronic mail is a principal use of Internet and is used in just about every enterprise.

REVIEW QUESTIONS FOR SECTION 1.1

1   Differentiate among communication, telecommunication, and telecommunications.

2   Give three definitions of information.

3   Distinguish between intercommunication and mass communication; discuss the limitations on their capacity.

4   Give a brief history of the telephone and telephone network.

5   What was the effect of the Modified Final Judgement (1984)?

6   List the technological developments of the 1990s.

7   What is the impact of the 1996 Telecommunications Act?

8   What was the origin of the facsimile machine?

9   Describe the origin of television.

10   Describe the origin of cable television. How does its original purpose differ from present applications?

11   Distinguish between videotelephone and videoconferencing.

12   For what purposes is electronic mail used?

1.2 DATA COMMUNICATIONS

Despite the efforts of mathematicians to make numbers continuous, counting is essentially a digital function. Married to 20th-century technologies, it has given rise to the computer and data processor together with a myriad of specialized terminals, and has created a new class of communication capabilities that facilitates sharing data amongst machines.

1.2.1 TELEGRAPHY

In the early years of the 19th century, Oersted, Henry, Gauss, and Weber, among others, developed the sciences of electricity and magnetism. One of the first applications of their work was the electric telegraph. In 1843, Samuel Morse and Alfred Vail reduced it to practice. By interrupting the electric current flowing in a wire strung between the cities, Morse demonstrated that a message could be sent from Washington, DC., to Baltimore, MD. Running their copper wires beside railroad tracks, telegraphers began exchanging messages over ever-increasing distances on the first real-time communication system. With the formation of the Associated Wire Service in 1846, and Western Union in 1856, even towns of modest size were able to receive news of important events shortly after they occurred. As a result, the Civil War was reported in timely detail in dispatches telegraphed to local newspapers.

The international promise of telegraphy was recognized in 1865. At the first International Telegraph Convention, the International Telegraphic Union (ITU) was formed. Its charter included the recommendation of operating procedures and standards so as to promote worldwide communication. In 1866, the completion of a transatlantic cable gave same-day access to the major cities of Europe. In 1886, Hertz discovered radio waves; in 1894, Marconi demonstrated the wireless telegraph; in 1898, ship-to-shore telegraphy was established; and in 1901, transatlantic wireless telegraphy came into being.

1.2.2 DEVELOPMENT OF DATA COMMUNICATION

Adopted by the railroads in the mid-19th century, telegraphs provided coded communication between control points (railroad stations) that was used for dispatching trains and supervising freight. Before long, the messages spread out to other enterprises, and centralized, quasi-real-time management of a distributed organization through sharing data became a reality. As the number of messages increased, the telegrapher's key, sounder, and message pad were replaced by a combination typewriter and printer. It prepared a paper tape image of the message to be sent and printed the messages received on a similar strip of paper.

In the 1950s, as electronic computers emerged from universities and research laboratories, updated versions of these machines, known as teletypewriters, were used to communicate with them. By the 1960s, an increasing number of commercial tasks produced a need to connect remote data terminals to central processors, and these processors needed to exchange information among themselves.

The development of a vigorous, general-purpose computing community produced demands for communication between remote machines. Ignored by the telephone companies, users began experimenting with various techniques. Led by the Advanced Research Projects Agency (ARPA) of the Department of Defense, government researchers pioneered packet-switched networks and procedures for a national network that could survive a nuclear attack. Led by IBM, industrial and commercial users adopted networks and procedures based on Systems Network Architecture (SNA) that provided a tightly managed network with minimum, predictable delays. Encouraged by the success of these activities, Specialized Common Carriers challenged the Bell System for the right to carry data signals over private facilities between major cities. At about the same time, the modern data terminal, with video display and keyboard, was introduced. Today, data terminals allow persons to communicate with each other, and with machines, and the digital coding they employ (data communication) permits the exchange of information between machines.

In the last half of the 20th century, the development and deployment of information technology created a demand for the exchange of data among data centers, and other facilities, that has revolutionized the business communication environment. A rapidly expanding number of jobs require communication to support the use of data terminals that are connected to host computers, database servers, and other equipment.

1.2.3 INTERNET, INTRANETS, AND EXTRANETS

Perhaps the most important development of the last part of the 20th century has been the establishment and growth of Internet. Founded on the pioneering work of ARPA, Internet uses packet-switching technology to interconnect personal computers, hosts, and servers in a global data network that facilitates the retrieval of information from anywhere and its use everywhere. At first restricted to government projects, the use of the facility has been expanded to provide ready access to commercial and residential users.

Internet has become a vehicle for disseminating information on every conceivable topic. Quick to grasp an opportunity to improve their internal dissemination and use of company information, firms have adapted Internet for their own communications. By establishing secure subnetworks called *intranets*, they have created an island of company information that can be accessed by authorized employees, no matter what sort of platforms they are using—and where they are located. Furthermore, by establishing secure links outside the firm, platform-independent information sharing can be extended to suppliers, vendors, and customers. Called *extranets*, these arrangements can help manage shared projects, encourage collaboration, and improve performance (as described in Section 1.2.4).

1.2.4 AREAS OF APPLICATION

In a 1988 study by the Office of Technology Assessment (OTA) of the United States Congress,[1] technologies for collecting, storing, manipulating, and communicating information are identified as having the potential to revolutionize the economy of the United States. Two areas of application singled out in the report are

- **Productivity**: they can increase the productivity of operations where real changes may never have been considered seriously
- **Economic networks**: they can link production systems together in ways that improve the performance of entire economic networks.

One (productivity) is concerned with improving the internal efficiency of economic units; the other (economic networks) is concerned with improving the ways in which economic units cooperate to produce finished goods. What the authors of the OTA study had in mind is that the information technologies have a decisive role to play in stimulating and maintaining a strong, expanding national economy in the information age of the 21st century. Telecommunications (for collecting and communicating information), databases (for storing information), and data processors and information systems (for manipulating information) will have a significant impact on the future success of business enterprises. As individual entities, and in combination with one another, they can contribute to increased margins and improved performance.

(1) Telecommunications and Automation

Modern business enterprises employ computers and computer-controlled machinery to sustain their competitive edge. They use them to control operations, increase productivity, incorporate custom features in their products, maintain quality standards, and provide accurate, timely responses to the demands of their markets. To continue to reduce the cost of units of output, and to improve the efficiency of operations, they must constantly increase the level of automation. What is more, as technical changes and competitive offerings diminish the overall life cycle of each product, there is a need to reduce development, engineering, procurement, and deployment times.

To promote timely decisions that have a high likelihood of being correct, a growing amount of data must be collected from many locations and processed into information that is distributed in a timely fashion. Achieving this performance depends on speedy collection and distribution facilities, information sys-

---

[1] U.S. Congress, Office of Technology Assessment, *Technology and the American Economic Transition: Choices for the Future*, OTA-TET-283, U.S. Government Printing Office, May 1988.

tems that organize the data in useful ways, and the free flow of comments and instructions among employees, suppliers, and customers. By providing voice, data, and video services over networks of increasing sophistication, telecommunication facilities contribute to the successful attainment of these goals and to the cost-effective operation of today's enterprises.

(2) Information Systems

An information system is a software entity that organizes data to produce information for the benefit of an enterprise. More specifically, information systems process data to create useful information for control and operating purposes. The types of systems needed to support each product or service depend on the positions they occupy in their respective life cycles. Thus, for a new product or service, information systems can identify markets and develop positioning strategies; for a mature product or service, information systems are used to improve manufacturing efficiency and match customer requirements with optional capabilities. Information systems can be classified as follows:

- **Evaluative**: one that provides information with which to plan future activities
- **Informative**: one that provides information with which to control present activities
- **Supportive**: one that provides (historical) information with which to operate the enterprise.

(3) Telecommunications and Information Systems

What sets one enterprise apart from another is its use of the combination of telecommunication facilities and information systems. Some of the applications that are improving the way in which enterprises conduct their business include the following:

- **Manufacturers** link together the computer-based tools they use to design, engineer, and manufacture products to create computer-integrated manufacturing (CIM). CIM is essential for organizations that must match the quality and turnaround time of global competitors. To make the producing process as efficient as it can be, these capabilities must be shared with their suppliers, distributors, and customers.
- **Financial institutions** broker global financial markets, serve multinational businesses, buy and sell currencies around the world, and sell financial services based on proprietary information packages. These activities are pursued over real-time networks that integrate many islands of information and expertise. To retain the attention of a dynamic, global customer base, it is necessary to cover the world with instant information, acceptances, and analysis.
- **Passenger airlines** depend on filling seats; their strategy is threefold:

- provide attractive fares and attractive schedules, and get them to travel agents ahead of competing airlines
- provide travel agents with special terminals linked to the airline's database so that the agents can sell and confirm tickets on the airline
- build customer loyalties through frequent flier, senior citizen, and other programs that are based on records stored in reservation systems.
- **Property** and **casualty insurance companies** link agents with proprietary programs and databases so that they can quote the best available terms to clients—immediately. Independent agents have the same access through special terminals. In this environment, underwriters must provide additional services to agents if they expect to retain their customers.

While the success of these applications is due in large measure to the information systems that they employ, the results would be diminished or negated without the immediacy of which telecommunication facilities are capable. Furthermore, the benefits they provide produce only temporary market leadership. To remain the leader, a stream of new services must be added to offset aggressive attempts by followers to regain market share. The right kind of telecommunications facilities can facilitate these actions.

1.2.5 LINKING PRODUCTION SYSTEMS TOGETHER

In the information age of the 21st century, the speed with which companies can do the following depends on the speed with which computer-based, telecommunications-assisted capabilities are adopted

- Serve large numbers of highly specialized markets by connecting together complex networks of producers around the nation or around the world to forge tighter (less costly and more timely) links between retail, wholesale, transportation, and manufacturing operations
- Concentrate production in areas (domestic or overseas) where labor skills, wages, business conditions, or living conditions are judged to be favorable—including the opportunity to move production activities (both goods and services) to any place in the world
- Control the activities of far-flung units of the enterprise to make the whole efficient and competitive
- Reduce travel among company locations for routine matters.

(1) Manufacturing Chain

Figure 1.1 shows an electronic goods manufacturing chain. Goods are manufactured in a sequence of operations beginning with raw materials and ending

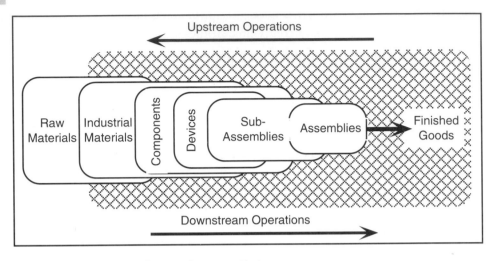

Figure 1.1 Electronic Goods Manufacturing Chain

Goods are manufactured in a sequence of operations beginning with raw materials and ending with the finished article. At each step, the input is transformed into what is needed for the next step.

with finished goods. At each step, the input is transformed into what is needed for the next step. Thus, raw materials are processed into industrial-grade materials. In turn, they are made into components, then devices, subassemblies, and assemblies to produce the final product. Separate companies may perform the steps one at a time, or a single company may perform several steps. At each step, the product may be dispatched to other manufacturing chains. For this reason, the shading denoting finished goods extends back up the chain to industrial materials. To complete the goods in a timely and cost-effective way, information is passed along the chain.

(2) Interenterprise Uses of Telecommunications

Figure 1.2 shows part of a manufacturing chain that produces electronic goods for consumers. Company A, a low-cost producer, links backward to two levels of suppliers to manage inventories and acquire confidence in the suppliers' abilities to deliver their products as required. Company B, an aggressive supplier of subassemblies, links backward to suppliers D, E, and F to assure their ability to support just-in-time manufacturing, and forward to company A to provide information support that will differentiate the product from others. The link from B to A represents a switching cost (to A) that may prevent A from changing suppliers in times of poor performance by B. At the component level, company D links forward to company B for the same reasons. Less aggressive than company B, company C links back only to supplier G to manage inventories.

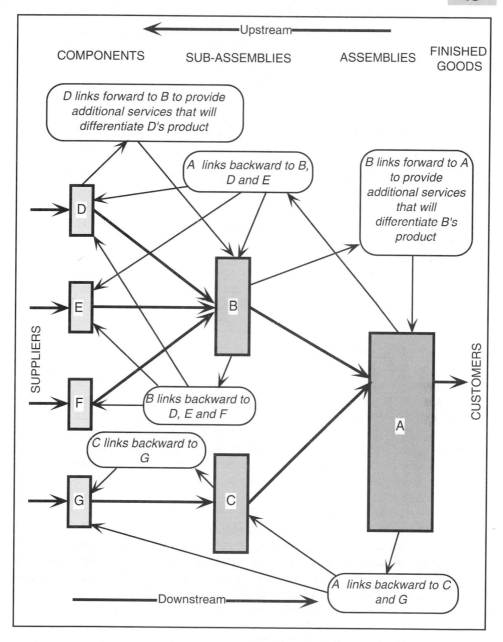

Figure 1.2 Interenterprise Uses of Telecommunication

Company A, a low-cost producer, links backward to suppliers to manage inventories. Company B, an aggressive supplier of subassemblies, links backward to suppliers, and forward to the customer. Company C, less aggressive than company B, links back to a supplier.

(3) Linking to Suppliers

The path to finished goods may lead through several firms that are responsible for a portion of the endeavor. To achieve the timely delivery of high-quality goods, the final downstream firm in the chain depends on the performance and skill of its upstream suppliers; in turn, they depend on their suppliers. To gather information with which to build confidence in their performance and the strengths of their commitments, each firm can extend telecommunications links to information nodes in their suppliers' internal networks—if they agree. The data they provide will permit the customer to evaluate progress on orders and to estimate the probability of on-time deliveries.

Moreover, once the link is established, the customer can send information to the supplier concerning future orders and schedules. Once suppliers and customer gain confidence in each other—and the system—the customer can take advantage of just-in-time operation as a way of reducing the cost of money for goods that might otherwise be held in the customer's inventory. In fact, with several telecommunications-based information systems, inventory costs can be reduced all the way upstream. Using customer-to-supplier linkages between the firms in the chain, the individual supplier to customer segments can operate more efficiently, and the entire chain is likely to achieve a higher percentage of on-time deliveries of higher-quality, less costly finished goods.

(4) Linking to Customers

One of the ways to create links forward is to place a terminal in the customer's office and support uses that the customer considers important. Not all of them need to be related to the supplier's product. For instance, if the product is a common use item, inventory management and automatic reordering might be included. In addition, the system could provide inventory services for related products—including those supplied by competitors—making it more difficult for the competition to persuade the customer to replace the terminal with one it supplies. This action may also encourage the notion that the convenience of automatic reordering could be extended to more products if they came from the same supplier. Fully exploiting the advantage of occupancy means finding creative ways to assist the customer and benefit the supplier.

In Figure 1.2, company B, an aggressive supplier, is attempting to build a relationship with A that will achieve a competitive advantage. Over the link to A, B provides information services. At worst, these services will differentiate B from other suppliers, and at best, they will create a relationship that prevents A from going elsewhere for B's products.

(5) Integrating the Chain

If the final supplier of finished goods is a significant customer of the other firms in the manufacturing process, it can use this fact to put links in place to connect it to the other firms. In this way, it is able to integrate data from the entire manufacturing chain—and communication does not have to be between manufacturing entities only. Engineering may collect data and exchange information with other engineering organizations in the chain; marketing may share information with the other marketing organizations, etc. Given cooperation, the more activities that can be coordinated, the greater the levels of efficiency that are likely to be achieved. Such networks will build relationships among suppliers and customers so that the entire manufacturing chain for a particular set of finished goods becomes the single, integrated, economic network referenced in the OTA study.

(6) Electronic Data Interchange

In industries such as transportation, automobiles, groceries, and pharmaceuticals and personal care products, dominant firms have built economic networks of the sort described and have standardized message formats and messaging procedures for themselves and their suppliers. The development of broader, general-purpose standards for electronic data interchange (EDI) is the task of industry associations coordinated by the ANSI Accredited Standards Committee X12. For buyer and seller transactions, this group has adopted a message architecture that accommodates a broad range of practices. It includes a feature set that can be tailored to different industries, and it is becoming the architecture of choice for many applications. EDI is most beneficial for industries with regular, repetitive buying, selling, and distributing functions.

1.2.6 STRATEGIC USES OF TELECOMMUNICATIONS

Telecommunications is one of many resources that are employed in running an enterprise. Normally, the level of management review of, and concern for, telecommunications activities will depend on the fraction of the corporation's total budget that is allocated to running expenses and capital acquisitions. When strategic results are expected, members of the top management team must get involved. Certainly, the preemptive strike described in Section 1.2.6(5) could not have happened without this level of support.

(1) Support of Marketing Strategy

Telecommunications networks can be used to pursue different information-driven marketing strategies. For instance, if the firm seeks to

- **Lower product costs** so as to become the low-cost producer (and market leader), the telecommunication network will emphasize linkages backward to suppliers, and suppliers' suppliers, so as to increase the efficiency of the manufacturing chain
- **Enhance product differentiation** to be the preferred source of supply, the telecommunication network will include linkages forward to the customer to provide information services that are not provided by competitors

- **Produce specialty products** that are tailored to particular market segments and to particular customers, the network will emphasize integrated operations so that custom parts and subassemblies can be produced with a minimum of difficulty.

(2) Building Barriers

By investing in electronic networks to link together all of the activities associated with a particular product, its markets, and its supporting materials, producers can manage the integrated community of firms that support them. Given the right tools for analysis and decision making, they can modify production schedules and adjust incoming supplies and outgoing distribution in an optimum manner to respond to changes in the marketplace and fluctuations in the world environment. Under the control of a dominant producer, the combination of a pervasive telecommunications network and powerful information systems tools can bind groups of firms together in a dependent way. If other producers wish to compete, they must establish a similar combination of telecommunications-based information systems capabilities—not an easy task. Thus, in building an integrated network, the producer has thrown up a significant barrier to would-be followers.

(3) Overintegration and Loss of Flexibility

Participation in these sophisticated arrangements entails more than just connecting to the network. Suppliers and distributors must share some of their proprietary information with the producer, and the producer must reveal proprietary plans and strategies. Thus, firms doing business as part of an integrated network become beholden to each other, and, after a period of confidence-building interaction, they become dependent on each other. Continuing success will bind them even more closely together. Eventually, they may lose the flexibility to pursue new implementations of existing products because the community they have created does not have the necessary expertise or incentives to shift from the mode in which they are operating. In this case, they are in danger of losing their dominant position to a less well established network that has greater flexibility to go after new opportunities.

(4) Retaining Customers

When a producer builds a telecommunications-based information system forward to a customer, it can offer information concerning product characteristics and applications that differentiate the supplier from its competitors. Properly thought out, these services can be important to the customer, yet inexpensive for the producer. A case in point is an original equipment manufacturer (OEM) that supplies a computer system for incorporation in a customer's product. On-line programming help, on-line maintenance guides, on-line applications news, and similar information can enhance the OEM's position with the buyer. The strategy can be particularly effective when the goods involved are complex and are likely to go through an applications evolution in their lifetime. With a linkage in place that is filled with useful information, the buyer is likely to find that switching to another supplier will have a cost—in time, in convenience, or in training. At times of stress in their relationship, the buyer may find the services to be a persuasive reason not to switch to another OEM's products. At times of price competition from other OEMs, the supplier may find the services provide some protection against the buyer substituting another's product.

(5) Preemptive Strikes

The first implementations of some of the examples cited earlier [in Section 1.2.4(3)] have been called preemptive strikes. The term describes the aggressive use of new technologies, or old technologies in new ways, to the advantage of the innovator, such that

- The move surprised the innovator's competitors
- The move was well received by the innovator's customers and attracted customers served by competitors
- The move reshaped the markets to the advantage of the innovator
- Competitors required significant time, and had to make substantial investments, to catch up.

Using the combination of telecommunications facilities and information systems, many companies have succeeded in attaining these ends.

A good example of a pre-emptive strike is the use made by American Airlines of telecommunications-based information systems.[2] In the 1970s, by building and exploiting an electronic reservation and ticketing system known as SABRE, American Airlines created a new playing field on which other airlines were forced to play. Once the computers and communication links were operational, and SABRE was available at the ticketing desks in airports, terminals were moved into travel agents' offices. The terminals created links forward to customers, assisting them and differentiating American from other airlines. At the same time, having terminals in agents' offices allowed American to provide the agents with promotional materials, and, unless the potential traveler had a strong preference for another line, encouraged the agents to *book American*.

In 1981, American announced a frequent flier bonus program. Designed to foster brand loyalty, it was a linkage forward to the ultimate customers—the travelers. Building a record of flights taken by full-fare passengers (typically business travelers) for which tickets were dispensed through SABRE, American awarded points for miles flown, kept track of the total, and converted the points into addi-

---

[2] For an extensive discussion of preemptive strikes, see Peter G. W. Keen, *Competing in Time, Using Telecommunications for Competitive Advantage*, updated and expanded (New York: Ballinger Publishing Company, 1988) 110–18.

tional tickets at the request of the flier. Its action met with acclaim from travelers and expanded the fraction of all frequent fliers whose carrier of choice was American Airlines.

Caught at a disadvantage, American's competitors were forced to respond on American's terms. Even those airlines with proprietary reservation systems had to spend a great deal of time and money adapting them to capturing information concerning the total distance flown by individual fliers. Because the record has nothing to do with selling a ticket to fill a seat, it posed a sizeable barrier to entry in the new marketplace of electronic ticketing and bonus programs. American continues to exploit the advantages of occupancy given by a terminal in the travel agent's office and the records of frequent fliers stored in SABRE's processors. This combination of telecommunications and information systems capabilities has helped American maintain a leading position in the airline industry.

1.2.7 INCREASING THE PRODUCTIVITY OF OPERATIONS

Internally, the ready availability of telecommunications resources has changed the way in which business is conducted. They have altered work requirements so that the skills possessed by persons presently on the job may not be adequate for the future, and they are changing the way in which work is organized. Intra-enterprise use of telecommunications has aided the rise of concepts of automated enterprises, automated offices, and automated factories.

(1) Person-Centered Applications

Two examples of ways in which telecommunications-based systems are used within the enterprise to improve the efficiency of operations are

- **Materials**: a materials specialist who keys in a part number that defines a manufactured subassembly, the quantity required, and the need date, may cause the
  - display of current inventory
  - ordering of critical parts from a supplier
  - issuing of a revised manufacturing schedule for the subassembly to a facility somewhere else.
- **Integrated circuits**: an engineer who inserts a description of the functions to be performed by an integrated circuit may cause the
  - design and simulation of a circuit
  - generation of the layout information for an integrated circuit chip to perform the task
  - development of the software instructions to adapt it to a specific set of operating circumstances

– development of a test suite that ensures the chip and software perform as specified.

Each of these examples reduces labor, speeds up activities, and increases the expectation of customer satisfaction. They substitute information processing, and the linking and transporting abilities of telecommunications facilities, for the labor of several persons. More than likely, the information systems in these examples will reside in different locations, and the tasks cited are enabled by the availability of telecommunications facilities.

(2) Machine-Centered Applications

Not all activities need be in direct response to actions by persons. At certain times, or under certain conditions, they may be initiated between machines operating under the control of programs designed to transfer information and perform specific functions.

- **Daily activity consolidation**: during the night, a machine at the headquarters of a supermarket chain will
    - call for information on the day's transactions from machines at branch locations
    - balance the total activity in the company
    - return consolidated records to each branch for use the next day
    - provide a summary of the day's activities to the operating managers when they arrive in the morning.
- **Process control**: periodically, a machine at an operations center that supervises a pipeline will receive status reports from machines at important nodes in the network. Should equipment failure or unusual operating conditions occur, it might transmit instructions to the controllers so as to mitigate the difficulty, or call for human intervention.

These examples reduce the amount of labor involved and increase the expectation of a higher-quality output. They are enabled by the availability of telecommunications facilities.

(3) Scope of Intra-Enterprise Telecommunications

Figure 1.3 depicts a manufacturing enterprise as a value-producing process surrounded by R&D, engineering, and marketing resources. It is coupled to the manufacturing chain by distribution channels and to executive and administrative functions by telecommunications-assisted information systems. The company's telecommunications-based networks collect data from all parts of the enterprise and distribute information to control and action points, as required. The shaded circle that extends over all resources and the distribution channels suggests the pervasive nature of their applications.

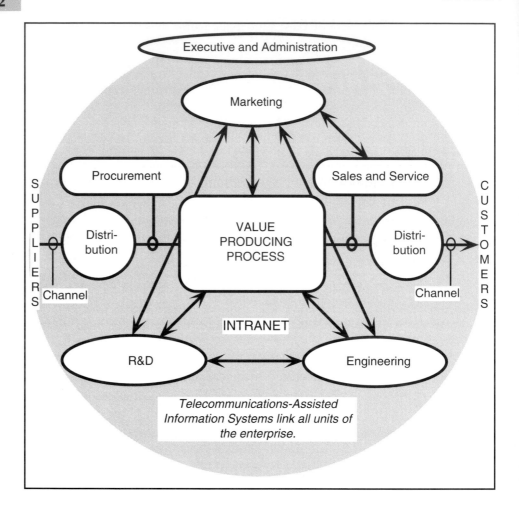

Figure 1.3 Representation of a Manufacturing Enterprise

The value-added process is surrounded by R&D, engineering, marketing, sales, and distribution. It is coupled to customers and suppliers by distribution channels, and to executive and administrative functions by telecommunications-assisted information systems supported by an intranet.

For firms whose operations are located in a limited area, the telecommunications facilities supply a full range of services over short pathways between information nodes. For firms whose operations are distributed across a continent, the telecommunications facilities supply a full range of services over local and long-distance pathways between information nodes. For firms that do not manufacture goods (hardware or software), the diagram can be modified to reflect the generation of value-added services.

1.2.8 AUTOMATED ENTERPRISE

Pervasive use of telecommunications-based systems, and other information technology, creates an environment in which more and more tasks become candidates for automation. As each company using information systems substitutes information technologies for human labor, it creates more information and further opportunities for automation. As each company developing information systems perfects new capabilities, it creates further opportunities for information technologies to substitute for labor. As a consequence, in the modern enterprise, the levels of automation, and dependence on information, increase inexorably. Companies that are caught up in information revolution in this manner make extensive use of telecommunications-based systems; they are called *automated enterprises.* I divide their activities into two categories: automated office and automated factory.

(1) Automated Office

In an automated enterprise, the use of information technologies to improve the effectiveness of persons performing office-based tasks has come to be known as *office automation*, and offices that adopt these services are known as *automated offices.* The specific functions performed in an automated office depend on the work to be done and on the capabilities of the equipment installed. Besides pervasive voice telephone services and desktop computing capabilities, automated offices incorporate telecommunications-based services that

- Achieve desk-to-desk, local, or remote transfers of text, data, and images as easily as voice information is transferred over the telephone
- Provide the ability to reach local or remote databanks, making central storage feasible and ensuring that all available, relevant data can be assembled and reviewed when needed
- Support electronic mail (text, voice, or graphics) and teleconferencing (computer, voice, or video) so that ideas can be exchanged and problems discussed with associates anywhere, thereby providing a higher level of expertise everywhere.

By fostering these activities, telecommunications-based systems merge the individual offices of the geographically distributed corporation into an extended electronic workspace in which to attend to the running of the enterprise.

(2) Automated Factory

A factory can be defined as a set of persons, machines, data processors, and other facilities that work together in a timely and cost-effective manner to produce goods. In an automated enterprise, the use of information technologies to improve the effectiveness of factory processes is known as *factory automation*, and factories that adopt these services are known as *automated factories.* The *goods* they manufacture may be hard goods (i.e., hardware), or soft goods (i.e., software or services).

Under the control of software systems supervised by operators, automated factories that manufacture and distribute hard goods employ computer-integrated manufacturing facilities. They consist of numerically controlled machines, robots, and other tools organized into flexible machining cells that may be served by automated materials handling devices. In addition, supervisory functions such as order entry, inventory control, scheduling, testing, etc., are integrated into the operation of the factory by information systems that collect data, produce information, and feed it to the proper equipment for processing or procurement.

Under the control of software systems supervised by operators, automated factories that manufacture and distribute soft goods employ generic programs, databases, graphical interfaces, expert systems, and other software procedures. They are used to produce information systems or services that meet a customer's requirements.

(3) Knowledge Workers

In many firms, most of the physical labor components of traditional jobs are performed by automated devices—leaving only the intellectual components to be performed by workers. As more of the remaining activities are assumed by information systems, the results are

- Steady increase in the technical skills demanded of a decreasing number of workers
- Decrease in the number of routine tasks performed by all employees
- Rapid decrease in the number of management persons engaged in collecting, analyzing, summarizing, and distributing information.

Overall, the total number of workers per million dollars of revenue is declining, and, for those jobs that remain, higher skills and/or different skills are required. Automated enterprise is creating a group of persons that does not fit the traditional work environment. Known as *knowledge workers*, they perform information-based tasks and communicate knowledge and expertise directly to manufacturing, operations, and other segments of the enterprise. They are the group of talented workers who understand the potential of new telecommunications facilities and information systems, and can apply them to achieve the objectives of the enterprise. Their tools are terminals and their workspaces are combinations of computers, telecommunication networks, and software-directed devices distributed throughout the enterprise.

(4) Changing Organizational Structure

The classical, hierarchical management structure is of little value when the enterprise is dependent on knowledge workers and automation. Knowledge is power, and, in many situations, the understanding of how to manipulate it makes the knowledge worker indispensable. In contrast, the manager who does not have this understanding is powerless—and expendable. The traditional hierarchical structure in which level is sacrosanct is giving way to loose-knit, ad hoc groups of specialists who form teams (task forces) to suit the requirements of the moment. They bypass regular management channels as they solve problems and perform other tasks. The result is a reduction in the number of manager levels and the elimination of many jobs that existed solely to summarize and interpret information and pass on directives. **Figure 1.4** conceptualizes these changes:

- **Contemporary firm**: in diagram A, a traditional organization is shown as a pyramid with layers occupied by officers, managers, and workers. Information flows upward through several layers to get to the chief executive, and instructions flow down to workers who produce the company's products. An important characteristic of this organization is that the person in charge (manager) is the person in control of an operation (group of workers). The corporation is under overall pressure from global competitors to increase productivity, product quality, and reliability. In addition, the corporation is under internal pressures from automation to reduce the number of middle managers, to reduce the number of low-skilled workers, and to increase the skill levels of the workers who remain.

- **Downsizing corporation**: in diagram B, the corporation is responding to these pressures. There are fewer managers and fewer workers. The organization requires higher levels of skills in all persons to exploit higher levels of automation.

- **Automated enterprise**: in diagram C, the corporation has reached the level of automation that makes it as an automated enterprise. There are still fewer persons involved, and at all levels there are knowledge workers who have risen to the challenge of applying new automation techniques for the good of the enterprise. Reporting to managers, they are in control of operations and are likely to be found in one, or more, of the task forces that guide the enterprise. The corporation's competitive edge will depend on the quality of its knowledge workers and on the teamwork displayed by managers and workers. This includes the understanding that the person in charge (manager) is no longer the person in control (knowledge worker) of an operation (group of workers).

Part businessperson, part computer scientist, and part engineer, *knowledge workers* need an up-to-date understanding of the spectrum of telecommunications. With it, they can implement electronic highways that allow information to be created *anywhere*, and used *everywhere*, without delay. It is for these persons that this book has been written.

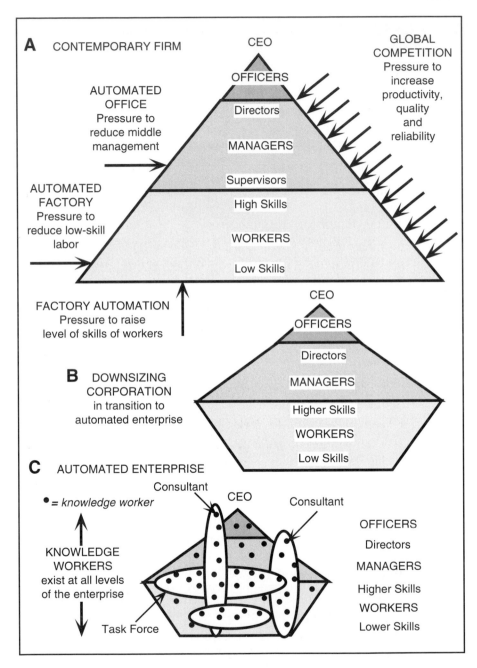

Figure 1.4 Conversion of a Contemporary Firm to an Automated Enterprise

Squeezed by global competition, contemporary firms increase automation, eliminate human tasks whose main purpose is to process data, upgrade workers' skills, create knowledge workers, and adopt a new management style.

REVIEW QUESTIONS FOR SECTION 1.2

1   Describe the origin of telegraphy. What did it contribute to the Civil War in the United States?

2   Describe the development of data communications.

3   What benefits will the adoption of computer-based, telecommunications-assisted capabilities bring to the information world of the 21st century?

4   What is Internet? What are intranets and extranets?

5   What two specific benefits does OTA expect technologies for collecting, storing, manipulating, and communicating information to bring to the economy of the United States?

6   In modern business, how do communication facilities promote timely decisions that have a high likelihood of being correct?

7   Define an information system. How are they classified?

8   Give examples of the use of telecommunications facilities and information systems to improve the way in which enterprises conduct their business.

9   For what productive purposes does an aggressive assembly operation communicate with suppliers and customers?

10 For what purpose have standards for Electronic Data Interchange been developed?

11 Explain how communications can support different marketing strategies.

12 Explain how, by building telecommunications-based information systems capability, a producer can place a significant barrier in the way of would-be followers.

13 What are the dangers in building too extensive telecommunications-based information systems capability?

14 How can a producer use telecommunications-based information systems to retain customers?

15 What is a preemptive strike?

16 Give two examples of the way in which telecommunications-based systems are used within an enterprise to improve the efficiency of operations.

17 What is meant by an automated enterprise?

18 What is meant by an automated office?

19 What is meant by an automated factory?

20 Who are knowledge workers?

21 Explain Figure 1.4.

1.3 TELECOMMUNICATIONS STANDARDS

Today, standards are a hotbed of activity as organizations challenge one another to design and build tomorrow's networks.

- **Standards**: documented agreements containing precise criteria to be used as rules, guidelines, or definitions of characteristics. Their consistent application ensures that materials, products, processes, and services are fit for their purposes.

1.3.1 WHY STANDARDS ARE VITAL TO THE INTRODUCTION OF NEW CAPABILITIES

Since telecommunications deregulation in 1984, and since the Telecommunications Act of 1996, competition among telecommunication providers in the United States has increased the sophistication of available facilities and produced a frenzy of activities in standards-making bodies. Mindful of the experience of product developers in other fields—that standardization is a prerequisite for low-cost implementation and market development—service providers, manufacturers, and users are working together. In bodies sponsored by national, regional, and international standards organizations, they are promulgating architectural and physical standards for the anticipated telecommunications requirements of the global information age. With them, manufacturers can produce the necessary equipment and users can pursue applications that build demand for the services they provide. An important consequence of the globalization of telecommunications is the emergence of consensus standards that combine North American and European practices.

1.3.2 PRINCIPAL INTERNATIONAL STANDARDS ORGANIZATIONS

From the middle of the 19th century, the standardization of operating procedures and equipment performance to promote international (global) telecommunication has been the concern of national governments.

(1) Global Standards Organizations

At the global level, standards are developed and maintained by three organizations

- **International Standards Organization** (ISO): a worldwide federation of national standards bodies. ISO is concerned with the development of standards and related activities for the purpose of facilitating the international exchange of goods and services, and developing cooperation in the spheres of intellectual, scientific, technological, and economic activity (http://www.iso.ch).

- **International Electrotechnical Commission** (IEC): concerned with promoting safety, compatibility, interchangeability, and acceptability of international electrical standards. IEC is the international standards and conformity assessment body for all fields of electrotechnology (http://www.iec.ch). In 1987, by merging their information technology committees, ISO and IEC formed a single body to pursue the development of standards that support the needs of the information industry.
  - *ISO/IEC Joint Technical Committee 1* (JTC1). Responsible for consensus standards in the area of information processing systems.
- **International Telecommunication Union** (ITU): concerned with the creation of standards that facilitate international telecommunication. Founded in 1865 as the International Telegraph Union, since 1947, ITU has been a *specialized* agency of the United Nations (http://www.itu.ch). Its objectives include
  - maintaining and extending international cooperation for the improvement and rational use of telecommunications
  - promoting development of technical facilities and their operation in order to improve the efficiency of telecommunication services and make them available to the public
  - harmonizing the actions of nations in pursuit of these goals.

Often, ITU's objectives are stated as—to *regulate* radio communications, *standardize* international telecommunications, and *develop* global networks.

In 1993, ITU was reorganized into three sectors that reflect these functions.

- **Radiocommunications Sector** (ITU-R): concerned with radio issues and the allocation of the electromagnetic spectrum. Work is accomplished through world and regional conferences and study groups, and the Radio Regulations Board promulgates regulations.
- **Telecommunications Standardization Sector** (ITU-T): develops and adopts *ITU-T Recommendations* designed to facilitate global telecommunication. Work is accomplished through world conferences and study groups.
- **Telecommunications Development Sector** (ITU-D): responsible for promoting the development of technical facilities and services likely to improve the operation of public networks. Work is accomplished through world and regional conferences and study groups.

Each sector is headed by a director who is supported by a staff bureau and an advisory board.

(2) Regional Standards Organizations

While following the lead role of ITU in global standardization, in various regions of the world, telecommunications organizations have banded together to form groups to develop regional standards when no international standards exist. In this way, they influence international standards so that they take account of regional requirements. Among such organizations are

- **European Telecommunications Standards Institute** (ETSI): created by the Committee for Harmonization of the European Conference for Post and Telecommunication (CEPT), and originally intended to serve the needs of the 12 states of the European Community (EC), ETSI now counts Eastern and Central European countries in its membership (http://www.etsi.fr). Its objective is to produce the technical standards that are necessary to achieve a large, unified European telecommunications market.

- **Telecommunication Technology Committee** (TTC): established in 1985 to develop and disseminate Japanese domestic standards for deregulated technical items and protocols, TTC is composed of representatives of Japanese carriers, equipment manufacturers, and users (http://www.ttc.or.jp).

- **Committee T1–Telecommunications** (ANSI-T1): established in the United States in 1984, Committee T1 is sponsored by the Alliance for Telecommunications Industry Solutions (ATIS).[3] Accredited by ANSI and approved by the FCC, it is designated an appropriate forum to address telecommunications networks standards issues (http://www.t1.org). Pledged to work toward consistent, worldwide telecommunications standards, but principally concerned with matters pertaining to the operation of domestic public networks, Committee T1 develops consensus standards and technical positions related to interfaces for United States and associated North American telecommunications networks.

1.3.3 U.S. STANDARDS ORGANIZATIONS

In the United States, the American National Standards Institute (ANSI) is the administrator and coordinator of the United States private sector voluntary standardization system.

---

[3] ATIS is sponsored by over 100 telecommunications companies to promote the development of national and international standards and operating guidelines.

(1) American National Standards Institute

Founded in 1918 by five engineering societies and three government agencies, ANSI does not develop standards itself but facilitates their development by establishing consensus among qualified groups of interested parties (http://www.ansi.org). ANSI is the sole U.S. representative to ISO and IEC. It performs its work through Accredited Standards Committees, such as the following:

- **X 3—Information Technology**: develops voluntary standards in the areas of information storage, processing, transfer, display, retrieval, and management

- **X 12—Electronic Data Interchange**: develops voluntary standards for interindustry applications

- **Committee T1—Telecommunications**: see Section 1.3.2(2).

(2) Other U.S. Standards Organizations

- **Electronic Industries Association** (EIA): a nonprofit organization that represents the interests of manufacturers and focuses on electronics, national defense, telecommunications, education and entertainment, and technical developments (http://www.eia.org). EIA provides a public forum for the discussion of national laws and policies.

- **Institute of Electrical and Electronic Engineers** (IEEE): a professional society, IEEE develops technical performance standards (http://www.ieee.org). The Institute pioneered in developing standards for local-area networks. The IEEE 802 committee of the IEEE Technical Activities Board includes the following subcommittees:

  - 802.1 High Level Interface
  - 802.2 Logical Link Control
  - 802.3 CSMA/CD Networks (Ethernet)
  - 802.4 Token Bus Networks
  - 802.5 Token Ring Networks
  - 802.6 Metropolitan Area Networks–Distributed Queue Dual Bus
  - 802.7 Broadband Technical Advisory Group
  - 802.8 Fiber Optic Technical Advisory Group
  - 802.9 Integrated Data and Voice Networks
  - 802.10 Network Security
  - 802.11 Wireless Networks
  - 802.12 Demand Priority Networks
  - 802.14 Residential Networks.

The results of the work of many of these groups are included in later chapters of this book.

- **National Institute for Standards and Technology** (NIST): an agency of the U.S. Department of Commerce (formerly the National Bureau of Standards) charged with providing an orderly basis for the conduct of business by maintaining measurement standards (http://www.nist.gov). NIST coordinates voluntary product standards and develops computer software and data communication standards for the federal government [Federal Information Processing Standards (FIPS)].

(3) Industry Forums

Producing and/or using organizations join in industry forums to promote the harmonization of specific technologies or the application or development of products. Often, their work is an important precursor for national, regional, or international standards. Some forums of interest to us are

- **ATM Forum**: an international nonprofit organization of producers and users committed to promoting the application of Asynchronous Transfer Mode technology through the development of interoperability specifications (http://www.atmforum.com)
- **Frame Relay Forum**: an association of producers and users committed to promoting the application of Frame Relay in accordance with national and international standards (http://www.frforum.com)
- **ADSL Forum**: an association of competing companies committed to the promotion of applications of Asymmetric Digital Subscriber Lines (http://www.adsl.com)
- **Telecommunications Information Networking Architecture Consortium** (TINA–C): an association of over 40 of the world's leading network operators, telecommunications equipment manufacturers, and computer manufacturers. They are committed to the definition, validation, and implementation of a *common* and *open* software architecture for the provision of telecommunication and information services (http://www.tinac.com).

REVIEW QUESTIONS FOR SECTION 1.3

1 Define standards.
2 Explain why standards are vital to the introduction of new capabilities.
3 Identify the principal global standards organizations.
4 Describe the three sectors of ITU.
5 Identify the principal regional standards organizations.

6   Describe the role of Committee T1–Telecommunications.

7   Identify the principal standards organizations in the United States.

8   Identify the other principal standards-creating bodies in the United States.

9   What is an industry forum?

CHAPTER

2

MODES OF
TELECOMMUNICATION

Before discussing the technology of telecommunication, it is helpful to consider the ways in which persons use telecommunication services. Their natural interests and prejudices have influenced the development of communications capabilities. In this chapter, the *softer* aspects of telecommunication are discussed.

2.1 ACHIEVING TELECOMMUNICATION

Telecommunication is achieved through the use of communication *channels* and communication *circuits*

- **Channel**: a one-way path between communicating entities
- **Circuit**: the combination of two channels, one in each direction, between communicating entities.

2.1.1 INFORMATION FLOW

There are three ways in which information can flow

- **Simplex** (SX): information flows in only one direction (a single channel is employed)

- **Half-duplex** (HDX): information flows in two directions, but only in one direction at a time. If a single channel is used, it must be *turned around* to allow information flow in the reverse direction. If a circuit is used, only one channel is active at a time.

- **Duplex** (DX), also known as **full-duplex** (FDX): information can flow in two directions at the same time. One of the following circuits may be employed
  - *symmetrical* —i.e., the channel capacity is the same in both directions
  - *asymmetrical*—i.e., the capacity of one channel (usually from source to receiver) is greater than the capacity of the other channel.

2.1.2 CONNECTION TOPOLOGIES

Channels and circuits are used to connect locations as follows

- **Point-to-point**: connects one location to one other location (e.g., a telephone connection); alternative nomenclatures are *one-to-one* and *intercommunication*

- **Point-to-multipoint**: connects one location to many other locations (e.g., a cable television connection that distributes programs from a head-end to thousands of homes); an alternative nomenclature is *one-to-many*. When the number of receiving locations becomes very large, we may use the description **point-to-all**. It is a connection that links one location to all possible locations (e.g., a broadcasting arrangement). Alternative nomenclatures are *one-to-all* and *mass communication*.

- **Multipoint-to-multipoint**: connects many locations to many locations (e.g., a conferencing arrangement or a local area network); an alternative nomenclature is *many-to-many*.

Figure 2.1 illustrates the relationships enumerated above. With regard to the multipoint-to-multipoint connection, three topologies—star, mesh, and bus—are shown. As far as they support the same communication capability, they are equivalent. If several stations send at once over such connections, the individual messages run together. In Figure 2.1, we show the procedure for communicating over a star. One station sends while all others receive; then another station sends while the others receive; and so on. This procedure works equally well for mesh and bus. In effect, the somewhat complex connections are reduced to a sequence of point-to-multipoint connections by a communication protocol.

2.1.3 COMMUNICATION PROTOCOLS

The existence of a connection is a *necessary*, but not *sufficient*, condition for communication. In addition, there must be procedures (called *protocols*) that organize and discipline the exchange.

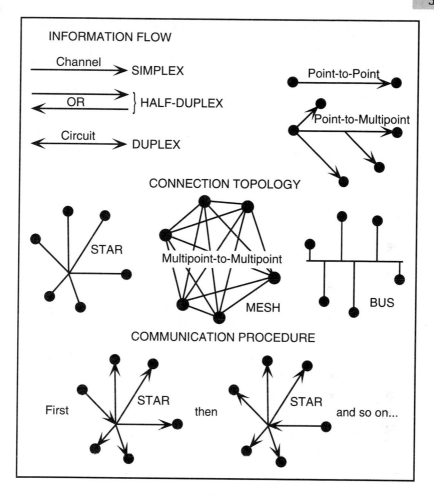

Figure 2.1 Examples of Information Flows, Connection Topologies, and a Procedure for Many-to-Many Communication

Connections are necessary but not sufficient prerequisites for communication. In addition, there must be an agreed-upon procedure as to how the participants will interact.

- **Communication protocol**: procedure that organizes and disciplines the process of sending messages so that one message does not corrupt another message, and all messages reach the intended receiver.

In data communication, a great deal is invested in procedures to ensure that messages are exchanged in an orderly fashion (see Chapter 8).

2.1.4 Types of Media

- **Medium**: agency that supports action-at-a-distance.

Persons who are remote from one another must make use of a medium to over-come interference from solid objects and span greater distances than the sound and reflected light waves of face-to-face encounters are able to do. This action requires conversion of the message to a form compatible with the medium employed. Broadly, based on the timeliness of the action they support, media are divided into two classes:

- **Record media**: by storing a limited amount of information for future refer-ence, record media make it possible to communicate *at a later time*. Further, if the storage medium can be transported readily, communication can occur *at another place, at a later time*. In addition, if the record medium can be dupli-cated in quantity, independent communication can occur with a large num-ber of persons, at a distance, at their convenience.
- **Real-time media**: make it possible to communicate at a distance, *at the same time*. Although segments from the information stream may be independently recorded using an appropriate record medium, no storage capability exists in a real-time medium. The price paid for real-time performance is that infor-mation present at one instant is destroyed by the information that succeeds it.

Record and real-time media may be used in combination to achieve *store-and-forward* operation. Messages are generated, sent to an intermediate location, stored, and forwarded to the recipient on demand. Electronic mail and voice mes-saging are store-and-forward services.

2.1.5 Electronic and Photonic Means

Electronic and photonic media, bearers, and transducers are used to implement telecommunication.

(1) Media

Examples of electronic and photonic media that are used to implement *record* com-munication are as follows:

- **Magnetostatic fields**: used to create records on tapes and disks
- **Electrostatic fields**: used to create records in electronic arrays
- **Mechanical or chemical imprints**: used to create holographic records.

Examples of electronic and photonic media that are used to implement *real-time* telecommunication are as follows

- **Electric currents**: flowing in telephone circuits, they enable signals representing the voices of talkers to pass to the telephones of listeners
- **Electromagnetic waves**: propagating through free space, or guided by coaxial cables, they carry television signals to millions of receivers or deliver digital signal streams to microwave relay points
- **Optical waves**: propagating along optical fibers, they carry high-speed streams of digital data between computer systems, or transport thousands of telephone calls between major telephone centers.

(2) Bearers

To implement real-time telecommunication, bearers carry electronic and photonic media. Thus

- **Copper wires** (*twisted* pairs) are bearers. They transport electric currents for the delivery of voice and low-speed data messages between parties separated by some distance.
- **Coaxial cables** are bearers. They guide electromagnetic waves to complete intercommunication connections that support very high-speed data communication between host computers, or distribute television signals to communities of users.
- **Free space** is a bearer. When excited by *omnidirectional* antennas it distributes electromagnetic waves to provide mass communications (e.g., broadcast television or radio programs). When excited by *directional* antennas, it transmits pencil beams of electromagnetic energy to complete intercommunication circuits.
- **Optical fibers** are bearers. They transport optical energy to complete intercommunication connections that support very high speed data communication between host computers or carry multiplexed voice and video signals between facilities.

(3) Transducers

- **Transducer**: device that converts messages in one medium to the same messages in a different medium.

So that signals may flow freely between sender and receiver, some transducers are found at the edges of a network where they perform the translation of user-oriented messages to network-oriented messages, and *vice versa*. Common conversions are speech-to-voice signals and keyboard-to-data signals. Examples of transducers that interface with users are as follows

- **Telephone instruments**: required at each end of a telephone circuit, telephone instruments convert sound waves to electrical currents, and *vice-versa*; they limit the message to the audio component of a face-to-face encounter

- **Keyboards** and **video display units**: required at the user's end of a connection to a database facility, they convert keyboard inquiries to digital signals and display the information retrieved

- **Television receivers**: required at the user's end of a television channel in order to receive television programs, television receivers convert complex signals to moving pictures and sounds

- **Facsimile machines**: required at each end of a communication circuit, facsimile machines convert printed material, diagrams, and pictures to electrical signals, and *vice versa*. They permit graphical information to be sent to a remote receiver.

Other transducers are found in the network where they link facilities to support the transmission of messages by dissimilar means. Common conversions are electric current to microwave or optical energy, and *vice versa*. Combinations of media, bearers, and transducers must be put together in a cost-effective way to provide communication channels whose characteristics are matched to the needs of specific services.

2.1.6 TELECOMMUNICATION NETWORKS

- **Telecommunication network**: an array of facilities put in place to provide connections to a large number of users so that they can exchange data, voice, or video messages.
  - From a *functional* standpoint, a telecommunication network carries messages from one user to another, on demand
  - From a *physical* standpoint, a telecommunication network consists principally of nodes (switches), links (transmission systems), and stations (terminals)
  - From a *theoretical* standpoint, a telecommunication network is a stochastic service system handling streams of random messages that seek routes between different pairs of users or among combinations of users. With many inputs and many servers, we might describe it as $M/E_k/n \to \infty$.[1]

The network provides as many routes as are required up to the limit of its resources. When the direct links are busy, the traffic will overflow to an alterna-

---

[1] Stands for a queueing model in which arrivals are Poisson-distributed, service times are Erlangian, and there is a large number of servers. D. G. Kendall, a British mathematician, introduced the notation system in 1951.

tive route formed by links that are not on the direct path. In this way, large networks are able to work around localized congestion or outages. When all routes are in use, further attempts to obtain service are refused, or the call requests are queued to await the release of facilities as calls are completed.

(1) Representations

Except for small installations, it is impossible to capture the full extent of a telecommunication network in a diagram. **Figure 2.2** shows two specialized views. In the top diagram, the network is represented by a space filled with links and nodes. As required, they are used to complete an exchange between two terminals. In most cases, the user is not aware of the routing through the network, only that a service is being implemented over the network *access* link. It is the responsibility of the network provider to see that the traffic is directed over as many internal links, and through as many nodes, as are needed to support the exchange between the originating terminal and the terminating terminal. If the exchange is duplex, this includes finding a return path so that the parties can interact appropriately.

The lower diagram illustrates some of the complexities of network access, including the range of providers that stand ready to carry the user's traffic. A major question is *Over which network can the information exchange be completed?* A closely related question is *What is the most efficient (usually meaning least cost) way to complete the task?* Before answering the questions, it may be as well to review the different categories of carrier:

- **Local exchange carrier** (LEC): company that terminates a collection of subscriber lines (the local loop) at a serving switch (the central office) or other routing device. Provides transport services directly between end-offices in a local access and transport area (LATA).

  - *Local Access and Transport Area.* A grouping of contiguous exchange areas served by a single LEC.

  Subscribers are treated equally, and facility costs are averaged over the subscriber base.

- **Interexchange carrier** (IXC): company that carries traffic between exchanges not located in the same LATA

- **Mobile carrier**:

  - company that operates terrestrial cellular radio facility to provide communication connections to (mostly) mobile subscribers in developed areas. Traffic is collected at a mobile telephone switching office (MTSO) that interfaces with LEC (and IXC) facilities

  - company that operates communication satellite system (geostationary, medium, or low Earth orbit—GEO, MEO or LEO) to provide communication connections to (mostly) mobile subscribers. Traffic may be delivered to terrestrial facilities.

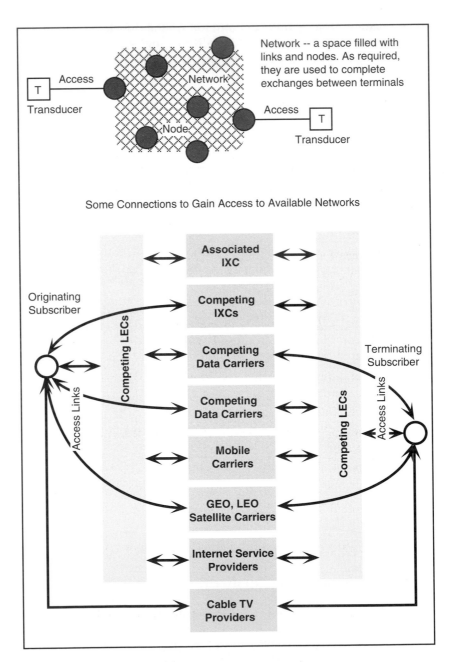

Figure 2.2 Two Views of a Telecommunication Network

In the upper diagram, the network is represented by a space filled with links and nodes. As required, they are used to complete an exchange between terminals. In the lower diagram, the network providers are shown. In many cases, the facilities of two or three carriers are needed to complete a connection.

- **Data carrier**: a value-added carrier that provides data communication services.
  - *Value-Added Carrier.* Besides transport, a value-added carrier provides additional services (e.g., error detection and correction).
- The carrier is likely to operate packet switches and lease transmission facilities from other carriers
- **Cable television carrier**: a cable television company that uses some portion of the cable bandwidth to provide data or voice services. May hand off traffic to other carrier.
- **Internet service provider** (ISP): a company that provides access to Internet facilities and supports data transport, information retrieval, and electronic mail services.

(2) Access Options

Setting the question of cost aside, I will enumerate some of the possibilities for access to the available networks:

- Route all calls to the end office of the serving local exchange carrier; the LEC will complete local calls and relay long-distance calls to an associated interexchange carrier or another IXC
- For long-distance calls, access the serving IXC facilities directly and bypass the LEC facilities
- For calls to mobiles
  - route calls through the serving LEC office to the mobile telephone switching office of a mobile carrier
  - use on-premises Earth station to access GEO satellite carrier
  - use mobile or portable equipment to access MEO and LEO satellite carriers.
- Route data calls through the serving LEC office to the central office of the serving data carrier
- Access the central office of the serving data carrier directly over private facilities
- Use private facilities exclusively.

The maximum bit rate depends on the facilities available and the acceptable error rate.

At home, individuals have an increasing number of choices. For many persons, access to half-duplex data, duplex voice, and simplex video (television) services is available from competing LECs and competing cable television providers. In turn, these carriers are likely to interconnect with an ISP, and on-line service providers, or perform these functions themselves. In addition, simplex data and television services may be provided by direct broadcasting satellite facilities, and data and voice services may be available from terrestrial mobile carriers or LEO satellite carriers.

(3) Message Handling

In what some call *plain old telephone service* (POTS), a path between the calling and called parties is established based on their telephone numbers. It is set up after the calling party enters the called number and before the terminating end office rings that number. That the right connection is made is confirmed when the called party answers. During the call, the circuit is maintained for the exclusive use of the parties—when they are talking and when they are not—and is subject only to the propagation delay of the transmission path and any delays in multiplexers or switches. At the end of the call, the facilities are returned to a common pool ready to be used to form a circuit that will complete some other call. This process can be described as *connection-oriented* with acknowledgment of the identity of the sending and receiving parties and confirmed receipt of understandable information. It is used for telecommunication when the messages exchanged cannot tolerate more than a short, constant delay. In addition, segments of the messages exchanged arrive in the sequence in which they were sent.

Data messages are handled in several ways depending on the following:

- How much delay they can tolerate
- Level of user interest in confirmation of delivery and message integrity
- Whether it is important that the message segments arrive in the sequence in which they were sent.

Figure 2.3 illustrates some of the options

- **Connection-oriented service**: a circuit is established between the sender and receiver (physical, as over a telephone network, or virtual or permanent virtual—see below—as over a packet data network). The receiver acknowledges receipt of message segments (packets) and is likely to report the detection of errors. The delay between packets may vary, but packets are delivered in the sequence in which they were sent. From the point of view of the data, the exchange is simplex; however, operation may be duplex or half-duplex, depending on the error-control procedures employed.
- **Connectionless service**: commonly provided over a packet network for short data messages. The receiver sends no acknowledgments. Packets carry originating and terminating addresses. At each network node, they wait their turn to be sent over a link that will get them closer to their destination. Subsequent packets are unlikely to follow the same path, so that the times they take to reach their destinations will vary and they may arrive out of sequence. The operation is simplex; any response is handled independently of the message received.

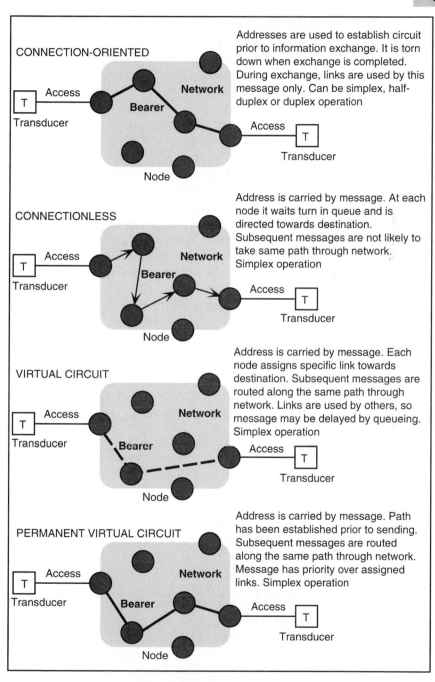

CONNECTION-ORIENTED

Addresses are used to establish circuit prior to information exchange. It is torn down when exchange is completed. During exchange, links are used by this message only. Can be simplex, half-duplex or duplex operation

CONNECTIONLESS

Address is carried by message. At each node it waits turn in queue and is directed towards destination. Subsequent messages are not likely to take same path through network. Simplex operation

VIRTUAL CIRCUIT

Address is carried by message. Each node assigns specific link towards destination. Subsequent messages are routed along the same path through network. Links are used by others, so message may be delayed by queueing. Simplex operation

PERMANENT VIRTUAL CIRCUIT

Address is carried by message. Path has been established prior to sending. Subsequent messages are routed along the same path through network. Message has priority over assigned links. Simplex operation

Figure 2.3 Types of Network Connections
Individual networks may support some, or all, of them.

- **Acknowledged connectionless service**: same as connectionless service except receiver confirms receipt of message.

- **Virtual circuit service**: commonly provided over a packet network for segmented data messages (packets) that are not sensitive to variations in intersegment delays, but must arrive at the destination in sequence. Delays between packets may vary. Each packet contains originating and terminating addresses that are used to ensure they are sent toward the destination over the same link as the previous packets were sent. An initial packet may be sent to establish the virtual circuit, or the necessary information may be stored with the routing information maintained at each node. The operation is simplex; any response is handled independently of the message received.

- **Permanent virtual circuit service**: commonly provided over data networks for longer messages that are not segmented, or segmented messages that must arrive in sequence and are sensitive to variations in intersegment delays. The path between sender and receiver is defined by information stored at the network nodes, and messages afforded this service receive priority over the designated links. The operation is simplex; any response is handled independently of the message received.

For data passed over data networks, the options range from unsupervised, but likely to be delivered in a reasonable time, connectionless service, to closely supervised, guaranteed delivery in a short time, permanent virtual circuit service. In between, acknowledged connectionless service provides some assurance for the sender that the message was received, and virtual circuit service provides a level of control and in sequence delivery that is adequate for many applications.

REVIEW QUESTIONS FOR SECTION 2.1

1 Define a channel and a circuit.

2 Define simplex, half-duplex, and duplex information flows.

3 Define point-to-point, point-to-multipoint, point-to-all, and multipoint-to-multipoint connection topologies.

4 Explain why the existence of a connection is a *necessary*, but not *sufficient*, condition for communication.

5 Define communication protocol.

6 Define medium, record medium, and real-time medium

7 What is store-and-forward operation?

8 Give three examples of media that are used to implement *record* communication.

9   Give three examples of media that are used to implement *real-time* telecommunication.

10   What is the function of a bearer?

11   Give four examples of bearers that implement real-time telecommunication.

12   What is the function of a transducer?

13   Give four examples of transducers that interface with users.

14   Where else are transducers to be found? What do they do?

15   Define a telecommunication network from a functional, physical, and theoretical point of view.

16   Enumerate the different categories of carrier available to carry user's traffic.

17   Enumerate the different combinations of carriers that can be used to complete an inter-LATA call.

18   What operational considerations determine the manner of handling a data call?

19   With regard to data traffic, define connection-oriented service, connectionless service, and acknowledged connectionless service.

20   With regard to data traffic, define virtual circuit service and permanent virtual circuit service.

2.2 PERSONAL COMMUNICATION

Person-to-person communication must take account of the human psyche and how it affects, and is affected by, the telecommunications environment. Among persons, communication occurs on a physical and intellectual level. Influenced by the emotional states of the participants, it involves messages that are expressed in spoken words, texts, and images (stationary or moving).

2.2.1 NATURAL COMMUNICATION MODE

The natural mode of person and person communication is caricatured in **Figure 2.4**. It can be described as two way, interactive, and *face to face*. For persons to engage in this style of communication, they must be close enough so that their speech may be heard and their body language may be seen. In practice, face-to-face communication occurs when the parties are within a few feet of each other. Pressure waves representing the words of the speakers are transmitted between the parties by gas molecules in the air; visible information is conveyed by light that is reflected from the parties and their visual aids.

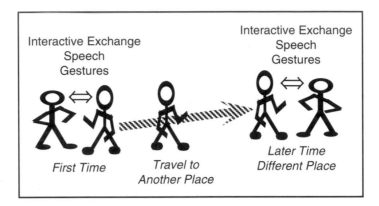

Figure 2.4 Two-Way, Interactive, Face-to-Face Communication Between Persons

This is the natural mode of communication between persons. By moving from place to place, messages are carried throughout a community.

Acoustic and optical waves are the physical media that make face-to-face communication possible. They accomplish action at a *short*-distance so that the process can proceed. In this arrangement, each party uses the full range of expressions of which they are capable—words and phrases emitted with appropriate emphasis, emotion, gestures, and body language. In addition, they may employ pencil and paper, or chalk and chalkboard, to illustrate points; use physical models or simulations displayed on computer terminals to demonstrate sequences; play audio or video recordings to reinforce perceptions; etc.

Once communication is completed, the participants may move to other places and use the information they have just gained in another encounter. In this way, information and ideas circulate within a community. The distance between the first and second meeting is whatever distance the traveler elects to travel, and the speed at which the information propagates is the speed of the traveler.

2.2.2 COMMUNICATION AT A DISTANCE

The essential property of a *medium* is its ability to produce action-at-a-distance. Persons who are remote from one another must make use of media to communicate. This action requires conversion of the natural message to a form compatible with the medium employed. In the conversion, much of the *richness* of face-to-face communication is sacrificed to obtain convenience and low cost. Nevertheless, for most purposes, telecommunication facilities provide acceptable extensions of personal communication capabilities so that persons can maintain communities of interest in increasingly mobile situations, and can embrace an expanding array of information sources.

By storing a limited amount of information for future reference, record media make it possible to communicate *at a later time*. Further, if the storage medium can be transported readily, communication can take place *at a distance, at a later time*; and, if the record medium can be duplicated in quantity, independent communication can occur with a large number of persons, at a distance, at their convenience. For example, after a book is written and printed, copies carry the author's ideas to a wide, contemporary audience that reads it at their convenience, and future readers can consult copies lodged in libraries for this purpose. By this mechanism, new generations of readers are able to build on the ideas, the ideals, and the discoveries of their forebears. In a similar fashion, audio and videotapes, and digital discs capture episodes of sights and sounds that can be reproduced in other places at later times. The caricatures in **Figure 2.5** illustrate the process of communication using a record medium.

Real-time media make it possible to communicate-at-a-distance, *at the same time*. Telephone, television, and radio are prime examples of communication systems that employ real-time communication media. Thus, so that persons who are separated from one another can discuss matters, the telephone can be used to connect them in pairs or conference groups. Moreover, as events happen, radio and television bring eyewitness reports to audiences as big as the world. From remote receivers, they deliver a continuous stream of information that mimics the observations of persons in proximity to what is going on. However, there is a price to be paid for real-time performance—the information present at one instant is destroyed by the information that succeeds it. Although segments from the information stream may be independently recorded using an appropriate record medi-

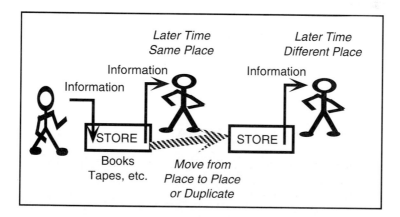

Figure 2.5 Communication Using a Record Medium

By storing information in books, or on tapes, a person can communicate with other persons at a later time. If the storage medium is transportable, and can be duplicated, it can be used to communicate with many persons, in many places, at their individual convenience.

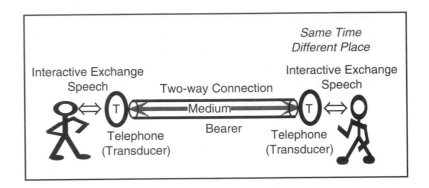

Figure 2.6 Communication Using a Real-Time Medium
The use of a real-time medium makes it possible to communicate at the same time, in different places.

um, no storage capability exists in a real-time medium. The caricatures in **Figure 2.6** show the process of telecommunication using a real-time medium. By way of illustration, the connection is divided into transducers (telephones), bearer, and medium.

2.2.3 THE SERVICES PLANE

Personal communication services cater to the exchange or distribution of messages that are associated with an extensive range of human interests. The richness of these services and the relationship between them can be diagrammed on the services plane.

(1) Message Components
Between and among persons, messages may employ symbols, sounds, or sights.

- **If they employ symbols,** they are likely to specify facts, to define entities, or to seek to inform. The information they contain is in the arrangement of letters, numbers, and shapes. If the arrangement is sequential, we have *one-dimensional* communication exemplified by the output of printers and video display terminals that support electronic mail services and other services that employ *text*. If the arrangement covers an area, we have *two-dimensional* communication exemplified by the output of facsimile machines that support graphics mail and other services that employ *graphics*.
- **If they employ sounds**, they are likely to discuss points, to describe situations, or to inspire feelings. The information is contained in the sequence and

intensity of the sounds and in their pitch. Together with the sequence and intensity of the elementary sounds, communication depends on time—unless the sounds are executed at the correct rate, the message is distorted and may be lost. This is *three-dimensional* communication. Three-dimensional messages are the products of telephone, radio, and similar entities that employ *audio*.

- **If they employ sights**, they are likely to present vistas, to witness events, or to assist with the visualization of ideas. The information is contained in the intensity of the images and in static or dynamic relationships among objects arranged in space. Thus, the messages contain two- or three-dimensional space, intensity, and time. The result is *four-* and *five-dimensional* communication. Four-dimensional messages are the products of cameras and displays that employ *video*. Five-dimensional messages may be the stuff of which *virtual reality* will be made.

(2) Richness

Using the concept of dimensions, an empirical measure called *richness* can be constructed

- **Richness**: a message is richer than another if it requires a more complex channel to deliver it. Channel complexity depends on the number of message dimensions.

Thus, video messages are richer than audio messages, audio messages are richer than graphics messages, and graphics messages are richer than text messages.

To put a measure to these statements, I have drawn a table in which bit positions represent dimensions. In **Figure 2.7**, bit position 1 is associated with one-dimensional communication (i.e., services that employ text). Bit position 2 is associated with two-dimensional communication (i.e., services that employ graphics). Bit position 3 is associated with services that employ audio, and bit position 4 is associated with services that employ video. The presence of a 1 in these positions defines the presence of the parameter. For instance, services that employ video are designated 1000; services that employ graphics are designated 10; etc. This notation permits us to construct designators for services such as television that employ both video and audio (1100), videoconferences that are augmented with facsimile (1110), telephone (100), and electronic mail (1). In addition, the notation permits us to interpret the designator as an empirical richness number. Thus, telephone is of richness 4 and electronic mail is of richness 1. We make use of these values later in this section.

(3) Asymmetrical Connections

Connections among pairs or groups of persons for a common purpose need not have the same richness in both directions. What is more, they are not constrained to assume the same topology in the forward and reverse directions. For

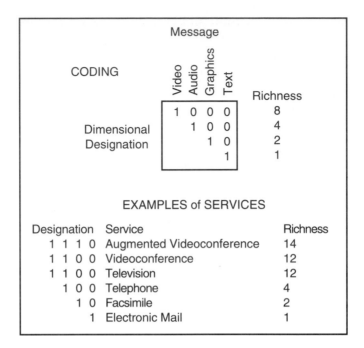

Figure 2.7 Richness of Text, Graphics, Audio and Video Messages, and Combinations

Richness is an arbitrary measure of complexity. One message is richer than another if a more complex channel is required to deliver it.

instance, **Figure 2.8** shows a concept for a telecommunication system that provides education at a distance (also known as distance learning). The sound and sight of the instructor, and the images employed by the instructor, are transmitted to all receiving sites, and, when authorized to do so, students at the sites may ask questions of the instructor. Under these circumstances, messages consisting of video, sound, and graphics (of richness 14) are distributed in simplex fashion from the instructor to the remote sites. When a student is given permission to ask a question, the site at which the student is located is connected to the instructor's site by a duplex voice connection (of richness 4). The discussion between student and instructor is retransmitted from the instructor's site to all other sites on the existing simplex connection. Thus, in one direction, the system employs a bearer that connects one to many and supports a medium that can transport messages of richness 14. In addition, on demand, the facility employs a bearer that connects one to one, duplex between the instructor and a receiving site, and supports a medium that transports messages of richness 4. When the duplex connection is made, that part of the richness 14, simplex channel that carries sound to the site at which the question originates, is disconnected (to prevent echoes and oscillations). It is replaced by local sound.

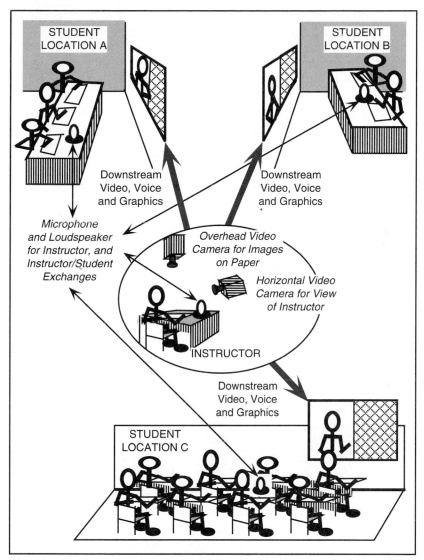

Figure 2.8 Distance-Learning Facility

(4) Directions and Information Space

The direction in which the more rich messages flow (i.e., instructor to receiving sites) is called the *downstream* direction and the direction in which the less rich messages flow (i.e., questioner to instructor) is called the *upstream* direction. Usually, one (perhaps more) communication mode is common to both directions. It is referred to as the common *information space*. In our tele-education example, the common information space is voice.

(5) Construction of Services Plane

Figure 2.9 shows a plane containing orthogonal axes that divide real-time media (right-hand side) from record media (left-hand side), and intercommunication (upper half) from mass communication (lower half). The four quadrants are used to segregate services according to the functions they perform. Further, if the

- Axes are marked off in richness in both directions from the crossing point
- The horizontal axis represents the richness in the upstream direction
- The vertical axis represents the richness in the downstream direction

then communication services can be positioned in the quadrants. This is shown in **Figure 2.10.** Because the richness of the connections in the downstream direction is always equal to or greater than the richness of the connections in the upstream direction, services can exist only in the upper 45° sector of the real-time media and intercommunication quadrant (upper right-hand side). Services that form the boundary line at 45° to the axes have the same richness in each direction; they share the maximum common information space. In the record media and intercommunication quadrant (upper left-hand side), the array of potential services follows a similar pattern—services exist only in the upper 45° sector. In the mass communication half-plane (lower half-plane), there can be no upstream messages and no common information space, so that all services are located along the vertical axis.

(6) Notation

The position a service occupies on the Services plane is defined by the richness of the downstream and upstream messages. If they employ record media, intercommunication services can be represented by a combination of two numbers (the richness of the downstream message followed by the richness of the upstream message) enclosed in square brackets, e.g., $[x,x]$, and, if they employ real-time media, in curly brackets, e.g., $\{x,x\}$. Mass communication services can be represented by a single number (denoting their richness) enclosed in the same styles of brackets, i.e., $[x]$ or $\{x\}$.

2.2.4 COMMUNICATION BY TELEPHONE

A significant fraction of all personal telecommunication is conducted over the telephone. The parties exhibit a sense of purpose and recognize that they are participating in a continuing process.

(1) Communication Among Persons

Persons are endowed with free will so they may communicate selectively. For senders, free will gives the freedom to attempt to initiate communication when it is convenient for them, and to seek to communicate with whoever

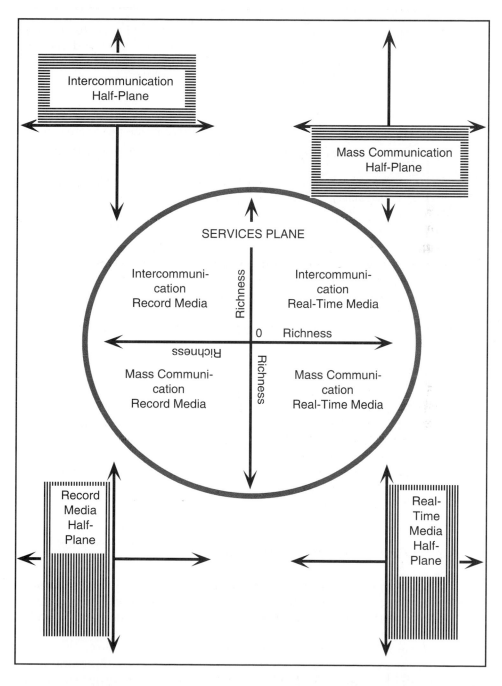

Figure 2.9 Development of Services Plane

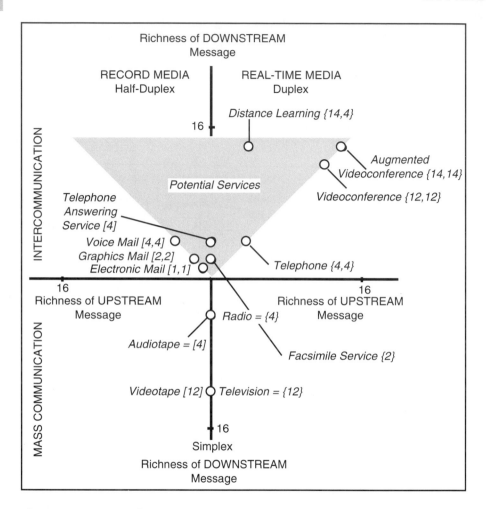

Figure 2.10 Services Plane

Downstream is the direction in which richer messages flow. Thus, services over half-duplex and duplex channels are limited to the triangular area in the intercommunication half-plane. Services over simplex channels are limited to the vertical axis of the mass communication half-plane.

appears to be able to satisfy their objectives. For receivers, it gives the right to refuse to communicate and to treat senders' messages in different ways. While some telephone messages demand immediate attention, others can be afforded less priority, some can be ignored completely, and some may be interrupted if the receiver loses interest or has other things to discuss.

(2) Using the Telephone

The activity associated with the use of the telephone occurs in three stages.

- **Physical stage**: establishing a path along which signals can flow between users. Under the supervision of software, procedures initiated by the originating user execute this task. It requires compatible terminals and network equipment.

- **Procedural stage**: ensuring that the information is in a common language and is exchanged between the right parties. At this stage, the parties check each other's authority to communicate and ensure that their information is exchanged in compatible formats. This task employs the routine intelligence of the users.

- **Intellectual stage**: managing the flow of information to create an understanding of the message. This stage is peculiar to person and person communication—it employs the users' intellectual capabilities to convert the information exchange into the messages that they wish to impart. Achieving a common understanding depends upon the intellectual processes employed by the users, and mutual comprehension is not always possible.

(3) Need for Judgment

For a telephone conversation to be completed successfully, there is need for a common understanding of the telecommunication process on the parts of the participants. Fortunately, the procedures are common knowledge so that the calling and called parties have a high expectation of creating the connection and of engaging in a productive communication session. That the protocols are relatively simple is due to the presence of intelligent beings at both ends of the connection. They exercise judgment—judgment to know that the connection is made, that the right person has been reached, when to speak and when to listen, whether the message is being received, and that the information provided has been understood. Of course, things can go wrong. The calling party may have the wrong number or may have the right number but enter it incorrectly. The called party may answer in a tongue that the calling party does not understand, causing one of the parties to change languages or to hang up. Before the calling party has completed the message, the called party may lose interest and terminate the call. The connection may be noisy, making it necessary for the parties to speak clearly, listen carefully, and repeat parts of their conversation. These are all situations in which judgment must be exercised. They include the possibility of starting over from the beginning.

2.2.5 COMMUNICATION BY VIDEO CAMERA

From time to time, video telephone service is suggested as a natural extension of voice telephone service that will benefit the users and exploit the rapidly expanding technical capabilities of public telephone networks.

(1) Videotelephone

Requiring both a video signal and a voice signal, major obstacles to the implementation of videotelephone have been the large bandwidth required by the visual information, and agreement on a common video format. However, ITU–T has standardized a coding technique for video telephone that provides communications-quality video over two 64-kbits/s digital channels—one for voice and one for compressed video. Meanwhile, taking advantage of even more efficient coding techniques (see Section 3.5.3), organizations such as AT&T, Northern Telecom, and NEC produce equipment that carries both voice and video frames over a single voice channel. In addition, individuals are using Internet to provide a variable-quality, simulated, one-on-one videotelephone service.

Reactions to videotelephone service are mixed and vary with time. After an initial fascination with the technology, continued use brings to the fore some of the limitations of the technique. Sitting in front of a camera attempting to maintain eye contact with the caller is a discipline we are not used to. What we are used to is the telephone, the freedom to lounge comfortably while talking, and to dress, or not to dress, as we choose. Eventually, there comes a realization that the videotelephone introduces restrictions that outweigh the advantages of being seen—at least for routine communications. What is more, there is no clear evidence that the addition of video to a telephone call provides more information than the callers can exchange by voice alone.[2]

(2) Videoconferences

To accommodate the large-screen requirements of videoconferences, ITU–T has recommended coding techniques that provide full-color, full-motion pictures and sound at 1.5 Mbits/s, and fractions thereof. In contrast to the video telephone, videoconferences are an acceptable mode of communication for many business purposes. With the miniaturization of many essential components, and the development of powerful video compression techniques, ad hoc videoconferences can be implemented between offices in most parts of the world, on demand.

2.2.6 COMMUNICATION BETWEEN PERSONS AND MACHINES

When messages are exchanged between persons and computers, data processors, file servers, and like machines, they are expressed in alphanumerics and graphics, and the communication session is governed by rules that have been programmed into the application. To communicate effectively, persons must abandon their nat-

---

[2] A. Chapanis, "Interactive Communication: A Few Research Answers for a Technology Explosion," *8th Annual Conference of the American Psychology Association, Toronto, Canada*, (Baltimore, MD: Johns Hopkins University Press, 1978).

ural mode of communication—free-flowing speech and gestures—and adopt the use of keyboards and video displays. With them, the user can select functions, enter commands, and provide whatever data may be involved in the transaction. In return, the user receives text and/or graphics, and cryptic requests for further actions. Provided the user is well trained in the operation of the terminal and understands the performance of the application being executed, such arrangements work well. They are the foundation of the knowledge-based activities being pursued by an increasing number of workers.

Key to communication mediated by data terminals is the software that executes the data exchange functions. If it is too complex, it is likely to be slow and cumbersome; too simple, and it might not protect the user from choices that cause gridlock and loss of messages. Like other machine-based operations, communication by data terminal can be described as user-friendly, user-hostile, or user-seductive.

- **User-friendly**: by most persons, the rules governing the operation of user-friendly services and equipment can be learned readily, and they cause little or no increase in anxiety level in users
- **User-hostile**: by most persons, the rules governing the operation of user-hostile services and equipment are difficult to learn, and they cause a significant increase in anxiety level in users
- **User-seductive**: most persons are fascinated by the services and equipment and are eager to use them; they dispel anxiety and produce pleasure in the user.

Early attempts at machine-based communication were user-hostile; to employ them required the user to subordinate the need to communicate to the need to present and handle messages exactly as the software demanded. With time, the importance of making them easy to use was recognized, and contemporary, data-terminal-mediated communications systems exhibit a substantial degree of user friendliness. Applying the descriptors to a broader range of services and equipment, for many persons, telephone, radio, and television are user-seductive, and most VCRs are user-hostile.

REVIEW QUESTIONS FOR SECTION 2.2

1 Describe the natural mode of person and person communication.
2 Why does communication at a distance require the use of a medium?
3 What capabilities does the use of a record medium bring to communication?
4 What capability does the use of a real-time medium bring to communication?

5 Explain how the dimensions of symbols differ from sounds, and sounds differ from sights.

6 Define the state of one message being richer than another message.

7 Discuss an empirical notation that can be used to describe the richness of a service.

8 Using the idea of richness, define the upstream and downstream directions of information flow.

9 What is common information space?

10 Explain the construction of the Services plane and describe the areas in which services can exist.

11 Explain the meanings of $[x,x]$, $\{x,x\}$, $[x]$, and $\{x\}$.

12 Persons are endowed with free will. What behavior might this property produce in persons using the telephone?

13 Describe the three stages of using the telephone.

14 Why does the successful completion of a telephone call require the exercise of human judgment?

15 List some common reactions to videotelephone service.

16 Describe the meanings of user-friendly, user-hostile, and user-seductive, terms that describe communication between persons and machines mediated by data terminals.

2.3 PERSONAL SERVICES

Some personal communication services are based on the transfer of information in the form of text and/or graphics. The information exists at the sender's station in physical documents or on video screens. Others preserve the message in electronic form and serve many of the non-real-time needs of businesses and households.

2.3.1 DOCUMENT SERVICES

At the receiver, the message appears as hard copy (i.e., on paper) or soft copy (i.e., on a video screen). The principal methods employed are facsimile and teletex.

(1) Facsimile

- **Facsimile** {2}: a communication service that captures images from paper (or a video screen), transmits them to a distant site, and reconstructs them. Facsimile is designated a real-time service (of richness 2) because the image is transmitted immediately as it is formed, and reproduced immediately as it is received.

In a facsimile system, the item that is copied is a document that may contain typed and printed matter, handwritten notes, and images in the form of graphs, diagrams, photographs, etc. Most facsimile systems scan the original document to detect black, white, and gray-scale patterns, convert the patterns into analog or digital signals, transmit them over a telephone connection to a remote site, and produce a replica of the original. Times to complete the transmission range from minutes for older equipment to seconds for newer equipment (see Section 10.1.3). Other facsimile systems are able to code, transmit, and reproduce replicas of color documents.

(2) Teletex

- **Teletex** [1]: a communication service that transports character-based messages to remote sites where they are printed out or displayed on a video screen. Teletex is designated a record service (of richness 1) because the message is composed and stored until the receiver is available, then transmitted and stored again until the recipient is ready.

In 1983, recognizing a demand to link sophisticated terminals, several telecommunications organizations inaugurated a service called Teletex to promote communication among electronic typewriters with storage capabilities and communicating word processors. At a Teletex terminal, texts are composed, edited, and stored, and the terminal may be used for local functions independent of the transmission and reception of messages. The basic character set includes all letters, figures, and symbols used in Latin-based languages, and the user may add national characters as required. Text to be transferred to another terminal is assembled in memory and transmitted automatically once a circuit has been established with the receiving terminal and a send instruction has been given by the receiver. Teletex allows each terminal to handle the forwarding and receipt of its own messages.

Teletex provides direct electronic document exchange between electronic typewriters, word processors, or personal computers equipped with sending and receiving storages and communications capabilities. Once a connection is completed, formatted documents are exchanged between the sending storage of the calling machine and the receiving storage of the called machine. Transmission speeds depend on the common capabilities of the machines involved and the channel chosen. Teletex employs an 8-bit code set for alpha numerics, graphics symbols, and control characters. In addition, the calling procedures require the exchange of 72-character reference information frames before the exchange of messages. The frames contain the called terminal identifier, the calling terminal identifier, date and time transmitted, and the document reference.

(3) Mixed-Mode Documents

Mixed-mode documents contain fields of text (that can be handled efficiently by teletex) and fields of graphics (that can be handled efficiently by facsimile). To preserve the ability of the recipient to edit and process them, they are exchanged in blocks that contain characters and blocks that contain bit-maps. This is accomplished by describing the document as a set of logical objects and as a set of physical objects that conform to a specific layout. The document's content, structure, and attributes are represented as a series of protocol elements that are exchanged between the transmitter and receiver.

2.3.2 MESSAGE SERVICES

Electronic mail services facilitate non-real-time exchanges of text, graphics, or voice messages. They transport messages point to point between persons, or distribute them within communities of interest. Called *email*, it has the advantage over voice telephone service in that no one needs to be there to answer and over normal mail in that delivery is made at electronic speed. They are store-and-forward message services in that messages are stored to await requests for delivery from the intended recipients. They are implemented as text mail, graphics mail, and voice mail.

(1) Text Mail

- **Text mail** [1,1]: an intercommunication record service of richness 1 with point-to-point (sometimes point-to-many) simplex topology. Composed at, and delivered by, data terminals, the text message may be stored in a central system of electronic mailboxes, or stored at one or another of the terminals serving the persons involved. When the intended recipients command, the messages are retrieved, forwarded (if necessary), and displayed on a video terminal or printed.

For many years, communicating teletypewriters and teleprinters have provided a text message service. Called *Telex* service, it began in the early 1900s and developed into a worldwide text communication service with some 2 million users. In its simplest form, the operator composed the message at the keyboard of the teletypewriter. This machine produced a strip of paper tape in which the message was encoded as punched holes. When ready, the operator contacted the intended receiving station and passed the punched paper tape through the reader on the teletypewriter. This produced a signal at some 30 to 50 bits/s in 5-bit Baudot code that mimicked the message entered by the operator. If direct communication was not possible, the operator sent the message to an intermediate station where a paper tape copy was made and used later to relay the message to its destination. Improved over the years by the introduction of automatic features, and later with electronics, it still provides slow but reliable service. In some parts of the world, Telex service is more readily available, and more dependable, than telephone service. In many countries, Telex is being upgraded to Teletex.

For those with a personal computer and Internet access, or a data terminal connected to a commercial data network, many software packages are available to implement electronic mail. With remarkable speed, these facilities enable users to send text messages to others with similar equipment. Usually, the service includes the ability to reply to incoming messages and to *file* messages in electronic storage, as appropriate.

(2) Graphics Mail

- **Graphics mail** [2,2]: an intercommunication record service of richness 2 with point-to-point (sometimes point-to-many) simplex topology. Entered in facsimile machines, graphics messages are reconstructed on similar machines that serve the communicating parties. Transmission between senders and receivers is rapid; usually, the messages are reproduced on paper. It is the storage medium that preserves the messages until the recipients retrieve them.

For more than 100 years, pictures have been sent over wires. In recent years, the desktop facsimile transceiver has become an important adjunct to business. Because ideographs are difficult to break down into a limited number of keystrokes for typewriter entry, graphics mail is particularly useful in countries where the written language employs them. In Japan, for instance, graphics mail is much more important than text mail, and the Japanese have pioneered in facsimile communication systems. All the features of store-and-forward operation can be employed. Delayed delivery, multi-address delivery, and mailbox delivery are of most interest to users. Desktop facsimile units are commonplace in industrial and commercial organizations. Connections are made on an ad hoc basis over telephone lines, or other circuits, as needs arise.

(3) Voice Mail

- **Voice mail** [4,4]: an intercommunication record service of richness 4 with point-to-point (sometimes point-to-many) simplex topology. Recorded using telephone connections to voice storage devices (tapes or solid-state memories), voice messages are delivered on commands from the addressed station. Delivery may be by the storage devices directly (answering machines at the recipients' telephones, for instance) or over telephone connections initiated by the recipients.

Voice mail records the spoken words of the senders, stores their messages, and delivers the sounds of their voices to the recipients when they request their messages. In addition, voice mail may

- Deliver general announcements
- Be used for message collection
- Furnish call-answering capability, both for no answer and line busy conditions
- Deliver messages automatically and keep trying until the message is accepted or refused
- Record both sides of a telephone conversation.

Control is exercised through the touch-tone buttons on the telephone set. The system prompts both sender and receiver, asking them questions to elicit essential information that they key in on the touch pad. Voice mail services are offered by a number of North American organizations. They route messages over telephone networks to central facilities where they are stored in electronic mailboxes until the recipients request they be forwarded, or they are handled under special arrangements made by the senders. In the public telephone network, advanced management support services incorporate versatile artificial voice announcement and instruction capabilities, and electronic private branch exchanges (PBXs) are available with voice message service facilities.

2.3.3 INFORMATION RETRIEVAL SERVICES

A major personal activity is the retrieval of information from distant databanks. Implemented originally as Videotex and Teletext, much of the demand today is for information over the World Wide Web.

(1) Videotex and Teletext

At the beginning of the 1970s, the intercommunication and mass communication authorities in the United Kingdom demonstrated systems that allowed large numbers of users to call for text or graphics information and display it in color on the screens of modified television receivers. Using a common display format, one service employs the telephone network in an interactive way to search through databanks containing thousands of pages of information; the other service supplies a hundred or more pages in the vertical blanking interval (VBI) of a broadcast television signal without affecting the television service. The first service was called Viewdata, and the second is called Teletext. Together, the two services were known as Videotex, with Viewdata being described as wired-Videotex and Teletext described as broadcast-Videotex. Similar systems were quickly developed in France and Japan, and a little later in Canada. In 1983, CCITT (now ITU–T) recommended the substitution of the name Videotex for Viewdata. Accordingly, the two services are now referred to as *Videotex* and *Teletext*.

Figure 2.11 shows the basic facilities required for the two services. In 1994, British Telecom discontinued its videotex service (service name Prestel). In France,

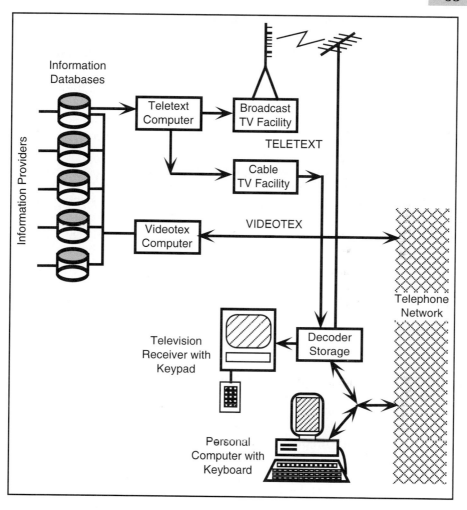

Figure 2.11 Videotex and Teletext Systems
Videotex provides encyclopedic information over the telephone network. Teletext provides an electronic magazine service; it is carried in unused lines in the television signal.

the communications administration continues to support its Videotex service (service name Minitel), which now supplies over 25,000 services.[3] In Europe, some 15 administrations provide Teletext. Pages from many of these services are available on the World Wide Web.[4]

---

[3] www.jou.ufl.edu/people/faculty/dcarlson/ehistory.htm
[4] www.demonco.uk/tssp/teletext/intro.htm

(2) Teletext—Electronic Magazine

In many ways, Teletext [2] adapts the capabilities of Videotex to a mass market. Like television, everyone can use Teletext at the same time without encountering a busy signal. In addition, it is cheaper to use, since it requires no telephone connection. These advantages are obtained by restricting the number of pages of information to that which can be made available in a reasonable time. It is the time the user is willing to wait before turning to more familiar sources such as newspapers, guides, and magazines. The challenge for the service provider is to create an electronic magazine that contains information that will be seductive, causing users to prefer Teletext to other sources of information.

The number of text characters or picture elements transmitted with each frame of the television picture controls the waiting time for page capture. In present broadcast systems, 4 lines (in the VBI) out of every 525 lines (in a complete picture) are used to carry mass Teletext information. With four to six pages of text transmitted each second, the average waiting time for a particular page in a 100-page magazine will be about 10 seconds. Dramatic decreases in search time can be achieved if all 525 lines are used. By dedicating an entire channel to Teletext information, consulting a 100-page electronic magazine is virtually instantaneous; and consulting a 12,000-page electronic magazine would take no more time than present Teletext. Several cable television systems devote one or more channels to Teletext.

Inserted within the envelope of the broadcast television signal, Teletext signals must conform to limitations set by the broadcast spectrum. Further, the time to transmit a single page must be as short as possible. Thus, while Teletext is compatible with Videotex, it uses some of the capabilities in moderation. Slowly, the individual national systems have been brought together into common specifications by ITU–T and other bodies. The current World System Teletext specification is compatible with previous standards (which are subsets of it).

(3) Competition

The ready availability of more familiar, alternative reference sources, and the lack of a clear need that Videotex and Teletext can assuage better than anything else, is a deterrent to their expansion. For instance, when the user seeks information normally contained in the newspaper, more than a 15- to 20-second wait at the terminal—while the receiver is switched on, warms up, and the connection is made (Videotex), or the proper page is seized (Teletext)—may be enough to discourage electronic searching in favor of leafing through the newspaper. However, such a wait is acceptable if the information can be obtained only by going to the public library.

(4) World Wide Web

To North American users, Videotex and Teletext are largely unknown—television is for entertainment, not serious information. Those who need reference materials have on-line searching capability using their personal computers, modems, and the many specialized computer-based information retrieval services available to them. Further, all of the public functions performed by Videotex, and many more, are available on the World Wide Web—the pervasive information retrieval service supported by Internet (see Section 12.5). With the opportunity to present text, colorful graphics, and audio and video on command, Webmasters are able to create composite information fields that suit the interests of the user. Their dissemination depends only on the ability of the receiver to support software of the requisite complexity. For many, such multimedia documents are user-seductive.

REVIEW QUESTIONS FOR SECTION 2.3

1   With regard to communication services, describe facsimile.
2   With regard to communication services, describe teletex.
3   Under what circumstances is electronic mail (a) better than the telephone and (b) better than surface mail?
4   Describe text mail, graphics mail, and voice mail.
5   What is Telex?
6   Describe Videotex and Teletext. Why is one called an electronic encyclopedia and the other an electronic magazine?

2.4 ELECTRONIC MEETINGS

Meetings vary in purpose, size, and format

- **Large forums**: papers are exchanged, graphics are used, and views and ideas are argued back and forth by persons whose ambitions and objectives may be at odds with one another
- **Small gatherings:** groups of like-minded persons in which the outcome depends solely on the persuasiveness of the spoken word, and the faith and trust each person has in the others.

Millions of meetings occur each working day. More than half of them last less than one hour, require no visual aids, and are between persons in the same building or the same facility. For some of the remaining meetings, participants travel from somewhere else to the meeting site. In doing so, they deprive their home locations of their presence, use up time in traveling, and may encounter stressful situations that impair judgment (e.g., jet lag, accidents, or terrorism). Insofar as electronic meetings provide an extended meeting space among remote sites, they can reduce the need for travel, save time, and contribute to the well-being of the individual. In short, electronic meetings should improve productivity.

2.4.1 Use of Electronic Media

With electronic media to provide an extended information space, meetings can be conducted so that those who participate are able to stay at their home sites. They do not have to travel to a designated point to share a common physical space. If a small number of sites are involved, the participants may interact freely over communication channels that are likely to be equally rich at all locations. If a large number of sites are involved, a chairperson must control interaction, and it is likely that one site is more important than the others are. Under this circumstance, the channels from the more important site to the other sites may be richer than the channels from them to the primary site.

The effect of limiting the dimensions of the exchange by introducing electronic media is not completely understood. What is known is that the degree of satisfaction with electronic meetings expressed by the participants depends directly upon the

- Richness of the channels employed
- Degree of familiarity of the participants with one another.

It varies inversely with the

- Complexity of the topics discussed
- Number of persons and/or locations involved.

From these reactions, providers of electronic meeting services conclude that

- For routine discussions among a few peers who are well known to one another, the richness of good-quality audio channels may suffice
- For unique discussions among several levels of management in several locations, the richness of full-color television may be insufficient
- For deliberations among technical persons, computer conferencing may be the ideal medium.

2.4.2 Range of Electronic Meetings

For intercommunication applications, the spectrum of electronic meetings extends from computer conferencing to videoconferencing (that may be augmented by graphics mail). **Figure 2.12** shows the set of electronic meeting alternatives.

- Computer conferencing [1,1]
- Audio conferencing {4,4}, which may be augmented by text or graphics {5,5 or 6,6}

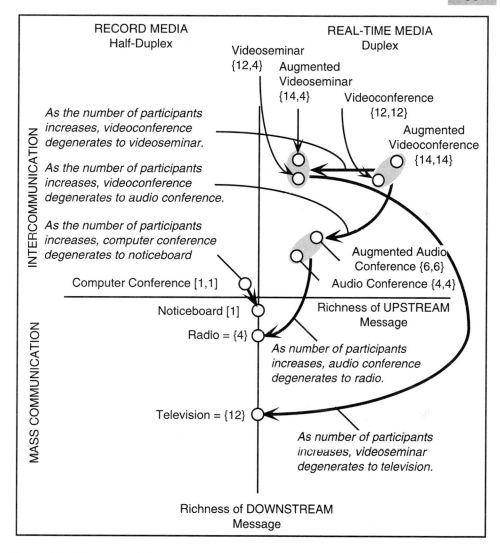

Figure 2.12 Spectrum of Electronic Meetings

As the number of participants, and/or locations, increases, individual participants are less able to use the return channel(s). The communication capabilities degenerate, causing services to migrate to less rich alternatives.

- Videoseminar {12,4 or 14,4}
- Videoconferencing {12,12 or 14,14}.

As the number of participants and/or locations increases, each participant is less able to use the return channel(s) and the communication capability degenerates.

The services plane can be used to visualize the migration of one service into another. For large numbers of participants

- Computer conferencing [1,1] degenerates to a computer notice board [1]
- Audio conferencing {4,4} degenerates to radio {4}
- Video conferencing {12,12} and augmented videoconferencing {14,14} may degenerate to
  - videoseminar {12,4 or 14,4}, and eventually to television {12}
  - audio conferencing {4,4 or 6,6} and eventually to radio {4}.

2.4.3 COMPUTER CONFERENCES

Computer conferencing is an ideal technique with which to support activities directed to producing a report or position paper. Participants can compose segments of the report, criticize one another's contributions, have access to resources at their home locations, and ponder issues without off-site distractions. In fact, computer conferencing may be an ideal medium for developing consensus without being swayed by zealous proponents of one position or another. To implement it, each participant has a terminal that is connected to a central computer equipped with conferencing software. To reach consensus on an assigned topic, participants pass messages to the entire group, a subset, or a single individual. The product of the conference activity is stored and made available to all participants as their work proceeds. The arrangement provides electronic interaction among a restricted group of individuals for a common purpose. The computer attends to the exchange of messages, the maintenance of files, and the assembly of the body of data that is the outcome of the computer conference. Like an electronic mail system, the participants do not have to be involved simultaneously. Unlike an electronic mail system, the participants share a common data space and message community that do not have to be defined each time and can operate on common information without having to build it each time.

Using Internet (or an intranet, see Section 12.8), persons with personal computers can engage in *desktop* conferences. Individual participants use their *browsers* to connect to a conference server with a multipoint control unit (MCU). They send data messages to one another that represent text and (possibly) diagrams. The MCU establishes the conference, sets up subnetworks so that factions can communicate privately among themselves, and generally supervises the participants (who joins, who withdraws, etc.). **Figure 2.**13 shows an arrangement for desktop conferencing using Internet.

2.4.4 AUDIO CONFERENCES

In a relatively primitive form, audio conferencing has been available for a long time. For groups small enough to gather around a desk, it can be implemented

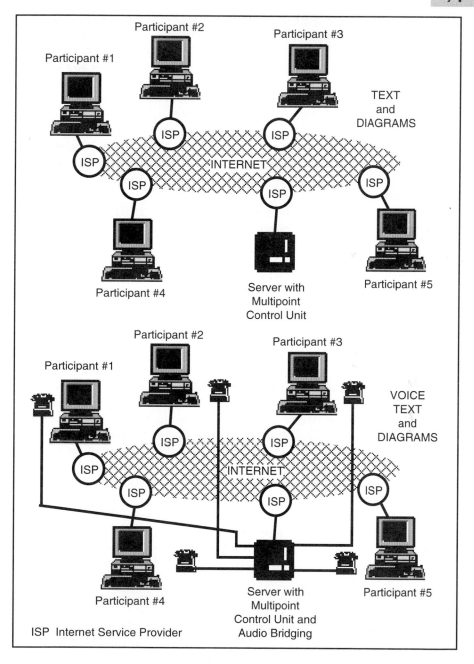

Figure 2.13 Desktop Conferencing Using Internet

In the upper diagram, the conferencing medium is text and possibly graphics. In the lower diagram, the participants employ voice also. In this instance, telephone connections are bridged at the multipoint control unit.

between two or more locations using speaker phones, or special conference telephones can be used in small conference rooms. For those served by modern telephone switches, a three-way conference (using telephones) can be established unaided, and the conference can be extended one station at a time to encompass larger groups (see Section 10.5.2). The call originator can connect conferences of around six persons, and, by calling each party in turn through the public switched network, larger groups of persons can be connected by an attendant. Very large groups of persons can call a *meet-me* bridge at a prearranged time for connection in a conference arrangement, or they may join a meeting already in progress. Usually, as a precaution against uninvited listeners, an attendant who has a list of those invited to participate supervises the connections. If privacy is important, the connection procedure may also include the use of a password. Unless there is a prearranged protocol administered by a strong chairperson, in large groups individual recognition and participation in the discussion is difficult, and most users must be content to listen.

For audio conferences, telephone connections are far from ideal. For one thing, the users hear the sum of whatever background noises is present on all the circuits. This includes both noises in the environment local to each user, as well as noise present on the circuits themselves. Tests have shown that relatively insensitive microphones worn by the users are better than one or two sensitive microphones that pick up sounds from an entire group at a particular conference location. Other tests have shown that the overall teleconferencing environment is improved by using a second channel to produce a stereo effect that assists the listeners to differentiate between persons who are speaking at the remote locations. In addition, multichannel sound has been reported to improve the conference ambiance. Above all else, participants seem to value good-quality sound.

For many purposes, it is helpful to augment voice conferences with text or graphics. When equipment or functions need to be described, or the relationships among variables need to be illustrated, there is no substitute for a picture; and when financial statements need to be analyzed, there is no substitute for properly prepared tables. The lower diagram in Figure 2.13 shows how this can be accomplished using desktop conferencing. Each participant uses the company PBX, or the public telephone system, to connect to the multipoint control unit. There, the voice signals are bridged together and returned to the participants.

2.4.5 VIDEOCONFERENCES

The concept of a videoconference between two sites is shown in **Figure 2.14**. For multilocation conferences, full video coverage of all rooms at all times requires that each location is connected to every other location by a separate video channel. For n locations, each location is connected to $(n-1)$ other locations so that there must be $n(n-1)$ video channels. Such an arrangement becomes expensive in a hurry!

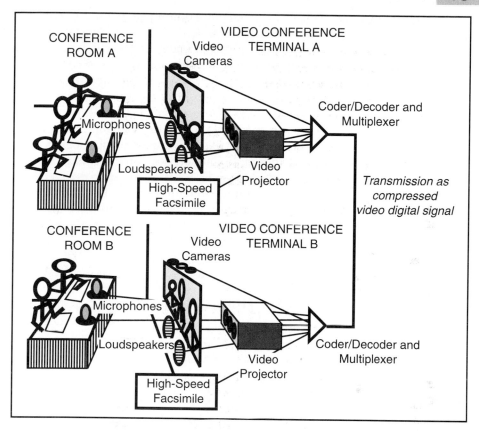

Figure 2.14 Implementation of a Videoconference

As an alternative, a video switch can be employed so that only the scene from the conference room in which the speaker of the moment is located is distributed to all other locations. This reduces the video connections to each location to one two-way connection, but limits the video space of the participants; they are required to look at the speaker and cannot look at others to judge reactions, etc.

For conferences among a number of locations, some observers find it confusing to see different scenes as the conversation shifts from site to site. In some systems, the video switch is voice activated. So long as the conversation is orderly, this technique works reasonably well—but if persons at different locations interrupt one another, particularly if they shout, the video may switch back and forth in a haphazard fashion. Alternatively, one speaker may capture the system by talking rapidly, and without pauses, so that no one else can speak. The convenience of voice-activated switching can be defeated unless participants adopt a cool, calm, permissive style of conversation.

Often, when conferences take place among a large number of locations, one location is more important than the others are—company headquarters for instance. Many meetings can be arranged using one-way video, audio, and graphics connections from headquarters to all sites, and an audio connection from each site to the chairperson at headquarters. This arrangement is similar to the distance-learning example discussed earlier. It is convenient when the primary site wishes to share information with the secondary sites and wants the secondary sites to ask questions about, or to react to, the information imparted.

If video is judged important, desktop conferencing (see Sections 2.4.3 and 2.4.4) can be augmented so that a relatively low quality image of the speaker is displayed at all workstations. However, it can become expensive because a camera is needed at each station, and it can be frustrating because the speaker must remain in the field of view. Furthermore, the MCU will have difficulty deciding whose image to distribute if several participants talk at once—or if no one speaks for some time.

REVIEW QUESTIONS FOR SECTION 2.4

1 Comment on the following statement: electronic meetings should improve productivity.

2 On what factors does the satisfaction of the participants depend?

3 Computer conferencing, audio conferencing, videoseminar, and videoconferencing are four modes of electronic conferencing. Use the services plane model to explain why, as the number of participants and/or locations increases, they become equivalent to a computer notice board, radio, and television.

4 Describe some alternative implementations for videoconferencing.

2.5 HUMAN FACTORS

Since the late 1960s, technical journals, professional magazines, and daily newspapers have extolled the future promises of *wired* cities, *paperless* offices, and *electronic* cottages. Driven by technology, a burgeoning array of messaging and meeting services that seemed to fit many business needs have been implemented. Yet, with some notable exceptions, communication in many industries continues in much the same way it did 30 years ago. Therefore, we are forced to conclude that there is more to personal communications than the availability of high-technology facilities.

2.5.1 ELECTRONIC MESSAGE SERVICES

Electronic message services have been readily accepted by scientific, professional, and commercial organizations. Usually, little difficulty is encountered in comprehending and employing them. Many users feel that they provide more effective, and more equal, communication with all levels of the operation, especially with those in remote locations.

(1) All Services

Electronic message services reduce the frustration associated with uncompleted telephone calls that result from line busy or party unavailable conditions. Satisfaction is gained from knowing that the message will be delivered at a time when the recipient is available. The store-and-forward feature is particularly effective when communication spans several time zones, or when colleagues work different hours.

(2) Text or Graphics Versus Voice

Text mail and graphics mail can be read from a video screen or printed copy. Specific points, and their context, can be studied readily, and at length. Time can be taken to compose replies, and they can be edited several times before sending. Long messages can be handled a piece at a time so that they present no more difficulty to the originator or recipient than a long letter.

Voice mail has an important advantage over text mail and graphics mail—it requires no terminal other than a touch-tone telephone. Voice mail can be sent and received from practically anywhere. What is more, voice mail delivers a message in the sender's voice with all the emphasis and color retained. However, if detailed consideration of specific points is required, the message must be listened to carefully and in its entirety. Because they require considerable attention, long voice mail messages impose hardships on the originator and the recipient. These points lead to the following conclusions:

- **Text** mail and **graphics** mail may be more suited to lengthy arguments, to position statements, and to technical matters
- **Voice** mail may be more suited to instructions and to short administrative messages
- For those more skilled in dictating than typing, **voice** mail is probably an easier service to employ than **text** mail
- For those more skilled in typing than dictating, **text** mail is probably an easier service to employ than **voice** mail.

Successful employment of any of these message systems requires that users monitor their mailboxes. No matter how user-friendly and speedy an electronic message service is, it will be to no avail if the messages are left unread or unheard.

(3) Saving Time and Increasing Participation

Saving time is the chief, direct contribution that electronic message services make to the business community. By eliminating the normal delay in letter mail, they accelerate the pace of decision making. Another direct contribution is opening participation in influencing decisions to all who have something to contribute. Electronic mail can have a positive effect on productivity, provided those who receive it have other productive things to do to fill up the time they save, and all *electronic* participants diligently study one another's contributions. However, it is unlikely that the number of productive thing to do will expand in sympathy with the use of electronic message services. Consequently, the diligent use of electronic mail services could result in a reduction of the need for so many professionals and management persons. One of Parkinson's celebrated conclusions is that work expands to fill the time, and consume the resources, allocated to it. With all users seeking to remain fully employed, electronic message services may generate more discussion, not less.

2.5.2 ELECTRONIC MEETING SERVICES

During the work day persons meet face to face to discuss sensitive, confidential, or personal problems; negotiate and reach agreements; solve problems; share information; or for any combinations of these reasons. In each case, the participants may belong to the same peer group or different peer groups, share a superior/subordinate relationship, or be total strangers.

(1) Observed Behavior

Many of these combinations have been studied in electronic meetings that use services of different richness. Analysis of the behavior of the participants leads to the following:

- Regional personnel perceive a greater need for, and utility of, electronic meeting services than do headquarters persons
- There is a substantial difference between the measured and perceived effectiveness of electronic meeting services (most users believe that video is an important element—problem-solving tests do not bear this out)
- An electronic meeting imposes a degree of formality that reduces extraneous discussion and concentrates the attention of the participants on the subject at hand (meetings are more topic-oriented and less person-oriented)
- Acquainted persons perform more efficiently than strangers do in an electronic meeting
- Acquainted persons perform more efficiently in an electronic meeting than in face-to-face meetings (some estimate as much as 30% shorter meetings)

- Groups in which no clear authority structure or previous acquaintance exists become more uncertain with regard to norms, values, and role expectations as the richness of the electronic meeting medium decreases

- So long as audio is included, both problem-solving and information-sharing meetings are relatively insensitive to the total richness of the meeting service employed

- As the richness of the meeting service decreases, many persons find it easier to say no or to argue for positions that they do not totally support

- For sensitive, confidential, or personal discussions, and for meetings whose purpose is to reach an agreement, the degree of success is sensitive to the richness of the communications channel (face-to-face meetings are best for these purposes).

(2) Importance of Voice

In university psychology departments, extensive testing of the speed with which problems are solved has shown that face-to-face meetings provide no significant advantage over communication by voice alone. Then *why have electronic meetings not substituted for face-to-face meetings and revolutionized the way in which business is conducted*? Some reasons may include the following:

- **Inappropriate application of results**: tests in psychology departments are conducted in a cooperative environment with the participants seeking to overcome the limits placed on them. Usually, they have no alternative means available to accomplish the assigned end. In business, cooperation is not necessarily present, and the alternative to an electronic meeting is to travel to a *real* meeting—a mode that is familiar and well understood, and that has worked in the past.

- **Imbalance of sensory channels**: audio conferencing has been described as talking to people in a dark room. This may be something we are willing to do with one person (by telephone), but not with several. In fact, the analogy of the dark room is not exact. Audio conferencing is an activity in which the participants talk to unseen persons while viewing their present surroundings. The field serving their ears and the field serving their eyes are not the same. For more than a person and person conversation, the participants find visual cues helpful in separating the speakers and may become confused when none is available in the immediate vicinity. Calling persons together in a conference room matches sound and sight and restores the coincidence of the inputs. This may be the reason many persons consider video to be an important component of an electronic meeting system intended for use in a variety of situations.

- **Richness of medium**: the richness of the medium affects the size of the group and the complexity of the subjects that can be handled effectively.

With an audio-only link {4,4}, more than a few persons discussing a complicated or unfamiliar topic can result in chaos. Graphics equipment increases the richness somewhat and may help to sort out conflicts and partition the problems. Computer conferencing [1,1] may assist with understanding the complexity at the expense of slowing the communication down and introducing an unfriendly interface. The richness of full-motion video {12,12} will make person identification easier, restore the balance between eyes and ears, and should make it possible to handle larger groups.

(3) Other Effects of Richness

The richness of the medium has other effects

- **Insincere support**: as the richness decreases, many persons find it easier to argue for positions they do not support. For this reason, for meetings in which the parties intend to discuss important issues, even full-motion video may not be entirely adequate.

- **Change in importance of participants**: as richness decreases the conference participants tend to equalize in importance. Thus, full-motion video allows easy identification of persons from headquarters by their manner, style, and dress. In an audio conference, none of the visible marks of rank is available. Attention is focused much more on eloquence, logic, and other vocal skills. A computer conference may even reverse the ranking, for successful participation depends on easy familiarity with terminals and protocols. This is more likely to be found in working-level persons than in management-level participants.

(4) User Benefits

If problem solving or information sharing is all that a meeting must achieve, it can be held using relatively simple electronic meeting facilities. Groups of acquainted, compatible persons who wish to solve a problem, or to share information, are likely to be able to use an audio-only link to achieve the results of a face-to-face meeting. This is the case for regular meetings of the kind at which project progress is reviewed or sales information is exchanged. Indeed, meeting electronically under these circumstances can do much more than a face-to-face meeting. Because many more persons can be present at each location, a better outcome is achieved through having those that really know what they are doing participate in the decision making. Important matters need no longer be left to traveling generalists, but can be discussed to any depth within the capabilities of the staff resident at each terminal location.

Those who use electronic meeting facilities report improvements in personal productivity and in the working environment. They cite decreases in the time away from work, greater job satisfaction, improved access to other persons, and more productive meetings. An important prerequisite appears to be a senior manager determined to break the face-to-face meeting habit. When used regularly, electronic meetings are shorter and more timely; more issues are raised and resolved, decisions can be made based on more correct information and less misinformation, and unproductive time is reduced. The results are said to be greater management visibility, improved communication throughout the corporation, improved morale, and increased cooperation.

(5) Intra- and Interorganization Meetings

Most of the applications of electronic meetings are reported to be within large institutions. This is not surprising, since a community of interest is already formed, the resources may be easier to obtain, and the location of the terminals is already established (each of the separated facilities). In addition, in most large organizations, there is a sense of belonging that makes employees who have never met not strangers, but unacquainted members of the same family. Conditions are such as to heighten the probability that an intra-organizational electronic meeting can be completed over relatively simple facilities (audio only or augmented audio) to the satisfaction of all participants.

However, what of interorganization electronic meetings? Inevitably, this implies more conferences among strangers, so that the richness of the medium must increase to have the same probability of satisfying all parties. More expensive facilities (full audio, video, and graphics capabilities) must be used. What is more, compatible terminal equipment is required so that one terminal can communicate with the other, and a switched video transmission network is necessary so that the expensive transmission facilities can be shared among a large number of users. Thus, interorganization electronic meetings are more complicated and more expensive than intra-organization meetings. For this reason, they have been slower to develop.

(6) Reducing Travel

Often, those charged with controlling expenses argue that the proper use of telecommunication facilities can reduce the amount of business travel. This is a persuasive statement—but it ignores several factors

- **Natural curiosity**: the major tool used in business and personal life to share information and to solve problems is the telephone. Usually, we reach for it without even considering travel as an alternative way to achieve our purposes. However, by expanding our knowledge of other places, the same instrument may stimulate our natural curiosity. The simultaneous achievement of more use of telecommunications and less travel can be contradictory objectives.

- **Complex objectives**: many meetings have complex objectives—and very often the stated purpose for the meeting is an excuse for accomplishing something else. Some attendees have special reasons for traveling to the meeting site, just as those who wish to have an electronic meeting may have special reasons to stay home. Some of their reasons will be official; some may be personal but public; and some may be personal but private. They influence the decision to travel or to communicate much more than whether the objectives can be accomplished over electronic meeting facilities.

- **Location of meeting rooms**: the decision to travel or to communicate is influenced by the location of the electronic meeting rooms. Within an organization, greater use of electronic meetings may occur between suburban locations and downtown headquarters than between suburban locations. This is particularly true in older metropolitan complexes, where going downtown means traffic snarls, uncertain parking, and the risk of vandalism. In this environment, travel problems exceed electronic meeting problems, and those working in the suburbs can be expected to use electronic meetings as often as the headquarters staff will allow them to. For the same reasons, a downtown public electronic meeting site will do little to encourage interorganizational electronic meetings unless it is conveniently located, provides adequate support services, and is easy to use.

So... *why have electronic meetings not revolutionized the way in which business is conducted*? It may be that, when a number of persons are concerned, the efficiency of doing business is not the only factor they consider.

REVIEW QUESTIONS FOR SECTION 2.5

1 What benefits do electronic message services provide?

2 What are the advantages/disadvantages of text or graphics mail versus voice mail?

3 Comment on the following statement: Electronic mail can have a positive effect on productivity, provided those who receive it have other productive things to do to fill up the time they save, and all *electronic* participants diligently study one another's contributions.

4 Describe the motivations and reactions of participants in electronic meetings.

5 Why have electronic meetings not substituted for face-to-face meetings and revolutionized the way in which business is conducted?

6 As the richness of the communication channels decreases, what human factors may come into play?

7 Comment on the following statement: If problem solving or information sharing is all that a meeting must achieve, it can be held using relatively simple electronic meeting facilities.

8   Why are most of the applications of electronic meetings within large institutions?

9   Comment on the following statement: What of interorganization electronic meetings? Inevitably, the richness of the medium must increase.

10 Comment on the following statement: Those charged with controlling expenses argue that the proper use of telecommunication facilities can reduce the amount of business travel. What factors do they ignore?

11 So... why have electronic meetings not revolutionized the way in which business is conducted?

3

BASEBAND SIGNALS

Signals are the manifestations of messages exchanged at a distance over telecommunication facilities. Their generation, transmission, reception, and detection are the fundamental tasks performed by communication systems. The messages consist of

- **Data**: as in file transfers between data processors
- **Text or graphics**: as in electronic mail, facsimile, or information retrieval from a remote database
- **Spoken words**: as in point-to-point telephone conversations or voice mail
- **Music**: as in broadcast programming
- **Television**: as in point-to-multipoint cable television, broadcast television, or videoconferences.

At the signal level, it is impossible to know for what purposes the messages are used. Accordingly, without reference to their operational context, we discuss them as data, voice, and video signals. To them, we must add those signals that interfere with and distort the wanted (message) signals. Collectively known as *noise*, it degrades the performance of all communication systems.

3.1 SIGNAL ANALYSIS

Signals can be measured. Therefore, in mathematical terms they are real and finite. However, they are not easy to describe, and their analysis challenges the ingenuity of engineers who would analyze them. To do this, they build models of a signal environment in which mathematical functions can be employed. The objective is to produce behavior that mimics the signals they wish to study. Among many properties, the variation of their values with time, and the degree to which their behavior can be described and predicted, are especially important.

3.1.1 VARIATION OF VALUES WITH TIME

Signals can be divided by the way in which their values vary over time

- **Analog**: a continuous function that assumes positive, zero, or negative values. Changes occur smoothly and rates of change are finite.

Analog signals are generated by many physical processes; they play a large part in the *transmission* of messages by electronic media. Indeed, over many *electrical* transmission systems, data signals are coded in analog form. Information is carried in the variations of amplitude, frequency, and/or phase of the signal.

- **Digital**: a function that assumes a limited set of positive, zero, or negative values. Changes of value are instantaneous, and the rate of change at that instant is infinite—at all other times it is zero.

An important subclass of digital functions is binary functions.

- **Binary**: a digital function that exists in two states only.

Binary signals are generated by computers and similar electronic devices. Binary techniques dominate the *processing* of messages, and digital signals are used exclusively in the transmission of messages by optical fibers. With binary signals, information is carried in the sequence of states (1s and 0s).

It is important to note the way in which information is carried is quite different in analog and digital signals. Digital functions can be converted to analog functions, and *vice versa*.

3.1.2 CERTAINTY WITH WHICH BEHAVIOR IS KNOWN

Signals may be divided by the degree of certainty with which their behavior is known; thus

- **Deterministic**: at every instant, a deterministic function exhibits a value (including zero) that is related to values at neighboring times in a way that can be expressed exactly.

The time at which a deterministic function will achieve a particular value is known, and the value the function will have at a particular time can be calculated. Consequently, deterministic signals are described by familiar mathematical functions.

- **Probabilistic**: a function whose past and future values are described in statistical terms.

While we may have a set of historical values, neither the value of a probabilistic function at a specific future time, nor the future time at which the function will have a specific value, can be known for sure. Future values are estimated from the statistics associated with past values and the assumption that future behavior is somehow connected to it. An important subclass of probabilistic functions is random functions.

- **Random**: a probabilistic function whose values are limited to a given range. Over a long time, each value within the range will occur as frequently as any other value.

With few exceptions, the signals in communication systems are probabilistic. They cannot be represented by classical functions, and we cannot calculate specific future behavior. To overcome this impasse, future values are estimated based on the statistics associated with past measurements and the *assumption* that future behavior will remain the same. In mathematical terms, we assume the process giving rise to the signals is *ergodic* and *wide sense stationary*. These terms are defined as

- **Ergodic**: statistical measures on sample functions of the random process in the time domain are equal to corresponding measures made on the ensemble of sample functions (the time average is equal to the ensemble average)
- **Wide sense stationary**: the expectation of the random process is constant, and the autocorrelation function depends only on the *time interval* over which it is evaluated (not the time *at* which it is evaluated).

They are powerful concepts whose application requires substantial mathematical skills.

3.1.3 SINUSOIDAL FUNCTIONS

Sinusoidal functions occur in the modeling of many simple, natural processes such as the motion of a pendulum and the vibrations of a structure. In the electric age, sinusoidal signals are present as

- Single-tone audio signals
- 60-Hz alternating current present in public power supply systems
- Carriers employed in radio and television broadcasting.

Basic to an understanding of sinusoidal functions are the notions of amplitude, frequency, and phase. To illustrate the meanings of these terms, we start with the model shown in **Figure 3.1**. Called an *Argand* diagram for its inventor, Jean-Robert Argand, an 18th-century Swiss mathematician, it consists of a phasor of amplitude *A* signal units (su) that rotates anticlockwise about the origin at a constant angular velocity and completes *f* rotations a second. For reasons we need not go into, the horizontal axis is known as the *real* axis, and the vertical axis is known as the *imaginary* axis. Angles are measured in relation to the horizontal, right-pointing, real axis (marked 0°). As the phasor rotates, its projection on the real axis is $A \cos\phi$, and on the imaginary axis, it is $A \sin\phi$.

Figure 3.2 introduces time as the third dimension. In this diagram, the tip of the phasor traces out the spiral path shown. When projected on to the plane containing the real axis and the time axis, the spiral of Figure 3.2 forms the cosine function shown in the time-plane diagram at the upper right-hand side of **Figure 3.3.** If one rotation of the phasor takes τ seconds, the repetition frequency *f* is $1/\tau$ times per second (i.e., $1/\tau$ Hertz). During each period, the phasor rotates through 360°, and the phase angle can be associated with position in the period as shown. These angles are repeated each period. In the time plane, $A \cos\phi$ becomes $A \cos 2\pi ft$—a periodic function that repeats in synchrony with the rotating phasor. The bottom diagram in Figure 3.3 shows the frequency-plane representation of the cosine function. It consists of a single line of amplitude *A* su situated at a frequency *f* Hz.

3.1.4 COMPLEX PERIODIC FUNCTIONS

When designing and testing communications equipment, the use of sinusoidal functions and their signal equivalents, single-frequency tones, simplifies calculations and measurements. However, real signals are more complicated than this. Fortunately, if they can be assumed periodic, they can be modeled by complex periodic functions formed from sinusoidal functions that comprise a Fourier series. The function with the lowest frequency (but not zero) is called the fundamental, and the other functions are called harmonics. They occur at integer multiples of the fundamental frequency.

Periodic functions are decomposed into sinusoidal functions with the *trigonometric* Fourier series and the Fourier *cosine* series (after their inventor, Jean Baptiste Fourier, a late 18th- and early 19th-century French mathematician). In addition, they may be converted into exponential functions with the *exponential* Fourier series. This technique introduces complex quantities and transforms the frequency domain into one that contains both positive and negative frequencies.

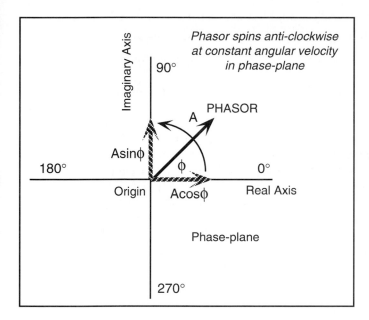

Figure 3.1 Argand Diagram Representation of Sinusoidal Function

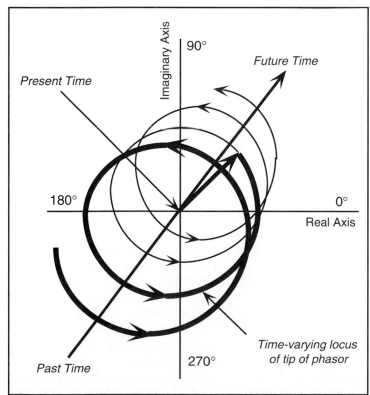

Figure 3.2 Three-Dimensional Model of Sinusoidal Function

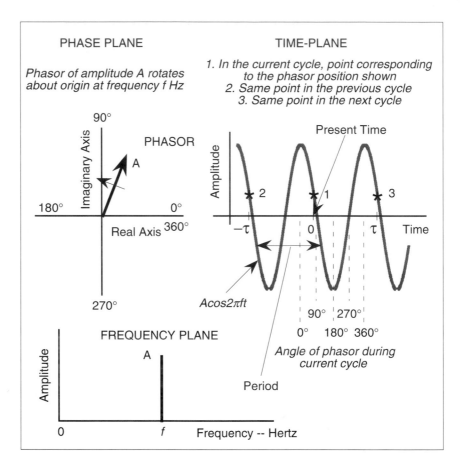

Figure 3.3 Phase-, Time-, and Frequency-Plane Representations of Sinusoidal Function

In the real world of signals, negative frequencies do not exist. In the make-believe world of Argand diagrams, a negative frequency is associated with a phasor rotating in a clockwise direction. The combination of positive and negative frequency values gives rise to what are called *two-sided* spectrums.

Figure 3.4 shows an example of a complex sinusoidal function. It is made up of a fundamental component at frequency f Hz of amplitude 1.25A su and a third harmonic component at frequency 3f Hz of amplitude 0.56A su. The fundamental component establishes the basic periodicity of the function; the third harmonic component provides periodic variations about the fundamental component. In the Argand diagram, a phasor of magnitude 1.25A su rotates at the fundamental frequency (f Hz), and a phasor of magnitude 0.56A su rotates at the third harmonic

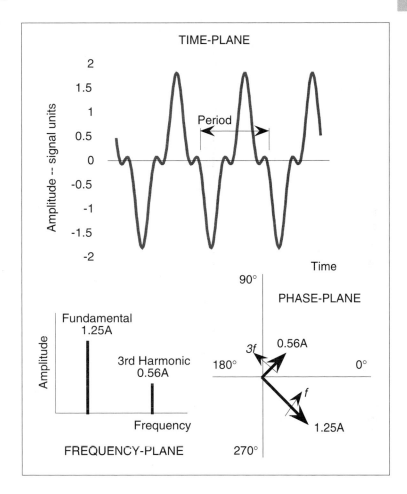

Figure 3.4 Complex Sinusoidal Function

frequency (3*f* Hz). The amplitude and phase of the resultant of these phasors are the amplitude and phase of the complex signal.

   Figure 3.5 shows another complex periodic function—a symmetrical pulse train. A binary digital function that changes from *A* to 0 and 0 to *A* at regular intervals, it contains an infinite number of harmonic components. Figure 3.5 shows the magnitudes of the components up to the ninth harmonic. That only odd harmonics are present results from defining zero time (i.e., *t* = 0) in the middle of a pulse so the train is an even function (i.e., values at ±*t* are identical). In the general case, this class of functions contains both odd and even harmonics. Further, because the function assumes only positive (+*A*) and zero values, it contains a dc-term (0 Hz, 0.5*A* su) that is the average value of the function.

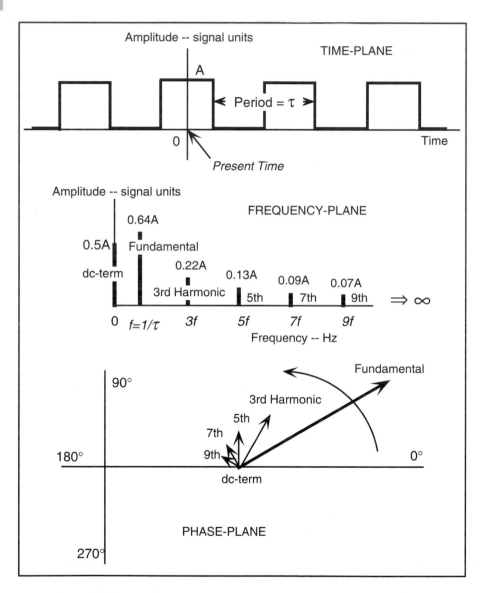

Figure 3.5 Phase-, Time-, and Frequency-Plane Representations of Symmetrical Pulse Train

In the phase plane, Figure 3.5 shows an Argand diagram in which the phasor representing the fundamental component rotates about the point $0.5A$. This is the offset created by the dc-term. The harmonic components rotate about the tips of the next lower components. Because each phasor rotates at a different multiple of the fundamental frequency, the diagram is a snapshot of the action; it is repeated f times a second.

3.1.5 Bandwidth

- **Baseband**: the frequency range associated with a message signal when first generated
- **Bandwidth:** the range of frequencies that just encompasses all of the energy present in a given signal. The signal may be the original message signal, in which case the bandwidth is equal to baseband, or the signal may be a processed version of the original message signal. In this case, bandwidth is likely to be greater than baseband.

Some of the power of the complex sinusoidal function shown in Figure 3.5 is carried by the fundamental component (0.78 signal watts, sw), and some by the third harmonic component (0.16 sw). To reproduce the signal exactly, these components must be transmitted to the receiver as they exist in the original signal. Thus, the bearer (and associated equipment) must carry the signal bandwidth without distortion. In the case of the complex sinusoidal function, the range of frequencies that just encompasses all the power present in the signal is from f to $3f$ Hz—i.e., a bandwidth of $3f$ Hz.

(1) Passband

A communication system must deliver the message signal exactly as it was sent. For this to happen, the passband of the system facilities must be sufficient to accommodate the bandwidth of the signal.

- **Passband**: the range of frequencies transmitted without distortion by a bearer and associated equipment.

Practical communication systems are endowed with finite properties, and this includes the frequencies transmitted by the bearers and associated equipment. Examples of their behavior are shown in **Figure 3.6**. Some will transmit energy from 0 Hz (dc) to gf Hz, but not at frequencies above gf Hz; they are said to exhibit *low-pass* characteristics and to have a passband of gf Hz. Others will transmit energy between mf Hz and nf Hz, but not at frequencies below mf Hz or above nf Hz; they are said to exhibit a *bandpass* from mf Hz to nf Hz, and to have a passband of $(n-m)f$ Hz. Yet others will transmit energy at frequencies greater than hf Hz, but not at frequencies below hf Hz: they are said to exhibit *high-pass* characteristics. In reality, because there will always be a limit to the frequencies they transmit, the concept of a high-pass system is artificial. In practice, communication systems are of two types—low pass and bandpass. If it is to be transmitted without distortion, what is important is the passbands of all the component parts of a transmission system must be greater than, or equal to, the bandwidth of the signal.

(2) Bandwidth > Passband = Distortion

When the bandwidth of a signal is larger than the passband of the communications system it is presented to, the signal that is transmitted is a distorted ver-

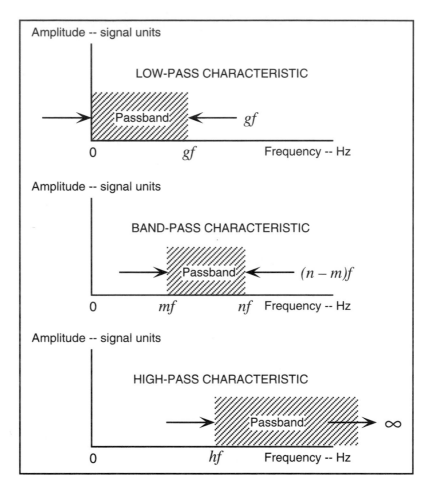

Figure 3.6 Passband Characteristics of Bearers and Associated Equipment

sion of the original. For the case of the symmetrical pulse train drawn in Figure 3.5, parts A, B, and C of **Figure 3.7** show the result of progressively diminishing the passband (as measured by the number of harmonic frequencies transmitted) of a low-pass bearer that carries the pulse train. Figure 3.7A shows the original pulse train. The second signal (B) is the signal received from a system that passes frequencies no higher than nine times the fundamental frequency. This system has a passband of $9f$ Hz, and transmits approximately 97% of the power in the original signal. Figure 3.7B shows the signal received from a system that passes frequencies no higher than five times the fundamental frequency. This system has a passband of $5f$ Hz and transmits approximately 96% of the power in the original signal. The fourth signal (D) is received from a high-pass system that passes fre-

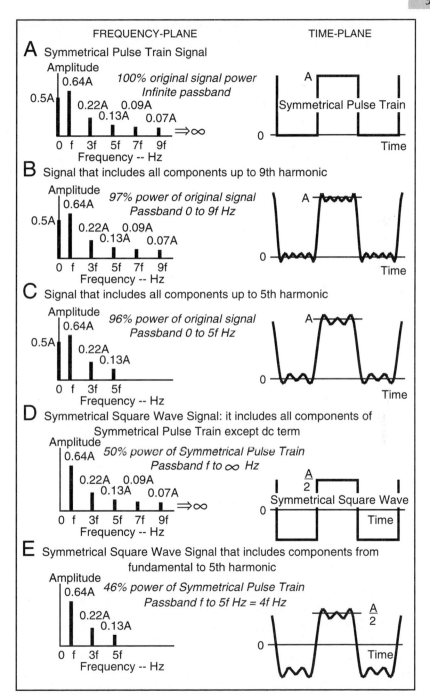

Figure 3.7 Effect of Limiting the Passband Available for Transmission of a Wideband Signal

quencies of f Hz, and above. Without the dc component, the pulse train becomes a symmetrical square wave with an average value of zero. Finally, Figure 3.7E shows the signal received from a bandpass system that passes frequencies from f Hz to $5f$ Hz. It has a passband of $4f$ Hz, and the signal is identical to C except for the elimination of the positive offset. With the same periodic features as the symmetrical pulse train, it requires only one-half of the power.

An important first point in our exploration of telecommunication signals is that the passband of the bearer must be large enough to accommodate the bandwidth of the signal. Thus, because coaxial cables have greater passbands than twisted pairs, they can carry signals of wider bandwidth. It will deliver an undistorted television signal (nominal bandwidth 6 MHz) to a point several miles away; this will not be the case if we try to send it over twisted pairs. Further, because optical fibers have a higher passband than electrical bearers do, optical fibers carry much higher speed datastreams than coaxial cables, microwave radios, or twisted pairs.

3.1.6 BINARY SIGNAL FORMATS

Some common examples of binary signal formats are shown in **Figure 3.8**. As drawn, they are current or voltage pulses that carry the bit sequence $\leftarrow$101100111000. For specificity, we describe them as current pulses.

- **Unipolar**: a 1 is represented by a current of $2A$ signal units, and a 0 is represented by a current of 0 signal units. Two implementations are possible
  - *nonreturn to zero* (NRZ). Currents are maintained for the entire bit period (time slot). In a long sequence in which 1s and 0s are equally likely, the power in a unipolar function is $\frac{1}{2}(2A)^2$ —i.e., $2A^2$ signal watts.
  - *return to zero* (RZ). Currents are maintained for a fraction of the time slot. In a long sequence in which 1s and 0s are equally likely, and current is maintained for one-half the time slot, the power in a return-to-zero unipolar function is $\frac{1}{2} \times 2A^2 = A^2$ signal watts—one-half of NRZ unipolar.

With NRZ operation

- Long strings of 0s produce periods in which no current is generated
- Long strings of 1s produce periods in which only positive current is generated
- When 1s and 0s are equally likely, the mean signal value is A signal units.

All three situations present problems to electronic receivers. When a constant current flows, or no current flows, no timing information is available and synchro-

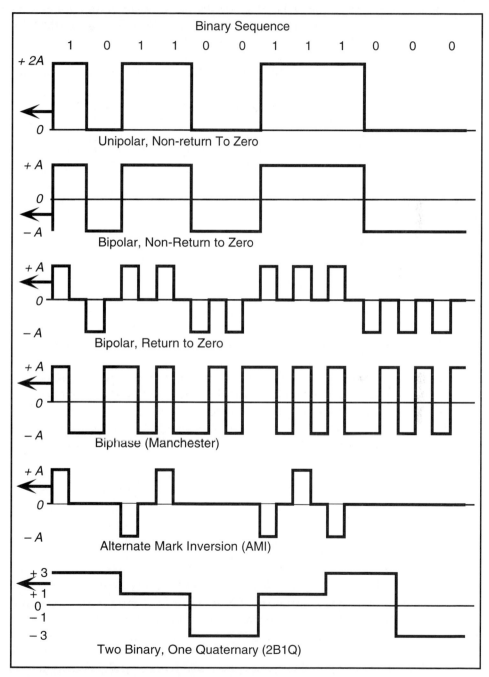

Figure 3.8 Examples of Binary Signal Formats

nization is impossible. When the current has a mean value (not zero mean), the receiver is required to follow long-term level shifts—this can be difficult.

With RZ operation

- Long strings of 0s produce periods in which no current is generated
- Long strings of 1s produce periods in which positive current flows for a fraction of a time slot and returns to zero between 1s, so the receiver can detect a change
- When 1s and 0s are equally likely, and the pulses are $T/2$ wide, the mean signal value is $A/2$ signal units.

Thus, RZ eliminates one of the receiver's problems—i.e., timing, but not long-term level shifts.

- **Bipolar**: a 1 is represented by a current of A signal units and a 0 is represented by a current of $-A$ signal units. Two implementations are possible
  - *nonreturn to zero* (NRZ). Currents are maintained for the entire time slot. In a long sequence in which 1s and 0s are equally likely, the power required to sustain an NRZ, bipolar stream is A^2 signal watts, i.e.—one-half the unipolar NRZ value.
  - *return to zero* (RZ). Currents are maintained for a fraction of the time slot. In a long sequence in which 1s and 0s are equally likely, and current is maintained for one-half the time slot, the power required to sustain a return to zero bipolar stream is $A^2/2$—one-half of NRZ bipolar—and one-fourth the unipolar NRZ value.

Long strings of 1s or 0s produce constant currents in NRZ bipolar; they are a problem for electronic circuits. The difficulties are largely eliminated with RZ bipolar because the receiver detects the current returning to zero in each pulse period. Moreover, when 1s and 0s are equally likely, the mean signal value is zero.

- **Biphase** or **Manchester**: a 1 is a positive current pulse of amplitude A signal units that changes to a negative current pulse of equal magnitude, and a 0 is a negative current pulse that changes to a positive current pulse. The changeover occurs exactly at the middle of the time slot. In a long sequence in which 1s and 0s are equally likely, the power required to sustain a biphase stream is A^2 signal watts. It is equal to that required by bipolar signaling; thus, it is one-half the power of the unipolar representation.

Manchester signaling is employed between equipment that operates at high speeds and must maintain close synchrony. It is a popular signaling technique for local-area networks. Long strings of 1s or 0s do not result in constant currents, and the signal is zero mean under all circumstances.

- **Alternate mark inversion** (AMI): 1s are represented by return-to-zero current pulses of magnitude A that alternate between positive and negative. 0s are represented by the absence of current pulses. In a long sequence in which 1s and 0s are equally likely, the power required to sustain an AMI function is $A^2/4$—one-half of RZ bipolar—and one-eighth of NRZ unipolar.

In a long sequence in which 1s and 0s are equally likely, the average pulse repetition rate of AMI signaling is one-half the bit rate (i.e., the same as unipolar). Because the polarity of the pulses alternates, virtually the entire power is contained within a bandwidth equal to the bit rate expressed in Hz (see Figure 3.9). With a pulse shape that approximates a raised cosine, AMI is used extensively in T–1 carrier systems for high-speed digital connections [see Section 6.2.1(2)].

- **Two binary, one quaternary** (2B1Q): four signal levels (± 3 and ± 1) each represent a pair of bits. Of each pair, the first bit determines whether the level is positive or negative (1 = +ve, 0 = −ve), and the second bit determines the magnitude of the level (1 = $|1|$, 0 = $|3|$). For the purpose of comparison, if we designate level 3 as A, in a long sequence in which 0s and 1s are equally likely, the power required to sustain a 2B1Q stream is $5A^2/9$ signal watts—approximately twice the power for AMI.

2B1Q signaling is employed to provide ISDN Basic Rate service (at 160 kbits/s, see Section 10.4) and ISDN digital subscriber loop services (see Section 6.3). For long sequences of 1s or 0s, or alternating 1s and 0s (i.e., 101010...) 2B1Q signaling produces constant currents, and synchronization is impossible.

Figure 3.9 shows (one-sided) frequency spectrums of random binary datastreams (i.e., 1s and 0s are equally likely) that employ bipolar NRZ, 2B1Q, AMI raised cosine, and Manchester formats.

- **Bipolar NRZ**: the full-width, zero-mean function is used as reference.
- **2B1Q**: because of the double-width time slots of 2B1Q, most of the power is present at lower frequencies, and the first zero occurs at one-half the frequency of the first zero of bipolar NRZ. Significant power is present in the sidelobes.
- **AMI Raised Cosine**: zeros occur at dc and at the zeros of bipolar NRZ. The sidelobes fall off rapidly, so only two lobes appear in Figure 3.10. Peak power density occurs approximately midway between the first two zeros.
- **Manchester**: because of the half-width pulse pattern of Manchester, most of the power is present at higher frequencies; zeros occur at dc and at the even zeros of bipolar NRZ. The peak power density occurs approximately at the first zero of bipolar NRZ. Significant power is present in the sidelobes.

Figure 3.9 shows a frequency progression from the lower frequencies in the double-width pulses of 2B1Q to the higher frequencies in the half-width pulses of Man-

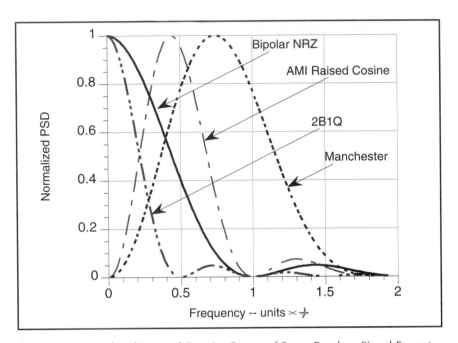

Figure 3.9 Normalized Spectral Density Curves of Some Random Signal Formats

chester. Because of their lower-frequency power density spectrums, 2B1Q and AMI Raised Cosine are employed in the bandwidth-limited environment of telephone connections. Manchester is employed in local-area networks and other applications where precise synchronization is important and bandwidth is available.

3.1.7 SCRAMBLING

For reasons noted above, certain signaling formats produce constant signal levels when carrying long strings of 1s or 0s (and alternating 1 and 0 in the case of 2B1Q). To avoid these situations, several bit-changing strategies can be employed. Those that are unique to specific facilities are discussed later [see Section 6.2.1(2)]. A general-use strategy is to *scramble* the datastream before producing the physical signal. This action breaks up long strings of the same symbol, or repeated patterns of the symbols, and makes the signal train more random. At the receiver, the scrambled train is descrambled to reclaim the original signal stream.

 Figure 3.10 shows a simple hardware scheme[1] that scrambles the data sequence. Before the datastream is transmitted, it is added to a stream derived

---

[1] J. E. Savage, "Some simple self-synchronizing digital data scramblers," *Bell System Technical Journal*, 37 (1958), 1501–42.

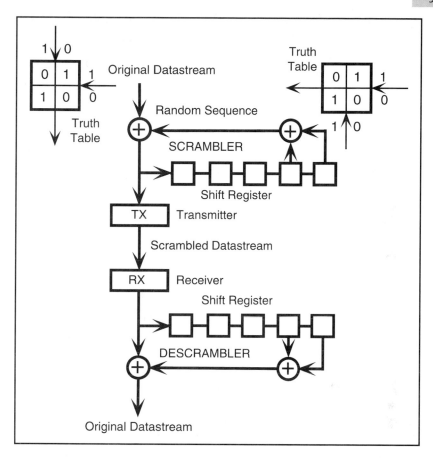

Figure 3.10 Self-Synchronizing Data Scrambler

Before transmission, the datastream is added to a stream derived from itself. As a result, the transmitted datastream is more random; long sequences of 1s or 0s are broken up, and repetitive patterns are changed. At the receiver, the operation is reversed.

from it by delaying it in a shift register and performing two exclusive-or operations. The result of using the scrambler is a new signal sequence that is related to the original, but is more random. Long sequences of 1s or 0s are broken up, and repetitive patterns are changed. At the receiver, the operations are repeated to produce the original datastream. The use of a five-stage shift register produces a scrambled sequence that repeats every 24 input bits. In practice, longer registers are used to achieve more thorough randomization.

In **Figure 3.11**, we show a table of the bits produced when an all-zeros datastream is scrambled (the initial condition of the shift register is 10101). As shown, instead of 24 0s in the input datastream, we have an irregular pattern of 13 1s and 11 0s—a pattern that contains a sufficient number of transitions to produce a useable signal.

The descrambler must restore the scrambled sequence to the original message datastream. In this case, applying the scrambled sequence to the descrambler must produce a stream of 0s. Using Figure 3.11 as a model, the reader can confirm the descrambler reverses the process and the output produced by the scrambled sequence is the original all-zeros stream.

REVIEW QUESTIONS FOR SECTION 3.1

1   What consequence can be drawn from the following statement: Signals can be measured.

2   Define an analog function.

3   Define a digital function.

4   Define a binary function.

5   Define a deterministic signal.

6   Define a probabilistic function.

7   Define a random function.

8   Give some examples of sinusoidal signals.

9   Describe an Argand diagram. How does it show amplitude, frequency, and phase?

10   How is a complex periodic function described by a Fourier series?

11   Define baseband, bandwidth, and passband.

12   What do you understand by the descriptors low-pass filter, and bandpass filter? Why is a high-pass filter a theoretical concept only?

13   Justify the following statement: When the bandwidth of a signal is larger than the passband of the communications system it is presented to, the signal that is transmitted is a distorted version of the original.

14   Describe the following binary signal formats: unipolar NRZ, unipolar RZ, bipolar NRZ, bipolar RZ, Manchester, AMI, 2B1Q.

15   Compare the frequency spectrums of the following binary signal formats: bipolar NRZ, AMI raised cosine, Manchester, and 2B1Q.

16   Why are signals scrambled?

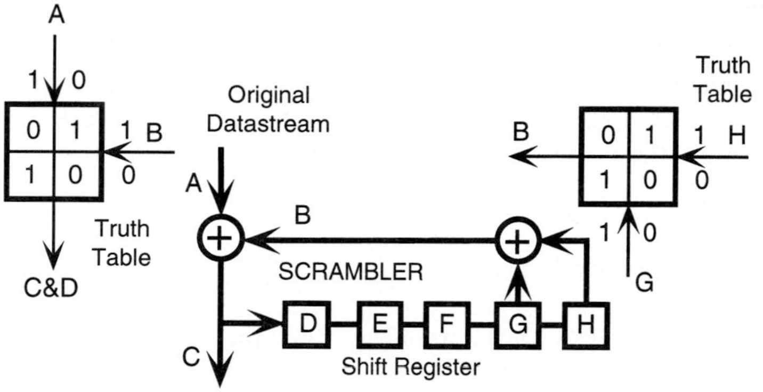

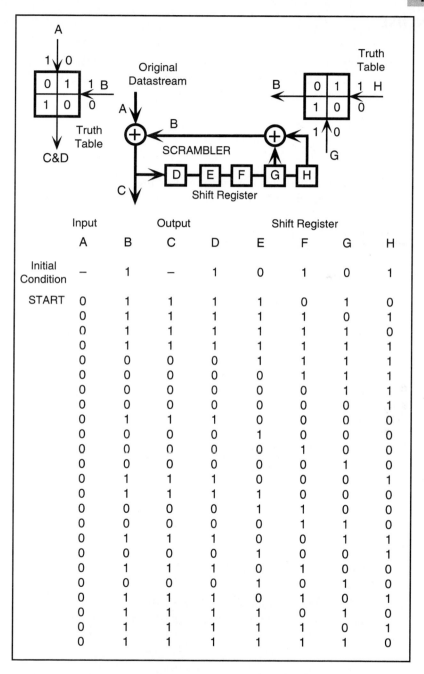

	Input		Output			Shift Register		
	A	B	C	D	E	F	G	H
Initial Condition	–	1	–	1	0	1	0	1
START	0	1	1	1	1	0	1	0
	0	1	1	1	1	1	0	1
	0	1	1	1	1	1	1	0
	0	1	1	1	1	1	1	1
	0	0	0	0	1	1	1	1
	0	0	0	0	0	1	1	1
	0	0	0	0	0	0	1	1
	0	0	0	0	0	0	0	1
	0	1	1	1	0	0	0	0
	0	0	0	0	1	0	0	0
	0	0	0	0	0	1	0	0
	0	0	0	0	0	0	1	0
	0	1	1	1	0	0	0	1
	0	1	1	1	1	0	0	0
	0	0	0	0	1	1	0	0
	0	0	0	0	0	1	1	0
	0	1	1	1	0	0	1	1
	0	0	0	0	1	0	0	1
	0	1	1	1	0	1	0	0
	0	0	0	0	1	0	1	0
	0	1	1	1	0	1	0	1
	0	1	1	1	1	0	1	0
	0	1	1	1	1	1	0	1
	0	1	1	1	1	1	1	0

Figure 3.11 Scrambling a Stream of 0s

3.2 NOISE

- **Noise**: unwanted signals that are added to the message signal in the generation, transmission, reception, and detection processes employed in effecting telecommunication.

3.2.1 SOURCES OF NOISE

In the real world of telecommunication, message signals are accompanied by noise. At the receiver output, the signal is a distorted replica of the signal the sender intended to send. The difference between the actual (corrupted) signal present at the output of the receiver and the intended (ideal) signal is called the *noise* signal. Noise consists of

- Signals picked up at the receiver such as
  - another modulated wave that interferes with the wanted signal
  - signals generated by natural, impulsive phenomena such as lightning, or by manmade impulsive sources, such as automobile ignition systems
  - galactic radiation
- Signals generated in the processes and equipment used to consummate telecommunication, such as
  - the random motion of electrons, ions, or holes in the materials that compose the (receiving) equipment
- Signals that represent errors such as
  - approximations made in processing the signal for transmission result in sending a signal that differs from the signal that the sender intended to send
  - errors made in detecting the signal at the receiver.

3.2.2 SIGNAL-TO-NOISE RATIO

In a communication system, the ratio of the power in the message signal to the power in the noise signal is known as the signal-to-noise ratio (SNR).

$$SNR = \frac{\text{Signal power } (P_2)}{\text{Noise power } (P_1)}$$

Because it makes calculations of overall performance easier, it is usual to express this ratio as a logarithm expressed in decibels:

$$(\text{SNR})_{\text{db}} = 10 \log_{10} \frac{P_2}{P_1}$$

Power ratios expressed in decibels are listed in **Table 3.1**. Good-quality communication systems achieve SNRs of 35 to 45 db, or better. Many persons have difficulty recognizing words when the SNR in a telephone system drops below 15 db.

Table 3.1 Power Ratios Expressed in Decibels

| | POWER RATIO | |
DECIBELS	Gain	Loss
0.0	1.000	1.0000
0.1	1.023	0.9772
0.2	1.047	0.9550
0.3	1.072	0.9333
0.4	1.097	0.9120
0.5	1.112	0.8913
0.6	1.148	0.8710
0.7	1.175	0.8511
0.8	1.202	0.8318
0.9	1.230	0.8128
1.0	1.259	0.7943
2.0	1.585	0.6310
3.0	1.995	0.5012
4.0	2.512	0.3981
5.0	3.162	0.3162
6.0	3.981	0.2512
7.0	5.012	0.1995
8.0	6.310	0.1585
9.0	7.943	0.1259
10	10	0.1000
20	100	0.0100
30	1×10^3	1×10^{-3}
40	1×10^4	1×10^{-4}
50	1×10^5	1×10^{-5}
60	1×10^6	1×10^{-6}

3.2.3 WHITE NOISE

When it is not possible to take account of the sources of noise individually, it is assumed they produce a single, random signal in which power is distributed uniformly at all frequencies. By analogy with white light (it contains all visible frequencies), this signal is called *white* noise. White noise has the property that different samples are uncorrelated, and, if the probability density function of the amplitude spectrum is Gaussian, they are statistically independent. White Gaussian noise is truly random. However, the spectral density is constant at *all* frequencies, so white noise is an unrealizable signal. Fortunately, in real situations, bandwidth is limited so that we can define

- **Band-limited white noise**: noise that has a constant power spectral density over a finite range of frequencies.

3.2.4 NOISE AND PROBABILITY OF ERROR

Noise corrupts the wanted signal and can produce errors in digital signals. Consider a stream of unipolar digital signals [denoted by $f(t)$] in which the levels are zero and A. The observed waveform $y(t)$ will consist of $f(t)$ and a noise signal $n(t)$—i.e., $y(t) = f(t) + n(t)$. **Figure 3.12** shows the 0 and 1 elements of a unipolar NRZ pulse train (signal C in Figure 3.7) and what it might look like when white Gaussian-distributed noise is present. Because the noise signal is random, it may add to, or subtract from, the pulse train signal and destroy the certainty of which level is present. If the decision threshold is positioned at $y = A/2$, so signals above it are identified as 1s and signals below it are counted as 0s, Figure 3.12 shows an occasion when the sum of noise and a 0 signal is greater than the threshold. If the signal is sampled at that instant, it will be read as 1 not 0, thereby creating an error.

    Figure 3.13 shows approximate values of probability of error and SNR for a bipolar train in which 1s and 0s are equally likely. The probability of error decreases rapidly as the signal-to-noise power ratio exceeds 10 db. It drops to around 10^{-8} when the ratio of the signal power to noise power is approximately 1,000.

REVIEW QUESTIONS FOR SECTION 3.2

1 What is meant by noise?
2 Enumerate some noise sources.
3 Define signal-to-noise ratio.

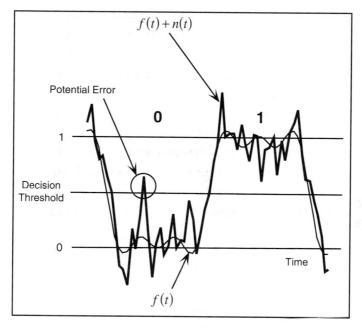

Figure 3.12 The Effect of Noise on Unipolar Pulses

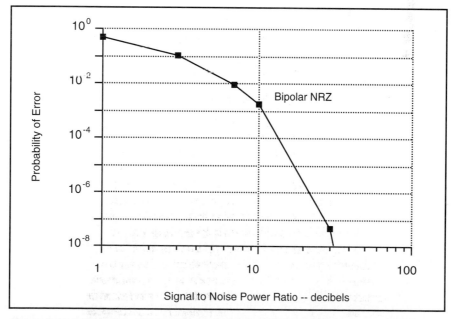

Figure 3.13 Probability of Error Due to Noise for Bipolar NRZ Random Pulse Train
The probabilities of error decrease rapidly as the signal to noise power ratio exceeds10 db.

4 Use Table 3.1 to find the SNR in decibels when the signal power is 10 watts and the noise power is 0.1 watts. If the signal power were 0.1 watts and the noise power were 10 watts, what would be the SNR in decibels?

5 Define white noise.

6 Define band-limited white noise. Why is this concept needed?

7 Explain Figure 3.13.

3.3 DATA SIGNALS

Between machines, information is exchanged by binary digits (bits). Since each bit is two-valued—either 0 or 1—per se, it can describe two situations (states). To describe more states, it is necessary to use the bits in sequences—to create unique codes—so a set of sequences can represent a range of information.

3.3.1 BINARY CODES

In order for two machines to communicate, they must employ sequences whose meanings are known to both. Two sequences in common use are

- **ASCII**: the *American Standard Code for Information Interchange* employs a sequence of 7 bits. Since each bit may be 0 or 1, ASCII contains 2^7, or 128, unique patterns. Some are assigned to represent the alphanumeric characters and symbols found in messages, and others represent the special characters required to control communication.

- **EBCDIC**: the *Extended Binary Coded Decimal Interchange Code* employs a sequence of 8 bits. Since each bit may be 0 or 1, EBCDIC contains 2^8, or 256, unique patterns. Some are assigned to represent the alphanumeric characters and symbols found in messages, others represent the special characters required to control communication, and the remainder may be used for purposes defined by the users.

Table 3.2 lists representative members of ASCII and EBCDIC. Examples are given of the decimal digits, upper- and lower-case letters, and some of the special characters used to control communication. The bit patterns are arranged so bit 1 (the least significant bit), is on the right-hand side of the member. It is transmitted first—a procedure that is opposite to digital voice octets, where the most significant bit is transmitted first [see Section 3.4.3(1)]. So there shall be no ambiguity in the meaning of ASCII and EBCDIC bit patterns, I use an arrow to show the direction of transmission, and thus identify the position of the least significant bit.

Table 3.2 Representative Members of ASCII and EBCDIC Code Sets

Alphas	ASCII	EBCDIC	Alphas	ASCII	EBCDIC
a	1100001 ⇒	10000001⇒	A	1000001 ⇒	11000001⇒
b	1100010	10000010	B	1000010	11000010
c	1100011	10000011	C	1000011	11000011
d	1100100	10000100	D	1000100	11000100
e	1100101	10000101	E	1000101	11000101
f	1100110	10000110	F	1000110	11000110
g	1100111	10000111	G	1000111	11000111
h	1101000	10001000	H	1001000	11001000
i	1101001	10001001	I	1001001	11001001
j	1101010	10001010	J	1001010	11001010
k	1101011	10001011	K	1001011	11001011
l	1101100	10001100	L	1001100	11001100
m	1101101	10001101	M	1001101	11001101
n	1101110	10001110	N	1001110	11001110
o	1101111	10001111	O	1001111	11001111
p	1110000	10010000	P	1010000	11010000
q	1110001	10010001	Q	1010001	11010001
r	1110010	10010010	R	1010010	11010010
s	1110011	10010011	S	1010011	11010011
t	1110100	10010100	T	1010100	11010100
u	1110101	10010101	U	1010101	11010101
v	1110110	10010110	V	1010110	11010110
w	1110111	10010111	W	1010111	11010111
x	1111000	10011000	X	1011000	11011000
y	1111001	10011001	Y	1011001	11011001
z	1111010	10011010	Z	1011010	11011010

Control	ASCII	EBCDIC	Numerics	ASCII	EBCDIC
SYN	0010110 ⇒	00110010⇒	0	0110000 ⇒	11110000⇒
SOH	0000001	00000001	1	0110001	11110001
STX	0000010	00000010	2	0110010	11110010
ETX	0000011	00000011	3	0110011	11110011
EOT	0000100	00110111	4	0110100	11110100
ENQ	0000101	00101101	5	0110101	11110101
ACK	0000110	00101110	6	0110110	11110110
NAK	0010101	00111101	7	0110111	11110111
DLE	0010000	00010000	8	0111000	11111000
ETB	0010111	00100110	9	0111001	11111001

3.3.2 CHARACTER TRANSMISSION

In some terminals, characters are generated at a keyboard and transmitted singly, one after the other. In other terminals, characters are run together to form a block of data.

(1) Characters Transmitted Singly

Given no two persons type at the same speed, and individual keystrokes may be separated by different time delays, for characters transmitted singly, the stream will be irregularly paced with spaces between each character. In some terminals, the characters are collected until a complete line of text is created, or the return key is pressed, causing the line to be sent as a burst of contiguous characters. Whether sent one by one as they are generated, or sent line by line as each line is completed, each character is framed by a start bit (0) and one or two stop bits (1 or 11). The start bit provides a reference from which the receiver measures time in order to sample the succeeding bits approximately in their center [see Section 8.3.1(1)]. The stop bit(s) provide(s) a positive indication of the end of the character. Called *character-framed data*, timing begins anew with each character received. Thus, the receiver does not have to maintain precise timing over more than 10 bits (7- or 8-bit character + stop bit, or 7- or 8-bit character + 2 stop bits). Because the characters are framed and sent without reference to any external timing source, this style of operation is known as *asynchronous*. **Figure 3.14** shows the datastream formats employed in transmitting characters singly.

(2) Characters Transmitted in Blocks

In more sophisticated situations, characters are run together without start and stop bits to form a block of data. Then the data block is framed by a start sequence and a stop sequence. The start sequence is called the *header*—it contains synchronizing, address, and control information. The stop sequence is called the *trailer*—it contains error-checking and terminating information. The entire data entity is called a *frame*, and the arrangement is described as *message-framed data*—i.e., the message data are framed by data devoted to the transmission process. In response to a command from *external* equipment, the frame is transmitted as a continuous stream. At the receiver, the header is used to signal the beginning of a frame and to synchronize the receiving equipment to cause all the bits in the frame to be sampled at exactly the same instant in the bit interval. For one frame in common use, precise timing must be maintained over 1,088 bits (5 characters in the header + 128 characters in the text + 3 characters in the trailer = 136 characters x 8 bits = 1,088 bits). Because transmission is paced by an external source, this style of operation is known as *synchronous*. **Figure 3.15** shows the datastream format used in transmitting characters frame by frame.

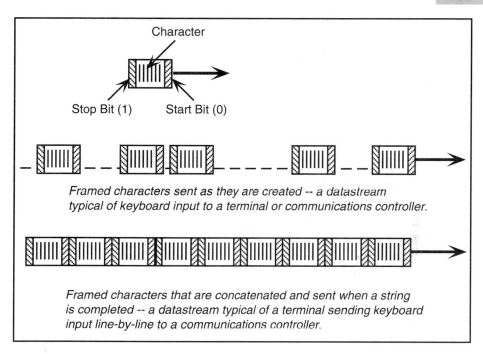

Figure 3.14 Transmission Formats for Characters Sent One-by-One, or String-by String
To ensure readability, each character is framed by a start bit and a stop bit. Individual characters or strings are sent when ready.

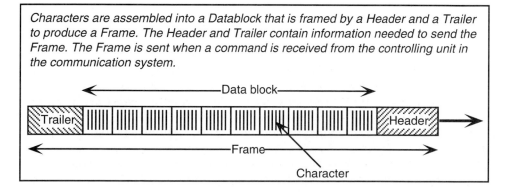

Figure 3.15 Transmission Format for Data Sent in a Frame

(3) Coding Efficiency

There is a significant difference in the coding efficiency of the two transmission methods

- **Character by character**: assuming an 8-bit character sequence, one start bit and one stop bit, the ratio of information bits to total bits is 0.8—i.e., the coding efficiency is 80%. With two stop bits, it is 73% efficient
- **Frame by frame**: assuming a 5-character header, a 128-character text (information characters), and a 3-character trailer, the ratio of information bits to total bits is 0.94. The coding efficiency is 94%.

Thus, under the conditions stated, when working character by character, the datastream is made up of 73% or 80% information bits and 27% or 20% overhead bits. When working frame by frame under the conditions stated, the datastream is made up of 94% information bits and 6% overhead bits.

(4) Frames, Packets, Envelopes, and Containers

An alternative nomenclature for a frame is *packet*. Invented for an early form of data network, it is the block of data (header + text + trailer) that is transmitted from a sender to a destination across a packet network (see Section 9.2). A later concept is an *envelope*. It transports *payload* between nodes in high-speed networks. The payload consists of packets, frames, and other digital sequences. Yet another concept is a *container,* which is used for even larger blocks of data that may be formed from several *data tributaries*. We discuss these concepts in Section 6.4.

3.3.3 CHARACTER-ORIENTED COMPRESSION

Data compression reduces the number of characters transmitted during a session so more information can be transmitted over a given facility. Data compression techniques are divided into those that employ character-oriented compression and those that employ statistical encoding. In this section, I describe character-oriented techniques.

By substituting special characters, or small groups of characters, for larger strings of characters that appear in a message, the length of the datastream can be made shorter than the original. Some of the techniques employed are

- **Null compression**: in datastreams that describe formatted documents, there are likely to be blank spaces (null characters) denoted by all-zeros characters (i.e., NUL = 0000000 in ASCII and 00000000 in EBCDIC). To reduce the number of zeros transmitted, a null-compression-indicating character is used followed by the number of all-zeros characters in the string. Thus, if the original character string is

$\Leftarrow$NAMEzzzzzzzzzzzzzzzzzzzzzzzzzzzzzzzzzCRSS#zzzzzzzzzzzzzzzzzzzzzzCR

where z represents the all-zeros or null character, and CR is carriage return, the compressed character string might be

$\Leftarrow$NAME©30CRSS#©20CR

where © is the null-compression-indicating character. The numbers 30 and 20 are the numbers of all-zeros characters in the two segments of the string. If they are expressed as 8-bit binary numbers (see below)—i.e., 00011110 and 00010100—the number of characters in the compressed string is 15. Since the number of characters in the original string is 61, the compression ratio is approximately 4:1.

- **Run-length compression**: a superset of null compression used to compress any string of repeating characters (not just zeros)
- **Pattern substitution**: common character pairs, character triplets, and entire words in a message are replaced by special characters. Thus, in the last statement, the word *character* appears three times. Represent it by ©. The statement will then appear as... common © pairs, © triplets, and entire words in a message are replaced by special ©s. In this case, the compression ratio is approximately 4:3.
- **Binary number substitution:** for decimal numbers of two digits and more, a significant character saving can be achieved by converting the numbers to binary. An 8-bit sequence taken from the code table can represent one decimal digit, or zero to 255 if full use is made of the binary representation. For 8-bit coding, this represents a compression of 2:1 for $10 \leq$ number ≤ 99, and 3:1 for $100 \leq$ number ≤ 255.

3.3.4 ENTROPY CODING

In this section, I describe character compression using statistical encoding.

(1) Information Theory

- **Information**: a commodity transferred from source to receiver over a communication channel; the value depends on the probability associated with the message being sent.

For technical purposes, *information* was defined by Claude Shannon[2] (a 20th-century American mathematician) in terms of the probability of a specific message being sent by a source that can send a finite number of different messages.

---

[2] Claude Shannon, "A mathematical theory of communication," *Bell System Technical Journal*, 27 (1948), 379–423, 623–56.

- **Self information**: when a source emits a sequence of symbols $x_1, x_2, \ldots x_j \ldots$ and each symbol x_j occurs with probability P_j, it conveys a quantity of information I_j given by

$$I_j = -\log P_j = \log\left(1/P_j\right)$$

The information conveyed by a symbol is equal to the logarithm of the inverse of the probability it will be sent. Thus

- If the same symbol is received all the time, the probability of being sent is 1 and the self-information is 0—i.e., $P_j = 1$ and $I_j = 0$
- If a symbol is received infrequently, the probability of being sent is close to zero and the self-information is high, i.e., $P_j \to 0$ and $I_j \to \infty$.

These relations mimic common experience. A message that is repeated endlessly is ignored; it provides no new information. On the other hand, attention is paid to a message that is received infrequently; it provides new information—perhaps very important information.

The unit of self-information is 1 *bit*. Unfortunately, this is also the term used for a binary digit. When confusion is possible, custom reserves the term *bit* for self-information and substitutes the term *binit* for binary digits. For a binary source equally likely to emit 0s or 1s

$$P(0) = P(1) = 1/2$$

and the self-information is

$$I_0 = I_1 = \log_2 2 = 1 \text{ bit}$$

Thus, the information conveyed by a binary symbol that is equally likely to be 0 or 1, is 1 bit. When 1s and 0s are not equally likely, the self-information of each member of the symbol class more likely to be sent is less than 1. Similarly, the self-information of each member of the symbol class less likely to be sent is more than 1.

(2) Entropy and Information Rate

- **Discrete memoryless source**: an information source that emits a steady stream of mutually independent symbols at a rate of r symbols/s. (The self-information of the symbols remain constant over time.)

- **Source entropy**: the average value of information per symbol emitted by a source. Denoted by $H(X)$, for a discrete memoryless source that emits a sequence of symbols $x_1, x_2, \ldots x_m \ldots$, the source entropy is

$$H(X)=\sum_{j=1}^{M}P_jI_j=\sum_{j=1}^{M}P_j\log\left(1/P_j\right)$$

where X is a discrete random variable having possible values $I_1, I_2, ...I_M$.

When the source only emits the same symbol, $P_j = 1$ and $H(X) = 0$, i.e., the average value of information per symbol is zero—a result that confirms a previous assertion. When the source is equally likely to emit any symbol, $P_j = 1/M$ and $H(X) = \log M$.

(3) Entropy Coding

If, from a set of M symbols, a discrete memoryless source emits symbols of length N_j binits and their probability of transmission is P_j, the average length $(\overline{N})$ of the codewords representing this set is

$$\overline{N}=\sum_{j=1}^{M}P_jN_j$$

If the length of the codewords (in binits) is proportional to the self-information of the symbols, we obtain $\overline{N}=CH(X)$ where C is a constant. If it was possible for $C = 1$—i.e., the length of the codeword in binits equals the self-information of the symbol in bits—we should obtain the minimum average length for the set of M symbols. However, we must use at least 1 binit to represent a symbol with close to zero self-information, so, in the general case, we cannot achieve the minimum average length. Nevertheless, by making the length of the codewords proportional to the self-information of the symbols they represent, we can produce efficient variable-length codes. Another way of expressing the procedure is we use the shortest codeword for the symbol used most, and the longest codeword for the symbol used least. One technique, known as Shannon-Fano coding, is described below.

A prerequisite for variable-length coding is unique decipherability. For the case of codewords of equal length, this requirement is easy to fill—start at the beginning and divide the symbol stream into equal length words $(\overline{N}=N)$. For symbol streams with variable-length coding, the task is more difficult. One approach is to require no member of the code set be the prefix of any other member.

(4) Shannon-Fano Coding

A *Shannon-Fano* code[3] is a code in which the codeword lengths increase as the symbol self-information increases, and no member of the code is the prefix of any other member. Consider eight symbols (S1, S2, ... S8) that have probabilities of

---

[3] R. M. Fano, "A Heuristic Discussion of Probabilistic Coding," *IEEE Transactions on Information Theory*, IT-9 (1963), 64–74.

being sent ranging from 0.24 to 0.03. They can be represented by the eight members of an equal length 3-bit code, or by codewords whose length is inversely proportional to the probability of them being sent. Specifically, suppose the eight symbols occur with probabilities

$$S1 = 0.03 \quad S2 = 0.17 \quad S3 = 0.08 \quad S4 = 0.12$$
$$S5 = 0.15 \quad S6 = 0.10 \quad S7 = 0.11 \quad S8 = 0.24$$

In **Figure 3.16**, we construct a Shannon-Fano code for these symbols, determine the entropy of the code, and compute the average number of binits per codeword.

To construct a Shannon-Fano code, the symbols are arranged in descending order of probability of being sent. The column is divided to split the sum of the probabilities in half—or as close to half as possible. To the symbols above the split, 0s are assigned, and to the symbols below the split, 1s are assigned. Each subdi-

Symbol	P_k	$P_k \log_2 \frac{1}{P_k}$	SHANNON-FANO CODING TABLE				Code	$P_k N$
S8	0.24	0.49413	0	0			00	0.48
S2	0.17	0.43459	0	1	0		010	0.51
S5	0.15	0.41055	0	1	1		011	0.45
S4	0.12	0.36707	1	0	0		100	0.36
S7	0.11	0.35029	1	0	1		101	0.33
S6	0.10	0.33219	1	1	0		110	0.30
S3	0.08	0.29151	1	1	1	0	1110	0.32
S1	0.03	0.15177	1	1	1	1	1111	0.12

Entropy = 2.8321 bits
Average Number of Binits per Codeword = 2.87
Coding Efficiency = 98.6%

Figure 3.16 Example of Shannon-Fano Code

vision is then divided again to have approximately half the sum of the probabilities above the dividing line and half below. Zeros and ones are assigned as before. The process continues until all symbols have been coded as shown in Figure 3.16. Therefore, the entropy of the code set is 2.83 bits, and the average number of binits per codeword is 2.87, so the coding efficiency is 98.6%. If we chose to code the symbols with an equal number of binits (3), the efficiency would have been 85.7%. Thus, for this particular set of symbols and statistical mix, Shannon-Fano coding improves coding efficiency by 15% over equal-length, 3-binit codewords.

However, is the code decipherable? To test it, consider the datastream

$$\Leftarrow 1101100011010011101101011000111100010111100100011101\ldots$$

We must match the bits in sequence against the eight members of the code set—i.e., S8 = 00, S2 = 010, S5 = 011, S4 = 100, S7 = 101, S6 = 110, S3 = 1110 and S1 = 1111.

- The stream begins with 1; this eliminates S8, S2, and S5. The second bit is 1; this eliminates S4 and S7. The third bit is 0; this eliminates S3 and S1, and matches S6. Thus, the first symbol is S6.
- Bits 4, 5, and 6 repeat 110, so the second symbol is S6.
- Bit 7 is 0, eliminating all but S8, S2, and S5. Bit 8 is 0 so the third symbol is S8.
- Bit 9 is 1, eliminating S8, S2, and S5. Bit 10 is 1, eliminating S4 and S7. Bit 11 is 0, making the fourth symbol S6.

The first four symbols in the message are S6 S6 S8 S6. I leave it to the reader to confirm that the entire message is

S6 S6 S8 S6 S4 S3 S6 S7 S4 S5 S6 S8 S7 S3 S2 S5 S7

That I can decode the datastream into a unique sequence of symbols demonstrates that the Shannon-Fano technique produces unique codewords.

3.3.5 TECHNIQUES FOR FACSIMILE

A letter-size document scanned with a horizontal and vertical resolution of 200 dots per inch (dpi) could generate 3.2 million bits. Transmitting this amount of data over a 56 kbits/s channel will take almost one minute. To achieve faster service, data compression techniques must be employed. Two techniques are in general use:

- **Modified Huffman code**: employs a combination of run-length coding and variable-length (entropy) coding. In any document scanned horizontally, there will be many sequences (runs) of all white picture elements (pels) and all black pels. In modified Huffman (MH) coding a run of length R pels is expressed as $R = 64m + n$, where, $m = 0, 1, 2, ... 27$, and $n = 0, 1, 2, ...63$. n and m are encoded according to tables of Huffman codes developed to suit a variety of documents. Average compression ratios of 8 are obtained with this technique.

- **Modified READ code**: in the Modified Relative Element Address Designate (READ) Code, run lengths are encoded by noting the position of change elements (i.e., where white changes to black, or black changes to white) in relation to a previous change element. The use of this coding procedure produces average compression ratios of from 12 to 15.

REVIEW QUESTIONS FOR SECTION 3.3

1 Explain the following statement: Between machines, information is exchanged by binary digits (bits). Since each bit is two-valued—either 0 or 1—per se, it can describe two situations (states). To describe more states, it is necessary to use the bits in sequences.

2 What are ASCII and EBCDIC? List their characteristics.

3 In a word formed from ASCII or EBCDIC, which bit is transmitted first?

4 What is done to delineate characters when they are generated at a keyboard and transmitted singly, one after the other?

5 What is done to delineate them when characters are generated at a keyboard and transmitted a line at a time?

6 When characters are run together to form a block of data, and turned into a frame, what information do the header and trailer contain?

7 Why is there a difference in coding efficiency when a message is transmitted character by character and frame by frame?

8 Discuss the following terms: frame, packet, envelope, payload, and container.

9 Explain the following data compression techniques: null compression, run-length compression, pattern substitution, and binary number substitution.

10 Define information and self-information. If the same symbol is received all the time, what is its self-information?

11 If the same symbol is received very rarely, what is its self-information?

12 Explain the difference between bit and binit.

13 What is a discrete memoryless source?

14 What is the entropy of a source?

15 Give a simple explanation of entropy coding.

16 Comment on the following statement: A prerequisite for variable-length coding is unique decipherability.

17 Give a simple description of how to devise a Shannon-Fano code.

18 Describe two techniques used to compress facsimile signals.

3.4 VOICE SIGNALS

The electrical equivalent of speech, voice signals carry sufficient usable information in a bandwidth of about 3 kHz to reconstruct conversations between people. Until quite recently, this narrowband signal set the performance requirements for telephone networks and influenced most of the developments in telecommunications.

3.4.1 SPEECH COMPONENTS

Speech consists of a sequence of elementary sounds uttered at rates up to 10 per second.

(1) Phonemes

Elementary speech sounds are called *phonemes*. Produced by bursts of air that originate in the voice box, acquire pitch by passing through the vocal chords, and excite resonances in the vocal tract, they result in the string of sounds we call speech. Because each person's vocal apparatus is different, the speech they produce is spoken in voices unique to the speakers.

(2) Frequency, Phase, and Amplitude

By means of a microphone, speech is transformed from sound pressure waves to electrical signals. At this point, the speech signal acquires technical descriptors: *frequency, phase,* and *amplitude*. Within reasonable limits, they correspond to the pitch, intelligibility, and intensity of the original sound. Of the three electrical parameters, phase appears to be the least critical. Significant phase shifts can be produced in the message signal without affecting the ability to recognize the speaker and understand the words spoken.

(3) Talkspurts

In a conversation, the participants share the task of talking. First one speaks, then the other. Occasionally, the talking activities will overlap as one interrupts the other. On average, each talks about 40% of the time. The remaining 20% represents the time neither party is talking. It is made up of pauses between words and sentences when the speaker is clearly going to continue, and pauses when one speaker has finished speaking and the other is preparing to respond. Considered from this perspective, speech signals consist of bursts of electrical energy interspersed with periods of no activity at all. More or less connected periods of vocal activity are called *talkspurts*.

(4) Electrical Parameters of Speech

Measurements on the speech signals produced by a large population of talkers indicate natural speech

- Contains energy between approximately 100 Hz and 7,000 Hz
- Has peak energy around 250 Hz to 500 Hz
- Has a dynamic range of some 35 or 40 db
- Contains an amplitude distribution that favors low signal levels and can be represented by a Laplace probability density function.

For transmission on telephone facilities, the signal energy is usually restricted to a range from 200 or 300 Hz to 3,200 or 3,400 Hz. On toll facilities, the value of the ratio of the power of the voice signal to the power of all interfering signals (signal-to-noise ratio) is generally expected to be 30 db, or better. For transmission on AM radio, voice energy is restricted to a frequency range from about 100 Hz to something less than 5,000 Hz. Only on FM radio is it possible to transmit the full range of voice frequencies.

(5) Speech Quality

The quality of the speech signal delivered to the human ear is a matter for the exercise of individual judgment. While not measurable in any absolute way, three levels can be defined that are recognizable by most persons.

- **Network quality:** contains the full range of frequencies without perceptible noise or distortion. It is suited to the speech requirements of radio broadcasting (particularly FM broadcasting) and television.
- **Communications quality**: contains a limited range of frequencies and may have discernible noise and distortion. The speech quality is adequate for easy telephone communication, including speaker recognition. This level includes toll-quality voice.
- **Synthetic quality**: contains sounds generated by an electronic analog of the human speech production mechanism. May be more intelligible than communication quality voice, but speaker recognition is missing.

Numerical values have been assigned to descriptions of speech quality. Called *Mean Opinion Score* (MOS), the values range from 1 to 5:

1 = Bad; cannot understand words or recognize the speaker
2 = Poor; can understand words but not recognize speaker

3 = Fair; understandable, but not very good quality

4 = Good; toll quality

5 = Excellent; equal to face-to-face communication.

The score for a particular sound is usually determined by the reactions of a panel of listeners. Roughly, network quality corresponds to MOS scores between 4 and 4.5; communications quality corresponds to MOS scores between 3.5 and 4; and synthetic quality corresponds to MOS scores between 2.5 and 3.5. (Use of the two scales is shown later in Figure 3.26.)

(6) Electrical Parameters of Music

Music is produced by instruments that emit essentially continuous sounds. The electrical signals they generate

- Contain energy between approximately 20 Hz and 20,000 Hz
- Have maximum energy between 200 Hz and 600 Hz
- Have a dynamic range of some 75 or 80 db.

AM radio programs contain speech and music in the frequency range from about 100 Hz to something less than 5,000 Hz. FM radio programs contain speech and music in the frequency range from about 100 Hz to 15,000 Hz. Neither facility is able to reproduce a dynamic range of 80 db. **Figure 3.17** shows some of the parameters of speech and music signals.

3.4.2 Digitizing Voice Signals

As digitalization is applied to all telecommunication facilities, the digital equivalents of analog signals become an important topic, particularly digital voice signals. Analog voice signals are changed to digital voice signals by sampling, and then expressing the sample values as binary numbers to produce a digital stream. Digital voice signals are changed back to analog voice signals by converting the binary numbers to amplitude modulated pulses (a procedure known as pulse amplitude modulation, PAM) that are applied to a reconstruction filter to produce an analog signal. Because of the dynamic range encountered, a technique known as companding is required so very loud signals do not overpower the system, and very soft signals are not lost in noise. Within limits, the reconstructed analog signal mimics the original analog voice signal.

(1) Band-Limited Signal

Telecommunication signals are limited to specific bandwidths by the equipment that processes them. Accordingly, we consider only band-limited signals—i.e., signals that contain energy in the range $0 < f \leq W$ Hz. Elsewhere the signal value is zero.

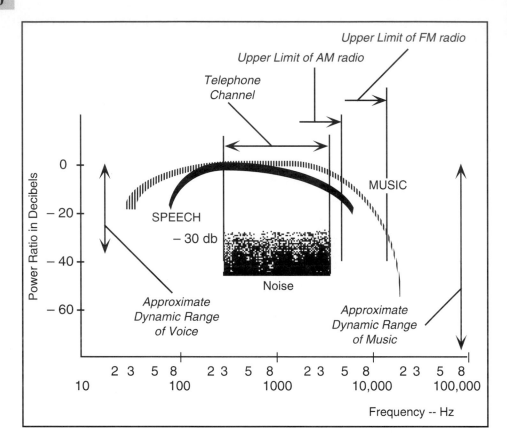

Figure 3.17 Some Properties of Speech and Music Signals

(2) Sampling

- **Sampling**: the process of determining the value of the amplitude of an analog signal at an instant in time.

Usually, sampling is done in a repetitive manner at equal time intervals. An *ideal* sampling function consists of a periodic train of Dirac delta functions. It consists of delta functions spaced apart by the sample time T_s. In the frequency domain, the ideal sampling function becomes a comb of delta functions of area f_s, separated from each other by f_s, where $f_s = 1/T_s$.

Figure 3.18 shows an example of these signals for the case of a square wave limited to the fundamental, third and fifth harmonics (based on Figure 3.7). In drawing the spectrum, we oversample (i.e., sample at 110% of the Nyquist rate).

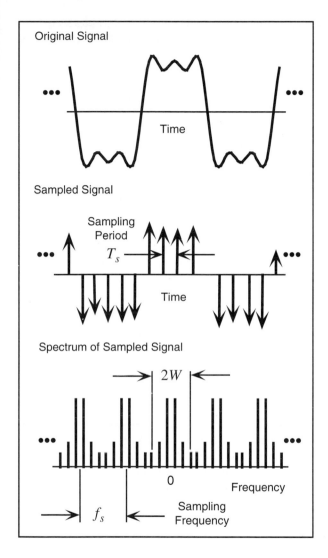

Original Signal

Time

Sampled Signal

Sampling
Period

T_s

Time

Spectrum of Sampled Signal

$2W$

0

Frequency

f_s

Sampling
Frequency

Figure 3.18 Sampling of a Periodic Waveform

The original signal is a square wave that is limited to the fundamental, third, and fifth harmonic components. It is sampled at 110% of the Nyquist rate. The spectrum of the sampled signal duplicates the spectrum of the original signal at intervals of $2W +$ 10% Hz.

- **Nyquist rate**: a sampling rate (f_s, in samples/s) exactly twice the highest frequency (W, in Hertz) in a band-limited signal $0 < f \le W$.

When $f_s \ge 2W$, sampling yields a set of delta functions that contains all of the information necessary to reconstruct the original, band-limited, signal. To demonstrate this result for the case of $f_s = 2W$, we imagine passing $S_\delta(f)$ through a low-pass filter that limits the frequencies to $\pm W$. Reference to Figure 3.18 shows this range just encompasses the frequency components necessary to reconstruct the original signal. Sampling at $f_s > 2W$ increases the interval between the repeated blocks of harmonic components in the bottom diagram in Figure 3.18; it does not change the internal relationship between the components.

(3) Aliasing

If $f_s < 2W$—i.e., the sampling rate is less than the Nyquist rate—the reconstructed signal will be a distorted version of the original with the energy that falls in the frequency range $f_s/2 < f \leq W$ Hz masquerading at lower frequencies. This effect is known as

- **Aliasing**: condition that exists when signal energy present over one range of frequencies in the original signal appears in another range of frequencies after sampling.

Figure 3.19 shows the effect on the spectrum of sampling the signal of Figure 3.18 at a rate 20% less than the Nyquist rate ($f_s = 1.8W$). Energy from the spectrums on either side of the baseband appears in the band $\pm W$. The result is the distorted signal shown in the lower diagram of Figure 3.19. To prevent aliasing, it is usual to *oversample* the band-limited signal (perhaps 110% to 120% of the Nyquist rate). To ensure the signal is band-limited, a low-pass filter of bandwidth W Hz may be included in the signal chain before the sampling operation.

3.4.3 QUANTIZATION

Using a process known as quantization, sample values are expressed in octets to produce digital signals. An octet is able to represent $2^8 = 256$ states. If they are used to describe the values of the train of impulses that is the sampled signal, we cannot describe each impulse value precisely; it must be rounded to the nearest binary value. Thus, the octets will only approximate the values of the real signal. This is known as

- **Quantization**: the process that segregates sample values into ranges and assigns a unique, discrete identifier to each range. Whenever a sample value falls within a range, the output is the discrete identifier assigned to the range.

By declaring the greatest sample values to be ± 1, the process can be normalized.

(1) Linear Quantization

Figure 3.20 illustrates linear quantization. The samples are represented by impulses whose values correspond to the amplitudes of the band-limited signals at the sample times. A set of uniformly spaced levels is shown. Each of the q levels is identified by an octet selected from those that can be made from an n bit binary code. When the levels span ± 1, the distance between each level is $2/q$. As

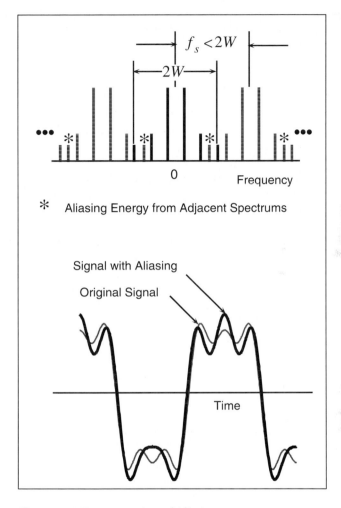

Figure 3.19 Demonstration of Aliasing
Undersampling the signal causes the spectrums of the sampled
wave to overlap.

the pulses arrive, they are assigned the octet that identifies the band in which they fall. The result is a series of octets that represents bands of sample values derived from the original signal.

Because voice signals contain many low-level samples, octets rich in 1s are assigned to the quantizing levels that correspond to them. In transmitting the octet, the most significant bit (the left-hand bit in Figure 3.20) is sent first. It defines whether the sample is positive (1) or negative (0).

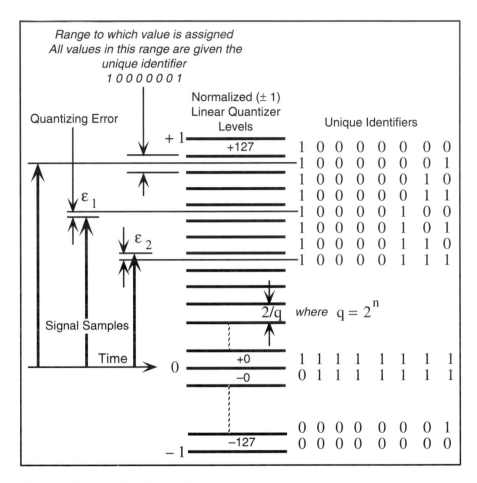

Figure 3.20 Principle of Linear Quantizing

Quantization is the process that segregates sample values into ranges and assigns a unique identifier to each range. The output is a series of octets that represent bands of sample values derived from the original signal. The octets are chosen so that the greater numbers of 1s occur for low-level sample values.

(2) Quantization Noise Signal

At the quantizer, the signal consists of two components, the quantized signal and a signal that is the difference between the quantized signal and the input signal. The latter is the amount by which the quantization process has degraded the output—it is the quantization *noise* signal. Assuming the quantization error is zero mean and uniformly distributed, and the quantizer is normalized with q levels, the magnitude of the quantization noise power can be calculated and the signal-to-noise ratio obtained. For a normalized input signal

$$S/N \leq 4.8 + 6.0n \quad \text{db}.$$

In **Figure 3.21** we plot the SNR experienced by a sinusoidal signal quantized in 256 equally spaced levels ($n = 8$). The signal power is expressed in db relative to a full-load signal (i.e., one that just sweeps through all 256 levels). From the graph, we see

- The quantization SNR experienced by a full-load signal is 52.8 db (i.e., a ratio of almost 200,000 to 1)
- An input signal with approximately one-hundredth (i.e., –20 db) of the power of a full-load signal will experience a quantization SNR of 32.8 db (i.e., a ratio of almost 2,000 to 1)
- An input signal with approximately one ten-thousandth (i.e., –40 db) of the power of a full-load signal will experience a quantization SNR of 12.8 db (i.e., a ratio of almost 20 to 1).

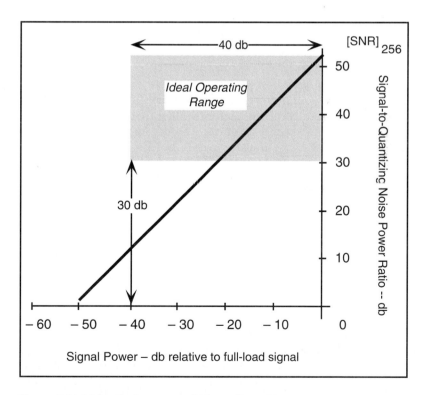

Figure 3.21 Noise Performance of Linear Quantizer
When the quantizing ranges are uniformly spaced, quantizing noise limits the ability of the quantizer to produce acceptable signal-to-noise power ratios at low levels of input signals.

Now, the dynamic range of voice signals can be as much as 40 db, and toll-quality voice requires a SNR of at least 30 db. Thus, a linear quantizer with 256 levels can handle only part of the required range with better than 30 db SNR.

3.4.4 COMPANDING

Speech signal samples show a preponderance of low amplitudes, and many samples have values close to zero. To improve the performance of a linear quantizer with respect to these low-level signal samples, some form of signal processing is required. One technique is known as *companding*.

(1) Distributing Signals Evenly

The sample distribution present in a voice signal can be approximated by a zero-mean Laplace distribution (after the late 18th-century and early 19th-century French mathematician Marquis Pierre Simon de Laplace). Shown in **Figure 3.22**, it is characterized by a peak at zero amplitude and a symmetrical exponential falloff. Most of the samples fall in levels close to zero where the effect of quantizing noise is greatest. Only a few samples will be encoded in the intervals approaching ± 1 where the effect of quantizing noise is least.

In the signal chain, to improve noise performance, processing is included that distributes signals approximately evenly across the levels of a linear quantizer. Known as companding, on the input side, it *compresses* the samples so the higher values are reduced with respect to the lower values; thus, with a fixed number of quantizing levels, more levels are available for lower-level signals. On the output side, it *expands* the samples so the compressed values are restored to their original values.

(2) μ-Law and A-Law Companding

Among the telephone administrations of the world, two companding functions are employed.

- **μ-Law companding**: used in North America and regions of the world that follow North American practices. If the amplitude of the input signal is denoted by x and the amplitude of the compressed output signal is $z(x)$, they are related by

$$z(x) = \mathrm{sgn}(x) \frac{\log_e(1+\mu|x|)}{\log_e(1+\mu)}$$

  where sgn(x) denotes the signum function, and $\mu = 255$. The positive quadrant of the compressor characteristic is shown in **Figure 3.23**

- **A-Law companding**: used in Europe and regions of the world that follow European practices.

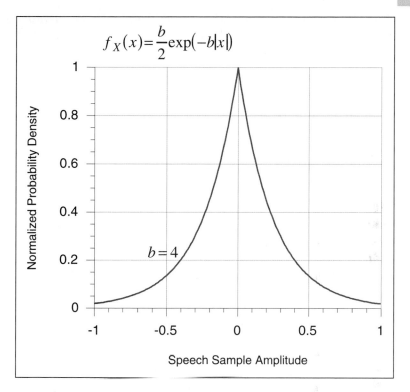

Figure 3.22 Laplace Distribution Density
The samples derived from speech signals show a Laplace distribution.

$$z(x) = \text{sgn}(x)\frac{A|x|}{1+\log_e(A)}, \quad 0 \le |x| \le \tfrac{1}{A}$$

$$= \text{sgn}(x)\frac{1+\log_e(A|x|)}{1+\log_e(A)}, \quad \tfrac{1}{A} \le |x| \le 1$$

where $A = 87.6$. For values of $|x| \le 0.01$ the relation is linear; for larger values, it is non-linear. For values of the input signal greater than 0.04 (i.e., 4% of full load), A-law and μ-law compander characteristics are practically the same.

Companding using one law, and restoring according to the other, will distort the signal.

In **Figure 3.24**, the signal-to-quantization noise ratio of a companded signal is plotted against the power of the input signal expressed in db relative to a full-

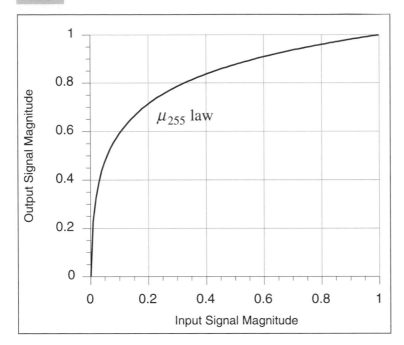

Figure 3.23 μ-Law Compressor Characteristic

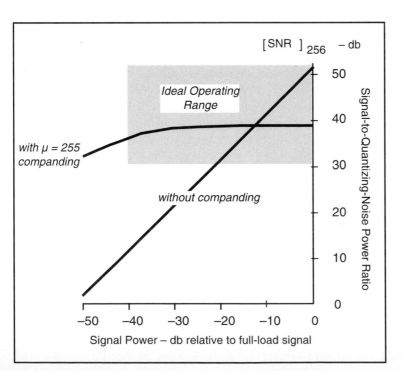

Figure 3.24 Companding Makes Operation Over the Ideal Operating Range Possible

load signal. Comparing this curve to the one obtained earlier for linear quantization, it is apparent that companding improves the SNR at lower signal levels and degrades it at higher levels. Companding is a solution to achieving better than 30 db SNR over a 40-db dynamic range.

3.4.5 RECONSTRUCTING THE SIGNAL

At the expander, a train of narrow pulses is created from the string of octets representing the compressed samples. Their amplitudes are determined by the quantizer level to which they were assigned multiplied by the inverse of the compressor characteristic. The pulse repetition rate is equal to the sampling rate. With an exact match between these two rates, the frequencies present in the original signal are duplicated. Next, this pulse train is applied to a filter (reconstruction filter) that passes frequencies $\leq W$ Hz. Provided sampling was conducted at, or above, the Nyquist rate, the signal at the output of the filter will be a close mimic to the original, band-limited signal. (From Figure 3.18, we see a low-pass filter of passband $\pm W$ Hz will isolate the spectrum of the input wave.)

In modern systems, the functions we have described have been combined and are implemented in digital signal processors (DSPs) or custom-integrated circuits. At the transmitter, the voice signal may be sampled and quantized in 4,096 (2^{12}) levels, in a linear fashion. Then, using a table stored in ROM, each 12-bit word is replaced in a biased fashion by one of 256 octets to produce an approximately uniform distribution of signal values. Usually, the companding characteristic is approximated by linear segments divided into equal quantizing levels.

3.4.6 PCM VOICE

Because telephone voice must withstand the rigors of worldwide service that may include several analog-to-digital, and digital-to-analog, conversions, the telephone administrations of the world adopted a robust coding scheme for voice signals. The system

- Employs a sampling rate of 8,000 samples/s
- Codes each sample in 8 bits to produce a signal with a bit rate of 64,000 bits/s
- Includes μ-law or A-law companding (ITU–T Recommendation G.711).

Called *pulse code modulation* (PCM), it provides communications-quality voice signals that are easily understood and readily attributable to particular speakers. Because it is based on sampling at 8,000 times per second, the analog signal reconstructed from the PCM stream cannot contain frequencies above 4,000 Hz.

(1) Decoding Noise

Signals that interfere with the digital signal stream can affect regeneration and produce erroneous values in the receiver. Inspection of Figure 3.22 shows the change in quantization levels due to an error in bit position k is 2^{k-1} levels— i.e., an error in the kth bit shifts the decoded level by $\varepsilon_k = \pm(2/q)2^k$. If P_ε is the probability of error in any bit position, it is possible to show that the decoding noise power is $N_D \cong 4P_\varepsilon/3$ signal watts.

(2) Signal-to-Total-Noise Power Ratio

The analog voice signal delivered by a PCM channel is corrupted by the sum of quantization noise and decoding noise so the total noise power N_T is

$$N_T = N_Q + N_D$$

In **Figure 3.25**, we plot values for the signal-to-total-noise power ratio for a range of probabilities of bit errors and power of a sinusoidal signal at the input to the quantizer.

- To achieve a channel with $[SNR]_T \geq 30$ db, and dynamic range ≥ 40 db, the probability of bit error must be greater than 1 bit in error in 100 million bits (1×10^{-8}). This value is hard to achieve on wire lines, although it is readily achievable over optical fibers.

- If we reduce $[SNR]_T$ to 15 db—a value many persons would describe as difficult to use—then a wire line with a probability of 1 bit in error in 100,000 bits (1×10^{-5}) can accommodate a dynamic range of approximately 35 db.

- If we limit the dynamic range to 20 db—a value practical in many situations—wire-line error rates as low as 1×10^{-4} will produce a $[SNR]_T$ of approximately 20 db.

Thus, the probability of error has a significant effect on the operating conditions that can be tolerated on a PCM channel.

3.4.7 LOWER BIT-RATE CODING

Because it allows more users to make simultaneous use of the limited channels available, lower bit-rate coding is of increasing importance in long-distance transmission systems, mobile radio systems, and personal telephone systems. By increasing the complexity of signal processing, communications-quality voice signals can be produced at rates significantly less than 64 kbits/s. The techniques employed are divided into *waveform coding* and *source coding*.

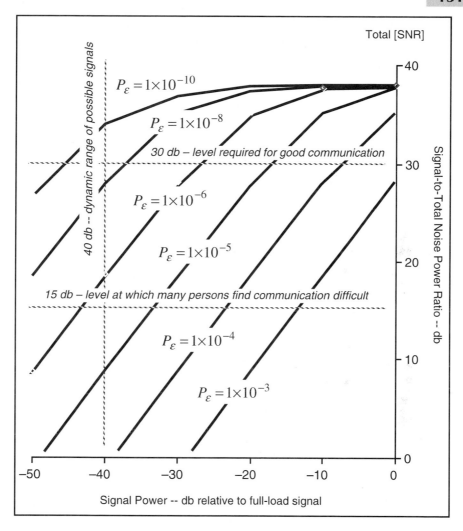

Figure 3.25 Effects of Errors on Performance of PCM Voice

(1) Waveform Coding

In waveform coding, speech is reduced to a string of bits by operating on the signal in one of two ways

- **Differential coding**: exploits the correlation between adjacent signal samples
- **Frequency domain coding**: divides the signal into a number of frequency bands and codes each separately.

(2) Differential Coding

If a band-limited voice signal is sampled at the Nyquist rate, the sample values will not vary greatly from one sample to the next (there is a high correlation between contiguous samples). Operating with the difference between adjacent samples, or samples and predicted values, reduces redundancy and permits the use of fewer quantization levels. The result is shorter codes and lower bit rates. At the receiver, adding up the reconstructed differences will produce sample values that mimic the original values (except for any dc component). Two important examples of differential coding are

- **Differential PCM** (DPCM): a coding technique that reduces redundancy between successive PCM samples. At the kth sampling time, we determine the difference between $s(kT_s)$ and $s[(k-1)T_s]$, the value at the $(k-1)$ th sampling time. The value is encoded and transmitted. By integrating these differences, the receiver is able to construct an approximation $[s'(kT_s)]$ to the transmitted signal. To make the process reliable, the transmitter also constructs $s'(kT_s)$. Should the transmitter detect unacceptable differences between its value of the receiver's predicted value, and the real signal at the transmitter, it sends a difference value based on $s(kT_s) - s'(kT_s)$ to make the receiver's value equal to the real value.

- **Adaptive DPCM** (ADPCM): a DPCM technique in which input signal samples are encoded in 14-bit linear PCM. Current samples are subtracted from predicted values derived from the weighted average of several previous samples and the last two predicted values. Differences are applied to a 15-level adaptive quantizer that assigns 4-bit codes. At the receiver, an inverse quantizer forms a quantized difference signal. Together with the output of an adaptive predictor, it is used to update the value of the estimated signal; 32 kbits/s ADPCM has been shown to perform better than 64 kbits/s PCM in noisy environments.

ITU-T has developed ADPCM recommendations (G.726, and G.727) for 4 kHz voice at bit rates of 40, 32, 24, and 16 kbits/s in public network applications. With minor modifications, ANSI-T1 has adopted these standards, placing particular emphasis on 32 and 16 kbits/s. In addition, ITU-T has developed an ADPCM recommendation (G.722) for 7 kHz voice at 64 kbits/s—i.e., a coding that carries the full range of voice frequencies in 64 kbits/s.

(3) Frequency Domain Coding

Because speech is not usually energetic across the entire frequency spectrum at the same time, it is possible to allocate more bits (number of quantization levels) dynamically to those bands in which the activity is significant, at the expense of those bands that are quiescent. Two techniques of this sort are subband coding (SBC) and adaptive transform coding (ATC).

- **Subband coding**: the input signal frequency band is divided into several subbands (typically four or eight). Each subband is encoded using an adaptive step size PCM (APCM) technique. SBC produces communications-quality voice at 9.6 kbits/s.

- **Adaptive transform coding**: the input signal is divided into short time blocks (frames) that are transformed into the frequency domain using techniques based on Fourier series. The result is a set of transform coefficients that describe each frame. These coefficients are quantized into numbers of levels based on the expected spectral levels of the input signal. The number of bits for any quantizer and the sum of all the bits per block are limited to a predetermined maximum. ATC produces communications-quality voice at 7 kbits/s.

(4) Source Coding

In source coding, speech sounds are identified and encoded. By transmitting a description of the speech and emulating the mechanism of the vocal tract in the receiver, it is possible to produce artificial speech at greatly reduced bit rates.

- **Vocoding**: in spectrum channel vocoders, the speech signal is divided into several separate frequency bands (e.g., 16 bands of 200 Hz each). The magnitude of the energy in each band is transmitted to the receiver together with information on the presence or absence of pitch. At the receiver, voiced or unvoiced energy is applied to a set of receivers that produce recognizable speech.

- **Linear prediction models**: speech analysis and synthesis are performed using models in which successive samples of speech are compared with previous samples to obtain a set of linear prediction coefficients (LPCs) used to reproduce the sound. In effect the transmitter listens to the input speech, calculates LPC strings, and transmits them to the receiver. It regenerates the sound using an artificial vocal tract. Reducing voice to a string of codes representing these sounds can produce a signal of the order of a few hundred bits per second. To achieve this rate, however, significant analysis of the signal is required, and it results in considerable delay.

In recent years, ITU-T has developed standards for 16 kbits/s, low-delay, linear-prediction voice coding; ANSI has developed standards for 7.9 kbits/s, vector-sum excited, linear prediction voice coding for digital mobile telephone; and the European digital mobile radio community has standardized a residual excited, linear prediction, voice coding technique at 7 kbits/s. In 1996, in G.729, ITU-T standardized 8 kbits/s voice for mobile radio applications, and extended it to include simultaneous voice and data (G.729A), and packet voice (G.729B). The voice quality is described as *wireline*, and rates a MOS of approximately 4.[4]

In addition, the Federal Telecommunications Standards Program has developed linear prediction voice coding standards at 2.4 and 4.8 kbits/s to serve the secure communications needs of the U.S. government.

(5) Processing Delay

In order to process the signal, it must remain within the encoding/decoding equipment for a finite time. The result is that the output train is displaced from the input train by a time delay. For 64 kbits/s PCM, this delay is the same as the sampling interval—i.e., 1/8,000 s, or 125 μs. For ADPCM, the average delay is close to 125 μs; for SBC, it may be 10 to 20 ms; and, for ATC and vocoders, it may be 50 ms (or more). Interactive, duplex communication becomes progressively more difficult as the magnitude of the delay extends into the tens of milliseconds region.

3.4.8 COMPARISON OF TECHNIQUES

In **Figure 3.26**, we summarize the performance of some of these techniques. The figure is divided into horizontal fields that embrace transparent (i.e., indistinguishable from the original), network, communication, and synthetic quality. Points are shown for telephone voice using 64 kbits/s PCM, 16 and 32 kbits/s ADPCM, and 8 kbits/s (G.729). For telephone voice, ADPCM at 32 and 16 kbits/s is judged by many to be more robust than the original 64 kbits/s PCM.

REVIEW QUESTIONS FOR SECTION 3.4

1 What are phonemes? How are they produced?
2 To what voice parameters do the frequency, phase, and amplitude of a voice signal correspond?
3 What are talkspurts?
4 Enumerate some electrical parameters of speech signals.
5 Describe three levels of output speech quality.
6 What is meant by sampling an analog signal?

---

[4] For a comprehensive discussion of the development of G.729, and the properties of some of the other voice standards, see *IEEE Communications Magazine*, 35, 9 (September 1997), 40–91.

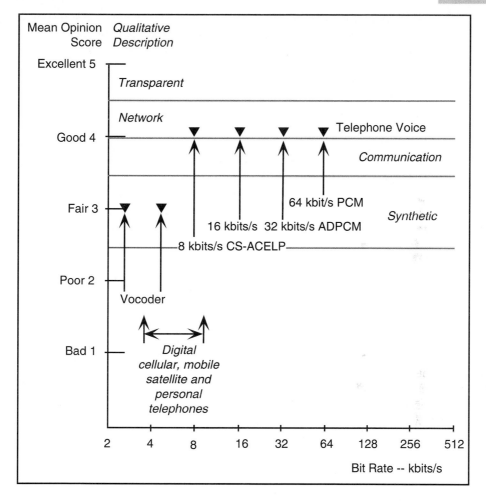

Figure 3.26 Comparison of Speech Coding Techniques

7 Define the Nyquist rate.

8 A band-limited signal is sampled above the Nyquist rate. Can the samples be used to reconstruct the original signal? Explain using Figure 3.18.

9 A band-limited signal is sampled below the Nyquist rate. Can the samples be used to reconstruct the original signal? Explain using Figure 3.18.

10 What is aliasing?

11 What is quantization?

12 What causes quantizing noise?

13 Explain Figure 3.21.

14 What is the purpose of companding?

15 Explain Figure 3.24.

16 How is the sampled signal reconstructed into a close mimic of the original analog signal?

17 Enumerate the main parameters of PCM voice.

18 Explain Figure 3.25.

19 Describe the principle of differential PCM.

20 Describe the principle of subband coding.

21 Describe the principle of vocoding.

22 Describe the principle of a linear prediction model.

3.5 VIDEO SIGNALS

The electrical equivalent of moving pictures, video signals are broad band signals. By employing complex processing, advanced coding, and compromising fidelity, video signals can be reduced to digital streams at rates as low as 64 kbits/s—the same as PCM voice.

3.5.1 PRODUCTION OF VIDEO SIGNALS

The production of video signals is essentially an analog process.

(1) Monochrome

While an electron beam scans it horizontally from top to bottom, scenes are projected on the optically active target of a camera tube. Each scene is converted to a signal composed of segments (that correspond to the horizontal lines) whose intensity is modulated in accordance with the details of the image. When the beam reaches the bottom of the target, it returns to the top to start again. In the time (known as the vertical interval) required for this action (known as vertical retrace) a small number of unmodulated lines are generated. They are used to transmit reference information to the receiver circuits, and to provide information to the viewer (closed captioning, for instance). In order to reduce the bandwidth required by the signal, yet provide a relatively flicker-free picture, the target scene is scanned in halves; all of the even lines are scanned and then all of the odd lines are scanned. The two frames (odd and even) make up a complete picture. In those countries with a power-line frequency of 60 Hz, the picture usually consists of 525 lines (495 lines picture; 30 lines vertical retrace), and a complete picture is pro-

duced 30 times a second. In countries with a power-line frequency of 50 Hz, the picture usually consists of 625 lines, and a complete picture is produced 25 times a second.

(2) Color

Color information is produced by separating the natural colors of the scene into red, green, and blue components, imaging them on separate targets, and scanning them with synchronized electron beams to produce three color signals. The red (R), green (G), and blue (B) signals are encoded to form a luminance (L) signal, and a chrominance signal that has in-phase (I) and quadrature (Q) components. They have the form

$$m_L(t) = 0.30m_R(t) + 0.59m_G(t) + 0.11m_B(t)$$
$$m_I(t) = 0.60m_R(t) - 0.28m_G(t) - 0.32m_B(t)$$
$$m_Q(t) = 0.21m_R(t) - 0.52m_G(t) + 0.31m_B(t)$$

Because of the repetitious scanning used to generate the color signals, the energy is concentrated at the line frequency (15,734.264 Hz) and its harmonics, with levels generally decreasing as the frequencies rise. The chrominance signals are modulated on to a carrier signal at 3.579545 MHz. The choice of this frequency (usually referred to as 3.58 MHz) causes the spectral lines representing the amplitude spectrum of the chrominance signal to fall between the lines representing the luminance signal. At the receiver, the signals are decoded to give

$$m_R(t) = m_L(t) - 0.96m_I(t) + 0.62m_Q(t)$$
$$m_G(t) = m_L(t) - 0.28m_I(t) - 0.64m_Q(t)$$
$$m_B(t) = m_L(t) - 1.10m_I(t) + 1.70m_Q(t)$$

3.5.2 ANALOG TELEVISION SIGNAL

In analog television, the signal consists of luminance, chrominance, and sound components. Television sound takes the form of a 50 Hz to 15,000 Hz baseband that frequency modulates a carrier signal of 4.5 MHz to produce a frequency deviation of 25 kHz. The upper diagram in **Figure 3.27** shows the frequency domain representation of a baseband analog television signal. To produce a modulated signal that will fit into the 6 MHz channels assigned to broadcast television service, National Television Standards Committee (NTSC) defines a picture carrier whose frequency is 1.25 MHz above the frequency of the lower boundary of the assigned channel. It is modulated by the composite signal to produce a vestigial-sideband, amplitude-modulated signal. This is shown in the lower diagram in Figure 3.27.

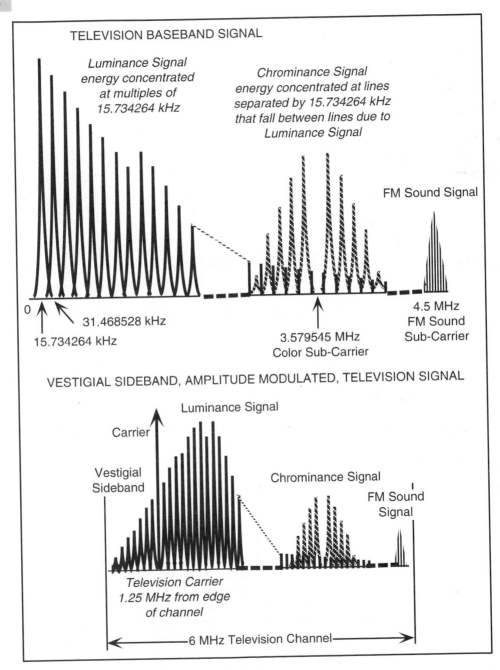

Figure 3.27 Some Properties of Television Signals

3.5.3 DIGITIZING VIDEO SIGNALS

Video pictures can be digitally encoded in three ways

- **Waveform encoding**: the video signal is reduced to a string of bits by sampling the original signal and encoding the values of the samples in some way (e.g., 8-bit PCM). Further processing may be used to reduce the bit rate (e.g., differential PCM).

- **Source encoding**: operates on the picture material in some manner to take advantage of correlations within frames, or between frames.

- **Entropy encoding**: uses statistical properties to implement variable-length coding and minimize the entropy of the coding source.

(1) Waveform Encoding

A color television signal can be digitized using the same waveform sampling procedures as for voice. For an NTSC system

- Taking 6 MHz as the maximum frequency to be reconstructed (analogous to 4 kHz for voice) and 256 quantization levels, the signal can be converted to a 96 Mbits/s, PCM bitstream

- Without jeopardizing picture quality, this rate may be reduced to 72 Mbits/s PCM by recognizing the video information is contained in 4.5 MHz, and the sound can be transmitted on a separate channel.

Reductions in these bit rates can be achieved by using differential PCM coding.

(2) Source Encoding

Source encoding of video signals begins by sampling and encoding the luminance and two chrominance signals and storing the digital information in a frame memory. Next, the digital image is divided into 8 x 8 or 16 x 16 pixel (i.e., picture element) magnitude blocks that are processed individually using a discrete-cosine transform (DCT) applied in the vertical and horizontal dimensions. The result is a series of values that represent the spatial frequencies contained in the image. When the entire frame has been processed in this way, a reconstructed image is stored in the frame memory. The succeeding frame is matched against this stored information, and the difference between the blocks is sent to the receiver where it is integrated with the previous block to form an estimate of the current block. In order to achieve a specific bit rate, odd or even fields may be discarded, and full frames may be eliminated.

Source encoding is used to produce video conference connections of varying fidelity at 1.536 Mbits/s, 768 kbits/s, and 384 kbits/s. Without degrading perfor-

mance to a level unacceptable to most users, because a typical videoconference consists of a few persons seated at a table, and a videotelephone call consists of a talking head, tradeoffs between picture quality and motion-handling capability are possible.

(3) Entropy Encoding

Encoding all symbols with equal-length binary words is only efficient if they have equal probability of being transmitted. Such would be the case with a source that emits randomly. However, in pictures, and with other practical source materials, there is a large component of correlated information. To create an efficient code, we need to encode symbols in a fashion that takes account of their self-information. In standards developed for still pictures and television frames, lookup tables based on common experience are provided that list the codes to be used for blocks of DCT coefficients formed from 8 x 8 pixel arrays. For still pictures, compressions of 4:1, or more, over equal-length coding, are achieved.

For motion pictures, four coded blocks form a *macro*block. Macroblocks in the current picture are compared with macroblocks in previous pictures to detect motion and create *motion vectors*—which are entropy-coded. The output rate from the encoder is maintained constant by a rate controller that adjusts the quantizing levels of the DCT coefficients and a buffer that smoothes out the bit rate.

In MPEG-2 (see below), motion compensation is a complex process that employs

- **Independent** (or **Intra-**) **frames**: known as I frames, they are coded without reference to other frames. They provide a reference for the calculation of motion vectors. A group of pictures (GOP) contains nine frames, one of which is an I frame.
- **Prediction** (or **Predictive**) **frames**: known as P frames, they are formed by using forward motion predictions from the previous I frame.
- **Bidirectional frames**: known as B frames, they are formed by using motion predictions from the previous I or P frame and the next I or P frame.

Because B frames use prediction from future as well as past frames, the frame sequence must be reordered before transmission to facilitate reconstruction of the original video stream. Thus, **Figure 3.28** shows a group of frames in the order they are created, and in the order they are transmitted, so the receiver has the information required to re-create the moving pictures. The choice of the number of P and B frames per I frame determines the quality of the transmitted picture and the final bit rate.

(4) Professional Group Standards Activities

The subject of video compression standards engages the attention of several professional groups:

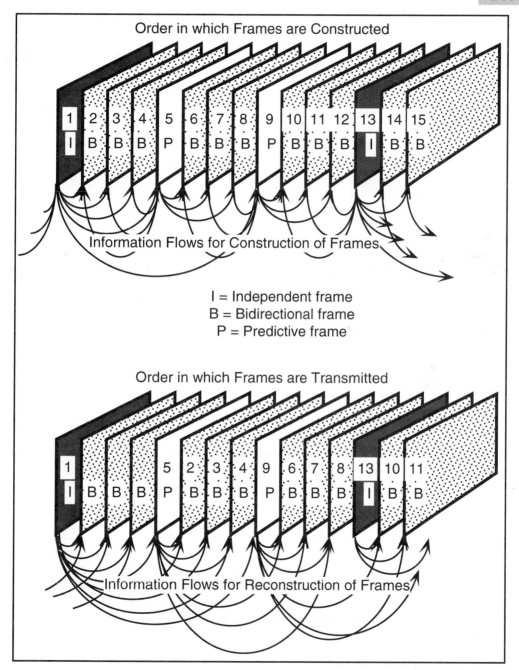

Figure 3.28 Construction of Sequence of Frames and Order in Which They Are Transmitted in MPEG-2

- **ITU–T**:
  - *H.261*. Also known as px64, standard for videotelephone ($p = 1$ or 2) and videoconference ($6 \geq p$) services over ISDN (see Section 10.4) circuits. Many of the techniques developed by H261 have been employed and extended by ISO MPEG.
  - *H.320*. Standard for videoconferencing in the range 64 kbits/s to 2.0 Mbits/s
  - *H.323*. Standard for LAN-based packet video applications
  - *H.324*. Standard for low-quality video applications.
- **ISO JPEG** (Joint Photographic Experts Group): concerned with developing coding standards for still pictures.
- **ISO MPEG** (Moving Picture Experts Group):[5] concerned with compression of full-motion video. MPEG has developed standards for various applications:
  - 300 kbits/s to 1 Mbits/s for videoconferencing applications
  - 1 to 2 Mbits/s for compact disc applications (includes fast forward, fast reverse, and random access features)
  - 2 to 4 Mbits/s for cable and satellite television
  - 4 to 9 Mbits/s for broadcast television
  - 7 to 15 Mbits/s for extended-definition television
  - 15 to 30 Mbits/s for high-definition television.

The work of ISO MPEG has resulted in the following standards:
- MPEG-1 video compression standard creates a 1.5 Mbits/s stream from relatively low definition sources. It reduces temporal redundancy with motion compensation, and spatial and perceptual redundancy with DCT coding and visually weighted adaptive quantization. It is an audiovisual coding standard directed to the storage and retrieval of multimedia information on a CD-ROM.
- MPEG-2 video compression standard creates a 1.5- to 20-Mbits/s stream from relatively high definition sources. Compatible with MPEG-1, it extends the redundancy reduction techniques to include adaptation to the characteristics of the video source and exploitation of the local contents of the video scene. MPEG-2 is directed to broadcast television applications, including high-resolution television.

---

[5] www.cselt.it/mpeg

- MPEG-3 was directed to high-resolution television applications. However, its development was overtaken by the success of MPEG-2
- MPEG-4, standardized in 1999, improves the error resilience of video coding for use in mobile radio applications.[6]

- **SMPTE** (Society of Motion Picture and Television Engineers):
  - *Task Force on Headers/Descriptors.* Concerned with the development of a universal header that can be attached to any video stream to provide the information required by the receiver to decompress the stream and reconstruct the image(s)
  - *Task Force on Digital Image Architecture.* Concerned with the development of digital standards that facilitate interoperation of high-resolution display systems.

3.5.4 DIGITAL MODULATION

Like other signals, the future format of modulated television signals is digital. Already, several direct broadcasting communications satellites beam digital channels to residential subscribers, and terrestrial broadcasting is not far behind.

(1) Standard Definition Television

Converting a standard (NTSC) analog color television signal to an 8-Mbits/s digital stream (using MPEG-2, for instance), and using QAM (see Section 4.2.2) to create a modulated wave, produces signals whose bandwidths are approximately 2.5 MHz (for 16-QAM), 1.9 MHz (for 32-QAM) and 1.6 MHz (for 64-QAM). Thus, using digital modulation and video compression, three high-quality TV signals can be accommodated in the 6-MHz channel allocated to standard television. What is more, using 32-QAM digital modulation, the same three TV signals can be delivered to a residence over a drop wire (twisted pair) up to 800 feet in length.

(2) High-Definition Digital Television

In 1998, in the United States, high-definition, compressed digital television (HDTV) made its debut as a public service. With MPEG-2 coding and a data rate of 19.42 Mbits/s in a 6-MHz channel, it employs 1,080 lines to produce a picture of 1920 x 1080 pixels with an aspect ratio of 16 x 9. An alternative, lower definition signal employs 720 lines to produce a picture of 1,280 x 720 pixels with the same aspect ratio. For broadcasting, the system uses 8-VSB modulation [see Section

---

[6] Raj Talluri, "Error-Resilient Video Coding in the ISO MPEG-4 Standard," *IEEE Communications Magazine*, 36 (1998), 112–19.

4.1.1(5)] with trellis coding. Standardized by the Advanced Television System Committee,[7] HDTV employs audio and video compression and specifies packet transport of audio, video, and data signals (see Section 10.6.2). The transport technology includes a stream of fixed-length (188 bytes) packets of data. Each packet is limited to one type of information—video, data, audio, or supervisory.

REVIEW QUESTIONS FOR SECTION 3.5

1. Describe the process of producing a color television signal.
2. Describe three ways in which video signal can be digitally coded.
3. Explain Figure 3.28.
4. Enumerate the distance and number of channels performance that can be obtained by using digital coding.
5. Give some parameters that describe high-definition television service.

---

[7] ATSC Technology Group (T.3), *Guide to the Use of the ATSC Digital Television Standard*, ATSC Document A/54, October 1995.

CHAPTER

4

MODULATED SIGNALS

By impressing baseband signals on high-frequency carriers, the action of *modulation* facilitates the long-distance transmission of data, voice, and video signals. Because they can be coupled more efficiently to the communication medium, the high-frequency modulated waves travel over greater distances than the original message signal can travel unaided.

- **Modulation**: a signal processing technique in which, at the transmitter, one signal (the *modulating* signal) modifies a property of another signal (the *carrier* signal) so that a composite wave (the *modulated* wave) is formed. At the receiver, the modulating signal is recovered from the modulated wave (this action is known as *demodulation*). The bandwidth of the modulated wave is equal to, or greater than, the bandwidth of the modulating signal.

The higher frequency (and shorter wavelength) enjoyed by the modulated wave means that it can be launched from, and received by, practical sizes of antennas, or conducted by moderate sizes of cables or waveguides. In addition, by creating sets of symbols of limited bandwidth, the action of modulation facilitates sending data over telephone lines and radio channels. Each symbol represents a specific sequence of bits, and the symbol set covers all possible combinations of bits. The maximum symbol rate is determined by the passband of the bearer and associated equipment.

4.1 ANALOG MODULATION

Analog modulation brings together a higher-frequency sinusoidal carrier and a lower frequency, analog message signal to create a modulated wave in which the parameters change smoothly, and rates of change are finite. The energy in the modulated wave is distributed at frequencies around the carrier signal frequency.

A sinusoidal carrier signal can be modulated by a message signal in three ways. Consider the signal $s(t)$ that is modeled as a cosine function—i.e.,

$$s(t) = A\cos(2\pi f t + \phi)$$

When serving as a carrier, it may be modulated by varying the amplitude A in sympathy with the message signal; in this case, we have *amplitude* modulation (AM). Also, it may be modulated by varying the frequency f or the phase angle ϕ; in these cases, we have *frequency* or *phase* modulation (FM or ΦM). Collectively, frequency modulation and phase modulation are known as *angle* modulation.

4.1.1 AMPLITUDE MODULATION

Amplitude modulated waves can take several forms—large-carrier AM (LCAM), double-sideband, suppressed-carrier (DSBSC) AM, single-sideband (SSB) AM, and vestigial-sideband (VSB) AM. To illustrate time domain behavior, a single-tone message signal $m(t) = \cos(\pi t)$, and a sinusoidal carrier signal $c(t) = 1.5\cos(5.3\pi t)$, are used. They are shown in **Figure 4.1**. In real life, the carrier will have a much higher frequency. The ratio of 5.3 to 1 for the frequencies of the carrier and message signals is chosen to make the diagrams easy to draw and understand.

(1) Large-Carrier AM

LCAM is the modulation technique employed by all radio stations broadcasting information and entertainment in the frequency range 535 to 1,605 kHz (the AM band). It has been specially crafted to make it possible to receive the message signal with inexpensive circuits—a prerequisite for the development of large-scale, consumer-oriented, public, broadcast radio services. Because the envelope of the modulated wave mimics the message signal, demodulation can be accomplished with a relatively simple envelope detector.

An LCAM wave, $s(t)$, is produced by multiplying the carrier signal $c(t)$ by $[1 + \mu m(t)]$ to give

$$s(t) = [1 + \mu m(t)] c(t)$$

Known as the *modulation index*, $\mu \leq 1$. It controls the *depth* of modulation. For the signals shown in Figure 4.1, the modulated signal becomes

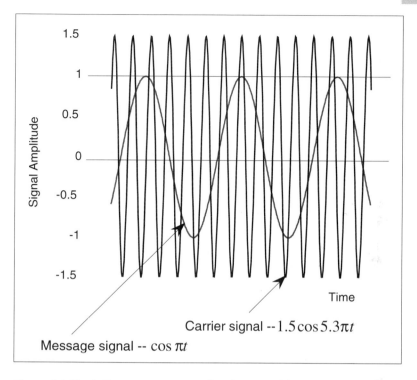

Figure 4.1 Single Tone Message Signal and Sinusoidal Carrier Signal Used To Construct Modulated Signal Diagrams

$$s(t)=1.5\big[1+\mu\cos(\pi t)\big]\cos(5.3\pi t)$$

and, when $\mu = 0.7$

$$s(t)=1.5\cos(5.3\pi t)+0.525\big[\cos(4.3\pi t)+\cos(6.3\pi t)\big]$$

As shown in **Figure 4.2**, the amplitude of the modulated wave swings from 0.45 to 2.55. (To aid visualization, the outline of the message signal is drawn on the modulated wave to emphasize that the envelope mimics it.) One-sided frequency plane diagrams are shown in **Figure 4.3**. The information sidebands are distributed about the carrier signal. In the upper half, we show the case of the single-tone message signal. In the lower half, we show the general case of a message signal $m(t)$ with spectral density $M(f)$. The modulated signal is *double-sideband* (i.e., the energy associated with the message signal is distributed on both sides of the carrier frequency) and contains a large carrier signal (i.e., there is a separate signal at the carrier frequency). If the baseband message signal is limited to the frequency interval $0 \le f \le W$, the modulated wave is limited to the band $f_c \pm W$, and the trans-

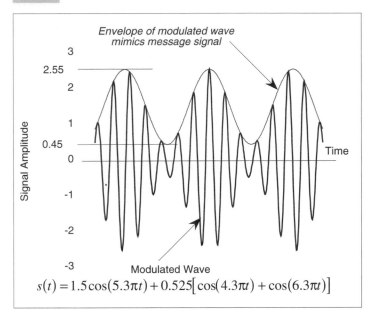

Figure 4.2 Large-Carrier Amplitude Modulated Wave

$$s(t) = 1.5\cos(5.3\pi t) + 0.525\big[\cos(4.3\pi t) + \cos(6.3\pi t)\big]$$

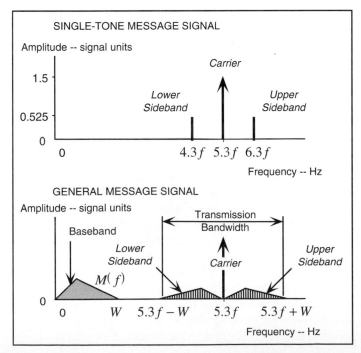

Figure 4.3 One-Sided Frequency Plane Representations of Large-Carrier Amplitude Modulated Wave

mission bandwidth is 2W Hz. Because of the presence of energy at the carrier frequency, less than 50% of the total power in the modulated wave is associated with the message-bearing sidebands. LCAM is a less efficient modulation process than the others described below.

(2) Double-Sideband Suppressed-Carrier AM

DSBSC is used in FM stereo broadcasting systems (the channel that carries the difference signal uses DSBSC, not FM), and as the first stage in the production of frequency-multiplexed signals (see Section 6.1.2).

Multiplying the carrier signal by the message signal produces DSBSC waves. Thus:

$$s(t) = m(t)\,c(t)$$
$$= 0.75\big[\cos(4.3\pi t) + \cos(6.3\pi t)\big]$$

The time-domain representation is shown in **Figure 4.4**, and one-sided frequency plane representations are shown in **Figure 4.5**. The upper half of Figure 4.5 shows the frequency distribution for a single-tone message signal. The lower half shows

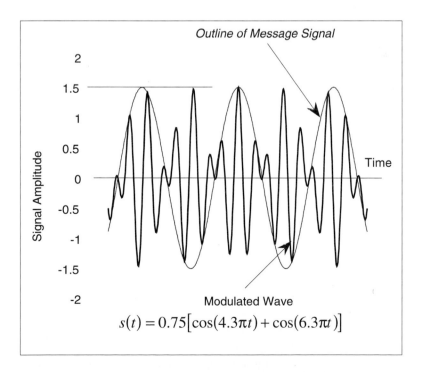

Figure 4.4 Double-Sideband Suppressed Carrier Modulated Wave

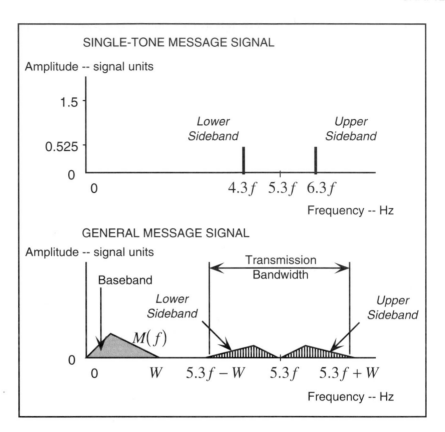

Figure 4.5 One-Sided Frequency Plane Representations of Double-Sideband Suppressed Carrier Amplitude Modulated Wave

the frequency distribution for the general case of a message signal $m(t)$ with spectral density $M(f)$ The spectral density is *double-sideband*—i.e., energy associated with the message signal is distributed on both sides of the carrier frequency, but there is no separate carrier component. If the message signal is band-limited to the frequency interval $0 \leq f \leq W$, the modulated wave is limited to the band $f_c \pm W$, and the transmission bandwidth is $2W$ Hz, the same as LCAM.

Without the powerful carrier signal component of LCAM, detection at the receiver is much more difficult. Usually, DSBSC (and other modulation schemes that eliminate the carrier component) employs *pilot* tones, or other special signals, from which they can reconstruct the carrier. All of the power in the wave is associated with the message-bearing sidebands; however, it still contributes two copies of the message.

(3) Quadrature-Carrier Amplitude Modulation

QAM is a bandwidth conservation procedure. Used in radio systems to carry digital signals [see Section 4.2.2(1)], it is employed to combine the in-phase and quadrature components of the luminance signal in color television [see Section 3.5.1(2)].

QAM is a special case of DSBSC modulation. Because sine and cosine functions are orthogonal, a DSBSC wave based on a cosine carrier can occupy the same transmission bandwidth with an independent DSBSC wave based on a sine carrier of the same frequency. The technique employs two independent message signals (m_1, m_2) that modulate two carrier signals that differ in phase by 90° but have identical frequencies. Thus:

$$s(t) = A_c \left[m_1(t) \cos(2\pi f_c t) + m_2(t) \sin(2\pi f_c t) \right]$$

The transmission bandwidth is $2W$ Hz, the same as LCAM and DSBSC—but two message signals share it.

(4) Single-Sideband AM

SSB is an efficient, bandwidth-minimizing, modulation technique that is used extensively in voice multiplexing systems (see Figure 6.3). All transmitted power is devoted to sending one copy of the message. Because the modulating technique produces sharp peak values when the message signal makes abrupt transitions, SSB cannot be used with digital signals. In addition, signals (such as television) that contain significant energy at frequencies close to 0 Hz [i.e., $m(t)$ contains important low frequencies] cannot use SSB.

When the higher-frequency sideband is employed, the modulated wave is described as upper-sideband (USB) SSB; when the lower-frequency sideband is employed, the modulated wave is described as lower-sideband (LSB) SSB. In the case of the sinusoidal signals defined in Figure 4.1, the waves are

$$s(t)_{LSB} = 0.75\cos(4.3\pi t) \; ; s(t)_{USB} = 0.75\cos(6.3\pi t)$$

One-sided frequency plane diagrams are shown in **Figure 4.6**. The upper half of Figure 4.6 shows the frequency distribution for a single-tone message signal. The lower half shows the frequency distribution for the general case of a message signal $m(t)$ with spectral density $M(f)$. If the message signal is band-limited to the frequency interval $0 \leq f \leq W$, the modulated wave occupies W Hz, and the transmission bandwidth is W Hz—i.e., one-half of that required by LCAM and DSBSC. In addition, all of the energy transmitted is devoted to a single copy of the message.

(5) Vestigial-Sideband AM

VSB is the modulation technique used for domestic television service broadcasting in the United States. Approximating the bandwidth efficiency of SSB, in VSB modulation one sideband is passed almost intact together with a *vestige* of the other sideband. This is achieved by a bandpass filter designed to have odd sym-

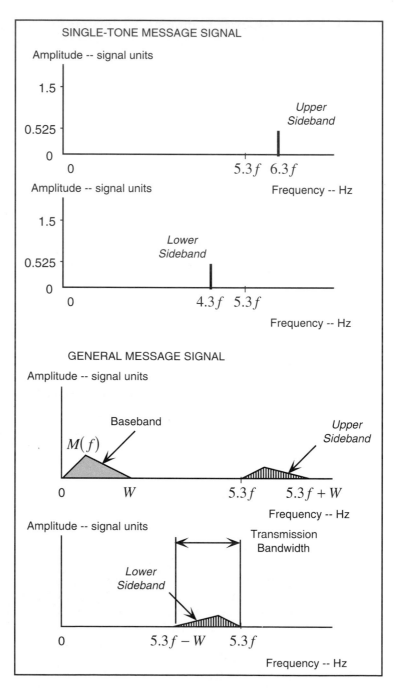

Figure 4.6 One-Sided Frequency Plane Representations of Single-Sideband Amplitude Modulated Wave

metry about the carrier frequency and a relative response of 50% at the carrier frequency. **Figure 4.7** shows how vestigial sideband operation is applied to an LCAM
signal. For equal total power, VSB provides less message power than SSB. However, the power at the carrier frequency simplifies signal detection and has made
the development of consumer-oriented television services possible.

4.1.2 ANGLE MODULATION

Angle-modulated waves exist in two forms—phase modulated and frequency
modulated.

(1) Phase Modulation (ΦM)

ΦM is employed to convey color (chrominance) information in NTSC television signals, and for other specialized functions. A ΦM wave is produced by varying the phase angle in sympathy with the amplitude of the message signal—i.e.,

$$s(t) = A_c \cos\left[2\pi f_c t + k_p m(t)\right]$$

k_p is the *phase sensitivity* of the modulator in radians per volt. The power in the wave
is $A_c^2/2$ signal watts, so that a ΦM wave is described as a constant power wave. As
modulation occurs, power is shared among the carrier and the side frequencies. The
total power remains constant at the value of the power in the unmodulated carrier
signal. The time-domain representation of a phase-modulated wave formed from

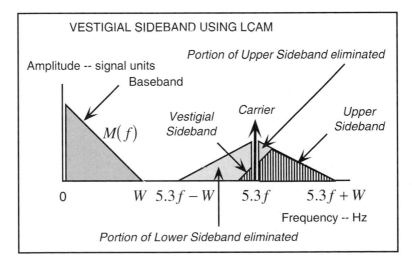

Figure 4.7 One-Sided Frequency Plane Representations of Vestigial-
Sideband Amplitude Modulated Wave

the single-tone message signal and the sinusoidal carrier signal of Figure 4.1 is shown in the lower diagram of **Figure 4.8**. Phase modulation is closely related to frequency modulation.

(2) Frequency Modulation

FM is used to implement point-to-point microwave-radio links and cellular mobile radiotelephone service. In addition, all radio stations broadcasting information and entertainment in the frequency range 88 to 108 MHz (FM band) employ it. Further, FM is employed in the sound segment of NTSC television signals.

An FM wave is produced by varying the frequency in sympathy with the amplitude of the message signal—i.e.,

$$s(t) = A_c \cos\left[2\pi f_c t + 2\pi k_f \int_0^t m(t)\,dt\right]$$

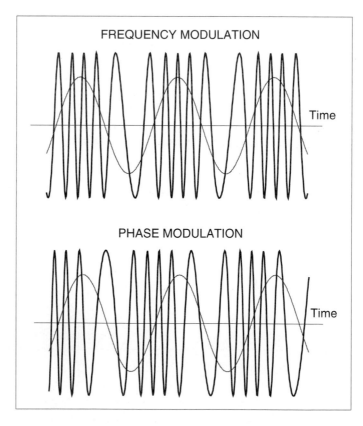

Figure 4.8 Time-Domain Representations of Frequency Modulated and Phase Modulated Waves

k_f is the *frequency sensitivity* of the frequency modulator in Hertz per volt. For a single-tone message signal of frequency f_n the expression becomes

$$s(t) = A_c \sum_{n=-\infty}^{\infty} J_n(\beta) \cos\left[2\pi(f_c + nf_m)t\right]$$

β is known as the *modulation index*. It is the frequency deviation in Hertz divided by f_n. $J_n(\beta)$ is the nth order Bessel function of the first kind with argument β. As the equations show, FM is a nonlinear process; even the case of single-tone modulation is complicated.

The time-domain representation of a frequency-modulated wave formed from the single-tone message signal and the sinusoidal carrier signal of Figure 4.1 is shown in the upper diagram of Figure 4.8. It differs from the phase modulation diagram (under it) in the location of the maximum frequency. To aid in visualization, the outline of the message signal is drawn on each wave. In phase modulation, the maximum frequency of the wave occurs at the positive going zero crossings of the message signal; in frequency modulation, the maximum frequency occurs at the positive peaks of the message signal.

Figure 4.9 shows spectral diagrams of single-tone-modulated FM waves. As the range of frequencies through which the modulated wave is changed increases, the bandwidth and the number of individual energy components increase. In theory, these side components extend to infinity. As a practical matter, they can be limited to the interval that just contains components whose amplitudes are $\geq 1\%$ of the amplitude of the unmodulated carrier. This is known as the *1% bandwidth.*

A convenient expression for calculating the approximate bandwidth is known as *Carson's Rule.* It states that, for small values of modulation index ($\beta < 1$), the spectrum is limited to the carrier and one pair of side frequencies, so that the bandwidth is $2f_m$ Hz. For large values of modulation index, the bandwidth is approximately $2\Delta f$ Hz, where Δf is the frequency deviation of the modulated wave.

An FM wave is a constant power wave. As modulation occurs, power is shared among the carrier and the side frequencies. The total power remains constant at the value of power in the unmodulated carrier signal.

(3) FM Radio

With a maximum permitted frequency deviation of 75 kHz, and a maximum audio signal frequency of 15 kHz, commercial, FM-band radio stations operate with a modulated wave whose bandwidth lies between 180 kHz and 240 kHz. As shown at the bottom of Figure 4.9, for operation in a spectrum-limited world, all FM systems contain a bandwidth-limiting filter to keep the radiation within authorized operating limits (200 kHz for FM band radio stations). In normal operation, the transmitted signal may be a bandwidth-limited version of the signal generated by the modulator.

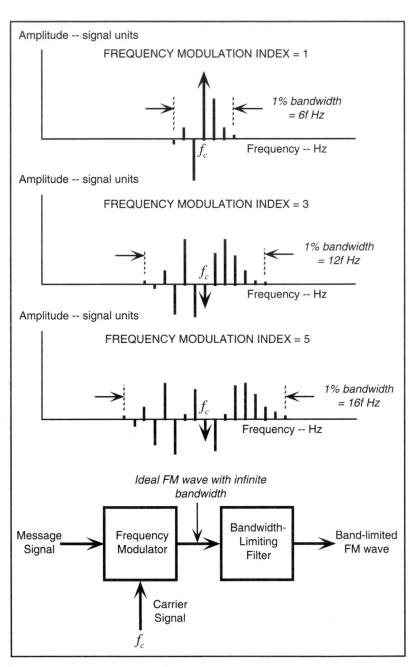

Figure 4.9 Frequency Plane Representations of Frequency Modulated Waves With Increasing Modulation Index

In order to limit the bandwidth excursions of an FM wave, it is usual to place a bandwidth-limiting filter at the output of the transmitter.

An important property of FM waves is their significant immunity to interference from noise. To a first approximation, this capability is derived from the fact that noise signals add to the wanted signal to change the amplitude of the wave; they do not change the frequency (the parameter that carries the information). In practice, the signal-to-noise ratio of FM can be 1,000 times (30 db) or more, better than LCAM. This is achieved by adding emphasis to the higher frequencies present in the message signal before modulation (known as preemphasis), and restoring the level during demodulation (known as deemphasis).

REVIEW QUESTIONS FOR SECTION 4.1

1 Given a sinusoidal carrier signal $s(t) = A\cos(2\pi ft + \phi)$, which parameters are varied to produce amplitude, phase, and frequency modulation?

2 List the basic forms of amplitude modulation.

3 LCAM is the modulation technique employed by all radio stations broadcasting information and entertainment in the frequency range 535 to 1,605 kHz (the AM band). Why is it used?

4 For the general case of a message signal $m(t)$ with spectral density $M(f)$, draw a one-sided frequency plane diagram that shows the distribution of energy for an LCAM wave.

5 If the message signal is limited to the frequency interval $0 \le f \le W$, what is the bandwidth of an LCAM wave?

6 Why is LCAM considered to be an inefficient modulation process?

7 For the general case of a message signal $m(t)$ with spectral density $M(f)$, draw a one-sided frequency plane diagram that shows the distribution of energy for a DSBSC wave.

8 If the message signal is limited to the frequency interval $0 \le f \le W$, what is the bandwidth of a DSBSC wave?

9 Describe QAM.

10 Why is QAM used?

11 For the general case of a message signal $m(t)$ with spectral density $M(f)$, draw a one-sided frequency plane diagram that shows the distribution of energy for an SSB wave. Draw the USB and LSB cases separately.

12 If the message signal is limited to the frequency interval $0 \le f \le W$, what is the bandwidth of an SSB wave?

13 Can SSB be used with digital message signals? Explain.

14 Describe VSB.

15 Name the two forms of angle modulation.

16   If an angle modulated wave exhibits constant power irrespective of whether it is modulated, how is information carried in the wave?

17   Describe the frequency-domain representation of a frequency-modulated wave formed from a single-tone message signal and a sinusoidal carrier signal.

18   What is meant by the 1% bandwidth of a frequency-modulated wave? Why is it necessary to define bandwidth in this way?

19   State Carson's Rule.

20   Comment on the following statement: An important property of FM waves is the significant immunity to interference from noise. Give a simple explanation for this performance.

4.2 DIGITAL MODULATION

Digital modulation brings together a high-frequency sinusoidal carrier signal and a digital datastream to create a modulated wave that assumes a limited number of states. The amplitude, frequency, and/or phase of the carrier are caused to change relatively instantaneously, and message signal energy is distributed at frequencies around the carrier signal frequency. Noise causes errors in the received message, and the fixed transmission bandwidth assigned to a particular channel sets an upper limit to the symbol rate.

- **Symbol Rate**: if symbols are generated at a rate of r per second to create a baseband signal with a bandwidth of W Hz, Nyquist showed that $r \leq 2W$. For a double-sideband modulated wave whose transmission bandwidth is B_T Hz, $B_T = 2W$, so that, $r \leq B_T$.

4.2.1 BINARY KEYING

The bits in the message stream switch the modulation parameters (amplitude, phase, and frequency) from one state to another. This process is known as *binary keying*.

- **Binary keying**: modulating action that creates two values of amplitude, phase, or frequency of a carrier signal in sympathy with the value of the bits in a binary signal stream. These actions are known as ASK (amplitude-shift keying), PSK (phase-shift keying), and FSK (frequency-shift keying).

(1) Binary Amplitude-Shift Keying

As shown in **Figure 4.10**, in binary amplitude-shift keying (BASK), the transmitted signal is a sinusoid whose amplitude is changed by *on-off* keying (OOK) so

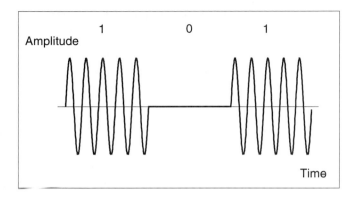

Figure 4.10 Time-Domain Representation of Binary Amplitude-Shift Keying

that the presence of the carrier signal corresponds to a 1, and no carrier corresponds to a 0. When signal one is present, the modulated pulse is

$$p_1(t) = \begin{cases} A\cos 2\pi f_c t, & \text{when } 0 < t \le T_b \\ 0, & \text{otherwise} \end{cases}$$

where T_b is the bit duration (in seconds). When signal zero is present

$$p_0(t) = 0$$

These are the pulses shown in Figure 4.10.

The energy in $p_1(t)$ is equal to $A^2 T_b/2$ signal joules, so that, provided 1s and 0s are equally likely, the average signal energy per bit $\overline{E_b}$ is one-half, or $A^2 T_b/4$ signal joules. The power spectral density—i.e., the distribution of power density (in signal watts/Hz) about the carrier frequency—is shown in **Figure 4.11**. The impulse function at the carrier frequency represents power due to the unipolar binary signal. With a symbol rate of r (symbols/s), the transmission bandwidth B_T is equal to r Hz, and most of the signal power is contained within the range $f_c - \frac{r}{2}$ to $f_c + \frac{r}{2}$ Hz. The main lobe of the signal spans the frequency range $f_c - r$ to $f_c + r$ Hz, and the power in the sidelobes drops off relatively slowly.

Because the spectrum extends well beyond the transmission bandwidth, *spillover* is an important concern with digital modulation. To reduce interference (interchannel interference), a bandpass filter may be applied to limit the energy of the pulses that form the datastream. By restricting the baseband signal to a frequency range that just encompasses the primary lobe of the spectral density diagram, around 90% of the datastream energy is available to the modulation process. Another bandpass filter is placed at the output of the transmitter.

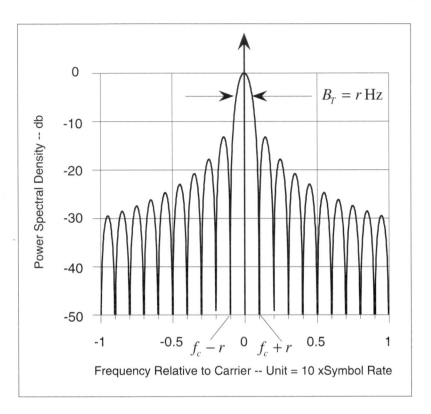

Figure 4.11 Power Spectral Density of ASK Signal-Carrying Random Datastream

(2) Binary Phase-Shift Keying

As shown in **Figure 4.12**, in binary phase-shift keying (BPSK), the transmitted signal is a sinusoid of constant amplitude whose presence in one phase condition corresponds to 1, and, in a state 180° out of phase with it, corresponds to 0. When signal one is present, the pulse can be described as

$$p_1(t) = \begin{cases} A\cos 2\pi f_c t , & \text{when } 0 < t \leq T_b \\ 0 , & \text{otherwise} \end{cases}$$

and when signal zero is present, the pulse can be described as

$$p_0(t) = \begin{cases} -A\cos 2\pi f_c t , & \text{when } 0 < t \leq T_b \\ 0 , & \text{otherwise} \end{cases}$$

These are the pulses shown in Figure 4.12. For obvious reasons, BPSK is also known as *phase-reversal* keying (PRK).

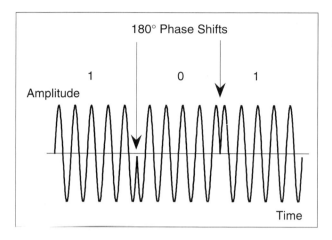

180° Phase Shifts

1   0   1

Amplitude

Time

Figure 4.12 Time-Domain Representation of Binary Phase-Shift Keying

The energy in $p_1(t)$ is equal to the energy in $p_0(t)$—i.e., $A^2T_b/2$ signal joules, and the signal power in the modulated signal is $A^2/2$ signal watts. The power spectral density has the same form as that shown in Figure 4.11 without the impulse at the carrier frequency. Thus, the percentage of the power devoted to signaling is greater in PSK than in ASK, and the error probability will be less for a given power level.

(3) Differential BPSK

Recovery of the datastream from a PSK-modulated wave requires *synchronous* demodulation—i.e., the receiver must reconstruct the carrier exactly so as to be able to detect phase changes in the received signal. *Differential* binary phase-shift keying (DBPSK) eliminates the need for the synchronous carrier in the demodulation process and simplifies the receiver. At the transmitter, the datastream is processed to produce a modulated wave in which the phase changes by $+\pi$ radians whenever a 1 appears in the stream; it remains constant whenever a 0 appears in the stream (i.e., $1 \rightarrow 1$ and $0 \rightarrow 1$, increase by $+\pi$; $1 \rightarrow 0$ and $0 \rightarrow 0$, no change). Thus, the receiver need only detect phase changes, not search for specific phase values. **Figure 4.13** shows a segment of a datastream (original), the changes in phase angle using the rules given above (relative phase angle), what the angles mean in terms of 1s and 0s in the transmitted datastream (processed datastream), and how the processed stream is interpreted at the receiver (demodulated datastream). To draw these sequences, I assumed that the bit before the bit beginning the original stream was 0.

(4) Binary Frequency-Shift Keying (BFSK)

As shown in **Figure 4.14**, in binary frequency-shift keying, the modulated wave is a sinusoid of constant amplitude whose presence at one frequency corresponds to 1, and whose presence at another frequency corresponds to 0. When signal one is present, the pulse can be described as

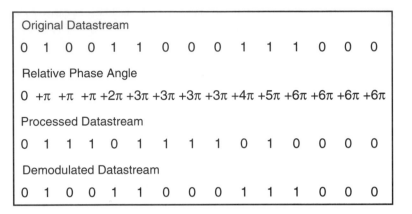

Figure 4.13 Datastream Processing for Differential BPSK

At the transmitter, the datastream is processed to produce a modulated wave in which the phase changes by +π radians whenever a 1 appears in the stream and remains constant whenever a 0 appears in the stream (i.e., 1 → 1 and 0 → 1, increase by +π; 1 → 0 and 0 → 0, no change). Thus, the receiver need only detect phase changes, not search for specific phase values.

$$p_1(t) = \begin{cases} A\cos 2\pi f_m t, & \text{when } 0 < t \le T_b \\ 0, & \text{otherwise} \end{cases}$$

When signal zero is present, the pulse can be described as

$$p_0(t) = \begin{cases} A\cos 2\pi f_n t, & \text{when } 0 < t \le T_b \\ 0, & \text{otherwise} \end{cases}$$

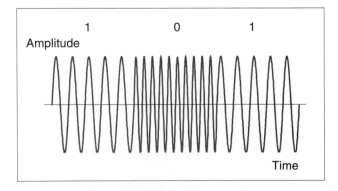

Figure 4.14 Time-Domain Representation of Frequency-Shift Keying

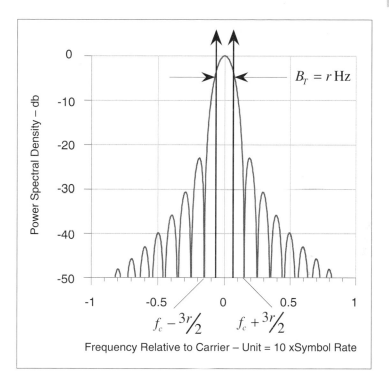

Figure 4.15 Power Spectral Density of FSK Signal Carrying Random Datastream

f_m and f_n are frequencies equally spaced on each side of the carrier frequency f_c (i.e., $f_m = f_c - f_d$ and $f_m = f_c + f_d$). These are the pulses shown in Figure 4.14. A BFSK wave is made up of two BASK waves.

The energy in $p_1(t)$ is equal to the energy in $p_0(t)$—i.e., $A^2 T_b/2$ signal joules— and the signal power in the modulated signal is $A^2/2$ signal watts. For binary FSK created by a random datastream and $f_d = r/2$, the distribution of power density about the carrier frequency is shown in **Figure 4.15**. The impulse functions are associated with the unipolar binary signals in the BASK component waves. They occur at the frequencies employed to represent the 1 and 0 states ($f_c \pm r/2$. The main lobe of the signal spans the frequency range $f_c - 3r/2$ to $f_c + 3r/2$ Hz, a range that is 50% greater than with BASK; however, the power in the sidelobes drops off more quickly. This particular version of BFSK is known as Sunde's FSK,[1] after its inventor E. D. Sunde, a 20th-century American engineer.

---

[1] E. D. Sunde, "Ideal pulses transmitted by AM and FM," *Bell System Technical Journal*, 38 (1959), 1357–1426.

How the symbol frequencies are generated has an important influence on the modulated signal. If the symbol frequencies are derived from two independent sources, discontinuities will occur when one symbol changes to another. Under this circumstance, the modulation is described as *discontinuous* FSK. If the symbol frequencies are derived from a single source by frequency modulation, discontinuities can be avoided and the modulation is described as *continuous phase* FSK (CPFSK). The amplitude of the modulated wave remains constant, and interchannel interference effects are avoided. In order to achieve continuity of phase across the boundary between symbols, the phase of the frequency representing the next symbol must be adjusted to compensate for a phase shift introduced by the frequency-keying process. Thus, CPFSK embodies both frequency shifts and phase shifts. In binary CPFSK, the phase shift between symbols is 90°.

(5) Minimum-Shift Keying

Also known as *fast* FSK and binary CPFSK, MSK uses half-sinusoidal data pulses and $f_d = r/4$—i.e., one-half the spacing used in Sunde's FSK. **Figure 4.16**

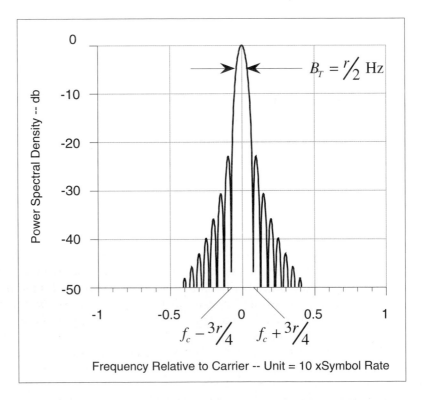

Figure 4.16 Power Spectral Density of MSK Signal Carrying Random Datastream

shows the distribution of power density about the carrier frequency. The impulses present in BFSK are eliminated, and the sinusoidal datastream waveforms produce smaller sidelobes. The main lobe spans the frequency range $f_c - 3r/4$ to $f_c + 3r/4$ Hz, a range that is 25% less than BASK and 50% less than BFSK. The power in the sidelobes drops off very quickly; more quickly than BFSK and much more quickly than BASK.

(6) Duobinary Encoding

The foregoing discussion of BFSK, CPFSK, and MSK confirms that the power spectral density (PSD) of a binary-modulated wave depends on the bit rate of the datastream. The greater the rate, the wider the PSD of the modulated wave. *Duobinary encoding* is a technique that, applied to the baseband signal, effects a reduction in the maximum frequency it contains and doubles the transmission capacity of the system. A specific embodiment of *correlative coding* (also known as *partial response signaling*), it adds intersymbol interference to the transmitted signal in a known way so that its effect can be interpreted in the receiver. In practical terms, duobinary signaling achieves a signaling rate of $2B_T$ symbols/s in a bandwidth of B_T Hz.

4.2.2 Quadrature Binary Modulation

A doubling of the symbol rate can be achieved by using the quadrature property of orthogonal functions introduced in Section 4.1.1(3). By dividing the data into two streams (even-numbered bits and odd-numbered bits) of symbols that are 2 bit periods long, and using them to produce binary shift keying of sine and cosine carriers (derived from the same source), we produce a composite wave containing $2r$ bits/s. It occupies r Hz, the same bandwidth as a single wave carrying r bits/s. Quadrature techniques are used with amplitude-shift keying and phase-shift keying. Generally, because of the larger bandwidth they require, they are not used with frequency-shift keying.

(1) Quadrature-Carrier AM

In QAM, bits are transmitted at rate $2r/s$ in the bandwidth required by binary ASK. In **Figure 4.17**, I show signal space diagrams for binary ASK and QAM. In polar form, they plot the possible signal states; amplitudes are measured from the zero point in any direction, and phase angles are measured about zero from the right-pointing axis. For binary ASK, the two signal points (1 and 0) are arranged on the horizontal axis. For QAM, signal points (1 and 0) occur on the horizontal and vertical axes. The set on the horizontal axis represents the signal states assumed by $\cos 2\pi f_c t$ when keyed by one of the datastreams, and the set on the vertical axis represents the signal states assumed by $\sin 2\pi f_c t$ when keyed by the other datastream. The four points 00, 01, 10, and 11 are the states assumed by the composite signal.

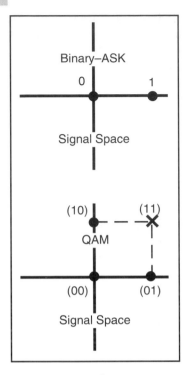

Figure 4.17 Signal-Space
Diagrams for Binary ASK
and QAM

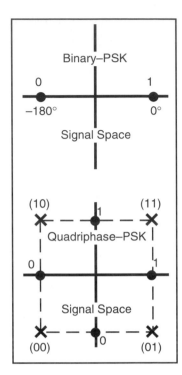

Figure 4.18 Signal-Space
Diagrams for Binary PSK and
QPSK

(2) Quadrature PSK

In QPSK (known as quadrature phase-shift keying, quaternary PSK, or quadriphase PSK) bits are transmitted at rate $2r$ per second in the bandwidth required by binary PSK. In **Figure 4.18**, I show signal-space diagrams for binary PSK and QPSK. For binary PSK, signal points (1 and 0) are arranged on the horizontal axis equidistant from the origin. The point representing 1 is used as the reference phase (0°), and the point representing 0 is 180° out of phase. For QPSK, signal points (1 and 0) are arranged on the horizontal and vertical axes. The set on the horizontal axis represents the signal states assumed by $\cos 2\pi f_c t$ when phase-modulated by one of the datastreams, and the set on the vertical axis represents the signal states assumed by $\sin 2\pi f_c t$ when phase-modulated by the other datastream. The four points 00, 01, 10, and 11 are the states assumed by the composite signal.

Phase-shift keying produces abrupt changes in phase that translate into substantial changes in the amplitude of the modulated wave on the output side of the bandwidth-limiting transmit filter. In a communication system, when passed through equipment with nonlinear characteristics (the power amplifier of a transponder in a communication satellite, for instance), these amplitude variations are suppressed—giving rise to additional spectral components outside the authorized bandwidth, and producing interchannel interference. Improved operation is achieved by delaying one of the datastreams by one *bit* period (one-half a symbol period) so that the two streams do not change states simultaneously. Known as *offset* QPSK (OQPSK), phase changes are limited to 90°.

4.2.3 M-ary Modulation

Binary keying produces two distinct signals (symbols) that represent the 1s and 0s in the message. In this case, the symbol rate is equal to the bit rate of the datastream. For a given transmission bandwidth (and thus a given symbol rate), by creating symbols that represent more than 1 bit, the data throughput is improved. For instance, if each symbol stands for 2 bits, the rate is doubled. To do this, different symbols are used for each of the four combinations 00, 01, 10, and 11, and the datastream is encoded 2 bits at a time. Similarly, if each symbol stands for 3 bits, the rate is tripled. To do this, different symbols are used for each of the eight combinations 000, 001, 010 011, 100, 101, 110, and 111, and the datastream is encoded 3 bits at a time. As the number of symbols increases, the penalty is more noise-related errors.

(1) M-ary Signaling

Creating four unique symbols is relatively easy; they may be formed from four amplitude values (quadrature ASK), four phase values (quadrature PSK), or four frequency values (quadrature FSK). To produce more complex symbols that represent larger groups of bits, the amplitude, phase, or frequency of the carrier can be used alone, or in combinations, with either a single carrier, or a coherent quadrature-carrier arrangement. Referred to as *M*-ary signaling, each symbol represents $\log_2 M$ bits, so that the bit rate is $r\log_2 M$ [r is the symbol rate (symbols/s)]. *M* has values of 2, 4, 8, 16, etc.

(2) Spectral Efficiency

To compare signaling schemes, it is convenient to define a quantity called spectral efficiency

*   **Spectral efficiency**: the ratio of the bit rate (in bits/s) of the modulating signal to the transmission bandwidth (in Hz) of the modulated signal—i.e., r_b/B_T bits/s/Hz.

For systems operating at the Nyquist rate, $B_T = r$, so that $r_b/B_T = \log_2 M$, and for values of *M* of 2, 4, 8, 16, etc., the ideal spectral efficiencies are 1, 2, 3, 4, etc., bits/s/Hz. For practical systems, the values are somewhat less.

(3) Comparisons

The selection of a digital modulation scheme is a complex tradeoff among bandwidth, symbol rate, bit rate, and permissible radiated power. Because noise affects the amplitude of the received signal, ASK techniques are inefficient in noisy environments. Because they are constant amplitude signals in which the information resides in parameters not greatly affected by amplitude perturbations, PSK and FSK perform better in noisy environments. As M increases, the transmitted power must increase to maintain the signal-to-noise ratio above an operating limit.

In practice, QPSK is widely used because it offers attractive power/bandwidth performance for acceptable levels of error. For several signaling schemes, **Figure 4.19** compares spectral efficiency and signal-to-noise power ratio for a constant probability of error of $P_\varepsilon = 10^{-5}$. The graph includes an upper limit to the performance of a communication system. Derived by Claude Shannon, it is independent of error probability.

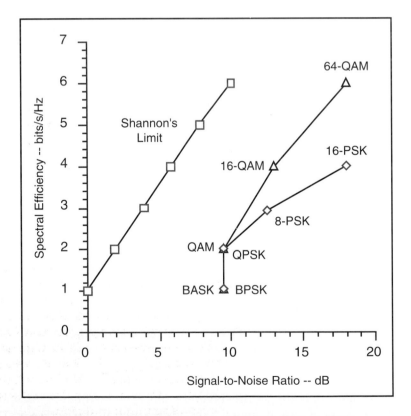

Figure 4.19 Spectral Efficiency and Signal-to-Noise Power Ratio Compared for Constant Probability of Error

REVIEW QUESTIONS FOR SECTION 4.2

1   Explain the difference between analog and digital modulation.

2   If symbols are generated at a rate of r per second to create a baseband signal with a bandwidth of W Hz, express r in terms of W.

3   For a double-sideband modulated wave whose transmission bandwidth is B_T Hz, express r in terms of B_T.

4   What is binary keying?

5   Describe BASK.

6   Explain why spillover is an important concern with digital modulation. What is done to reduce its effect?

7   Describe BPSK. Why is it known as phase-reversal keying?

8   Why is DPSK employed?

9   Describe the manner in which DPSK detects 1s and 0s.

10   Describe BFSK. Explain how it is related to BASK.

11   Distinguish between discontinuous and continuous FSK.

12   Describe MSK. How is the narrow bandwidth achieved?

13   What is the effect of duobinary encoding?

14   Explain the following statement: A doubling of the symbol rate can be achieved by using the quadrature property of orthogonal functions. How can this be done?

15   Sketch the signal-state diagram for QAM. What does it mean?

16   Sketch the signal-state diagram for QPSK.

17   Why is offset QPSK employed? How is it implemented?

18   Comment on the following statement: For a given transmission bandwidth (and thus a given symbol rate), by creating symbols that represent more than 1 bit, the data throughput is improved. How is this done?

19   In M-ary signaling, if the symbol rate is r symbols/s, what is the bit rate?

20   Define spectral efficiency.

4.3 SPREAD SPECTRUM MODULATION

First used by the military because it is hard to detect and almost impossible to jam, spread spectrum modulation (SSM) is employed in global positioning systems (GPSs), mobile telephones, personal communication systems (PCSs) and very small aperture terminal (VSAT) satellite communication systems. SSM is a technique in which the transmitted wave occupies a much larger bandwidth than the minimum bandwidth required to transmit the information it carries (i.e., $B_T >> r$). Commonly, this is accomplished through the use of a high-rate, pseudo-noise sequence as a *spreading* function.

Because spread spectrum signals use a wide frequency band, the *average* signal power per unit bandwidth (in watts/Hz) can be very low. Further, when employed in conjunction with families of orthogonal spreading functions [CDMA, see Section 6.5.2(3)], SSM provides a greater level of multiple access than other approaches [FDMA and TDMA, see Section 6.5.2(1&2)].

4.3.1 CHANNEL CAPACITY

For a signal-to-noise power ratio S/N, Shannon showed that a channel of bandwidth B Hz can transport (a maximum of) C error-free bits/s given by

$$C = B\log_2(1 + S/N)$$

Figure 4.20 shows the nature of this relationship. When operating under very noisy conditions (i.e., $S/N << 1$), it shows that it is possible to transport C error-free bits/s by adjusting the channel bandwidth so that $B = 0.693C(N/S)$ Hz. By way of

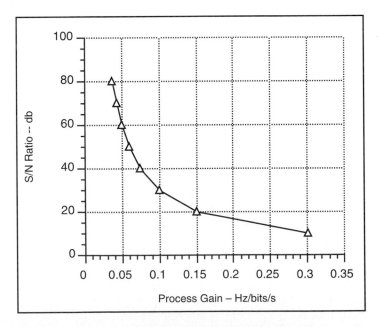

Figure 4.20 Process Gain and Signal-to-Noise Ratio

example, suppose we wish to send 9.6 kbits/s when $S/N = 0.1$. This expression tells us we must use a modulation technique that spreads the signal over a bandwidth of approximately 66.5 kHz. Further, if we wish to send 9.6 kbits/s when $S/N = 0.01$, we must use a modulation technique that spreads the signal over a bandwidth of approximately 665 kHz.

4.3.2 PROCESS GAIN AND JAMMING MARGIN

The relationship $B = 0.693C(N/S)$ tells us we can achieve error-free transmission of signals well below the noise level if we have a wideband channel and a modulation scheme that makes use of the bandwidth. Spread spectrum modulation fulfills these requirements. It is convenient to define two quantities:

- **Process gain**: denoted by G_p, the ratio of the output signal-to-noise power ratio to the input signal-to-noise power ratio
- **Jamming margin**: denoted by M_j, the difference between the noise level at the receiver and the received signal level that will assure the desired value of $(S/N)_{out}$.

Thus, if we have a system with $G_p = 30$ db and we require $(S/N)_{out} = 10$ db, the jamming margin is 20 db—it can operate in a noise environment that is 20 db greater than the signal level. The relationships among output signal level, noise level, jamming margin, process gain and received signal level are shown in **Figure 4.21**.

By way of example, suppose we wish to send 7 kbits/s when $S/N = 0.1$, and the output signal-to-noise ratio at the receiver is 10 db. Using the nomenclature of Figure 4.21, the received signal power is –20 db with respect to the noise level so

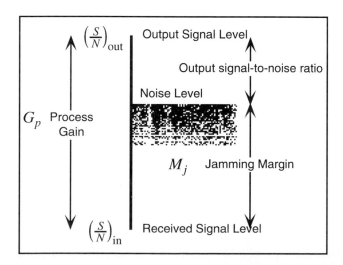

Figure 4.21 Output Signal Level, Noise Level, Jamming Margin, Process Gain, and Received Signal Level

that the jamming margin is 20 db. The output signal-to-noise ratio is 10 db; hence, the process gain is 30 db. Finally, with $C = 7$ kbits/s, and $G_p = 1,000$, the bandwidth of the system is 7 MHz.

4.3.3 PSEUDO-NOISE SEQUENCE

The spreading signal must

- Have a very wide spectrum
- Be randomly distributed so as not to interfere with the existing radio environment
- Be capable of duplication so that the receiver can generate the same spreading signal sequence for detection purposes.

These requirements can be satisfied by a logic-generated sequence known as

- **Pseudo-noise** (*pn*) **sequence**: a periodic binary sequence with a noiselike distribution of 1s and 0s in which the number of 1s is equal to the number of 0s plus or minus 1. The distribution of multiple occurrences of 1s or 0s is close to the distribution present in a random binary stream. Generated by a feedback shift-register arrangement, the parameters of the sequence are
    - *length*. Depends on the number of stages in the shift register. Commonly expressed in *chips* (see below).
    - *composition*. Depends on the initial state of the register, the number of feedback connections, and the logic in the feedback loop(s).
    - *period*. Depends on the length (in chips) and the rate at which the register is driven (in chips per second).
- **Chip**: the output of a code generator during one clock period. The name given to each element of a *pn* sequence.
- **Maximum-length** (*m*) **sequence**: a *pn*-sequence produced by an *m*-stage feedback shift register whose period is $2^m - 1$ chips.

To illustrate the generation of *pn*-sequences, **Figure 4.22** shows a three-stage shift register in which the state of the second stage is combined with the output state to provide feedback to the first stage in accordance with the truth table shown. Initially, suppose the stages are set to contain 1, 0, 0. On the first shift command, the sequence 1, 0, 0 is shifted to the right so that the state of stage 2 becomes 1, stage 3 becomes 0, and 0 is present on the output line. With inputs of 0 and 1, the feedback logic produces a 0 that is entered in stage 1 to make the register read 0, 1, 0. On the second shift command, this sequence is shifted to the right so that stage 2 becomes 0, stage 3 becomes 1, and 0 is present on the output line. With inputs of

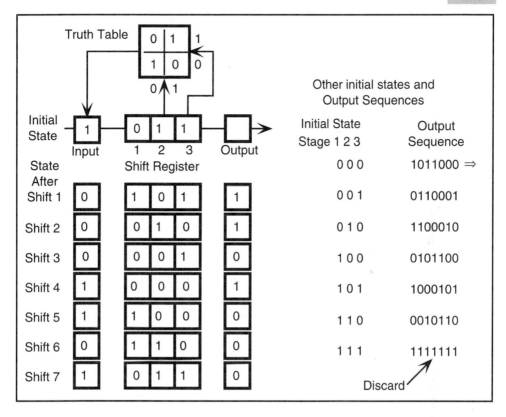

Figure 4.22 Generation of *pn* Sequences

0 and 0, the feedback logic produces a 1 that is entered in stage 1 so that the register reads 1, 1, 0, ..., and so on. On the seventh shift command, the register stages return to 1, 0, 0, and the process repeats.

The simple arrangement of Figure 4.22 produces a repetitive code made up of the seven chips 1 1 0 1 0 0 0. Different initial conditions produce different results. In Figure 4.22, for the eight possible input situations, we list the sequences they will produce. Ignoring the all-1s case, we have seven sequences that contain three 1s and four 0s. It is important to emphasize that they are unique to the feedback arrangement shown. Changing the logic to exclusive-or produces a set of seven 7-chip sequences in which there are four 1s and three 0s, and an all-0s sequence (which is discarded). These examples mimic the properties of the general *m*-sequence produced by a shift register of *n*-stages.

- **Length**: $2^m - 1$ chips per sequence
- **Number**: $2^m - 1$ different sequences
- **Composition**: (number of 1s) = (number of 0s) ± 1.

An immense range of register sizes and feedback connections has been explored to find useful codes of almost any length. In Section 11.2.4(2), I describe the codes used in cellular mobile telephone applications.

4.3.4 SPREADING TECHNIQUES

To create a spread spectrum signal, a *pn*-sequence is employed to convert the (narrowband) information signal to a very wideband modulated signal. Two basic techniques are used.

(1) Direct-Sequence Spread Spectrum

In *direct sequence* (DS) *spreading* the information signal modulates a *pn*-sequence to produce a signal that extends over the wideband spectrum of the digital sequence. In turn, this signal is used to modulate a sinusoidal carrier (PSK, for instance). The speed of the digital sequence is known as the *chip* rate. In DS applications, it may range from 1 Mchips/s to 100 Mchips/s. The DS signal occupies the entire spectrum space, the power density is low everywhere, the appearance is noiselike, and the signal level can be significantly below ambient noise. By way of example, a 1-watt, information-bearing signal of bandwidth 10 kHz has an average power per unit bandwidth of 100 µwatts/Hz. If it is spread over 10 MHz using a 10-Mchips/s *pn*-sequence the spread signal has an average power per unit bandwidth of

$$1 \text{ watt}/10 \text{ MHz} = 0.1 \text{ µwatts/Hz}$$

Thus, the average power in 10 kHz will be 0.1% of the transmitted power (a reduction of 30 db).

At the receiver, the spread spectrum signal is modulated a second time by the *pn* sequence to recapture the original information signal. The use of this spreading code during demodulation spreads interfering signals and reduces their effect on the wanted signal. The number of simultaneous users is limited by the cross-correlations of their spreading sequences. By restricting knowledge of the *pn* sequence to the transmitter and receiver, the communicators can be assured of a significant level of privacy.

(2) Frequency Hopping Spread Spectrum

In *frequency-hopping* (FH) *spreading* the information signal modulates a sinusoidal carrier to produce a narrowband-modulated wave. Subsequently, the center frequency is changed by mixing the modulated wave with individual frequencies selected by a *pn* sequence (the spreading signal). As a result, the transmitter *hops* from one frequency to another over a wide frequency band. Hopping rates may be as high as 100 khops/s. The FH signal occupies a narrow slice of spectrum for a very short time before shifting to another position in the spectrum.

While the average power density across the operating spectrum is low, it is high in the slice that happens to be occupied and must be above the noise level for the signal to be detected.

For the same 1-watt, information-bearing signal of bandwidth 10 kHz, spread in a bandwidth of 100 MHz, the signal can occupy any one of 10,000 slices (without overlap). If the hopping rate is 10,000 times per second, on average, a slice will be selected once per second. Each time it is selected, the signal will dwell on it for 100 µs so that it is present 0.01% of the time. The average power density is

$$1 \text{ watt}/100 \text{ MHz} = 0.01 \text{ µwatts/Hz}$$

However, the power is concentrated at one frequency or another, and is 1 watt in the occupied slice.

To reclaim the information signal, the receiver must know the frequency hopping sequence of the transmitter. When several cooperating transmitters are operating, multiple signals may appear at the same time in the same spectrum slice. This creates interference and errors that set a limit to the number of channels that may be operated simultaneously. As before, by restricting knowledge of the hopping sequence to the transmitter and receiver, the communicators can be assured of a significant level of privacy.

(3) DS and FH Systems

Figure 4.23 shows the principles incorporated in direct sequence and frequency-hopping systems:

- **DS spreading**: the *pn*-sequence is employed at the first modulator in the transmitter to produce a wideband spread signal and at the last demodulator in the receiver to *despread* the wideband signal. At the second modulator in the transmitter, the spread signal modulates the carrier to produce the direct sequence spread spectrum signal. In the wideband channel, this signal is distributed across the channel and picks up noise and interference from other transmitters. At the first demodulator in the receiver, the carrier is removed to yield the spread signal and noise. The final demodulator in the receiver despreads the wanted signal.

- **FH spreading**: the *pn* sequence is used to create a pseudo-random set of frequencies that are mixed with the modulated carrier to achieve frequency hopping. In the wideband channel, the hopping signal is narrowband. It picks up noise and interference from other transmitters that may occasionally be on the same frequency.

In **Figure 4.24** we provide waveform information to illustrate the use of a *pn* sequence in the direct spreading application of Figure 4.23. At the transmitter a data signal modulates a *pn* sequence. In turn, the composite digital signal is used

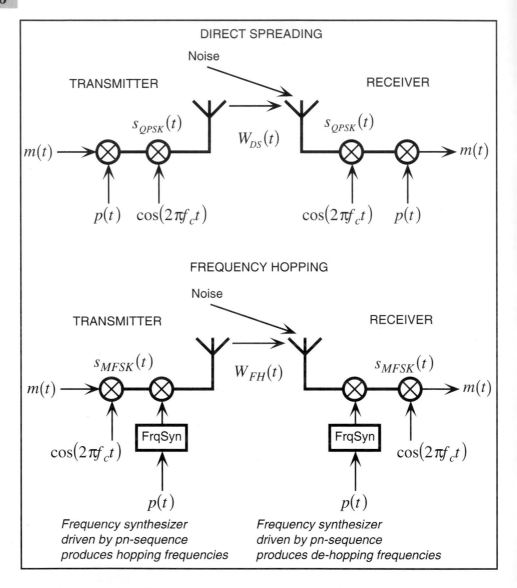

Figure 4.23 Principles of Direct Sequence and Frequency-Hopping Spread Spectrum Systems

to create BPSK of a sinusoidal carrier. At the receiver, the composite digital signal is reclaimed. On multiplication by the original *pn* sequence, it yields the original data signal. Should interfering signals be present, multiplying them by the *pn* sequence spreads their spectrums and reduces their negative effect. A prerequisite for successful operation is that the spreading signal and the carrier signal at the receiver are synchronized with their counterparts in the transmitter.

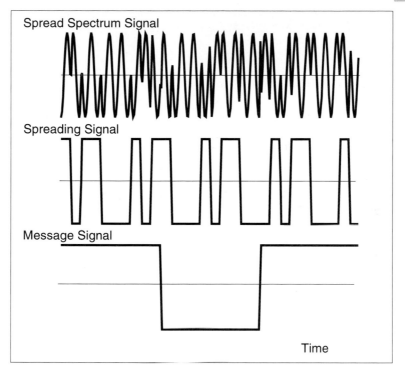

Figure 4.24 Formation of Direct-Sequence Spread Spectrum Signal

REVIEW QUESTIONS FOR SECTION 4.3

1 What does Shannon's channel capacity expression tell us about operating in a very noisy environment?
2 Define process gain.
3 Define jamming margin.
4 List the requirements for a spreading signal.
5 Define a *pn* sequence.
6 Define a chip.
7 Define an *m*-sequence.
8 Describe the generation of a direct-sequence spread spectrum signal.
9 Describe the generation of a frequency-hopping spread spectrum signal.

5

BEARERS

By effecting *action at a distance*, electromagnetic energy carries messages between communicating entities to make real-time telecommunication possible. Described by *Maxwell's* equations (after James Clerk Maxwell, a 19th-century Scottish physicist), signal energy propagates under the influence of oscillating electrical and magnetic fields. At relatively low frequencies, the energy is manifest as electric currents and potentials in wire cables. At higher frequencies, it is manifest as electromagnetic waves in coaxial cables and free space. At very high frequencies, it is manifest as infrared light in optical fibers. The bearers—wire cables, optical fibers, coaxial cables, and free space—are indispensable elements in the provision of telecommunication services.

5.1 WIRE CABLES

The first significant telecommunications bearer was a copper wire running between Baltimore, Maryland, and Washington, DC. In 1843, Samuel Morse and Alfred Vail used it to demonstrate the *electric telegraph*. Since then, telegraph and telephone companies have consumed a sizeable fraction of the world's copper supply to connect 500 million customers to their networks. Despite challenges from optical fibers, cellular radios, and communication satellites, the dominant bearer remains copper wire, twisted in pairs, in plastic insulated cables. In this section, I describe some of the characteristics and limitations of modern wire cables.

5.1.1 Transmission Phenomena

In a cable, each pair of wires behaves as a transmission line that must be terminated by its characteristic impedance if reflections are to be avoided.

(1) Characteristic Impedance of Ideal Transmission Line

An ideal transmission line is one without losses—losses due to the resistance of the conducting wire members and the leakance (conductance) between them. The upper diagram in **Figure 5.1** shows an equivalent circuit representation in which the line consists of an infinite ladder network of identical small sections containing incremental capacitors (ΔC) and inductors (ΔL). The line begins at the left and extends to infinity on the right.

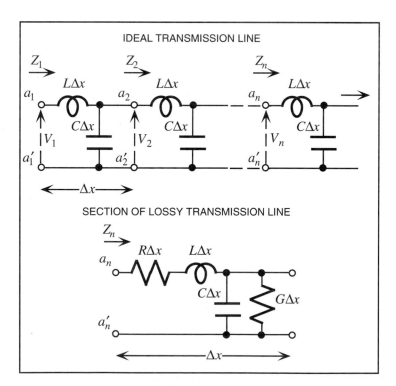

Figure 5.1 Equivalent Circuits for Ideal and Lossy Transmission Lines

If the line is broken at any point and terminated by a resistor equal to the resistance presented by the severed line, a signal at the input cannot distinguish it from an infinite line. Furthermore, because the line is infinite, no matter where the line is cut, the value of this terminating resistance will be the same. Known as the *characteristic impedance* of the line and denoted by Z_0, the value of the terminating resistance that turns a finite line into a circuit that mimics an infinite line is $\sqrt{L/C}$ ohms.

Over an ideal (loss-free) line, the magnitude of the signal is not attenuated as it travels along it, and the phase angle of the signal varies along the line depending on the signal frequency. This phase shift is manifest as a wave that propagates with velocity $1/\sqrt{LC}$ meters/s.

(2) Lossy Transmission Line

The lower diagram of Figure 5.1 shows a section of transmission line that includes the resistance (R) of the conducting members and the conductance (G) between them. R, L, C, and G are the fundamental electrical parameters of the line, and are called *primary* constants.

(3) Reflected Waves

At a point x, if a current i (Amps) flows in the conductors under the influence of a voltage v (Volts) between them, then

$$\partial v/\partial x = -\{Ri + L\,\partial i/\partial t\} \; ; \; \partial i/\partial x = -\{Gv + C\,\partial v/\partial t\}$$

For sinusoidal voltages, the solutions are

$$V = A\exp(-px) + B\exp(px) \; ; \; I = \frac{1}{Z_0}\{A\exp(-px) - B\exp(px)\}$$

p is the *propagation* constant and Z_0 is the characteristic impedance; both are independent of length. (They are called *secondary* constants.)

The negative exponential terms [i.e., $A\exp(-px)$] represent a wave traveling from source to load, and the positive exponential terms [i.e., $B\exp(px)$] represent a wave traveling from load to source. The constants A and B are determined by the values of the terminations at each end of the line.

If the source impedance is Z_S and the load impedance is Z_L, on a line of length l, at the load, the ratio of the reflected (i.e., directed to the source) voltage wave to the forward (i.e., directed to the load) voltage wave is

$$\vartheta_R = (Z_L - Z_0)/(Z_L + Z_0)$$

In addition, the ratio of the reflected current wave to the forward current wave is

$$\iota_R = (Z_0 - Z_L)/(Z_0 + Z_L)$$

These expressions show that, to prevent reflections at the load end, the load impedance must equal the characteristic impedance of the line. A similar result can be developed for the source end. For reflection-free operation then, both the source impedance and the load impedance must be equal to the characteristic impedance of the line.

(4) Skin Effect

At low frequencies, the current flowing in a conductor is distributed uniformly over its cross-sectional area. As the operating frequency is raised, the distribution is disturbed, the current density falls at the center and begins to increase in the outer layer of the conductor. At very high frequencies, the current flows along the *skin*. While the exact physical mechanism is difficult to describe, most analysts conclude that the ratio of the resistance at a high frequency to the resistance at dc is proportional to the square root of the frequency. Thus, signals that contain a range of frequencies will suffer frequency-dependent distortion; the higher-frequency components will be attenuated to a greater degree than the lower-frequency components.

(5) Twisted Pairs

Telephone wires are pervasive.[1] Often called *twisted pairs* because of the way in which they are constructed, they carry voice and data signals from customers' premises to local switching and routing centers. Each loop is fashioned by twisting together a pair of plastic insulated, copper wires. To limit interference [crosstalk, see Section 5.1.2(1)] between them, neighboring *twisted* pairs are constructed with different pitches (twist lengths). The pairs are stranded in units of up to 100 pairs, and cabled into plastic-sheathed cores. Common cable sizes range from 6 pairs to 2,700 pairs. Cables are laid in ducts under roadways and sidewalks, plowed into the earth, or suspended from telephone poles.

The primary constants, R, L, C, and G, depend on the gauge of the wire, the ambient temperature, the materials employed to construct the cable, and the operating frequency. Consequently, the characteristic impedance varies significantly, and it is impossible to terminate the lines exactly. Compromises have been reached that use 900 ohms on predominantly fine gauge (i.e., 26 and 24 AWG) circuits, and 600 ohms on predominantly coarse gauge (i.e., 22 and 19 AWG) circuits. On private lines that transport high-speed data signals, terminations of 135 to 150 ohms are used.

---

[1] For a comprehensive discussion of the characteristics of twisted pair cables see Whitham D. Reeve, *Subscriber Loop Signaling and Transmission Handbook—Digital*, (New York: IEEE Press, 1995), Chap. 7.

5.1.2 IMPAIRMENTS

Twisted pair cables are subject to several impairments that distort the signals they carry. Four major examples are described below.

(1) Crosstalk

Electromagnetic coupling between twisted pair circuits causes a signal in one to produce a disturbance in the other. Called *crosstalk*, it contributes to the noise background in the receiving circuit. **Figure 5.2** shows two parallel pairs that are terminated by their characteristic impedances. If distances are small compared to the wavelengths of the operating frequencies, their interaction can be characterized by the mutual inductance and capacitances shown. A voltage from the source will produce a current around a,a'. In turn, through the action of the inductive coupling represented by the mutual inductance M, this current will induce a voltage that drives a current around the circuit b,b'. In addition, the source voltage drives a current around the shunt circuit a,b,b',a' through the two halves of the capacitance C_U. Between b and b' the current divides into equal parts that traverse the circuit in different directions. From the model, two forms of interference, or crosstalk, are apparent

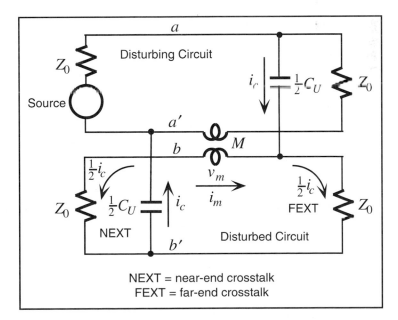

Figure 5.2 Equivalent Circuit for Calculating Crosstalk between Parallel Circuits

- **Near-End crosstalk** (NEXT): crosstalk whose energy flows in the opposite direction to that of the signal energy in the disturbing channel—i.e.

$$i_{NEXT} = i_c - i_m$$

Because the currents are out of phase, the effects of mutual inductance and capacitance reinforce each other.

- **Far-End crosstalk** (FEXT): crosstalk whose energy flows in the same direction to that of the signal energy in the disturbing channel—i.e.

$$i_{FEXT} = i_c + i_m$$

The effects of mutual inductance and capacitance oppose each other.

Thus, when parallel transmission paths are involved—such as pairs of conductors in the same cable—NEXT is usually a more serious problem than FEXT. In addition, for high-impedance circuits, capacitive coupling is the significant contributor to crosstalk; for low-impedance circuits, inductive coupling is the significant contributor to crosstalk.

(2) Echoes

In telephone plant, many existing arrangements impair transmission integrity. Often, they represent compromises made many years ago that have become embedded in operations and are too expensive to remove. The impedance mismatches they produce give rise to echoes. Three common impairments are

- **Multiple wire sizes**: because space is limited in well-developed wire centers, cables connected to the distributing frames employ 26 AWG twisted pairs. On longer loops, it is necessary to increase the wire size once or twice to keep the loop resistance within prescribed limits. The change in wire gauge produces an imbalance that can create reflections.
- **Two- to four-wire conversion** (2W/4W): a single twisted pair that carries signals in two directions serves most subscribers. At a point in the loop, or at the serving office, the channels are separated on two twisted pairs, one for each direction of signaling. This conversion is achieved with a *hybrid* transformer, or an electronic device. Almost impossible to balance exactly, these arrangements create reflections.
- **Bridged taps**: in the outside plant, some loops may have been used to connect to several terminating points at different times. In the process of rearrangement, the pair may have been left connected to some of them so that there are unused, open-circuited, connected cable pairs. They represent shunting impedance at the point of connection and cause reflections.

As discussed in Section 5.1.1(3), signal energy is reflected back to the source in proportion to the mismatch between the characteristic impedance and the load impedance. **Figure 5.3** shows the 2W/4W transitions that occur when a telephone circuit is completed across a four-wire network. Echoes are generated at impedance mismatches due to the hybrid transformers (and possibly other impairments). On speaking, the speaker receives echoes from the near end and from the far end. If the circuit is long enough, and the echoes are energetic enough, the far-end echo is discernible and annoying to the sender.

Subjective testing has established that echoes are less likely to annoy speakers when the delay between launching the energy and receiving the reflected wave is short, and more likely to annoy speakers as the delay becomes longer. To alleviate the annoyance, echoes must be attenuated; the longer the delay, the greater the attenuation that is required. **Figure 5.4** shows the general relationship between round-trip delay and the loss required to prevent speaker discomfiture. Echo conditions are at their worst when the energy is reflected from the far-end of a circuit that includes significant propagation delay (such as a trans-continental circuit). Fortunately, electronic devices known as *echo suppressers* are available. **Figure 5.5** shows the principle of such a device. At the sending end, they create an inverted echo signal that cancels the echo signal received from the impedance mismatch point(s).

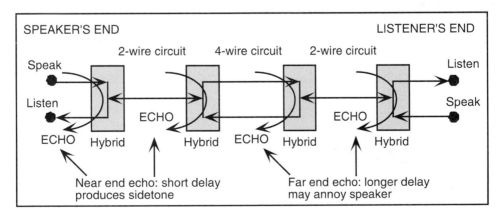

Figure 5.3 Signal Reflections Caused By Impedance Mismatches at Hybrid Transformers in a Telephone Circuit

Echoes are generated at impedance mismatches. On speaking, the speaker receives echoes from the near end and the far end. If the circuit is long enough, and the echoes are energetic enough, the far end echo may make continuous speech difficult.

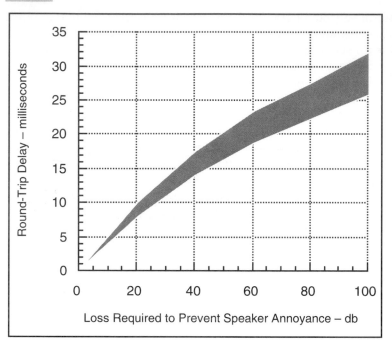

Figure 5.4 Relationship Between Round-Trip Delay and Loss Required to Prevent Speaker Annoyance

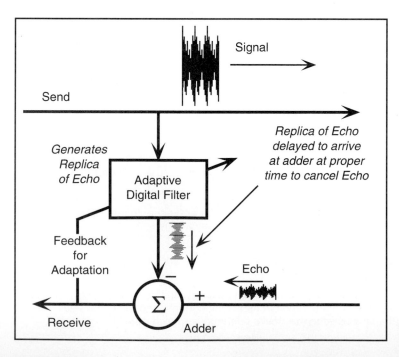

Figure 5.5 Principle of Echo Canceller

(3) Equalization

Because their performance varies with frequency [see Section 5.1.1(3)], wire and cable transmission channels distort high-speed pulse signals. The effect can be minimized by *equalization*, a technique in which the received signals are passed through a filter that approximates the inverse of the distortion of the channel. **Figure 5.6** shows the principle.

The bandwidth and pulse shape of the transmitted signal is set by the characteristics of the transmit filter (Π_{tx}). During transmission over the channel, the frequency distribution of the energy is modified by the characteristics of the channel (represented by the channel filter Π_{ch}). At the receiver, energy passes through the front end of the receiver, which is represented by a third filter (Π_{rx}). If the transmit and receive filters are matched so that, in the absence of channel distortion, the signal is received perfectly, a fourth filter (Π_{eq}), the equalizer, is needed to correct the effects of the channel filter. If the response of the channel filter is

$$\Pi_{ch}(f) = |\Pi_{ch}(f)| \exp\{j\theta(f)\}$$

where $|\Pi_{ch}(f)|$ is the amplitude response and $\theta(f)$ is the phase response, to compensate for the effect of the channel, the response of the equalizer filter must be

$$\Pi_{eq}(f) = \frac{1}{|\Pi_{ch}(f)|} \exp\{-j\theta(f)\}$$

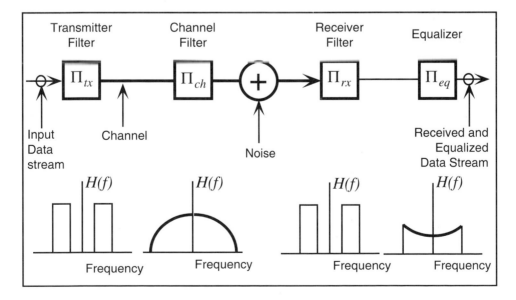

Figure 5.6 Principle of Channel Equalization
The equalizer filter compensates for the distortion introduced by the channel filter.

Figure 5.7 Eye Diagram

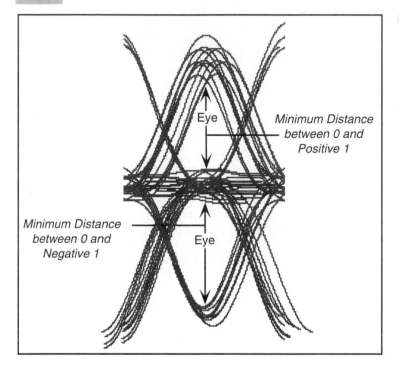

A transversal filter with a finite number of taps can implement the equalizer. In low-speed operation, the filter coefficients are adjusted manually at the time of setup. In high-speed operation, adaptive equalization[2] is employed; the coefficients are periodically checked and adjusted automatically throughout the life of the connection.

Exchanging a known sequence of pulses between the transmitter and receiver trains the filter. During the training sequence, the filter characteristics are adjusted to produce a pulse train that the receiver can decode correctly. The training process may last from a few seconds up to as long as a minute. When training is completed, the filter is able to correct (most of) the channel-induced distortion suffered by pulses sent over the channel.

To assess the performance of the equalization process, it is customary to view the waveforms at the receiver. On an oscilloscope whose time sweep is synchronized with the signaling rate, it is possible to produce a display that superimposes all of the waveform combinations over adjacent signaling intervals. Called an *eye* diagram, **Figure 5.7** shows the sort of display produced by a random raised cosine pulse AMI train. The clear areas in the middle are the *eyes*. They show the signal differences between 1s and 0s for the particular installation being measured. Poor

---

[2] R. W. Lucky, "Automatic equalization for digital communication," *Bell System Technical Journal*, 44 (1965), 547–88.

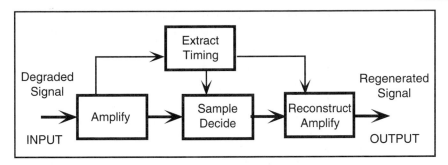

Figure 5.8 Principle of Regenerator (Repeater)

equalization is shown by smaller eyes; good equalization corresponds to larger eyes. Eye diagrams are a convenient way to assess the performance of a digital transmission channel, and to evaluate the efficacy of equalization.

(4) Regeneration

Another strategy that overcomes pulse distortion and achieves higher signaling rates is to reduce the distance over which the pulse stream is transmitted before processing it. This is done by introducing regenerators (also known as repeaters) at regular intervals along the transmission path. Their purpose is to read the pulse stream before it degrades substantially, generate a new stream, and pass it on to the next unit. **Figure 5.8** shows a functional model of a regenerator. It consists of sections that amplify, extract timing, sample and decide, and reconstruct and amplify. With a degraded signal as input, it produces an output signal that is almost the equal of the signal present at the beginning of the transmission chain.

In general applications, the signal train will be irregular and will not contain a pulse in each time slot. To generate a regular repetitive timing signal, the incoming signal is used to excite a resonant circuit that is tuned to the pulse repetition rate. The output of the resonant circuit is a sinusoidal signal synchronized with the fundamental frequency of the signal train. It is used to generate a square wave from which timing signals can be derived; they control the sampling of the received signal.

The timing signal may not match exactly the timing used to generate the pulse train for the first time. The difference between the actual timing signal and an ideal signal is known as *jitter*. In a string, the resonant circuit in each regenerator reduces incoming jitter from its upstream neighbor and passes the remainder along with a component of its own to its downstream neighbor. If jitter averages more than a small fraction of a symbol time, it produces errors in a datastream. Jitter can be divided in two parts:[3]

---

[3] P. Bylanski and D. G. W. Ingram, *Digital Transmission Systems* (Stevenage, UK: Peter Peregrinus, 1976), 205–15.

- **High-frequency jitter**: caused by resonant circuit misalignment and other effects, it is attenuated rapidly. In a long string, the peak high-frequency jitter does not exceed four times the high-frequency jitter in a single unit.
- **Low frequency jitter**: caused by threshold detector misalignment and other effects, it is accumulated. In a string of n units, the final value will be n times the low-frequency jitter in a single unit.

In a chain of regenerative repeaters, the probability of error is directly proportional to the number of units. While the average, voice-grade, twisted-pair telephone line (with equalizers) can support a signaling rate of 2,400 baud (i.e., 2,400 symbols/s), a twisted-pair line with regenerators every 6,000 feet will support a T-1 channel with a signaling rate of 1.544 Mbits/s.

REVIEW QUESTIONS FOR SECTION 5.1

1 Define the characteristic impedance of an ideal transmission line.
2 What must be done to prevent reflections from the load end and the source end of a transmission line?
3 Describe skin effect.
4 What is done to twisted pairs in a cable to reduce crosstalk?
5 What values of terminating resistances are used on fine-gauge loops, coarse-gauge loops and private-data loops?
6 What causes crosstalk in twisted pair cables?
7 Define NEXT and FEXT. Which is more troublesome?
8 What causes echoes in a telephone circuit? Give examples.
9 What does delay have to do with the annoyance produced by echoes?
10 What effect is minimized by equalization? How is equalization achieved?
11 How is an equalizer trained?
12 What is an eye diagram? For what is it used?
13 Why are regenerators used?
14 Describe the major parts of a regenerator.
15 What are the causes of high-frequency jitter and low-frequency jitter?
16 At the end of a long string of regenerators, what is the magnitude of high-frequency jitter?
17 At the end of a long string of regenerators, what is the magnitude of low-frequency jitter?

5.2 OPTICAL FIBER

An optical fiber is a strand of exceptionally pure glass with a diameter about that of a human hair (125 microns = 0.005 inch). The refractive index varies from the center to the outside in such a way as to guide optical energy along its length.

5.2.1 TRANSMISSION PHENOMENA

Several types of fibers are recognized in the scientific and engineering communities. The predominant design in communications applications is *single-mode fiber*. Shown in **Figure 5.9**, in such a fiber the central core of elevated refractive index glass is < 10 microns in diameter. A significant (and essential) fraction of the optical energy travels in the cladding. Because its velocity is slightly higher than the energy in the core, conditions are right to support single-mode propagation. With a refractive index of 1.475, the velocity of energy in the core is approximately 2×10^8 meters per second (i.e., approximately two-thirds the velocity of light in free-space).

(1) Comparison With Copper Wire

Transmission of information over an optical fiber has several fundamental advantages over transmission along a copper wire.

- Optical fibers are insulators; they provide electrical isolation between transmitter and receiver.
- Optical energy is not affected by lower-frequency electromagnetic radiation; optical communication can occur in noisy electrical environments without degradation.
- When launched properly, all of the optical energy is guided along the fiber; adjacent fibers do not interfere with one another; there is no fiber to fiber crosstalk.
- Optical frequencies are very high compared to any conceivable message bandwidth; they can be used to transport wideband signals.

Optical fibers have disadvantages, too.

- Unlike electrical energy in a twisted-pair cable that propagates along the conductors in two directions at once, optical energy propagates in one direction only along the fiber.
- Optical fibers do not conduct electricity; thus, it is impossible to power equipment (such as telephones or repeaters) down the fiber. When required, electrical power is sent over wires placed in the cable sheath.
- Microbends and other mechanical insults increase fiber loss.

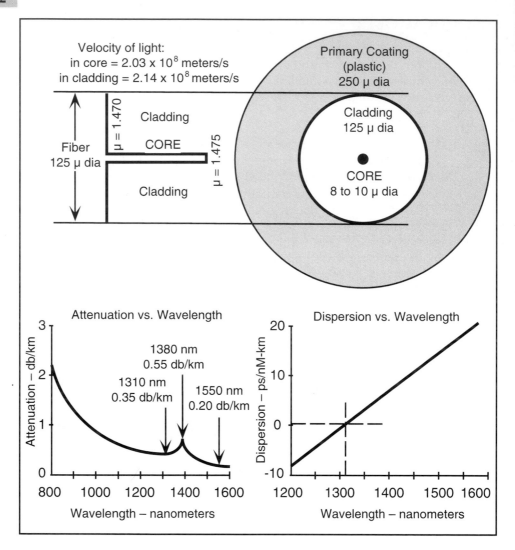

Figure 5.9 Properties of Single-Mode Optical Fiber

(2) Mechanical Properties, Cables, and Connections

The mechanical properties of optical fiber are difficult to measure. Glass is a brittle solid that does not deform plastically to relieve local stress—it shatters. The strength of a given length of optical fiber is determined by the most severe flaw present. For a 1-μm flaw, it is estimated that failure occurs around 80,000 pounds per square inch (psi). This is somewhat higher than the tensile strength of hard-drawn copper wire (60,000 to 70,000 psi). During manufacture, the fiber is subjected to a proof test, coated with a polymer coating (primary coating) to provide abrasion protection, and coated with a second polymer coating (buffer coating) to provide additional protection. Cables are produced with as few as 2 fibers up to 200 fibers. As noted above, most cables contain a few wire pairs for service use, and a tensile strength member to minimize fiber stress during installation. Cables are finished off with a plastic jacket and, if environmental conditions require it, steel armor.

The success with which fiber-to-fiber connections are made depends critically on the radial geometry of each fiber and on the preparation given to the fiber ends. Radial geometry is built into the fiber during manufacturing. The closer the fibers conform to same-size concentric circles, the better the connection that can be made. Referring to Figure 5.9, for each fiber, the core eccentricity (e) must be minimized, and the core and fiber ellipticity measures (c_1/c_2 and f_1/f_2) must equal one. For the two fibers, the diameters must be equal. As for the mating fiber ends, they must be cleaved square or polished so that they make intimate contact with each other.

Permanent connections are called *splices*. They are made by

- Aligning the fibers with reference to their outer surfaces and employing a coaxial connector structure to hold them in place. With some connectors, adjustments can be made to obtain the lowest splice loss after installation is made.
- Aligning the fibers with reference to their outer surfaces and cementing them into a structure to hold them in place.
- Employing a gaseous arc to fuse the ends together. During fusion, the surface tension of the molten glass assists in aligning the fiber ends.

Figure 5.9 lists representative values of losses due to these techniques. While fusion splicing produces the lowest loss connection, it cannot be used in hazardous environments, and requires a skilled operator and significant capital investment. In the 20 years since the first trial installations, the cost of optical fibers has reached parity with wire systems. The greatest concern of those engineering contemporary installations is with the joints and connections, not with the glass medium itself.

(3) Optical Properties

Single-mode fibers are used with solid-state lasers and photodetectors and operated at wavelengths of 1,310 or 1,550 nanometers. The lasers are switched on and off to produce unipolar pulses of intense infrared energy. Figure 5.9 shows the attenuation and dispersion per kilometer of a single-mode fiber as a function of the optical wavelength. At 1,550 nanometers, the fiber has a minimum attenuation of around 0.2 db/km, and a dispersion of some 17 ps/nm-km. At 1,310 nm, the fiber has zero dispersion and an attenuation of 0.35 db/km. Early installations were designed to operate in the minimum dispersion region. Their capacity is limited by a combination of fiber loss and the bandwidth of electronic repeaters. Later installations operate in the minimum attenuation region. With optical amplifiers, system operation has moved to the 1,550-nm region exclusively.

(4) Optical Amplifier

Optical amplification is achieved (amplification but not regeneration) using erbium-doped fiber. Called EDFAs (erbium-doped fiber amplifiers), they provide uniform gain across a wide spread of wavelengths centered on 1,550 nm, the wavelength of minimum attenuation. Using EDFAs, spans of 120 km (75 miles) between amplifiers can be employed. However, beyond 40 km, intersymbol interference produced by dispersion becomes a factor, and dispersion compensation is required.

Figure 5.10 shows the principle of an optical amplifier. The upper diagram identifies the component parts. The lower diagram shows an arrangement that employs erbium-doped fiber as the active medium. It is terminated by optical isolators. The pumping energy at 980 or 1,480 nm is injected through a WDM [wavelength division multiplexing, see Section 5.2.1(5)] device so that the signal energy

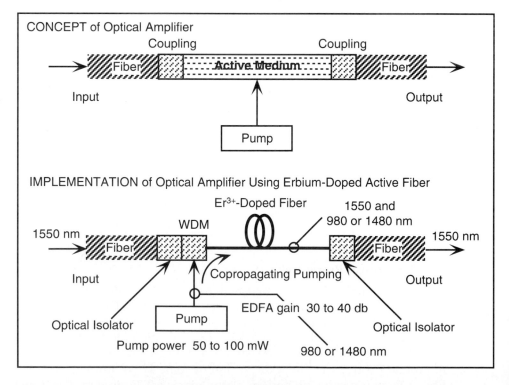

Figure 5.10 Principle of Erbium-Doped Fiber Amplifier

(at 1,550 nm) and the pumping energy flow in the same direction in the erbium-doped fiber section (this is known as copropagating pumping). In an alternative construction, the signal energy and the pumping energy flow in opposite directions (counterpropagating pumping).

Amplification depends upon the erbium ions in the active medium being excited to a higher energy level by the pumping radiation. Once there, they are metastable, and in times usually less than a millisecond they will spontaneously decay back to the ground state while emitting radiation. However, while in the metastable state, should the erbium ion be stimulated by a photon whose wavelength is between 1,530 and 1,560 nm, it emits a photon that exactly mimics the parameters of the exciting photon (i.e., same direction, phase, and wavelength) as it returns to the ground state. Because a single-signal photon can stimulate many erbium ions, the result is an amplified stream of coherent photons at the signal wavelength. Metastable erbium ions that are not stimulated by a signal photon spontaneously decay to the ground state. In so doing, they emit photons in the 1,530 to 1,560 nm region that have random directions and phases. This incoherent radiation constitutes amplifier noise. With a pump power of 50–100 mW, the amplifier produces a gain of 30–40 db. Current EDFAs show a relatively flat gain characteristic from approximately 1,540 nm to 1,565 nm, and significant gain from 1,530 to 1,570 nm.

(5) Wavelength Division Multiplexing

To make full use of their bandwidth, several optical carriers are transmitted simultaneously in the same fiber. Called *wavelength division multiplexing*, (WDM), the wavelengths of the carriers can be spaced about 2 nm apart. Current practice employs up to 32 carriers, with the expectation that this can be upgraded to 64 carriers in the near future. Crosstalk is a major concern in WDM. Interference is produced by fiber nolinearities that scatter the optical energy of the carriers, and by imperfections in network equipment.

5.2.2 TERRESTRIAL OPTICAL FIBER TRANSMISSION SYSTEMS

Optical fibers are used in point-to-point applications, and arrangements in which access at intermediate points is limited. In areas in which underground ducts are used, the substitution of fiber cables for copper cables frees significant space for future expansion.

(1) Synchronous Digital Network

Operating companies have developed a flexible, multipurpose network architecture for optical fiber transmission systems. Known as Synchronous Digital Network (SONET), it uses add and drop multiplexers, digital cross connects, and other terminals to provide point-to-point, hub and ring connections. They are being used extensively to expand interexchange, carrier serving area, and business customer routes. I describe the multiplexing and network aspects of SONET in Sections 6.4.1 and 10.2.3(7).

(2) Fiber Exhaust

In the early 1990s, recognizing the superiority of fiber technology, long-distance carriers installed optical fiber backbones in their networks. In addition, as prices came down, and performance went up, major feeder routes were converted. Most installations employed one optical carrier per fiber.

In 1996 and 1997, unexpected growth in demand for data circuits threatened to *exhaust* the capacity of the long-distance networks. Carriers reported most of their fibers *lit* and they were faced with the necessity of adding cables, or seeking other technical ways of meeting demand.[4] Fortunately, wavelength division multiplexers and erbium-doped fiber amplifiers were available to increase the number of optical channels per fiber. Consequently, most carriers have added WDMs and EDFAs to their existing cables to provide a major increase in capacity. Starting with 4-channel wavelength division multiplexers, 32-channel WDM devices are now in service.[5] Because one EDFA can amplify all of the wavelengths present in the fiber, their use has simplified equipment problems and eliminated electrical to optical conversions enroute. With improvements in electronic multiplexing speeds so that bitstreams of 10 Gbits/s (STM-64) are possible, a single fiber can transport up to 400 Gbits/s.

The sudden growth in demand for long-distance transport has stimulated the development of other devices to convert multiwavelength, point-to-point transmission systems into optical networks. Attention is focused: on optical add-drop multiplexers that permit numbers of channels to be dropped at a site without demultiplexing the entire optical stream; on optical cross connects; and on optical switches. A question occupying much attention is *protection*: how to recover capacity should a fiber route be interrupted. Most networks have adopted a ring on ring configuration that permits rapid rerouting at the optical level.

5.2.3 UNDERSEA OPTICAL FIBER TRANSMISSION SYSTEMS

The maturing of fiberoptic cable technology, the perfection of optical amplifiers and wavelength division multiplexing, and the adoption of synchronous digital hierarchy (SDH) standards (see Section 6.4.2) have led to the installation of SDH-based, optical fiber, high-capacity, digital, undersea cables and equipment.[6] Arranged in networks that span oceans and encompass continents and smaller

---

[4] John P. Ryan, "WDM: North American Deployment Trends," *IEEE Communications Magazine*, 36 (2) (February 1998), 40–44.

[5] Robert K. Butler and David R. Polson, "Wave-Division Multiplexing in the Sprint Long Distance Network," *IEEE Communications Magazine*, 36 (2) (February 1988), 52–55.

[6] Patrick R. Trischitta and William C. Marra, "Applying WDM Technology to Undersea Cable Networks,"*IEEE Communications Magazine*, 36 (2) (February 1998), 62–66.

regions, they provide fiber highways that encircle the globe. With an anticipated useful life of 25 years, the transmission networks are designed to require no more than one *ship repair event* in that time.

(1) TAT-12/13 Cable

Completed in 1996, TAT-12/13 (TAT is an acronym for Transatlantic Telephone) connects Europe and North America over four optical-fiber circuits. The four undersea cable segments are arranged in a ring with two landings in Europe (one in the United Kingdom, the other in France) and two on the East Coast of the United States, as follows:

- Shirley, New York ↔ Green Hill, Rhode Island (162 km, no undersea repeaters)
- Green Hill, Rhode Island ↔ Lands End, United Kingdom (5,913 km, 133 undersea optical repeaters spaced 45 km apart)
- Lands End, United Kingdom ↔ Penmarc'h, France (370 km, 4 undersea optical repeaters spaced 74 km apart)
- Penmarc'h, France ↔ Shirley, New York (6,321 km, 140 undersea optical repeaters spaced 45 km apart).

Each cable segment contains four fibers equipped to carry digital traffic at a rate of 5 Gbits/s (2 x STM-16). The fibers are divided into two pairs: a *service* pair that transports noninterruptible, bidirectional traffic between the cable stations, and a *restoration* pair that provides the ability to bypass a transmission fault in any segment by carrying traffic in the opposite direction around the ring. When all service pairs are operating normally, the restoration pairs may be used to carry lower-priority (i.e., interruptible) traffic giving a maximum capacity of 20 Gbits/s between Europe and North America. WDM is expected to be added before 2000.

(2) TPC-5 Cable

Completed in 1996, TPC-5 (TPC stands for Trans Pacific Cable) connects Japan and North America over four optical-fiber circuits with a combined bidirectional capacity of 20 Gbits/s. The six undersea cable segments are arranged in a ring with two landings in Japan, two on the West Coast of the United States, and two on islands in the Pacific Ocean, as follows:

- Ninomiya, Japan ↔ Coos Bay, Oregon (8,600 km)
- Coos Bay, Oregon ↔ San Luis Obispo, California (1,170 km)
- San Luis Obispo, California ↔ Keawaula, Hawaii (4,200 km)
- Keawaula, Hawaii ↔ Tumon Bay, Guam (6,580 km)
- Tumon Bay, Guam ↔ Miyazaki, Japan (2,920 km)
- Miyazaki, Japan ↔ Ninomiya, Japan (1,050 km).

Optical repeater spacings range from 33 km to 85 km. Each segment contains four fibers equipped to carry digital traffic at a rate of 5 Gbits/s (2 x STM-16). The fibers are divided into two pairs: a *service* pair that transports noninterruptible, bidirectional traffic between the cable stations, and a *protection* pair that provides the ability to bypass a transmission fault in any segment by carrying traffic in the opposite direction around the ring. When all service pairs are operating normally, the protection pairs may be used to carry lower-priority traffic. WDM is expected to be added before 2000.

(3) China–US Cable System and Atlantic Crossing–1

Across the Atlantic and Pacific Oceans, demand for channels is estimated to overwhelm the capacities of TAT-12/13 and TPC-5. Two new transoceanic ring networks are planned. Called China–US Cable System and Atlantic Crossing–1 (AC–1) Network, they will take advantage of the most advanced WDM technology available to provide the maximum capacity possible over the longest distances.

(4) Fiber Link Around the Globe

FLAG (Fiberoptic Link Around the Globe) cable network connects 12 countries using 27,000 km of cable (approximately 26,000 km undersea and 1,000 km overland). Providing a 10 Gbits/s SDH-based fiber circuit, it links the European end of TAT-12/13 (at Lands End, United Kingdom) and the Asian end of TPC-5 (at Miura, Japan). The cable segments form a modified daisy chain with landings in Europe, Africa, and Asia.

Each segment contains four fibers that are equipped to carry digital traffic at a rate of 5 Gbits/s (32 STM-1s). The fibers are divided into two pairs: a *local* pair that appears at each cable station and an *express* pair that bypasses branch cable stations.

(5) Asia Pacific Cable Network

APCN (Asia Pacific Cable Network) connects nine countries using 11,500 kms of cable. Employing a 10 Gbits/s SDH-based fiber circuit, 13 cable segments are arranged in three trunk and branch sections as follows:

- Japan ↔ Hong Kong with branches to Korea and Taiwan
- Taiwan ↔ Malaysia with branches to Philippines, Hong Kong, and Singapore
- Thailand ↔ Indonesia with branches to Malaysia and Singapore.

Optical repeaters are spaced from 73 km to 90 km apart. APCN serves as a regional network and as a feeder network for interregional networks (such as TPC-5 and FLAG) to provide connectivity to other areas of the world.

(6) Africa Optical Network

By far the largest network so far, Africa ONE (Africa Optical Network) will encircle the entire African continent with undersea fiber cable arranged in a trunk and branch architecture. As planned, Africa ONE will connect some 40 countries using 40,000 km of cable to provide a combined capacity of 40 Gbits/s (16 x STM-16s); 30 Gbits/s is dedicated to normal service traffic and 10 Gbits/s is used for protection, restoration, or lower-priority traffic. The trunk cable employs wavelength division multiplexing (WDM) with eight wavelengths per cable. It connects a ring of branching units (containing optical add/drop multiplexer, ADM, equipment) and regional hubs (called central offices). In turn, the branching units are connected to cable stations situated at more than 50 landing points on the African mainland and contiguous islands. Within its region, each central office assigns a specific optical wavelength to the cable stations. Of the eight optical wavelengths available on each cable, four wavelengths are assigned to cable stations, and four wavelengths are assigned to traffic between central offices. Thus, using both cables, a central office can communicate with up to eight stations. At the branching unit serving a specific station, the optical ADM adds and drops traffic associated with its wavelength to/from the trunk and directs it to/from the cable station. In this way, each nation using Africa ONE is guaranteed a measure of security over traffic originating and terminating within its borders. Initially, all cable stations will be equipped with from one to four STM-1 (155 Mbits/s) circuits. Eventually, they will be equipped with at least one STM-16 (2.5 Gbits/s) circuit.

(7) Other Networks

Other fiberoptic cable networks are under construction, or planned.

- **Atlantis-2, Columbus-3, and Americas-II**: three connected systems that will form a ring around the South Atlantic Ocean and connect countries on four continents.
- **SEA-ME-WE-3**: extends from Germany to Singapore connecting 16 countries en route.
- **Pan American Cable System**: a 7,300 km trunk and branch undersea cable system configured in four logical rings. With landings in seven countries in South America and the U.S. Virgin Islands, it will have connections to transatlantic and trans-Pacific networks so as to provide links with North America, Europe, and Asia.

REVIEW QUESTIONS FOR SECTION 5.2

1. Describe an optical fiber.
2. What is the velocity of the energy in the core of a single-mode fiber?
3. Enumerate the advantages of optical fibers with respect to copper wires.
4. Enumerate the disadvantages of optical fibers with respect to copper wires.

5 Why are the mechanical properties of optical fibers difficult to measure?

6 Comment on the following statement: The success with which fiber-to-fiber connections are made depends critically on the radial geometry of each fiber and on the preparation given to the fiber ends.

7 List three ways in which splices are made.

8 What is the advantage of operating at 1,310 nm? What is the advantage of operating at 1,550 nm?

9 What is an EDFA?

10 What is WDM?

11 Why is WDM used in all major networks today?

12 List the major undersea optical-fiber transmission systems. State which continents they connect.

5.3 FREE SPACE

Free space is the name given to the volume that surrounds the Earth; it is the bearer that supports the propagation of radio waves for communication purposes.

5.3.1 CHARACTERISTICS OF RADIO WAVES

The behavior of radio waves depends on their frequencies—and thus their wavelengths. [Wavelength = c/f where c is the velocity of light (187,000 miles/s or 300,000 km/s) and f is the frequency of the wave (Hz).]

(1) Lower Frequency Limit

In theory, radio waves can be generated at any frequency. In practice, a lower limit is set by the size of the antenna needed to launch the wave into free space. For instance, at a frequency of 60 Hz, the wavelength is 5,000 km (3,125 miles), a distance greater than the distance between Boston and San Francisco. Thus, a quarter-wavelength antenna for a 60 Hz wave will be 1,250 km long. However, at 1 MHz, the wavelength is 300 meters, and at 1 GHz, the wavelength is 30 cm. For these frequencies, antennas come in manageable sizes.

(2) Effect of Earth's Atmosphere

The impact of the sun's rays on oxygen molecules in the upper atmosphere creates an ionized layer called the *ionosphere*. Its thickness varies from day to night, and from the equator to the polar regions. The presence of this dynamic conducting layer produces a range of propagation effects. At frequencies

- Around **100 kHz**, the combination of ionosphere and Earth's surface guides electromagnetic energy around the Earth. Signals in this region are used for worldwide broadcast services.

- From around **1 MHz** up to around **30 MHz**, energy begins to travel in straight lines and is reflected by the ionosphere. Signals in this region include those used by AM broadcast stations and short-wave radio services. As the height of the ionosphere rises and falls with incident sunlight, the signals illuminate some areas and *skip* others.

- Above **50 MHz**, energy is neither reflected by the ionosphere nor conducted along Earth's surface, but propagates in straight lines. Under this circumstance, the receiver must be in the *line of sight* of the transmitter.

- Above **10 GHz**, intense rainfall attenuates signals significantly.

- Above **20 GHz**, water and oxygen molecules in the atmosphere become significant energy absorbers. Communication is limited to *windows* between the absorption bands.

For the most part, telecommunication facilities employ *microwave* frequencies—i.e., frequencies between around 1 GHz to 20 or 30 GHz. In this range, the receiver must be in the *line of sight* of the transmitter.

5.3.2 ADMINISTRATION OF ELECTROMAGNETIC SPECTRUM

Unless guided, or contained by shielding, electromagnetic energy radiates freely in all directions. When used to effect communication between two parties, the energy can spill over and interfere with others who are trying to communicate. For this reason, specific frequencies are assigned to specific stations, and part of the action of establishing communication involves adjusting (tuning) the equipment at the receiving station to the frequency used by the transmitting station. Other stations within receiving distance of the first pair of stations will employ different frequencies so as not to interfere with them, or to be interfered with by them.

In principle, selecting suitable frequencies should be a simple task. In practice, because of the large number of potential users, and the fact that certain frequencies may propagate around the world, frequency assignments must be made carefully, in full compliance with international agreements.

(1) Frequency Allocations and Assignments

Specialized terms are used to distinguish between all the frequencies allocated to a particular service, and the particular frequency assigned to a specific operator who provides the service.

- **Frequency allocation**: the set of frequencies allocated to all stations offering a specific service. Frequency allocations are administered by the Interna-

tional Telecommunications Union (ITU) with the help of information gathered at periodic, regional and worldwide Administrative Radio Conferences (RARCs and WARCs). In **Table 5.1**, we indicate some of the frequency allocations in the United States.

Table 5.1 Some Frequency Allocations in the United States

ONE-WAY RADIO SERVICE

- **AM broadcasting**: 107 channels, 10 kHz wide, 535 - 1605 kHz.
- **FM broadcasting**: 100 channels, 200 kHz wide, 88 - 108 MHz.
- **Television broadcasting**: 83 channels, 6 MHZ wide, in four frequency bands, 54 - 72, 74 - 88, 174 - 216, 470 - 806 MHz. Also, broadcast satellite services at 12.2-12.7 GHz (downlink) and 17.3-17.8 (uplink).

TWO-WAY RADIO SERVICE

Fixed Service: two-way radio service between fixed points
- **Common carrier point-to-point microwave radio service** (CCMRS): two-way, point-to-point, radio service in 40 bands between 928.0125 and 19620 MHz.
- **Multipoint distribution service** (MDS): provides one-way, point-to-multipoint, radio service (up to 6 MHz channels) and limited return capabilities at frequencies around 4, 6, 11, 12, 13 and 23 GHz..
- **Fixed satellite service** (FSS): voice, data and video common-carrier services in C-band (3.700-4.200 GHz downlink, 5.925-6.425 GHz uplink) and Ku-band (11.7-12.2 GHz downlink, 14.0-14.5 GHz uplink).
- **Digital electronic message service** (DEMS) and Digital termination service (DTS): two-way, fixed service for exchange of digital data. 2.5 and 5 MHz channels in frequency bands 10.55 - 10.68, and 17.7 - 19.7 GHz.

Mobile Service: two-way radio service between mobile stations, or between a mobile station and a fixed station
- **Air traffic control** (ATC): frequencies allocated in four bands, 118 - 121.4, 121.6 - 123.075, 123.6 - 128.8, and 132.025 - 135.975 MHz.
- **Cellular radio service** (CRS): 832 frequency pairs spaced 45 MHz apart with 30 kHz channel spacing allocated in one band (base to mobile), 869.040 - 893.970 MHz, and one band (mobile to base), 824.040 - 848.970 MHz. Divided between wire-line and non-wire-line carriers.
- **Wireless network service**: WARC'92/ITU–R allocated 1880 to 2025 MHz and 2110 to 2200 MHz for use by wireless networks. Two subbands -- 1980 to 2010 MHz and 2170 to 2200 MHz -- are designated for mobile satellite links. FCC has allocated frequencies in the 1850 to 1970 MHz and 2130 to 2200 MHz bands for personal communication services.

- **Frequency assignment**: the specific frequency on which a particular station must operate to provide the service to which the frequency is allocated. In the United States, frequency assignments are made by the Federal Communications Commission (FCC) for nongovernment operations, and by the National Telecommunication and Information Agency (NTIA) for government operations.

(2) Operating Licenses

The electromagnetic spectrum is a finite resource. It is administered by governments to ensure that as many persons as possible are able to benefit from its use. To use a radio transmitter, an operating license must be obtained. As well as a specific frequency assignment and authorization to provide service(s), operating licenses usually include restrictions as to permissible radiated power and type of modulation and may limit times of operation and other matters. Usually, a prerequisite for obtaining such a license is a demonstration by the applicant that existing facilities will not suffer significant interference.

5.3.3 PROPAGATION MODELS

The physical mechanisms that affect radio-wave propagation are quite diverse. Generally, communication situations require two kinds of propagation model.

- **Large-scale**: provides an approximate value for the average signal strength at a receiver located a large distance from the transmitter when there is an unobstructed line of sight between them.

This model describes propagation between microwave radio relay stations or an Earth station and a communication satellite, as if they are in free space. Under this circumstance, the energy drops of as the square of distance. In engineering terms, the energy decreases by 20 db for every 10-fold increase in distance. Thus, if a wave at 1 mile from the antenna in free space has a power level of 0 db, at 10 miles from the antenna it has a power level of –20 db.

- **Small-scale** or **fading**: characterizes rapid fluctuations in signal strength between a fixed transmitter and a moving receiver when the line of sight is obstructed.

This model describes propagation between a fixed radio site and a mobile receiver in an urban environment. Measurements in the range 30 MHz to 3 GHz show that the average path loss is proportional to the fourth power of distance. Thus, if a wave at 1 mile from the antenna in an urban environment has a power level of 0 db, at 10 miles from the antenna it has a power level of –40 db. The presence of

buildings, structures, and other impediments produces a 100-fold increase in attenuation over the free-space value.

At the top of **Figure 5.11**, I compare path loss in free space (power of 2) and in an urban environment (power of 4). Figure 5.11, shows another effect of the urban environment. As the mobile unit moves, it receives radiation directly from the base station antenna. In addition, it receives radiation that is reflected from

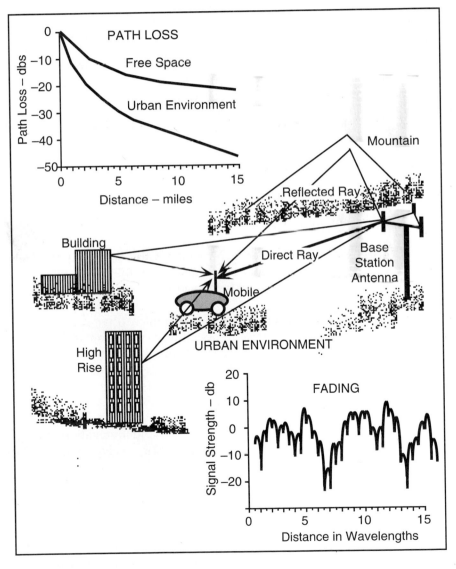

Figure 5.11 Urban Radio Environment

buildings, mountains, and other features of the environment. Because the path they travel is longer than the direct path, these rays arrive out of phase with the main signal. Moreover, as the unit moves through the environment, the relationship between the reflecting objects and the mobile change, and the phase differences between the signals vary. The result is the strength of the received signal changes in the fashion simulated in the lower graph. Significant *fades* can occur at approximately half-wavelength intervals as the unit moves forward.

REVIEW QUESTIONS FOR SECTION 5.3

1　What is free space?
2　What is the wavelength of a radio wave whose frequency is 100 MHz?
3　Why is it not practical to consider using 60 Hz as a frequency for radio communication?
4　What is the ionosphere? How is it formed?
5　As its frequency increases from 100 kHz to 100 GHz, how does the propagation mode of a terrestrial radio wave change?
6　With respect to international control of radio transmissions, define the term frequency allocation.
7　With respect to international control of radio transmissions, define the term frequency assignment.
8　Why is an operating license needed to operate most radio transmitters?
9　Distinguish between large-scale and small-scale, or fading, propagation models. For what purposes are they used?

5.4 CELLULAR RADIO

Cellular radio has shown how to uncouple terminals from fixed service points. No longer must a telephone or data terminal be attached to a cable to send and receive information. Cellular radio has made affordable mobile communications possible for the public.

5.4.1 FREQUENCY REUSE

By deploying more wires, wireline communications (i.e., communication at a distance over wires) can easily accommodate a growing number of users. In principle, the number of possible channels is without limit. On the other hand, wireless communication is executed in a limited signal space (spectrum). To keep up with demand, radio users must make better use of the allocated frequencies. The success of present cellular radio systems is based on frequency reuse.

- **Frequency reuse**: the principle of using a set of frequencies to communicate in one location, and simultaneously using the same frequencies to support independent communication in another location.

For 50 years, *frequency reuse* has been employed in terrestrial microwave radio relay systems in which focused beams of microwave energy distributed about a specific carrier frequency are sent from one station to another. The same carrier frequency can be used in other beams so long as individual receivers only *see* their transmitter. Successful operation depends on providing sufficient physical separation between entities that use the same frequencies to prevent one from receiving signals from another. With mobile units (mobiles), we use a band of frequencies in a limited area (known as a *cell*) and arrange that other areas in which the same frequencies may be in use are separated by enough distance so that the energy is properly attenuated.

5.4.2 SYSTEM CAPACITY

Figure 5.12 shows a cellular radio building block. Seven hexagonal cells are arranged in a *cluster*. At the center of each cell is a transmitter and receiver combination that uses a specific set of frequencies to communicate with mobiles moving within the cell. In the situation shown, seven sets of frequencies are employed so that the cells within the cluster do not interfere with one another.

Outside the seven-cell cluster, the frequencies are reused. **Figure 5.13** shows how the cluster fits with other clusters so that cells operating at different frequencies separate cells operating at the same frequencies. Other common arrangements use clusters with 4 or 12 cells.

(1) Ideal System

If radio resources are shared equally among all cells, the number of circuits that are available in each cell (k_{CE}) is

$$k_{CE} = B_{CL}/Nb_{ch}$$

where N is the number of cells in a cluster, B_{CL} is the radio bandwidth assigned to the cluster, and b_{ch} is the bandwidth required for each two-way circuit. For the cluster, the number of circuits is

$$k_{CL} = B_{CL}/b_{ch}$$

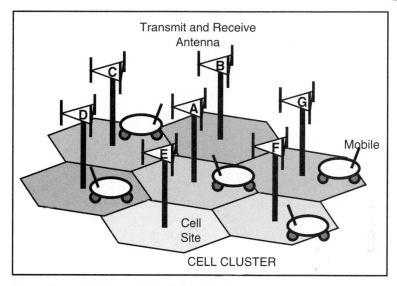

Figure 5.12 Cellular Radio Building Block

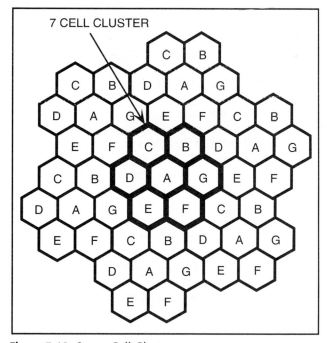

Figure 5.13 Seven Cell Clusters
The same frequencies are used in each cluster.

and, if the cluster is duplicated M times to form a system, the number of circuits in the system is

$$k_{SYS} = MB_{CL}/b_{ch}$$

Under ideal conditions, this is the number of simultaneous users that the system can support. Thus, system capacity is proportional to the number of clusters and the total allocated bandwidth, and inversely proportional to the bandwidth required by each circuit. In practical systems, not all of these circuits will be available to users; perhaps as many as 10% of them will be used for system control purposes.

(2) Influence of Cluster Size

To illustrate the influence of cluster size on system capacity, consider a specific system (AMPS, see Section 11.2.1). It operates in two frequency bands: 824 to 849 MHz mobile to cell site, and 869 to 894 MHz cell site to mobile. A circuit consists of a pair of 30-kHz simplex channels, one in the 824- to 849-MHz band and the other separated by 45 MHz in the 869- to 894-MHz band. From this information, $B_{CL} = 50$ MHz and $b_{ch} = 60$ kHz, so that the capacity of a cluster is 833 circuits.

Further, suppose the cell size is fixed and the system is to cover an area of 100 cells. Then

- If a cluster of 4 cells is employed

$$N = 4, \quad M = 25, \text{ and } k_{SYS} = 20,833 \text{ circuits}$$

- If a cluster of 7 cells is employed

$$N = 7, \quad M \cong 14, \text{ and } k_{SYS} = 11,666 \text{ circuits}$$

- If a cluster of 12 cells is employed

$$N = 12, \quad M \cong 8, \text{ and } k_{SYS} = 6,666 \text{ circuits}$$

- If a cluster of 19 cells is employed

$$N = 19, \quad M \cong 5, \text{ and } k_{SYS} = 4,166 \text{ circuits}$$

Thus, for a constant size cell and a fixed serving area, the smaller the cluster size, the greater the system capacity. Optimizing the coverage of a mixed urban, suburban, and rural area may lead to small clusters in the city and large clusters in the country.

(3) Signal-to-Interference Noise Power Ratio

If a mobile is distant x_0 from the transmitter in its cell, and x_I from the nearest interfering transmitter, the signal-to-interference (SIR) noise power ratio—i.e., the ratio of the power received from the cell site transmitter (P_0) to the power (P_I) received from the interferor—is

$$SIR = P_0/P_I = (x_0/x_I)^4$$

In fact, there are several first layer interferors, and many more farther than one layer away. In **Figure 5.14**, the capacity of a 100-cell system and the SIR for multiple interferors is plotted against the number of cells in a cluster. In common practice, an SIR of 18 db, or greater, is believed necessary for satisfactory reception. Reducing the cluster size to four cells makes for marginal operating conditions.

(4) Cells and Microcells

Cell size cannot be reduced without limit. It must be matched to the anticipated speeds of the mobiles. The smaller the cell, and the faster the mobile, the

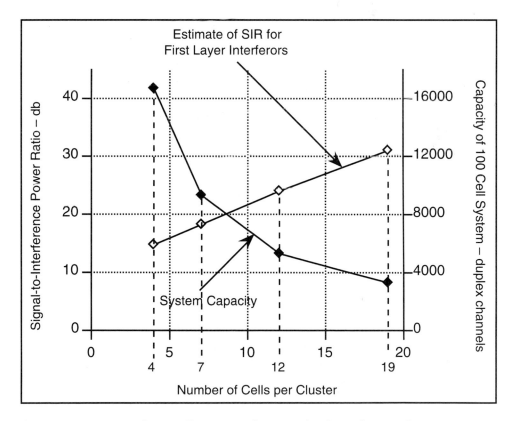

Figure 5.14 Capacity of 100-Cell System and Magnitude of Signal-to-Interference Noise Ratio as a Function of Size of Cell Cluster

more often the base stations will have to *hand off* to the next cell as the mobile traverses the system. (Hand off is a procedure we discuss in Section 11.2.) It requires time and system resources and can produce a discontinuity in the message stream. The result is that cells for systems that serve high-speed vehicles (e.g., cars on an interstate highway) may be several miles in diameter, while cells that serve pedestrians may be a thousand feet, or so, in diameter. For a given range of mobile speeds, there is a practical minimum cell size and thus a maximum limit to the system capacity that can be achieved by decreasing the size of the cells.

In Figure 5.13, a cellular radio system is shown as consisting of regular hexagonal cells. In the real world, the cell boundary is defined by the following:

- Continuum of points at which the power received from the transmitter in the cell is just equal to the minimum level at which satisfactory detection is achieved by the mobile receiver (i.e., the probability of detection is say $\geq 90\%$). This level will depend on the signal-to-noise ratio due to interferors (and possibly other disturbers).
- Maximum power of the mobile transmitter. It must be heard at the cell site receiver.
- Maximum speed at which mobiles traverse cells.
- Any anomalous propagation conditions or geographical features.

Establishing a cell size is a complicated process that requires full definition of the purpose, location, environment, and equipment employed.

REVIEW QUESTIONS FOR SECTION 5.4

1 What is meant by frequency reuse? Explain why it is the key to cellular radio.

2 Draw a seven-cell cluster. Indicate the frequency band allocated to each cell. Draw a second seven-cell cluster that abuts the first. Indicate the frequency bands allocated to these cells.

3 List the factors on which system capacity depends, and state the relationship.

4 What effect do cluster sizes have on system capacity?

5 Comment on the following statement: a SIR of 18 db, or greater, is believed necessary for satisfactory reception. What does this imply for 4-cell clusters?

6 List the factors that determine real cell size.

5.5 COMMUNICATION SATELLITES

Communication satellites orbit Earth; some perform as *repeaters* in the sky,[7] others perform as *switches* in the sky. They communicate with units on the ground, and may communicate among themselves (satellite to satellite). In the simplest case, the spacecraft electronics captures the signals from Earth, converts them to lower frequencies, and transmits them back to Earth so that a message originating from a single point in the field of view of the satellite can be received at a second point on Earth's surface in the field of view of the satellite. The relayed message can be broadcast to the entire area, in which case it can be received by everyone, or dispatched to a specific area by a spot-beam antenna on the spacecraft. In this case, only those within the illuminated area can receive it.

Commonly, satellite communication is used to provide the radio-relay of data, voice, and video messages between widely separated points. In addition, it is used to broadcast data, voice, and video messages to particular areas of the world. The broad geographical coverage and multiple connectivity that a satellite provides require that access be carefully managed so that a large number of Earth stations can be interconnected through the same satellite simultaneously.

Many satellites contain multiple *transponders*—i.e., amplifying and frequency-changing devices that operate over a portion of the frequency band assigned to the satellite. Channels may be pre-assigned (i.e., fixed allocation) for routes with heavy traffic (or for private usage), may be assigned on demand (demand-assigned multiple access, known as DAMA), or may be seized by Earth stations when the need arises (random access).

5.5.1 PROPERTIES

Some properties of communication satellites are illustrated in **Figure 5.15**.

(1) Orbits

Figure 5.16 plots height above the surface of Earth against the number of revolutions per day required to sustain the satellite in a circular orbit. At 22,300 miles above the surface, it must complete one revolution a day. If it is in the equatorial plane, this orbit is described as *geostationary* because, from Earth, the satellite appears stationary.

- **Geostationary Earth orbit** (GEO): some 22,300 miles above the equator, a path on the equatorial plane in which satellites rotate about Earth's axis at

---

[7] They are also described as *bent pipes* ; they receive energy and redirect it to the ground.

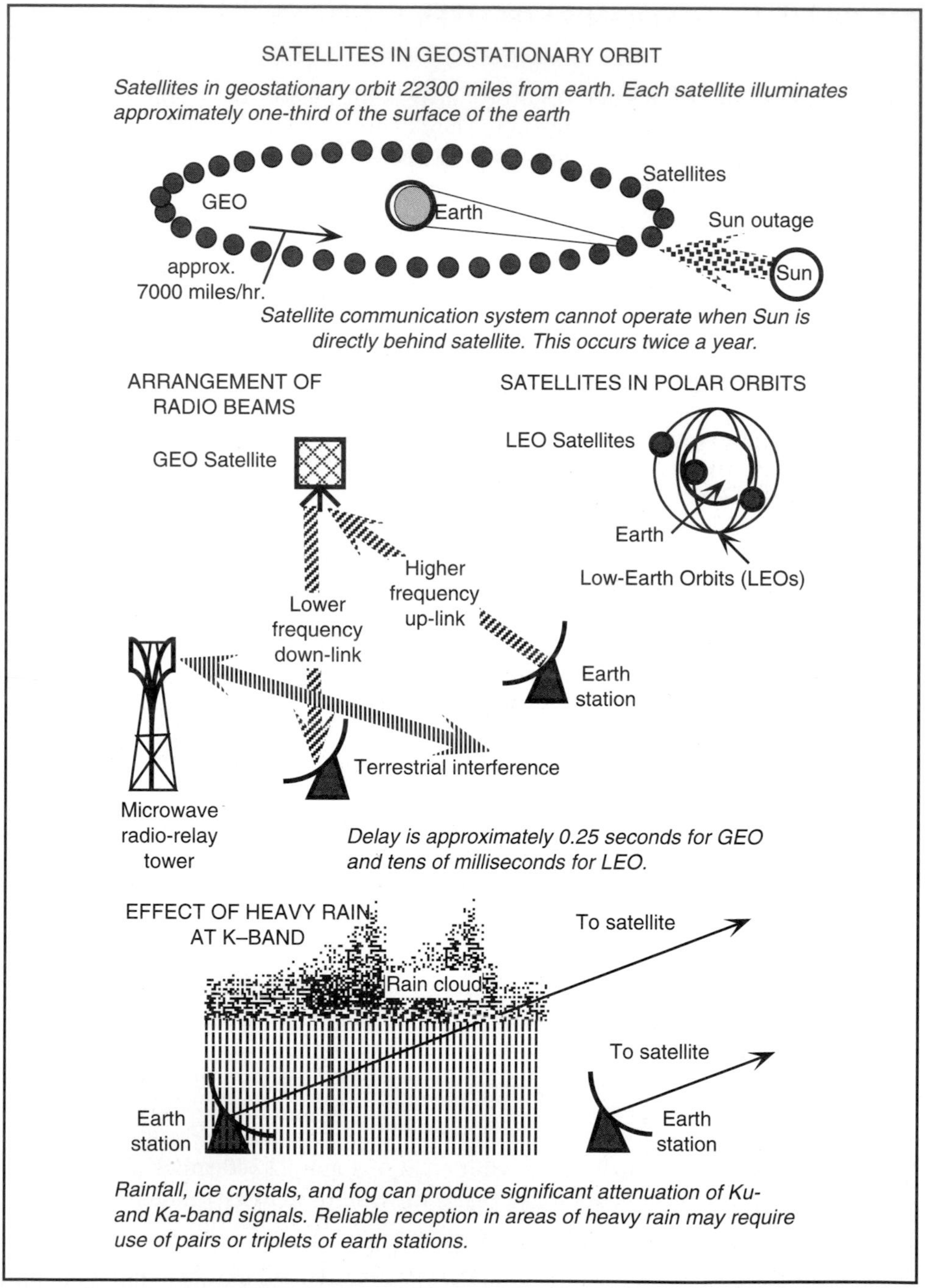

SATELLITES IN GEOSTATIONARY ORBIT

Satellites in geostationary orbit 22300 miles from earth. Each satellite illuminates approximately one-third of the surface of the earth

Satellites

GEO

Earth

Sun outage

approx.
7000 miles/hr.

Sun

Satellite communication system cannot operate when Sun is directly behind satellite. This occurs twice a year.

ARRANGEMENT OF
RADIO BEAMS

SATELLITES IN POLAR ORBITS

GEO Satellite

LEO Satellites

Earth

Low-Earth Orbits (LEOs)

Higher
frequency
up-link

Lower
frequency
down-link

Earth
station

Terrestrial interference

Microwave
radio-relay
tower

Delay is approximately 0.25 seconds for GEO and tens of milliseconds for LEO.

EFFECT OF HEAVY RAIN
AT K–BAND

To satellite

Rain clouds

To satellite

Earth
station

Earth
station

Rainfall, ice crystals, and fog can produce significant attenuation of Ku- and Ka-band signals. Reliable reception in areas of heavy rain may require use of pairs or triplets of earth stations.

Figure 5.15 Satellite Communications Environment

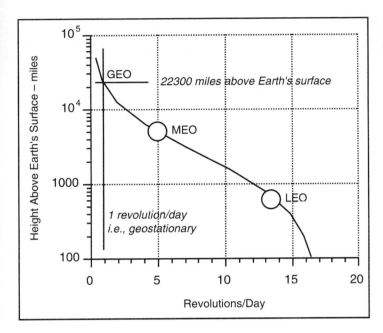

Figure 5.16 Number of Revolutions per Day to Sustain Orbit Height

the same angular velocity as Earth (i.e., 360° or 2π radians in 24 hours). They illuminate approximately one-third of the surface of Earth.

In geostationary orbit, the satellite travels at approximately 6,900 miles per hour, and illuminates a circular area approximately 7,000 miles in diameter. When positioned between approximately 90° and 130° West longitude, GEO satellites can cover the 50 states of the United States, and when positioned between approximately 60° and 90° West longitude they can cover the 48 continental states. GEO satellites deployed by major communications organizations for intercontinental and intracontinental trunking applications are spaced at intervals of arc of ≥ 2°. When used to provide direct television broadcasting, they are spaced ≥ 9° apart. Other orbits are

- **Polar Earth orbit**: a path on the polar plane—one that passes through the center of Earth and contains the poles
- **Inclined Earth orbit**: a path on a plane that passes through the center of Earth situated between the polar and equatorial planes
- **Medium Earth orbit** (MEO): a path about 6,000 miles above the surface of Earth. The orbit is likely to be polar or inclined and the footprint is around 5 to 6,000 miles in diameter.
- **Low Earth orbit** (LEO): a path a few hundred miles (400 to 900 miles) above Earth's surface. The orbit is likely to be polar or inclined, and the footprint is around 1,000 to 3,500 miles in diameter.

(2) Transmission Delay

Due to the distance the energy must travel, there is a time delay between sending a signal from a transmitter on Earth and receiving the signal back on Earth. For GEOs, it is approximately 0.25 second; for MEOs, it is 60 to 70 milliseconds; and for LEOs it is less than 10 milliseconds. **Figure 5.17** shows values of the minimum sender to receiver transmission delay as a function of height above the surface of Earth.

(3) Interference among Satellites

To avoid signal interference among satellites, both space and frequency diversity are employed.

- **Space diversity**: satellites are physically separated from one another. Either, they occupy different orbits, or, if in the same orbit, they occupy different positions in the orbit.
- **Frequency diversity**: satellites assigned to the same orbit positions operate in different frequency bands.

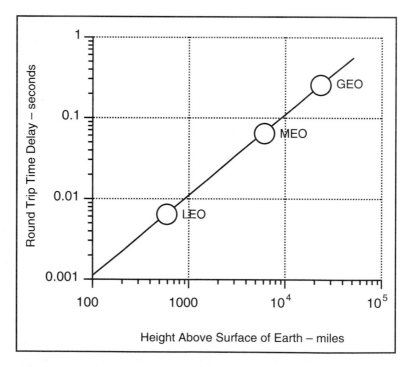

Figure 5.17 Round-Trip Transmission Delay and Satellite Height

(4) Frequency Allocations

- **GEO satellites**: employ frequencies in C band (around 4 GHz, space to Earth, and 6 GHz, Earth to space) and Ku band (around 11 and 12 GHz, space to Earth, and 14 GHz, Earth to space). Additional frequencies are allocated in Ka band (around 18 GHz, space to Earth, and 30 GHz Earth to space).

At 4 and 6 GHz, interference from terrestrial microwave networks that share the same frequency bands may restrict opportunities for siting Earth stations. At the higher frequencies (Ku and Ka bands), the absence of terrestrial facilities competing for the same frequencies makes the selection of sites for Earth stations easier. However, atmospheric attenuation due to rainfall, ice crystals, and fog can be significant. During heavy rainstorms in the eastern and southern regions of the United States, pairs or triplets of stations may be required to assure reliable signal reception. Based on statistics describing the extent of local storms, they are sited so that at least one station is likely to be in the clear at all times.

- **MEO and LEO satellites**: allocations exist from as low as 150 MHz (149.9 to 150.05 MHz) to as high as 30 GHz. Important bands for satellite provided personal communication services exist at 312 to 315 MHz (downlink) and 387 to 390 MHz (uplink), 1,610 to 165 MHz (downlink) and 2,483.5 to 2,520 MHz (uplink), and 1,980 to 2,010 MHz (downlink) and 2,170 to 2,200 MHz (uplink). Other allocations exist in S, C, and Ka bands.

(5) Power and Footprints

Most spacecraft employ power supplies that are charged from solar cells. In the design of the communications system, care is taken to minimize the power load presented by the electronics. Because atmospheric attenuation is greater at the higher frequencies, the frequency that requires the greater power is likely to be used by the Earth station transmitter; thus, the downlink frequency is less than the uplink frequency. Another way of saving downlink power is to employ shaped-beam antennas that concentrate the radiated power on specific areas of the global surface. For instance, geostationary satellites intended to carry traffic between continents separated by an ocean employ antennas that concentrate the radiated energy on the landmasses and not the ocean. For domestic coverage of the entire United States, GEO satellite antennas illuminate the 48-state continental landmass and provide additional spot beams focused on Alaska, Hawaii, and Puerto Rico. In the 48 states, spot beams may be employed to serve the major population centers.

Restricting the radiated energy to areas surrounding terrestrial receivers allows better use of satellite power, provides some privacy by limiting the area in which the signals can be received, and permits the available radio frequency spectrum to be reused. In each spot beam, the entire assigned bandwidth can be available for use provided adequate levels of isolation are maintained between the beams. This is accomplished by using an antenna with a radiation footprint that is a maximum in its intended service area and a minimum in areas served by other spot beams.

To use these techniques with MEO and LEO satellites, the spacecraft antennas must be steerable so as to track locations on Earth as the spacecraft passes overhead. Usually, systems that furnish communication services to mobile users employ a down-looking circular radiation footprint. Because the angle subtended by the service area is larger, the gain of the spacecraft antenna is significantly less than that of a GEO satellite.

(6) Received Power

The total loss in a free-space path from Earth to GEO satellite to Earth can be close to a daunting 400 db. This must be overcome by gain in the system components. Typical values are

Gain of transmitting antenna	=	+ 70 db
Gain of spacecraft receive antenna	=	+ 30db
Gain in spacecraft amplifier	=	+ 80 db
Gain of spacecraft transmit antenna	=	+ 30 db
Gain of receiving antenna	=	+ 50 db

With a 100 watt transmitter in the sending Earth station the received power may be around–84 dbm, or approximately 3×10^{-12} watts. Improved transmitters, higher-gain antennas, and the use of spot beams will permit GEO satellites to support hand-held terminals and mobile operations [see Section 11.1.1(5)]. For a LEO satellite operating at 1.6 GHz (uplink and downlink) and orbiting at a height of 750 km, the total path loss is 300 db.

5.5.2 GEO Satellites—International Telecommunications Satellite Consortium

INTELSAT (International Telecommunications Satellite Consortium) a consortium of national communications satellite organizations owns and operates over 20 spacecraft in geostationary orbit. Managed by COMSAT (Communications Satellite Corporation), a company created by act of Congress in 1962, INTELSAT provides repeater-in-the-sky services to most of the countries in the world for international (and some domestic) telecommunications. Approximately 300 Earth stations are owned and operated by local administrations and operating companies.

(1) Service Regions

INTELSAT satellites are deployed in four regions that cover the globe:

- **Atlantic Ocean Region** (AOR): serves most of the Americas, Western Europe, India, and all of Africa (11 spacecraft)
- **Indian Ocean Region** (IOR): serves Eastern Europe, Africa, the Middle East, India, Southeast Asia, western Australia, and Japan (5 spacecraft)
- **Asia Pacific Region** (APR): serves Eastern Europe, Russia, and other C.I.S. countries, Asia from India to Japan and Australia (1 spacecraft)
- **Pacific Ocean Region** (POR): serves Southeast Asia to Australia, all of the Pacific, and the western parts of Canada and the United States (4 spacecraft).

(2) Spacecraft

INTELSAT has four generations of spacecraft currently in orbit. They are

- **INTELSAT V** and **INTELSAT VA**: operating in C and Ku bands, IS Vs provide 12,000 two-way telephone circuits and two television channels, and IS VAs provide 15,000 two-way telephone channels and two television channels. With a design life of seven years, the majority of these spacecraft have exceeded this mark.
- **INTELSAT VI**: operating in C and Ku bands, IS VIs provide 24,000 two-way telephone circuits and three television channels. With digital circuit multiplication equipment (DCME), they provide up to 120,000 two-way telephone circuits. With a design life of 13 years, IS VIs are the largest commercial spacecraft ever built. The high-power spot beams provide a variety of voice, video, and data services. IS VIs have a number of transponders that are interconnectable by static on-board switches, or by an electronic subsystem to provide satellite-switched, time division multiple access (SS/TDMA).
- **INTELSAT VII** and **VIIA**: operating in C and Ku bands, IS VIIs provide 18,000 two-way telephone circuits and three television channels. IS VIIAs provide 22,500 two-way telephone circuits and three television stations. With DCME, IS VIIs provide 90,000 two-way telephone circuits, and IS VIIAs provide 112,500 two-way telephone circuits. Series VII spacecraft can be reconfigured in orbit to vary coverage capabilities in response to changing traffic patterns and service requirements.
- **INTELSAT VIII** and **VIIIA**: operating in C and Ku bands, Series VIII spacecraft are designed to meet the needs of Pacific Ocean region users for improved C band coverage. They incorporate sixfold, C band frequency reuse and increased C band power.

In addition, INTELSAT operates a specialized spacecraft in the Atlantic Ocean region.

- **INTELSAT K**: launched in 1992, and operating in Ku band, INTELSAT K provides service to international broadcasters. Its 16, 54 MHz transponders can be configured to provide 32 television channels.

5.5.3 GEO Satellites—International Maritime Satellite Organization

Established in 1979 to provide worldwide mobile satellite communications for the maritime community, INMARSAT (International Maritime Satellite Organization) has 79 member countries and operates global mobile satellite services on land, at sea, and in the air. To cover the globe, four Inmarsat-3 generation GEO satellites each employ a global beam and spot beams. Forty Earth stations in 31 countries connect satellite users to national networks.

5.5.4 MEO and LEO Systems

Constellations of medium Earth orbit and low Earth orbit satellites are being deployed to provide continuous global coverage to connect mobile and stationary terminals. The systems are in various stages of development, implementation, and operation. In Sections 11.1.2 and 11.1.3, I discuss them in more detail.

- **Ellipso**: a 17-satellite MEO/LEO constellation. Seven satellites in a ring on the equatorial plane provide coverage of lands on the equator and in the southern hemisphere. Ten additional satellites are arranged in two highly elliptical orbits to provide coverage of the densely populated areas of North America, Europe, and Asia.

- **ICO**: a 10-satellite MEO constellation orbiting in two 45° planes at 10,354 km from Earth, ICO offers mobile telephone service to users from existing terrestrial cellular systems who travel to places where cellular coverage is incomplete or does not exist.

- **Globalstar**: with a 48-satellite LEO constellation orbiting in eight planes inclined 52° at 1,410 km from Earth, Globalstar will offer wireless telephone and other telecommunication services worldwide to users equipped with hand-held or vehicle-mounted terminals (including airplanes). The constellation provides single-satellite coverage between ± 70° latitude, and at least two-satellite coverage between 25° and 50° north and south latitude.

- **Iridium**: with a 66-satellite LEO constellation orbiting in six polar planes at an altitude of 750 km, Iridium offers global personal satellite-based communications with hand-held terminals. Initial operation began in November 1998. Manufactured by Motorola and associated companies, it uses GSM architecture.

- **Teledesic**: originally designed to operate with a 840- to 924-satellite LEO constellation orbiting in 21 polar planes at an altitude of 700 km, Teledesic has been repositioned in a higher orbit with 288 satellites in 12 polar planes. It will offer broadband global services to mobile terminals at data rates from 16 kbit/s to 2.048 Mbit/s.

REVIEW QUESTIONS FOR SECTION 5.5

1 Justify descriptions of communication satellites as repeaters in the sky, bent pipes, and switches in the sky.

2 Describe a geostationary orbit, polar Earth orbit, inclined Earth orbit, medium Earth orbit, and low Earth orbit.

3 What magnitude of transmission time delays will you encounter when using GEO, MEO, and LEO satellites?

4 Describe two ways to avoid signal interference among satellites.

5 List the communications frequencies allocated to GEO, MEO, and LEO satellites.

6 Why use spot beams?

7 Explain the following statement: the total loss in the free-space path from Earth to (geostationary) satellite to Earth is a daunting 394 db.

8 What is INTELSAT? What does it do?

9 What is INMARSAT? What does it do?

10 List the names of the groups of MEO and LEO satellites that are being deployed to provide continuous global support of hand-held and mobile terminals.

6

MULTIPLEXERS
AND SUBSCRIBER LINES

In Recommendation G.902, ITU–T divides a telecommunications network into an *access* network and a *core* network. Shown in **Figure 6.1**, individual stations (clients) are connected to the access network. The core network provides communications services, and transports signals among access networks. Service nodes reside on the edges of the core network. They consist of switches, servers, and routers that provide the services required by clients. The interface between a client and the access network is known as the UNI (user network interface), and the interface between a service node and the access network is known as the SNI (service node interface). Network resources are managed by a network management system (TMN, see Section 8.6.8).

Over the access network, the stations send and receive various kinds of signals. To lower the cost of communication services, providers share transmission links among as many users as possible. The techniques they use range from twisted-pair cable connections that carry several channels, to the *laissez-faire* approach of *Aloha* on radio channels—send when ready, and repeat if necessary.

6.1 MULTIPLEXED SIGNALS

Because transmission costs increase with distance, on all but the shortest connections, a bearer should carry as many signals as possible. Performing this feat requires two actions

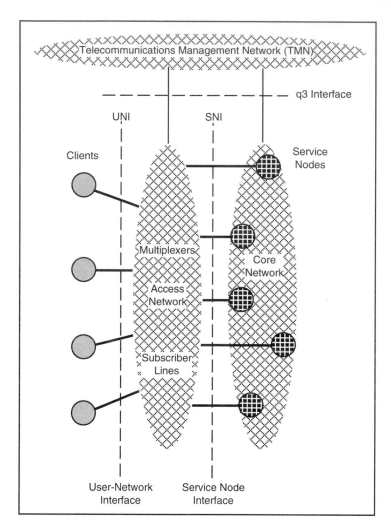

Figure 6.1 Components of a Telecommunications Network

- **Multiplexing**: action of placing several signal streams on a single bearer
- **Demultiplexing**: action of recovering individual signals from a multiplexed signal stream.

6.1.1 MULTIPLEXING TECHNIQUES

In the belief that demultiplexing is self-evident once multiplexing is understood, my principal focus is on multiplexing.

(1) Basic Multiplexing Techniques

Three basic techniques are used to place several channels on a single bearer.

- **Frequency-division multiplexing** (FDM): to each channel, a fraction of the bandwidth of the bearer is assigned so that the channels exist at the same time, at different frequencies, on the same bearer. A channel is defined by its center frequency, and its bandwidth, or by its upper- and lower-frequency limits.
- **Time-division multiplexing** (TDM): to each channel, the entire bandwidth of the bearer is assigned for very short periods of time (time slots) so that the signals exist in series on the same bearer. A channel is made up of a sequence of time slots. Time-division multiplexing is said to be
  - *synchronous* if the time slots are assigned to each channel in turn in a regular sequence paced by the system
  - *statistical* if the time slots are assigned to signals as they arrive at the multiplexer. They are read out at a speed determined by the multiplexed line.
- **Code-division multiplexing** (CDM): to each channel, the entire bandwidth of the bearer is assigned so that the signals exist simultaneously on the bearer. User's messages are differentiated by modulating them with orthogonal codes. The receiver uses these codes to demodulate the individual message signals. Principally, CDM is used with radios and free-space bearers.

(2) Special Multiplexing Techniques

The basic techniques can be combined in different ways to satisfy special multiplexing requirements. For instance, a channel can be time-divided, with signals introduced alternately from the ends.

- **Time-Division Duplex**: messages are sent back and forth, a time slot at a time. In sequence, the channel is available to
  - End A for a time slot (when A sends to B)
  - Neither end for a short guard time
  - End B for a time slot (when B sends to A)
  - Neither end for a short guard time
  - End A for a time slot (when A sends to B)
  - Neither end for a short guard time
  - And so on.

The purpose of the guard time is to compensate for the propagation delay along the connection.

In addition, FDM and TDM may be combined so that each frequency divided channel is further divided into time slots

- **Frequency- and Time-division multiplexing** (FD/TDM): the bandwidth of the bearer is divided up to produce frequency-divided channels. Then, each channel is time-divided to produce time slots. A channel is made up of a sequence of time slots in a frequency divided channel.

Figure 6.2 provides a conceptual view of FDM, synchronous TDM, and FD/TDM. To illustrate the make up of a channel, I have shaded the bandwidth that is FDM channel 9, the time slots that are TDM channel 9, and the time slots that are FD/TDM channel 9–9.

Communication satellites and mobile cellular systems employ FDM, TDM, FD/TDM, and CDM. Described in Section 6.5.1, CDM is a spread-spectrum technique. Like TDM, it can be used in combination with FDM as FD/CDM.

(3) Channel Banks and Multiplexers

Equipment that conditions and combines analog voice signals into FDM or TDM signals is known as a *channel bank*

- **Channel bank**: device that conditions and multiplexes analog voice signals using frequency division or time division
  - *FDM.* When transmitting, the channel bank combines each voice signal with one of a set of carrier frequencies to produce a comb of single-side-band (SSB) modulated signals; when receiving, it demodulates each SSB signal to obtain the voice signals
  - *TDM.* When transmitting, the channel bank performs quantization, coding, μ-law compression, and multiplexing; when receiving, it performs demultiplexing, decoding, μ-law expansion, and reconstruction of the voice signals.

Equipment that combines already-digitized signals (digital voice or data) into a TDM signal is known as a *multiplexer*.

- **Multiplexer**: device that employs time division to transport several digital signals on the same bearer
  - when transmitting, the multiplexer reduces the pulse width and increases the bit rate of incoming signals to match the parameters of the available time slots
  - when receiving, the multiplexer restores the pulse width and reduces the bit rate of the received signals to match the original signal.

6.1.2 FREQUENCY-DIVISION MULTIPLEXING

For many years, frequency modulated (FM) microwave radio systems carried frequency-division multiplexed voice signals in medium- and long-haul telephone

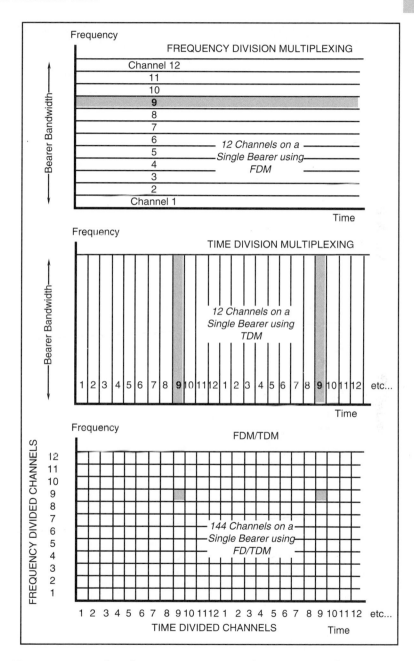

Figure 6.2 Principles of Frequency-, Time-, and Frequency- and Time-Division Multiplexing

In FDM, a channel is defined by a frequency band. In TDM, a channel is defined by a series of time slots. In FD/TDM, a channel is defined by a frequency band and a series of time slots.

transmission applications. Later, to increase the capacity of long-haul systems, single-sideband (SSB), amplitude modulated (AM) microwave systems were employed. Now, the combination of optical fibers and digital techniques has displaced FDM in favor of TDM. Nevertheless, because of the amount of FDM equipment in place, a significant fraction of the world's voice traffic will continue to use these techniques into the next century.

Figure 6.3 shows how 12 analog voice channels are combined in a *channel bank* to form a composite, multiplexed signal. To produce a double-sideband sup-

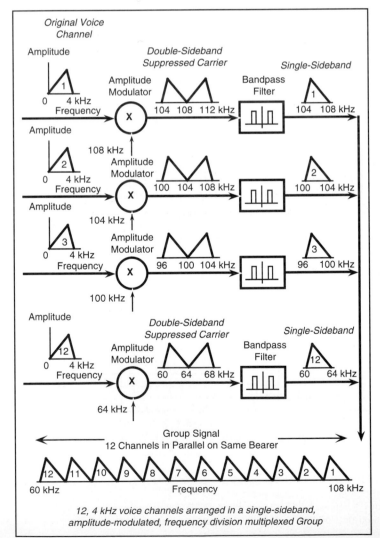

Figure 6.3 Frequency-Division Multiplexing of 12 Voice Channels

Twelve channels form a group. They exist as lower sideband, amplitude-modulated signals on a single bearer.

pressed carrier (DSBSC) signal [see Section 4.1.1(2)], each voice channel modulates a carrier chosen from the set 64, 68, 72, ... 108 kHz. When passed through a filter, only the lower sideband (LSB) remains. As shown at the bottom of Figure 6.3, because each LSB spans a different frequency range, they may be added (frequency multiplexed). This arrangement of channels is known as a *group*.

Figure 6.4 illustrates how groups are combined in steps to become part of a signal containing 600 (or more) channels. In turn, these signals are used to modulate carriers in radio relay systems. Of course, a complete transmission system consists of two multiplexing and demultiplexing arrangements—one for each direction of transmission.

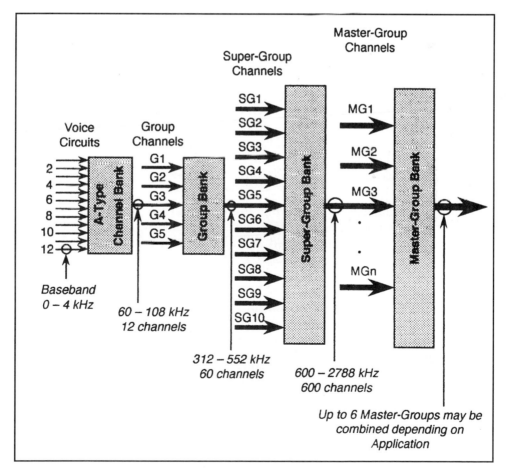

Figure 6.4 Frequency-Division Multiplexing Hierarchy

Five groups are combined to form one supergroup. Ten supergroups form a mastergroup. Up to six mastergroups may be combined to create a higher level of multiplexing.

6.1.3 Synchronous Time-Division Multiplexing

Synchronous time-division multiplexers combine n lower-speed, digital signals on higher-speed single bearers. Whether signals are present or not, time slots on the bearer are regularly allocated to each input channel at a rate determined by the multiplexing equipment. To ensure that they are synchronized with other equipment in the network, a stable clock is incorporated in all major multiplexers [see Section 9.6.4)]. Demultiplexing restores the n lower-speed channels. To avoid congestion, the speed of the multiplexed line must be equal to the sum of the speeds of the input channels, *plus* the speed of the framing channel.

Time division multiplexers operate at three levels

- **Octet-level multiplexing**: octets from each input line are interleaved on the common bearer in time slots reserved for them. After each sequence of octets (one from each input line) has been serviced, the multiplexer adds a bit to define the frame. A delay of one octet time (i.e., the time required to transmit 8 bits at the input speed) occurs in the passage of each datastream from input line to multiplexed output line.

- **Bit-level multiplexing**: bits from each input line are interleaved in turn on the common bearer in a regular fashion. Timing is more critical than at the octet level, but throughput delay can be reduced. In the public-switched telephone network, bit-level multiplexing is used to build higher-speed multiplexed datastreams from lower-speed multiplexed datastreams (see Section 6.2.3).

- **Frame-level multiplexing**: frames from each input line are interleaved on the common bearer. Frame-level multiplexing (which is likely to be asynchronous) is performed by data concentrators (see below) and packet switches (see Section 9.2.2).

Octet interleaved multiplexing is simple and efficient. It employs 8-bit segments of the datastreams in a fashion already engineered into computers, and other digital equipment. In contrast, bit-interleaved multiplexing is more complex. Success requires that all datastreams be exactly synchronized so that all bits arrive at exactly the right time to join the multiplexed stream. To provide a cushion, it is usual to include a buffer of a few bits in each feed. Even so, occasionally, there will be timing differences that prevent a bit being available when required.

In order to maintain proper frame structure and line speed, the multiplexer *stuffs* a bit into the stream. In addition, it adds overhead bits that tell the demultiplexer which bit is bogus. At the demultiplexer, the stuffed bit is ignored. Such measures produce timing irregularities in the demultiplexed output stream that are described as *jitter* or *wander*. They can lead to transmission errors.

Bit-interleaved multiplexing is used in the upper levels of the T-carrier hierarchy (see Section 6.2.3). When they were developed, the lack of fast processors and inexpensive memory made it the only practical implementation.

Figure 6.5 illustrates multiplexing and demultiplexing functions and shows the time relationship among octets on the lower-speed lines and octets in the datastream on the higher speed line. In a dynamic process, the multiplexer collects octets from the input lines, reads them out at the higher-speed of the multiplexed line, and places each in a time slot that is allocated to it on a regular basis. After all time slots have been serviced once, the multiplexer adds a framing bit. If one or more of the incoming lines are inactive, one or more of the slots on the multiplexed line will be empty.

At the receiving end, the demultiplexer uses the framing bits to synchronize activity and locate the beginning of a frame. It collects the octets from the multiplexed line and distributes them to the lower-speed output line corresponding to the input line from whence the octet came. Throughout the process, frame-level synchronization must be maintained so that the demultiplexer can assign the received octets to the correct output channels. Because the incoming octets are processed at a rate, and at a time, determined by the multiplexer, the action is called *synchronous* multiplexing.

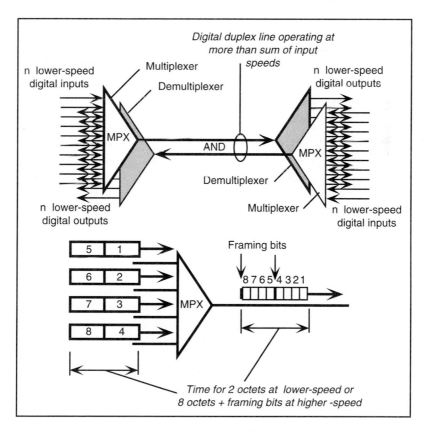

Figure 6.5 Principle of Synchronous Time-Division Multiplexing

6.1.4 Statistical Time-Division Multiplexing

The irregular nature of data messages, and the fact that, on average, the two channels that support a conversation are only 40% active, has produced statistical multiplexers (SMPXs). First applied to voice signals on transatlantic cables, statistical multiplexing is used now with any class of bursty traffic that is carried over even moderately expensive facilities.

(1) Bursty Data Traffic

Consider a terminal consisting of a keyboard and video display that is connected to a host computer. During periods of input activity, characters may be generated at an average of three per second (corresponding to 30, six-character words per minute). They are transmitted asynchronously as bit-framed characters [see Section 3.3.2(1)] at 300 bits/s. Under these circumstances, the line is occupied only 10% of the time. Between periods of input activity there may be periods of thinking during which nothing is transmitted, making the long-term average occupancy significantly less than 10%. If a cluster of terminals is involved, and a synchronous time-division multiplexer is employed to carry the datastreams to the host, only a small percentage of the time slots allocated to any terminal may be occupied. Obviously, this is a very inefficient use of the transmission facility.

The bursty nature of many digital streams has stimulated the development of statistical multiplexing techniques. Only when data are present at an input port do SMPXs allocate a time slot on the multiplexed line. Arrival times at the multiplexer, not what input lines they are on, determine the order in which octets are sequenced. As they are placed on the multiplexed line, the multiplexer adds an identifier so that the demultiplexer will know to which output line the octets are to go. To reconstruct the original datastreams, this information is vital—but it adds overhead bits to the process. In principle, some, or all, input lines can operate in bursts faster than the multiplexed line operates. The guiding requirement is the average number of input bits per second, plus the average number of address bits per second added by the multiplexer, must be less than the speed of the multiplexed line.

(2) Statistical Multiplexing

In **Figure 6.6**, I show an example of a simple, four-line, SMPX. Eight-bit characters, framed by start and stop bits, are presented on lines operating at 300 bits/s. During multiplexing, the framing bits are stripped off; then 2 bits are added to each character to identify the input line from which it came. (Two bits because there are four channels.) Because the characters are generated irregularly, the average bit rate on each input line is much less than 300 bits/s. The multiplexed line runs at 600 bits/s.

As each character arrives at the SMPX, the multiplexer allocates a time slot to it and places it on the common bearer. If two, or more, characters arrive at the same time, the SMPX handles them in a consistent manner—e.g., it may take them

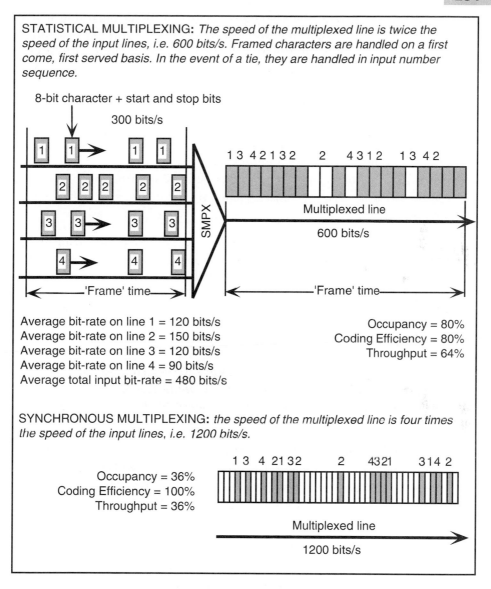

STATISTICAL MULTIPLEXING: *The speed of the multiplexed line is twice the speed of the input lines, i.e. 600 bits/s. Framed characters are handled on a first come, first served basis. In the event of a tie, they are handled in input number sequence.*

8-bit character + start and stop bits

300 bits/s

1 3 4 2 1 3 2   2   4 3 1 2   1 3 4 2

SMPX

Multiplexed line

600 bits/s

'Frame' time

'Frame' time

Average bit-rate on line 1 = 120 bits/s
Average bit-rate on line 2 = 150 bits/s
Average bit-rate on line 3 = 120 bits/s
Average bit-rate on line 4 = 90 bits/s
Average total input bit-rate = 480 bits/s

Occupancy = 80%
Coding Efficiency = 80%
Throughput = 64%

SYNCHRONOUS MULTIPLEXING: *the speed of the multiplexed line is four times the speed of the input lines, i.e. 1200 bits/s.*

1 3 4 2 1 3 2   2   4 3 2 1   3 1 4 2

Occupancy = 36%
Coding Efficiency = 100%
Throughput = 36%

Multiplexed line

1200 bits/s

Figure 6.6 Comparison of Statistical Multiplexing and Synchronous Multiplexing

in the sequence, line 1, line 2... line 4. The second character must wait while the first is regenerated at the higher speed and transferred to a time slot on the multiplexed line. Under the circumstances shown in Figure 6.6, the occupancy of the multiplexed stream is 80% (16 out of 20 slots are filled), and throughput with respect to framed characters is 480 bits/s.

The addition of 2-bit identifiers to the 8-bit characters results in a coding efficiency of 80%. In conjunction with occupancy of 80%, this produces an overall efficiency of 64% in the multiplexed stream—i.e., throughput with respect to information bits is 384 bits/s.

(3) Comparison With Synchronous Multiplexing

If a synchronous multiplexer is presented with the traffic drawn in Figure 6.6, it will strip off the start and stop bits from each octet and, to demarcate the frames, will add one bit after every four octets. The multiplexed line will run at four times the speed of an input line plus the speed of the framing channel.

In four time slots there are potentially 32 information bits + 1 framing bit. Thus, the speed of the multiplexed line is

$$1{,}200 \times 33/32 = 1{,}237.5 \text{ bits/s.}$$

In 44 time slots, 16 are occupied. Thus, the occupancy is 16/44 = 36%, and the number of frame markers is 11. The maximum number of bits in 44 time slots is 352 + 11 = 363. In fact, the total number of information bits in 44 time slots is 128. Hence, throughput with respect to information bits is

$$1{,}237.5 \times 128/363 = 436 \text{ bits/s}$$

Throughput efficiency is 436/1,237.5 = 35%.

In contrast, if the line speed of the SMPX described in Section 6.1.4(2) were 1,200 bits/s, it would be able to handle twice as many lines of bursty data traffic. Care should be taken not to extrapolate the results too far. They are illustrative and true only for the particular situation postulated in Figure 6.6—including the statistics of the input traffic.

(4) Average Wait for Service

A statistical multiplexer can be modeled as an M/D/1 queueing system. **Figure 6.7** shows the M/D/1 queueing model of the SMPX shown in Figure 6.6. M/D/1 is shorthand for a queueing system in which the following exist:

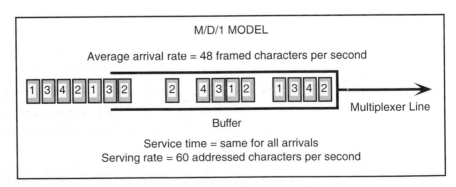

Figure 6.7 Statistical Multiplexing as a Queueing Model

- The arrival process is a Poisson distribution
- The service function is the same for all customers
- A single server handles arrivals on a first-come, first-served, basis.

For an M/D/1 queue, the average time waiting for service is

$$\overline{W} = \lambda/[2\mu(\mu - \lambda)] \text{ seconds}$$

where λ is the average number of framed characters arriving per second and μ is the average number of addressed characters transmitted per second.

In Figure 6.6, $\lambda = 48$ framed characters per second (i.e., 480 bits/s) and $\mu = 60$ addressed characters per second (i.e., 600 bits/s). Substituting in the expression introduced above, the average time waiting for service is 0.0333 second.

Figure 6.8 plots values of the average time waiting for service against the speed of the multiplexer line. At 1,200 bits/s, the average delay waiting for service is only 0.00222 second. As the speed of the multiplexer line slows toward 480 bits/s (the average total input bit rate), the delay increases dramatically.

Average results derived from the M/D/1 model are relatively simple to obtain and serve to illustrate the general behavior of a statistical multiplexer. However, the representation of the input process as a Poisson distribution (an infinite number of sources generating short messages at random) is a serious distortion of reality. This compromise is made to simplify the analysis.

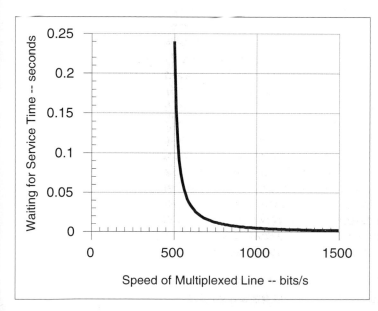

Figure 6.8 Waiting for Service Time as a Function of Speed of Multiplexed Line

When the line speed approaches the average speed of the arriving characters, the time spent in the queue increases dramatically.

(5) Average Delay Between Source and Destination

Because of the possibility of echoes, knowledge of the average delay between source and destination is important when voice signals are being handled and may be important when data are involved. To illustrate the problem, let's find the average delay on line 1 in Figure 6.6.

The framed characters are assembled in buffer storage at the input to the SMPX at the speed of line 1 (10 bits at 0.00333 second = 0.0333 second). They are read out as addressed characters at the speed of the multiplexed line (10 bits at 0.00167 second = 0.0167 second) in a first in, first out (FIFO) manner. If the multiplexer is empty when a framed character appears on line 1, it passes through with only the delays for assembly and read out (a total of 0.05 second). If the multiplexer is serving a framed character when the next one arrives, it must wait in the queue. [The average waiting time was derived in Section 6.1.3(4).]

If propagation delays are negligible, the total time for a framed character to pass from source to destination over the statistical multiplexer is

- Time for source DTE to send character on line 1 and assemble in buffer storage in SMPX = 0.0333 second
- Waiting for service time = 0.0333 second
- Time to read from source SMPX on to line (i.e., service time) = 0.0167 second
- Time to read from destination SMPX to destination DTE = 0.0333 second
- Total delay = 0.1167 second.

In contrast, to send a framed character from source to destination using TDM (i.e., no waiting for service time) requires 0.1167 − 0.0333 = 0.0834 second. Obviously, there is a tradeoff between transit time and transmission efficiency.

(6) Synchronous Transmission

SMPXs can be configured to use synchronous transmission on the common bearer. To do this, the offered traffic is organized into a frame in which the text portion contains the traffic sequence and their identifiers. Such a frame is shown in **Figure 6.9.** With four input lines, the multiplexer of Figure 6.5 can use a 2-bit code to identify them. In practical applications, these identifiers may be 4 or 5 bits long to allow 16 or 32 input lines to be handled at once.

(7) Data Concentrator

The principle of a data concentrator is shown in **Figure 6.10**.

- **Data concentrator**: a frame-buffering and multiplexing device that buffers frames received from several lower-speed lines until they can be sent to a central site on a single higher-speed line.

Data concentrators are likely to perform as primary sites with respect to inputs from DTE devices (polling, error checking, requesting resends, etc.) and as sec-

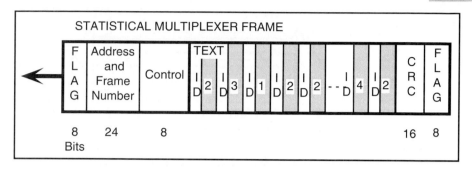

Figure 6.9 Statistical Multiplexer Frame for Synchronous Transmission

ondary sites with respect to the hosts they serve (responding to polls, resending frames, etc.).

6.1.5 INVERSE MULTIPLEXING

Earlier, I stated that a synchronous time-division multiplexer combines n lower-speed digital signals on a higher-speed bearer. In many ways, inverse multiplexing may be considered the reverse of this activity.

- **Inverse multiplexing**: creating a higher speed, point-to-point digital circuit from several independent, lower-speed digital channels that exist between the originating and terminating locations.

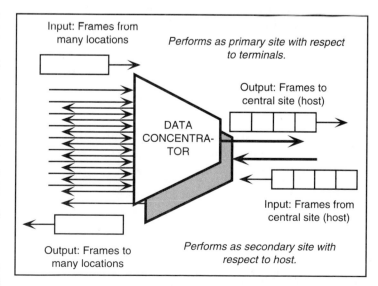

Figure 6.10 Principle of Data Concentrator

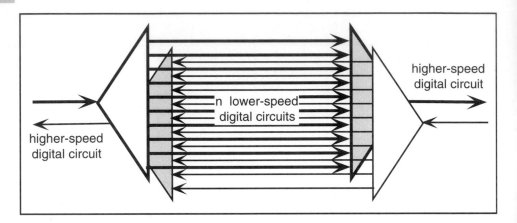

Figure 6.11 Principle of Inverse Multiplexer

The principle of inverse multiplexing is shown in **Figure 6.11**. Inverse multiplexers are used to send composite, higher-speed signals (such as 384 kbits/s videoconference signals) over lower-speed channels (such as 6 x 64 kbits/s channels). Inverse multiplexing is employed when a channel that matches, or exceeds, the higher-speed signal is not available, or when the higher-speed signal must be sent to a location served by lower-speed, switched lines. The inverse multiplexer establishes the lower-speed connections, segments the higher-speed datastream, and sends the data segments over the lower-speed lines to the destination. At the receiving location, an inverse demultiplexer receives the segments and reassembles them into the higher-speed stream—an action that may require reordering some segments and compensating for different channel delays. As carriers deploy higher-speed switched services, the need for inverse multiplexing diminishes.

REVIEW QUESTIONS FOR SECTION 6.1

1 Give two reasons why multiple signals must be carried on individual bearers.
2 Define multiplexing and demultiplexing.
3 Describe three basic techniques used to create several channels on a single bearer.
4 Describe two other techniques that are used to create several channels on a single bearer.
5 What is a channel bank?
6 What is a multiplexer?

7 Describe how 12 analog voice channels are combined by a *channel bank* to form a composite, multiplexed signal.

8 Distinguish among, and compare the performance of, octet-level multiplexing, bit-level multiplexing, and frame-level multiplexing.

9 Define synchronous multiplexing.

10 Describe statistical multiplexing.

11 Compare the performance of synchronous and statistical multiplexers.

12 How can you calculate the average wait for service at a statistical multiplexer?

13 When using statistical multiplexing, how can you calculate the average delay between source and destination?

14 How does a statistical multiplexer use synchronous transmission?

15 What is a data concentrator?

16 What is inverse multiplexing?

6.2 T–CARRIER MULTIPLEXING

T–carrier multiplexers produce composite signals that contain 24, 48, 96, 672, 2,016, and 4,032 digital voice signals. In addition, provided certain restrictions are observed, they transport data signals. The operating speeds of T–carrier equipment extend from 1.544 Mbits/s to 274.176 Mbits/s in four levels. While there were good reasons for the development of the entire family, today, the first (T–1, 24 voice channels, 1.544 Mbits/s) and third (T–3, 672 voice channels, 44.736 Mbits/s) levels dominate the T–carrier environment. Accordingly, while we will describe all members, we will emphasize T–1 and T–3.

6.2.1 T–1

In its original form, a T–1 system

- Employed channel banks that processed 24 analog voice signals to become 24, 64 kbits/s, PCM signals
- Octet multiplexed them to produce a bitstream of 1.544 Mbits/s
- Carried the signal on conditioned, four-wire circuits.

Figure 6.12 shows the format of a T–1 frame. It contains 24 octets that represent samples from 24 voice signals and a single framing bit that is used for synchronizing. In total, there are (8 x 24) + 1 = 193 bits per frame. Because each voice signal is sampled 8,000 times per second, the frame must repeat 8,000 times per sec-

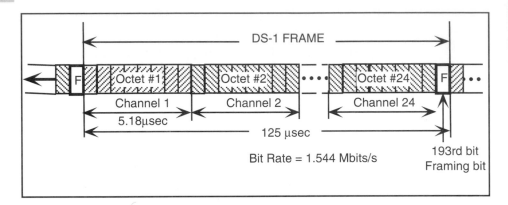

Figure 6.12 DS–1 Frame Format

A frame consists of one octet from each of 24 channels, together with a framing bit. There are 193 bits in a frame.

ond. Thus, the output bit rate is 193 x 8,000 = 1.544 Mbits/s. The signal at the output of the multiplexer is described as DS–1—i.e., digital signal level 1. The digital input signals are described as DS–0 signals—i.e., digital signal level 0.

(1) System Components

Figure 6.13 shows the general arrangement of the components required to implement a traditional, copper-based T–1 facility between a customer's premises and a carrier's central office. Digital signals are sent and received by the customer's digital PBX that includes equipment that performs the multiplexing and CSU [channel service unit, see Section 7.1.1(2)] functions. Twenty-four DS–0 signals are multiplexed to produce a DS–1 signal on the network side of the CSU. Connections are made over two pairs of wires with intermediate repeaters spaced at intervals of approximately 6,000 feet. The first and last repeaters in the T–1 line are spaced ≤ 3,000 feet from the central office and the customer's premises.

Because of the timing jitter that accumulates in long sequences of repeaters [see Section 5.1.2(4)], the maximum distance over which T–1 is used is 50 miles. Ninety-five percent of the time, T-1 facilities provide an error rate of 1 bit in error in 10^6 bits. Using 400 pairs, 200 T–1 systems can operate over a single, 900-pair cable.

(2) DS-1 Signals

DS–1 employs alternate mark inversion (AMI, see Section 3.1.6) signaling. Ones are signified by alternating positive and negative pulses, and 0s are signified by zero level. Because they are represented by the absence of pulses, long runs of 0s may cause the repeaters to lose synchronization. To maintain reliable working, engineering practices require that, on average, the signal stream must contain at least one 1 in 8 bits (12.5% 1s), and there cannot be more than 15 consecutive 0s.

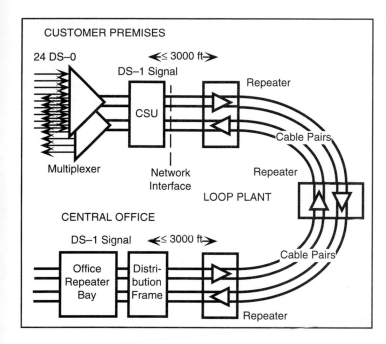

Figure 6.13 Outline of T–1 Facility Connecting Customer Premises to Carrier's Central Office

Twenty-four duplex DS–0 signals are combined into one duplex DS–1 signal.

The way in which this is achieved depends on whether the T–1 is employed exclusively in carrying voice, or in carrying data, or voice and data.

- **Voice only**: during active periods, the PCM quantizing procedure achieves the required *ones-density* automatically (see Figure 3.20). During idle periods, the transmitter sends the octets 00000010 (bit 7 forced) or 00011000 (bits 4 and 5 forced). In unassigned channels, the octet 11111111 is sent.

- **Data only,** or **Voice and data**: in unassigned channels, the transmitter sends the octet 00011000. The 0 in the eighth slot identifies it as a code introduced by the carrier. In assigned channels, substituting an arbitrary code in the middle of a user's datastream is unacceptable. Instead, one technique forces bit 8 in all user data time slots to 1. Therefore, the user data rate drops to 56 kbits/s—but the data *can* contain unlimited sequences of zeros. This is the technique employed in DDS [Digital Data Service, see Section 13.3.2(1)]. Another technique is tailored to providing a full 64 kbits/s *clear* channel. Phantom 1s are inserted into any sequence of eight 0s in such a way that the receiving equipment can detect and remove them. To do this, the pulse representing a phantom 1 is inserted in the same polarity as the previous (genuine) 1. This is a *violation* of the normal AMI sequence and is recognized by the system.

The most common scheme used to introduce bipolar variations is known as

- **Bipolar with 8 zeros substitution** (B8ZS): in a sequence of eight 0s, a bipolar violation is inserted at the fourth and seventh positions. Each violation is followed by a 1 in the next position (fifth and eighth positions). The receiving equipment is programmed to accept the sequence 000V10V1 as 00000000 so that the operation is transparent to the user. The upper signal waveform in **Figure 6.14** shows the AMI format for the sequence 011100000000011. The lower signal waveform shows the application of B8ZS. The sequence becomes 0111000V10V1011—a pattern that the receiver can detect and correct. Use of B8ZS removes the need to introduce special codes and other techniques to satisfy the 1s-density requirement, and eliminates the no more than 15-zeros requirement in the user's data.

A scheme that does not employ bipolar violations to achieve a full 64 kbits/s clear channel is known as

- **Zero-byte time-slot interchange** (ZBTSI): the transmitter maintains a buffer that is filled by successive groups of four frames (96 octets) from the T–1 stream. In the buffer, octets with all zeros are removed, and the remaining nonzero octets are consolidated at the back of the buffer. The first octet space at the front of the buffer is filled with a 7-bit binary number that gives the location (in the sequence of 96 octets) of the first all zeros octet. The eighth

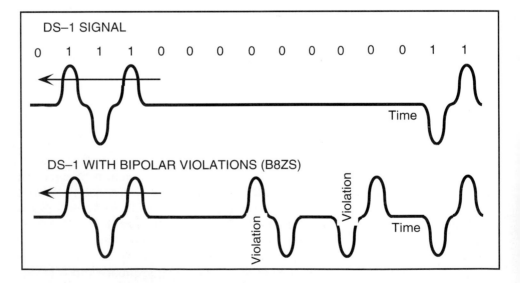

Figure 6.14 DS–1 B8ZS Signaling

Bipolar violations are introduced to prevent the occurrence of strings of eight 0s in the signal stream. The receiver recognizes the violations and restores the 0s.

bit in the octets is used to indicate whether there are more all zeros octets in the sequence. The second octet space is used to indicate the position of the second all zeros octet, and so on. (There are as many spaces as all zeros octets.) At the receiver, all zeros octets are reintroduced in the positions indicated. Because the processing delays the stream by 1.5 milliseconds, ZBTSI is less popular than B8ZS. Use of ZBTSI removes the need to introduce special codes and other techniques to satisfy the ones-density requirement, and eliminates the no more than 15 zeros requirement in the user's data.

(3) Framing and Channel Signaling—Voice Only

Over time, the framing of the DS-1 signal stream has evolved from the simple 193rd bit approach illustrated in Figure 6.11 to the use of sequences of framing bits to provide control patterns and error information. Two common arrangements are

- **Superframe format** (SF): also called D4 superframe format, 12 DS-1 frames form a superframe that contains 2,316 bits and takes 1.5 milliseconds to transmit at the DS-1 rate. The 12 framing bits form a pattern (100011011100) that is used for synchronization. In addition, they are divided into two sequences. The bits in the odd-numbered frames provide the sequence 101010; it is used by the receiving equipment to establish, maintain, or reestablish frame boundaries. The bits in the even-numbered frames provide the sequence 001110; it is used by the receiving equipment to identify frames 6 and 12. (Frame 6 occurs on a 1 preceded by three 0s—the first zero is the 12th bit from the previous frame—and frame 12 occurs on zero preceded by three 1s).

- **Extended superframe format** (ESF): 24 DS-1 frames form an extended superframe that contains 4,632 bits and takes 3 milliseconds to transmit at the DS-1 rate. The 24 framing bits are used to
  - synchronize terminal equipment (F-bits)
  - secure multiframe alignment (F-bits)
  - provide a 4,000 bits/s data link (FDL, facility data link) that carries application-dependent information and is used to acquire historical facility performance data (D-bits)
  - provide means of checking the error performance of the system (C-bits).

Figures 6.15 and **6.16** present views of SF and ESF. Twelve 193-bit frames are stacked one on the other to produce the SF diagram, and 24, 193-bit frames are stacked to produce the ESF diagram. Seen in this way, overhead bits and user octets are arranged in columns that represent individual bitstreams. To read the serial DS–1 stream, start in the left-hand upper corner, proceed along frame 1, return to the left-hand side and continue along frame 2, etc., until you reach the bottom, right-hand corner. The sum of the 24-octet bitstreams (i.e., the 24 DS–0 sig-

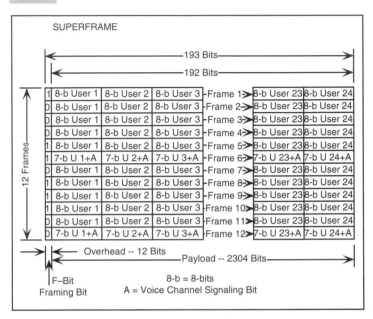

Figure 6.15 T–1 Superframe Format

A superframe consists of 12 T–1 frames. Coding is introduced into the sequence of framing bits.

nals) is called the *payload*. In frames 6 and 12 of SF, and frames 6, 12, 18 and 24 of ESF, bits are robbed from the user to provide a 1,333 bits/s signaling channel for each voice channel. Used in pairs (SF and ESF), they can report 4 states, and used in quadruples (ESF only), they can report 16 states. Messages include *idle, busy, ringing, no-ring* and *loop open*. Frame alignment is crucial to interpreting multiplexed streams. Loss of frame alignment renders the entire information stream unreadable.

(4) Framing and Channel Signaling—Data Only

For data-only applications, the general formats of SF and ESF are as described above except that per channel, robbed bit signaling is eliminated in favor of common channel signaling. Customers have access to all but the 24th channel that is reserved for synchronizing and signaling chores [see Section 6.2.1(5)].

(5) ESF Error Checking

The error-checking capability of ESF employs a 6-bit cyclic redundancy check (CRC). At the sending end, the polynomial formed by the 4,632 superframe bits is divided by a 7-bit polynomial (1000011) to produce a 6-bit remainder (the frame check sequence, FCS). It is transmitted in the 193rd bit of frames 2, 6, 10, 14, 18, and 22. At the receiving end, the CRC calculation is repeated. If the FCS produced at the receiver matches the FCS received from the sender, no error was introduced by transmission. The CRC process and the bit patterns in the framing information are used to define a set of events that describe the error performance of the link. Some of them are

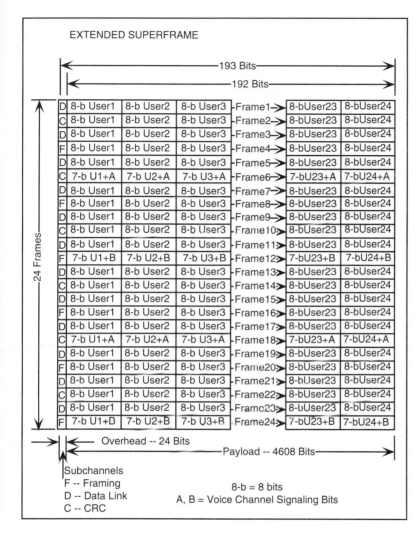

Figure 6.16 T–1 Extended Superframe Format

An extended superframe consists of 24 T–1 frames.

- **Out-of-frame** (OOF) **event**: when 2 out of 4 consecutive framing bits are incorrect

- **CRC-6 error event**: when the received FCS does not match the FCS calculated at the receiver

- **ESF error event**: an OOF event, CRC-6 event, or both

- **Errored second** (ES): any second in which one, or more, CRC-6 events are detected

- **Bursty second** (BS): any second in which from 2 to 319 CRC-6 error events are detected

- **Severely errored second** (SES): any second in which from 320 to 333 CRC-6 error events are detected, or one OOF event occurs

- **Failed seconds** (FS) **state**: state achieved after 10 consecutive SESs have occurred. The state remains in effect until the facility transmits 10 consecutive seconds that are not severely errored.

Figure 6.17 shows the general arrangement of a T-1 facility operating with ESF. Channel Service Units (CSUs) in the ESF facility store the number of error events detected. Using the facility data link, controllers poll the CSUs for data that are used to establish the level of performance being achieved. This is important information for billing, and for scheduling maintenance.

6.2.2 EVOLUTION OF T-1 SYSTEMS

Originally developed to multiplex and transport voice signals, T-1 systems have evolved to general-purpose, voice and data, digital multiplexing, and transmission systems.

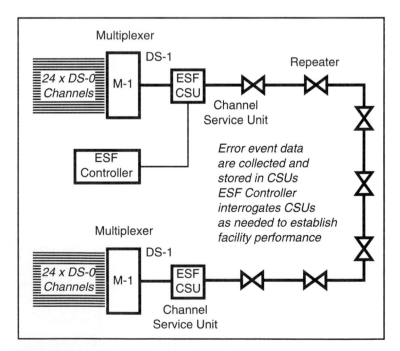

Figure 6.17 T–1 Facility With ESF Error Checking

ESF contains a CRC that checks for transmission faults. This enables the carrier to measure the level of performance of the link.

(1) Common Types

Traffic carried by T–1 systems includes

- Digital voice at 64, 32, or 16 kbits/s on DS-1 channels
- Digital data at 56 kbits/s or multiples of 64 kbits/s (i.e., n x 64 kbits/s, where $n = 1, 2$, etc.) on DS-1 channels
- Subrate data at 2.4, 4.8, 9.6, or 19.2 kbits/s on DS-0 channels.

Accommodating these signals on DS-1 lines requires a range of multiplexers. Some of them are

- **M24**: original T–1 system. Customer access = 24 voice channels. Per channel, signaling uses 8th bit robbing in every sixth frame. Framing uses 8,000 bits/s. Output = DS–1.
- **M44**: customer access = 44 voice channels that are converted to 44, 32-kbits/s voice channels. They occupy 22 DS–0 channels. The remaining two DS–0 channels are used for common channel signaling. Framing uses 8,000 bits/s. Output = DS–1.
- **M48**: customer access = 48 voice channels that are converted to 48, 32-kbits/s channels. They occupy 24 DS–0 channels. Per channel signaling uses bit robbing. Framing uses 8,000 bits/s. Output = DS–1.
- **M88**: customer access = 88 voice channels that are converted to 88, 16-kbits/s voice channels. They occupy 22 DS–0 channels. The remaining two DS–0 channels are used for common channel signaling. Framing uses 8,000 bits/s. Output = DS–1.
- **M96**: customer access = 96 voice channels that are converted to 96, 16-kbits/s channels. They occupy 24 DS–0 channels. Per channel signaling uses bit robbing. Framing uses 8,000 bits/s. Output = DS–1.
- **M24/64**: customer access = 23, 64-kbits/s data channels. Employs B8ZS. Uses one 64-kbits/s channel for data synchronization, alarms, and signaling. Framing uses 8,000 bits/s. Output = DS–1. Known as *clear channel capability* (CCC) or *clear 64* (i.e., the customer has full use of the 64 kbits/s channel), it is used with ESF.
- **M24/56**: customer access = 23, 56-kbits/s data channels. Output = DS–1. The bit layout for a 56 kbits/s data superframe is shown in **Figure 6.18**. The eighth bits in the 23 user time slots are set to 1 to ensure meeting the ones-density requirement. The 24th time slot is used for
  - data synchronizing
  - alarm channel for reporting failure in the outgoing direction (bit 6, the Y bit)
  - 8 kbits/s data channel for use by the digital transmission surveillance system (bit 7, the R bit) employed in digital data service (DDS).

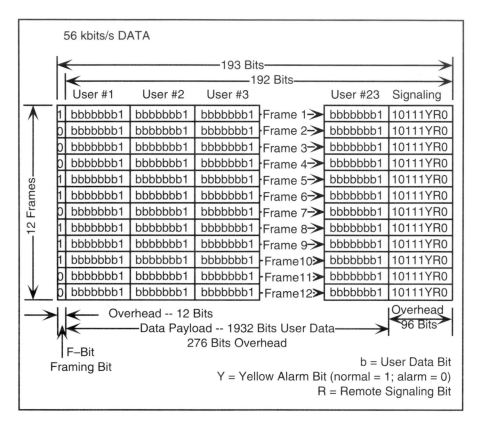

Figure 6.18 56-kbits/s Data Superframe Format
The eighth bit in each octet is set to 1 to prevent strings of eight or more 0s in the data. Consequently, only 7 bits per octet are available for customer data. In addition, the 24th channel is reserved for carrier uses.

(2) Subrate Data

In DDS [see Section 13.3.2(1)], subrate data (2.4, 4.8, 9.6, and 19.2 kbits/s) are multiplexed on DS-0 channels that are then multiplexed with other DS-0 channels on a DS-1 line. To connect

- A **single subrate station** to a M24/56 multiplexer, bytes are repeated as necessary to match the sender's speed to the speed of the DS–0 line. In this case, it is known as a DS–0A signal.

- **Several subrate stations** over the same DS–0 channel, subrate signals are multiplexed in a subrate multiplexer (SRM) to produce a stream of octets that contains a subrate synchronizing bit (bit 1) and 6 data bits (bits 2 through 7). The eighth bit is set to 1 to ensure meeting the 1s-density requirement. Thus, the 64 kbits/s signal contains 48 kbits/s of customer subrate data, 8 kbits/s of subrate synchronizing, and 8 kbits/s of ones. It is described as a DS–0B signal.

Some typical SRMs are

- **M5**: customer access = 5, 9.6-kbits/s data channels. Output = DS–0B.
- **M10**: customer access = 10, 4.8-kbits/s data channels. Output = DS–0B.
- **M20**: customer access = 20, 2.4-kbits/s data channels. Output = DS–0B.

SRMs are used in combination with M24/56 multiplexers to produce DS–1 signals. An example is a combination of 23 M5 multiplexers feeding an M24/56 multiplexer

- 23 **M5** + **M24/56**: customer access = 115, 9.6-kbits/s data channels to 23 M5 SRMs. Output from each M5 = DS–0B. Input to M24/56 multiplexer = 23 DS–0B. Output = DS–1. The bit layout for a 56-kbits/s, subrate data super-frame is shown in **Figure 6.19.**

(3) Unchannelized T–1

Some data are not organized in octets. For instance, the output of a video codec may be a continuous stream of bits composed of variable length code words (entropy coding, see Section 3.3.4). In addition, a customer may combine subrate data in one continuous stream to improve on the 71.5% efficiency associated with the subrate data superframe in Figure 6.19. For transmission over a DS–1 line, the bitstream can be divided into segments of 192 bits to which framing bits are attached. Since neither the sending nor the receiving equipment makes use of the boundaries that define channels, the technique is described as *unchannelized*. Incompatible with public networks, unchannelized operation is confined to private connections. It is the user's responsibility to ensure that the bitstream complies with T–1 transmission requirements.

6.2.3 T–CARRIER HIERARCHY

Called the T-carrier hierarchy, the facilities shown in **Figure 6.20** combine DS–0 signals into progressively higher speed streams.

(1) Hierarchy

The four levels of the hierarchy are designated T–1, T–2, T–3, and T–4. In addition, there is an intermediate level, T–1C between T–1 and T–2, and another intermediate level, T–4NA between T–3 and T–4. In conformity with the DS–0, DS–1 nomenclature introduced for the input and output signals of T–1, the output signals are designated DS–1C, DS–2, DS–3, DS–4NA, and DS–4, and the multiplexers are designated

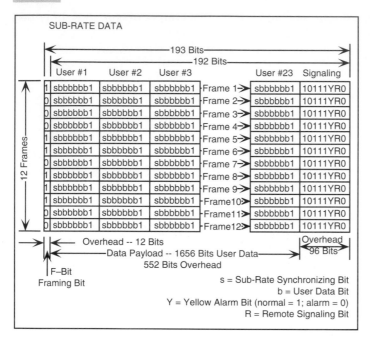

Figure 6.19 Subrate Data Superframe Format

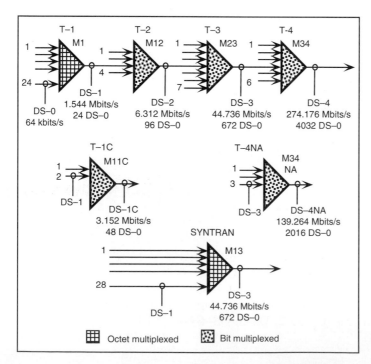

Figure 6.20 T–Carrier Hierarchy

Only the first level is octet multiplexed. All other levels are bit multiplexed, except SYNTRAN—the T–3 synchronous system used in private networks.

- **M1**: combines 24 DS–0s into one DS–1, also known as M24, M24/56, and M24/64.
- **M1C**: combines 2 DS–1s into one DS–1C
- **M12**: combines 4 DS–1s into one DS–2
- **M13** combines 28 DS–1s into one DS–3
- **M23**: combines 7 DS–2s into one DS–3
- **M34NA**: combines 3 DS–3s into one DS–4NA
- **M34**: combines 6 DS–3s into one DS–4.

For the multiplexing process to succeed, DS–0 octets must be formed in 125 μseconds. Further, they must arrive at the first-level multiplexer (M1) in synchrony with octets from other sources so that they can be octet-multiplexed into DS–1 frames. In turn, individual frames of the DS–1 stream must be formed in 125 μseconds. They must arrive at the second-level multiplexer (M12) in approximate bit-synchronism with frames from other DS–1 sources, and their speeds must be equalized before being bit-multiplexed into a DS–2 frame. Proceeding to the third level, DS–2 frames must arrive at the third level multiplexer (M23) in approximate bit-synchronism with frames from other DS–2 sources, and their speeds must be equalized before being bit-multiplexed into a DS–3 frame, etc. The activity is described as

- **Plesiochronous multiplexing**: a multiplexing process in which signals are generated at nominally the same digital rate and their significant instants occur at nominally the same time.

With signals homing on the multiplexers from many locations, maintaining frame or bit-synchronism is no easy task. Nevertheless, so that each octet shall arrive at its destination without error, the beginning of each frame must be identified unambiguously, the octets they contain must be demultiplexed exactly, and the bits decoded perfectly. Providing satisfactory service to all of the 4,032 users of a T-4 facility allows no room for errors.

(2) T–1C

T–1C converts a T–1 transmission link to one that transports twice the number of bits per second without changing the physical transmission arrangements. It operates over the same wire pairs with duobinary signaling [see Section 4.2.1(6)], the same repeater spacing (6,000 feet), and the same limitation ($\approx$ 50 repeaters). Using 300 pairs, 150 T–1C systems can operate over a single, 900-pair cable. This is a 50% increase over the performance of T–1. The DS–1C signal is formed by interleaving two DS–1 streams, bit by bit, and adding a control bit after every 52 bits (26 from each DS–1). To maintain the ones-density, one DS–1 stream is logically inverted before multiplexing. In addition, because the input DS–1 streams are likely to be operating at slightly different speeds, extra pulses (stuffing pulses) are inserted in specific locations in each stream to equalize their rates and raise the composite to 3.09 Mbits/s. With the insertion of one control bit for every 52 DS–1 bits, the final speed is 3.152 Mbits/s. At the receiver, the stuffing bits are removed before demultiplexing the DS–1C stream into two DS–1 streams.

(3) T–2

In M12, the DS–2 signal is formed by interleaving four, DS–1 streams, bit by bit, and adding a control bit after every 48 payload bits. To maintain the ones-density, bits in DS–1 streams 2 and 4 are logically inverted before multiplexing. In addition, because the DS–1 streams are likely to be operating at slightly different speeds, stuffing pulses are inserted in specific time slots in each stream to equalize their rates and raise the composite to 6.183 Mbits/s. With the insertion of one control bit for every 48 DS–1 bits, the final speed is 6.312 Mbits/s. At the receiver, the stuffing bits are removed before demultiplexing the DS–2 stream into four DS–1 streams. Because T–2 requires special low-capacitance cable and repeaters spaced at approximately 3-mile intervals, it is not widely deployed.

(4) T–3

In M13, the DS–3 signal is formed from 28 DS–1 signals by bit-interleaving groups of four DS–1s to create seven DS–2 signals as described above. In turn, the seven DS–2 signals are bit-interleaved to create one DS–3 signal. In the conversion of seven DS–2 to DS–3, a control bit is added for every 84 payload bits, and bits are stuffed in specific slots to equalize the speeds of the individual streams and achieve a DS–3 speed of 44.736 Mbits/s. The overhead functions associated with M13 consume 1.504 Mbits/s. DS–3 employs bipolar signaling and may be transmitted over coaxial cables, microwave radios, or optical fibers. DS–3 frame formats are known as traditional DS–3 (also called M23), C-bit parity, and SYNTRAN (called synchronous M13). The first two are asynchronous techniques (i.e., actions are driven by the state of the signal stream). The third is a synchronous technique (i.e., actions occur in synchrony with an external timing source). They are described as follows

- **Traditional DS–3**: the DS–3 signal is partitioned in frames containing 4,760 bits. Called M-frames, they consist of seven subframes of 680 bits each. The subframes are divided into eight blocks of 85 bits. In these blocks, the first bit is a control bit and the other 84 are payload bits. Each frame, then, includes 54 overhead bits that are used to perform frame alignment, subframe alignment, performance monitoring, alarm and status reporting, and other functions. **Figure 6.21** shows the format of a DS–3 frame and identifies some of the uses of the control bits. To adjust the speeds of the seven DS–2 signals, in the last block of the seven subframes, a single bit can be stuffed into one, some, or all of them. Setting the three C bits to 1 in the subframe in which it occurs indicates the presence of a stuffing bit. Because the DS–3 rate is not an integral multiple of the DS–1 rate ($44.736 \div 1.544 = 26.974$), and bits

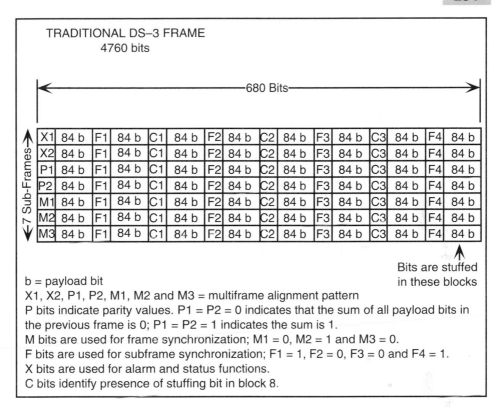

Figure 6.21 Traditional DS–3 Frame Format

are stuffed as needed, individual DS–1 frames *slide* relative to the DS–3 frames. Thus, to demultiplex a specific DS–1 signal, the entire stream must be demultiplexed.

- **C-bit parity**: the format of the traditional DS–3 frame is preserved. However, the seven DS–2 signals created in the multiplexing process are generated at exactly 6.306272 Mbits/s and a stuffing bit is inserted in every subframe. The result is a DS–3 signal at exactly 44.736 Mbits/s. Because they are not needed to report the presence or absence of stuffing bits, the 21 C bits can be used for other purposes. For instance, the three C bits in subframe 3 indicate the parity status of the preceding frame at the origination point of the stream. At the receiver, comparison of the locally calculated value (P1, P2) with this value indicates whether an error has occurred. Other bits are used to provide data links, alarms, and status information.

- **SYNTRAN**: the format of the traditional DS–3 frame is changed, and 699 M frames are combined into a synchronous superframe. The DS–1 frames have fixed locations within the superframe (699 M-frames occupy 74.375 milliseconds—the time required for 595 DS–1 frames). The principal purpose is to provide simple, reliable access to all DS–0 and DS–1 signals contained in the SYNTRAN DS–3 signal. To do this, *octets* from 28 DS–1 channels of exactly 1.544 Mbits/s are multiplexed directly to DS–3. No stuffing bits are used—a circumstance that frees approximately 66 kbits/s for other overhead functions. Because each M-frame contains 588 octets, and a full set of frames (i.e., 28) from the 28 DS–1 input signals requires 672 octets, a 125-µs segment of the input signals occupies more than one M-frame. Figure **6.22** shows how the 672 octets might occupy parts of two frames. The C bits, and the bandwidth no longer occupied by stuffing bits, are used for data link, CRC error checking, and other purposes.

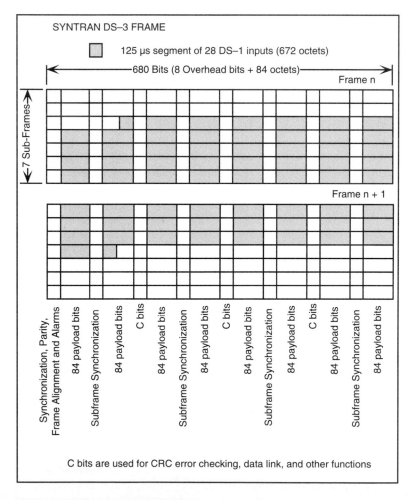

Figure 6.22 SYNTRAN DS–3 Format

Principally, traditional DS–3 is used by carriers in public networks, and SYN-TRAN DS–3 is used in private networks.

(5) T–4

In M34, six DS–3 signal streams are combined into a DS–4 stream that operates at 274.176 Mbits/s to create T–4, a high-density multiplexing system that is the natural extension of the T–carrier family. However, it is little more than a definition. An intermediate step, DS–4NA has been standardized as a 139.264 Mbits/s stream that is formed from three DS–3s. With SONET and SDH available (see Section 6.4), it is likely that neither DS–4, nor DS–4NA, will see much use in the future.

(6) Comparison of Levels in North America, Europe, and Japan

Five-level multiplexer hierarchies have been standardized in Europe and Japan. **Figure 6.23** shows how the number of DS–0 signals that North American, European, and Japanese multiplexers carry varies with level and region. To make this comparison, M4NA has been placed at level 4 and M4 has been placed at level 5. At level 1, the three are synchronous, and in Japan and North America, the num-

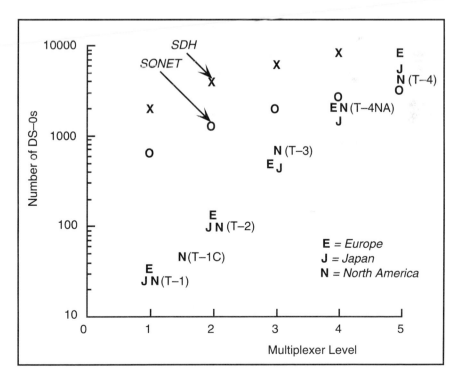

Figure 6.23 Comparison of North American, European, and Japanese Multiplexing Hierarchies

ber of DS–0s they carry is the same (24). In Europe, the first level (E–1, 2.048 Mbits/s) multiplexes 32 DS–0 channels. Of these, 30 channels carry customer traffic, and 2 channels are used for supervision. At level 2, both Japanese and North American multiplexers carry 672 DS–0 channels. However, in Japan, octet multiplexing is employed and the process is synchronous, while in North America, bit multiplexing is employed and the process is asynchronous. In Europe, the second-level bit multiplexes 128 DS–0s. At level 3, with the exception of SYNTRAN M13, the multiplexers employ bit multiplexing, carry different numbers of DS–0s, and are asynchronous. At levels 4 and 5, the multiplexers employ bit multiplexing, carry different numbers of DS–0s, and are asynchronous. The lines at the top of the chart indicate the numbers of DS–0s multiplexed by different levels of SONET and SDH (see Section 6.4).

REVIEW QUESTIONS FOR SECTION 6.2

1   Describe the original form of a T–1 system.
2   Describe the format of a T–1 frame. How many bits does it contain?
3   Justify the output bit rate of 1.544 Mbits/s.
4   Describe the repeater spacing in a T–1 system.
5   What is the maximum distance over which T–1 is used? Why?
6   How many T–1 systems can operate over a 900-pair cable?
7   What pulse format does T–1 use?
8   What does the term ones-density mean with respect to T–1 signaling?
9   How is the ones-density requirement maintained when a T–1 system carries voice only, carries data only, and carries voice and data?
10   Describe B8ZS and ZBTI. For what are they used?
11   Describe the superframe format for voice channels. What uses are made of the framing bits?
12   Describe the extended superframe format for voice channels. What uses are made of the framing bits?
13   If the channels are used for data, what modifications are made to SF and ESF formats?
14   Explain how the 6-bit CRC is obtained in ESF.
15   How is the error performance of a T–1 link with ESF described?
16   Name some common signals carried on a T–1 system.
17   Distinguish among M24, M44, M48, M88, and M96 voice multiplexers.
18   Distinguish between M24/64 and M24/56 data multiplexers.
19   Distinguish between DS–0A and DS–0B signals.

20 Distinguish among M5, M10, and M20, multiplexers.

21 What is unchannelized T–1? What are the responsibilities of the user?

22 List the six levels of the T–1 hierarchy. How are their output signals designated? How are the multiplexers that form the six levels designated?

23 Define plesiochronous multiplexing.

24 How is ones-density maintained in T–1C?

25 How is ones-density maintained in T–2?

26 What are the components of the 6.312 Mbits/s stream in T–2?

27 Why is T–2 not widely deployed?

28 What are the components of the 44.736 Mbits/s stream in T–3?

29 What pulse format does T–3 employ?

30 Distinguish among traditional DS–3, C-bit parity, and SYNTRAN formats. Why, and where, are they used?

31 Compare multiplexing levels in North America, Europe, and Japan.

6.3 CUSTOMER ACCESS LINES

For many years, telephones have been connected to the local switch by means of pairs of wires called *customer loops*; today, most telephones are still attached in this manner, as are many personal computers and other digital devices.

6.3.1 OUTSIDE PLANT

As far as possible, customers are grouped into convenient areas that contain an interface (or cross-connect) frame. Called the *serving* area, it is shown in **Figure 6.24**. Customer loops are constructed from three types of cables. On the switch side of the serving-area interface (SAI), connections are made from the central office (CO) distribution frame to SAIs by means of feeder cables (wire pairs or optical fibers). On the customer side of the SAI, connections are made through distribution cables to service access points (SAPs) such as pedestals. The final connection from the pedestals to the premises is made by drop cables. All the facilities situated between the distribution frame of the serving office and the customer's premises constitute *outside plant* (OSP).

(1) Resistance Design

Close to the switch, the feeder cables contain many wire pairs. As a consequence, to conserve duct space, and reduce cost, they employ small diameter wires (26 AWG). In the outlying feeder cable sections, the number of wire pairs

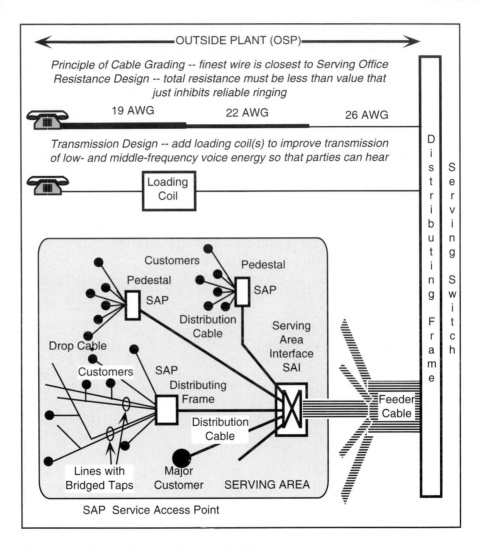

Figure 6.24 Wire Facilities in the Outside Plant

Outside plant is the name given to all facilities between the customer and the distributing frame in the carrier's serving office.

decreases, and the diameter of the individual wire pairs may be increased (e.g., to 22 and 19 AWG). The combination of wire sizes is chosen so that the total resistance of the loop does not exceed 1,100 to 1,300 ohms, the value that allows sufficient current to flow from the serving office battery for ringing. Known as *resistance design*, this technique ensures reliable signaling to customers situated at a considerable distance from the switch (30,000 to 50,000 feet).

(2) Transmission Design

Since there is no point to being able to signal if the parties cannot understand each other, inductance is added to long loops to reduce the voice band attenuation of the connection. Known as *transmission design*, the technique inserts loading coils at fixed intervals to improve the transmission of low- and middle-frequency voice energy. In **Figure 6.25**, loop length is plotted as a function of use (business or residential) and the percentage of loops that are no longer than this length. Approximately 10% of residential loops exceed 18,000 feet and are loaded. Very few business loops require loading.

To provide satisfactory service to the user, a loop must satisfy both resistance and transmission design parameters. The concepts of resistance and transmission design become obsolete as digital loop carrier equipment (see below) is installed, and individual physical loops are replaced by channels contained in multiplexed streams.

(3) Two-Wire and Four-Wire Connections

Most connections in the OSP are *two-wire* (2W)—i.e., the user's terminal is connected to the CO by a pair of wires that constitutes a duplex circuit and carries information in both directions. In some situations, two pairs of wires, one for each direction of transmission are employed. *Four-wire* (4W) connections are used when the distance is great enough to require repeaters or when the existence of two signals on the same bearer will cause them to interfere. Exclusively, on the

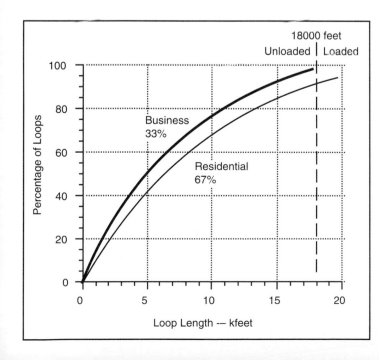

Figure 6.25 The Average Percentage of All Loops Shorter Than a Given Length

Approximately one-third of all loops connect to business customers, and two-thirds connect to residential customers.

network side of the CO, four-wire connections are employed. The 2W/4W transition is made on the subscriber line interface in the switch [see Section 9.6.1(2)] in the CO. In addition, a 2W/4W transition is made in the telephone instrument. A *hybrid* transformer is used to derive the talking and speaking connections (2 x 2W) for the microphone and earphone from the 2W circuit to the central office (see Figure 10.1). Today, although many channels are carried by radio and optical fiber, they are still described as two-wire or four-wire connections.

(4) Subscriber Line Bandwidth

It is usual to think of telephone lines as narrowband. Actually, it is the digitizing process, not the physical line that limits voice to 4 kHz. A subscriber loop, per se, can support a distance–bit-rate product of around 70 Mbits/s x 1,000s of feet. **Figure 6.26** shows data from several digital subscriber line systems (see Section 6.3.3) matched to a distance–bit-rate product of this value.

6.3.2 DIGITAL LOOP CARRIER

Even though they could increase the number of channels and improve speed, for many years, because of the marginal reliability of electronic devices, it was anathema to use active devices in customers' loops. However, the proven reliability of

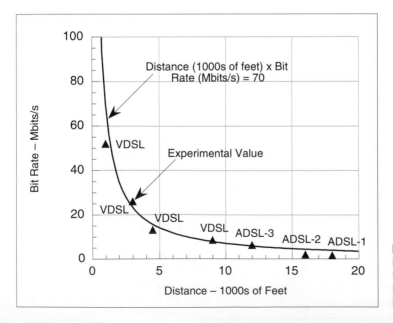

Figure 6.26 Distance–Bit-Rate Performance of Unshielded Twisted Pair Based on Digital Subscriber Line Performance

solid-state devices, and the digitalization of the network, changed this situation and forced the introduction of digital loop carrier systems.

- **Digital loop carrier** (DLC): a carrier system that provides time-division multiplexing on subscriber loops and operates between the user's premises or the serving area interface and the serving central office.

Also known as *digital channel banks* (D–channel banks) and *subscriber loop carrier* (SLC) systems, DLCs provide remote monitoring capabilities, perform concentration, and permit more channels to be implemented over existing loops. Because they operate in the loop plant, DLCs provide standard support functions for the user's terminal—i.e., battery feed, overvoltage protection, ringing, supervision, coding and filtering, hybrid, and testing (known by the acronym BORSCHT).

DLCs permit the use of digital technology in the feeder and distribution plant and the substitution of optical fibers for wire cables. In principle, they transport multiplexed digital voice and data signals between central office terminals (COTs) located at the serving office, and remote terminals (RTs) that provide digital services to a distinct geographical area known as a *carrier serving area* (CSA).

A typical digital loop carrier arrangement is shown in **Figure 6.27**. Each COT and RT handles four T-1 facilities (or one T-2 facility) to provide 96 DS–0 channels. To serve a major concentration of business subscribers, 28 T-1 facilities (a T-3 facility) may be multiplexed on optical fiber to provide 672 DS–0 channels. They are fed through seven COT/RT pairs. In some digital switches, the COT is incorporated within the switching electronics.

Contemporary digital loop carrier systems are

- **D4–channel bank**: multiplexes 48 or 96 individual voice or data channels over DS–1, DS–1C, or DS–2 facilities
- **SLC–96**: multiplexes 96 individual voice or data channels over four DS–1 facilities or a DS–2 facility. Includes the option to concentrate 48 channels on each DS–1 facility.
- **D5–channel bank**: multiplexes 96 individual voice or data channels over four DS–1 facilities or a DS–2 facility. Includes a common control subsystem that manages up to 20 units and supports fiber connections to the user's premises.
- **Automated digital terminal equipment system** (ADTS): multiplexes individual voice and data channels over DS–1, DS–2, or DS–3 facilities. ADTS provides
  - improved monitoring and maintenance functions
  - routing capability
  - clear–64 and ESF capability
  - fiber connections to the user's premises
  - SONET connection support.

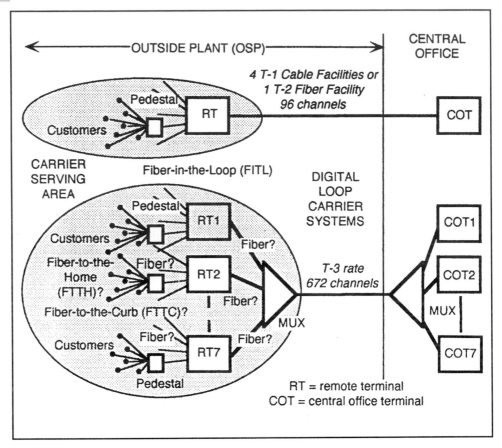

Figure 6.27 Example of Digital Loop Carrier Facilities

6.3.3 DIGITAL SUBSCRIBER LINE

The original digital subscriber line system was developed to provide ISDN Basic Rate access (see Section 10.4.2). Since then, demand has grown for subscriber access at higher speeds for other purposes. The result is there are now at least five different digital subscriber line systems with overlapping characteristics.

(1) ISDN Digital Subscriber Line

Called DSL (Digital Subscriber Line), it provides duplex service over a single, unshielded twisted pair (UTP) of 24-gauge, copper wire. DSL uses echo cancellers at each end of the line. With a bit rate of 160 kbits/s (160 kbits/s = 2 x 64 kbits/s B-channels + 16 kbits/s D-channel + 16 kbits/s control), it operates up to

18,000 feet. To achieve this performance, 2B1Q (2 binary, 1 quaternary) coding is used (2B1Q signal format is shown in Figure 3.8). In the 160 kbits/s signal streams

- 144 kbits/s are available for message transport (2B + D channels)
- 12 kbits/s are used for framing and synchronization
- 2 kbits/s are used for CRC error checking
- 2 kbits/s are used for a supervisory data channel.

Extensions of the technology are used to provide

- **Digital-added main line** (DAML): implements a second circuit over a single pair of wires for circuit expansion purposes
- **Universal digital channel** (UDC): implements six digital circuits on a single pair of wires.

Examples of the use of 2B1Q-configured DSLs are given in **Figure 6.28**.

(2) High Data Rate Digital Subscriber Line

Called HDSL (High data rate Digital Subscriber Line), it provides duplex service over two unshielded twisted pair at 1.544 Mbits/s (DS–1). Each pair operates at 784 kbits/s and employs echo cancellers to achieve duplex transmission. With 24-gauge copper pairs, service is limited to 12,000 feet—the nominal limit of the CSA (see Figure 6.27). HDSL provides customers with *repeaterless* T–1 service from the CSA interface over pairs that do not require special engineering.

(3) Single-Line Digital Subscriber Line

Called SDLS (Single-line Digital Subscriber Line), it provides duplex service over one unshielded twisted pair at 1.544 Mbits/s and a 4-kHz telephone channel. With 24-gauge copper pairs, service is limited to 10,000 feet. Because it uses only one pair, SDLS is suited to residential connections.

(4) Asymmetric Digital Subscriber Line

Called ADSL (Asymmetric Digital Subscriber Line), it transports simplex 1.5, 6.3, or 13-Mbits/s channels downstream, a duplex 16 to 640-kbits/s channel, and a 4-kHz duplex voice channel. Originally intended to carry 1, 2, or 4 television signals encoded according to MPEG2 digital video compression standards [see Section 3.5.3(4)], ADSL can be used in many applications that require a broadband, high-speed channel from network to user. Three levels of service are possible.

- **ADSL–1**: operates at 1.544 Mbits/s over UTP loops of 24-gauge copper wire up to 18,000 feet. Supports one MPEG2 video channel, a duplex data channel, and a duplex voice channel.

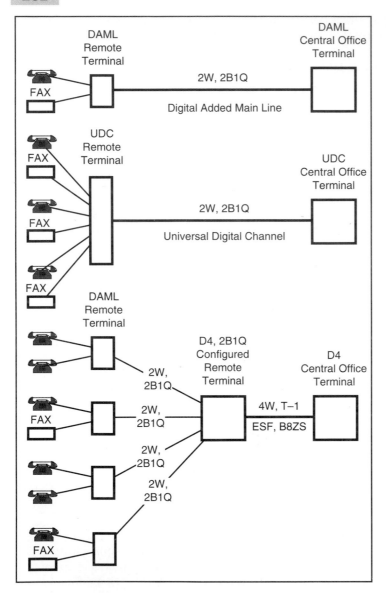

Figure 6.28 Typical Applications of ISDN Digital Subscriber Lines and T–1 Carrier

ISDN DSL uses 2B1Q pulse format.

- **ADSL–2**: operates at 6.312 Mbits/s over UTP loops of 24-gauge copper wire up to 12,000 feet. Supports two MPEG2 video channels, a duplex data channel, and a duplex voice channel.

- **ADSL–3**: operates at 9 Mbits/s over UTP loops of 24-gauge copper wire up to 9,000 feet. Supports four MPEG2 video channels, a duplex data channel, and a duplex voice channel.

ADSL enables the local exchange carrier (LEC) to offer switched television service to its subscribers. **Figure 6.29** shows the principle of what is called *video dial tone*. For all three versions, fibers are employed to transport the digital signals to the ADSL interfaces at the SAI.

(5) Very High Data Rate Digital Subscriber Line

Called VDSL (Very high data rate Digital Subscriber Line), it transmits downstream at 12.96, 25.82, or 51.84 Mbits/s [STS–1, see Section 6.4.1(1)]. An upstream digital channel has speeds between 1.6 Mbits/s and 51.84 Mbits/s, and there is a 4 kHz duplex voice channel. Over a 24-gauge copper pair, service is limited to

- 4,500 feet at 12.96 Mbits/s
- 3,000 feet at 25.92 Mbits/s
- 1,000 feet at 51.84 Mbits/s.

The very high speed data channel incorporates forward error correction [Reed-Solomon coding, see Section 7.4.2(5)] to guard against burst errors, and employs ATM format (see Section 9.4.1).

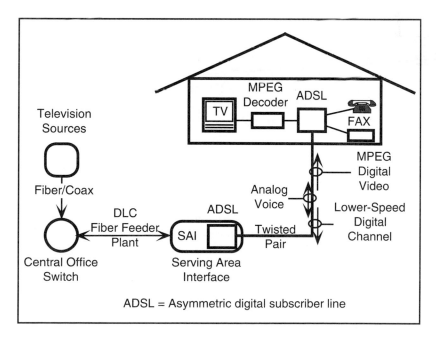

Figure 6.29 Principle of Video Dial Tone Residential Connection

6.3.4 Voice, Data, and Television

The majority of households in the United States is connected to two networks:

- A **switched** network over which they receive local voice and data transport services from the serving LEC
- A **nonswitched** network over which they receive television from the serving CATV company.

With the continuing evolution of the regulatory environment, the possibility of combining the two networks is being explored. No easy task, it requires the reconciliation of dissimilar technologies—the distribution of television signals over cable is an analog, FDM-based activity, while the distribution of interactive voice and data signals is a digital, TDM-based activity. In the near future, they must remain so, for a consumer-oriented television service must be compatible with existing receivers, and local voice and data transport must be compatible with digital distribution facilities. The result is a hybrid network that contains both optical fiber (a point-to-point, digital bearer) and coaxial cable (a point-to-many, analog bearer).

(1) Hybrid Fiber and Coaxial Cable Network

Figure 6.30 shows a *fiber and coax* network that provides duplex voice and data circuits and simplex broadband channels. It employs optical fibers for signal transport to local distribution points (analogous to serving area interface points) where the digital, TDM signal streams are converted to an FDM stream and distributed to individual homes over shared coaxial cables. In each home, an interface device separates television from voice and data.

Two-way operation over cable is achieved by allocating a portion of the spectrum for upstream signals. Usually, it is the band 5 MHz to 30 MHz. In this circumstance, 30 MHz to 54 MHz serves as a guard band, and downstream signals occupy the cable passband above 54 MHz. Unfortunately, in many areas, there is significant interference in the band 5 to 30 MHz.

The return path serves all those who want to use duplex services. Using FDM, the bandwidth available to individual users may be quite small and the signal power in active channels must be adjusted to prevent overloading in the cable amplifiers. Using TDM, operation is vulnerable to interference that can affect all the channels in the stream. In addition, the digital stream is subject to distortion by echoes and intersymbol interference [see Section 5.1.2(2)]. Probably, the best strategy to provide a large number of residences with reasonable levels of service quality is to use FD/TDMA (see Section 6.5.1). In each frequency band, the number of interferors will be smaller; however, keeping total signal power within the linear range of the cable amplifiers remains a problem.

(2) Other Implementations

Other ways of implementing multimedia customer access are

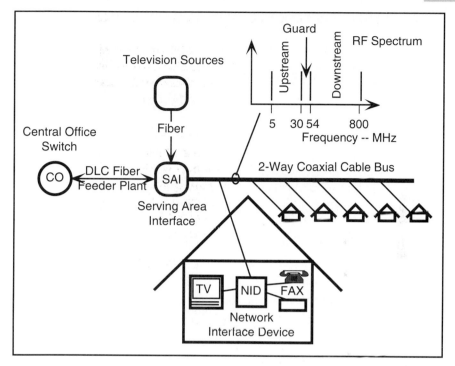

Figure 6.30 Principle of Fiber and Coax Network Providing Data, Voice, and Video Services to Residential Customers

- **xDSL**: provides high-speed Internet access and video transmission downstream over existing wire facilities
- **FTTC/VDSL**: a combination of fiber to the curb (FTTC) and VDSL (which operates over short distances) to provide wideband access downstream with limited upstream capability
- **FTTH**: fiber to the house (FTTH) provides the possibility of wideband access in both directions
- **Satellite with wire return**: provides downstream wideband signals over satellite (video programs or high-speed Internet downloading, for instance), with narrowband control signaling upstream over existing wire facilities
- **Local Multipoint Distribution System**: a radio system operating in the gigahertz region that provides television services to multiple sites.

As high-definition digital television becomes available [see Section 3.5.4(2)], and B–ISDN equipment is deployed (see Section 10.6.2), each residence can have a single signal stream that will provide all services. The techniques listed above may be interim measures to solve residential broadband access.

REVIEW QUESTIONS FOR SECTION 6.3

1   Identify the three types of cables used in the outside plant. Describe the points between which they are deployed.

2   What is the purpose of resistance design?

3   What is the purpose of transmission design?

4   Why are the concepts of resistance and transmission design becoming obsolete?

5   What is meant by two-wire and four-wire connections? Where are they likely to be found? Where are 2W/4W transitions found?

6   Why are telephone lines described as narrowband? What is the truth of the matter?

7   Define the term digital loop carrier. What other names are used for this equipment?

8   What does the acronym BORSCHT stand for?

9   Describe a typical digital loop carrier arrangement.

10   List, and describe, some contemporary DLCs.

11   What is a digital subscriber line system? Distinguish between DLCs and DSLs.

12   Describe the performance of ISDN DSL. For what other applications is it used?

13   Describe the performance of HDSL.

14   Describe the performance of SDSL.

15   Describe the performance of ADSL–1, –2, and –3. For what application was it developed?

16   Describe the performance of VDSL.

17   Describe a hybrid network that delivers data, voice, and video to residences over optical fibers and coaxial cables.

18   How is two-way operation over coaxial cable arranged? Discuss the strategies that can be used to provide return service to a number of users.

19   Sketch some other possibilities for broadband residential access.

6.4 SONET AND SDH

In the United States, spurred by the need for ever higher bit rates, and understanding the advantages of optical fibers, carriers are implementing synchronous optical networks (SONETs). At the same time, ITU–T has defined a synchronous digital hierarchy (SDH) that transports digital signals at gigabit speeds over microwave radios and optical fibers. Fortunately, the frames for both systems are compatible, and one is a multiple of the other. In this Section, I describe their multiplexing capabilities.

6.4.1 SYNCHRONOUS OPTICAL NETWORK

SONET standards define a set of optical interfaces for network transport. First promulgated as United States national standards by ANSI/ECSA, and now incorporated in ITU-T standards, they include electrical and optical speeds, and frame formats. As is to be expected of a more recent system that employs newer technologies, the features available in SONET make it superior to T–3 and T–4 systems. They include improved transport performance, ability to identify subsidiary streams with certainty, international connectivity without format conversion, and enhanced control and administrative functions.[1]

(1) SONET Transport Speeds

The electrical signal hierarchy has N-members

- **Synchronous transport signal level 1** (STS–1): with a basic speed of 51.84 Mbits/s, STS–1 signals are designed to carry DS–3 (synchronous or asynchronous) signals, or a combination of DS–1, DS–1C, and DS–2 signals that is equivalent to DS–3
- **Synchronous transport signal level N** (STS–N): with speeds that are multiples of STS–1—i.e., n x51.84 Mbits/s (where n may assume any integer value between 1 and 255)—STS–N signals are created by interleaving N STS–1 signals, octet by octet. For various reasons, the values N = 3, 12, 24, 48, 96, and 192 are preferred.

Corresponding to the STS signal hierarchy, the optical signals transmitted over the fiber facility are

- **Optical carrier level 1** (OC-1): the optical equivalent of STS–1
- **Optical carrier level N** (OC-N): the optical equivalent of STS–N

Synchronous transport signals and optical carriers are manifestations of the same user data in different media. **Figure 6.31** shows their relationship with each other and SONET multiplexing and transmission devices.

---

[1] For a comprehensive discussion of SONET, see Walter Goralski, *SONET: A Guide to Synchronous Optical Network* (New York: McGraw-Hill, 1997).

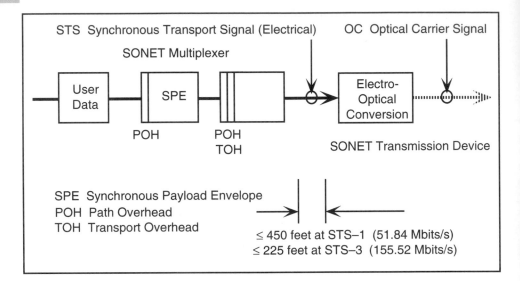

Figure 6.31 Illustrating the Relation Between SONET Electrical Signals and SONET Optical Carrier Signals

At STS–1, the electrical equipment can be no more than 450 feet from the optical equipment. At STS–3, the distance is no more than 225 feet.

(2) STS–1 Frame

To achieve compatibility with existing systems, synchronous transport signals are divided into fixed frames of 125 μs duration. An STS–1 frame consists of 6,480 bits or 810 octets. Shown in **Figure 6.32**, the octets can be arranged in a frame of 9 rows of 90 columns (with one octet in each location) that are divided into

- **Transport overhead**: 9 rows x 3 columns (27 octets) are occupied by transport overhead (TOH) information. It is made up of 9 octets (3 rows by 3 columns) of *section* overhead (SOH) and 18 octets (6 rows by 3 columns) of *line* overhead (LOH).
- **Synchronous payload envelope**: the remaining 9 rows x 87 columns (783 octets) are called the STS–1 synchronous payload envelope (SPE). Of the 783, 9 octets (one column) are reserved for *path* overhead (POH) information.
- **STS–1 payload**: within the SPE, 774 octets (9 rows x 86 columns) are payload.

Some of the functions performed by the overhead streams are

- **Section overhead**: defines and identifies frames, and monitors section errors and communication between section terminating equipment

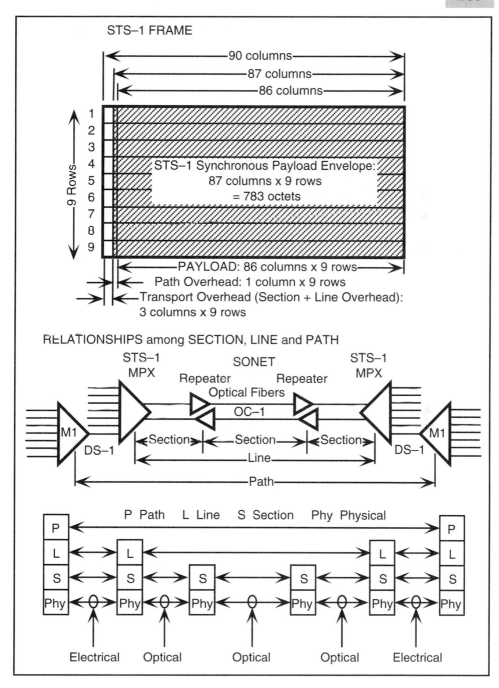

Figure 6.32 STS–1 Frame and Relations Among Section, Line, and Path

- **Line overhead**: locates first octet of SPE (pointer function), and monitors line errors and communication between line-terminating equipment
- **Path overhead**: verifies connection, and monitors path errors, receiver status, and communication between path-terminating equipment.

The lower part of Figure 6.32 shows the physical meaning of section, line, and path. The STS–1 line is deployed between STS–1 multiplexers. It consists of several *sections* that are bounded by repeaters or the multiplexers that terminate the line. On the *line*, the STS–1 signal may contain several tributary signals. In Figure 6.32, I use the example of a DS–1 signal. A *path* for this signal exists between the terminating M1 multiplexers. There are as many paths as there are tributaries in the STS–1 signal. Overhead octets are included in the STS–1 frame to supervise and manage events associated with the transport of user data at the section, line, and path levels.

At the bottom of Figure 6.32, I have drawn the four levels of a connection over SONET. In addition to the physical level, they are

- **Section layer**: uses the section overhead (SOH) to manage the transport of STS frames over the physical path between section terminating equipment. Performs error monitoring, framing, and signal scrambling.
- **Line layer**: uses the line overhead (LOH) to manage the transport of an entire SONET payload (that may be embedded in several STS frames) over the physical span between two SONET devices. Performs multiplexing and synchronization to create or maintain SONET payloads
- **Path layer**: uses the path overhead (POH) to communicate end to end and provide SONET services between customer premises equipment. Converts customer data into a format compatible with the line layer.

In a dynamic situation, the SPE does not necessarily start in column 4, nor must it remain within one frame. This has an impact on the position of the path overhead octets. **Figure 6.33** shows the organization of octets in columns 1, 2, and 3 of two STS–1 frames. The path overhead information floats across the frame to the column in which the SPE starts. The important point is that one octet of POH occurs immediately before every 86 octets of payload.

(3) Tributary Signals

SONET uses octet multiplexing to create a higher-speed stream from lower-speed tributary signals

- **Tributary signal**: a lower-speed signal consisting of a synchronized stream of octets that is incorporated in a higher-speed multiplexed stream.

Payloads whose sizes are less than an STS–1 payload (i.e., < 774 octets) are carried in special structures. Known as *virtual tributaries* (VTs), they consist of a

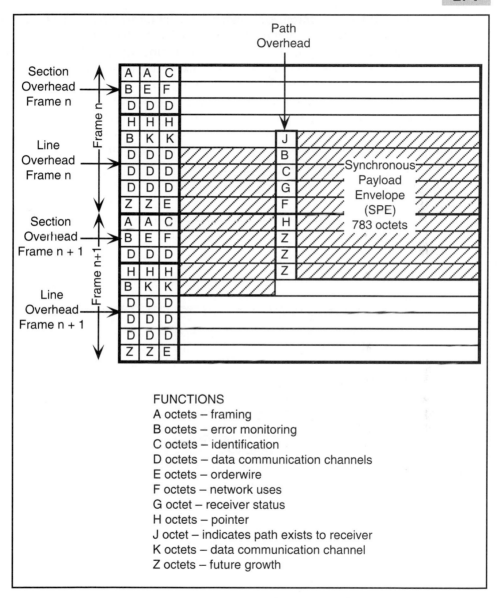

Figure 6.33 SONET Synchronous Transport Envelope Contained in Two STS–1 Frames
Path overhead designates the beginning of the payload. It moves to the column in which the SPE starts.

fixed number of octets that occupy 9 rows x *n* columns in the SPE. For instance, VT1.5, the virtual tributary for DS–1, consists of 27 octets (9 rows x 3 columns, equivalent to 1.728 Mbits/s). They are divided into 2 octets of VT overhead, 24 DS–0 octets (or 192 bits of unchannelized data), and, to preserve the information contained in SF or ESF framing sequences, 1 octet for framing. For European applications, VT2, the virtual tributary for E–1, consists of 36 octets (9 rows x 4 columns, equivalent to 2.304 Mbits/s). VT3, the virtual tributary for DS–1C, consists of 54 octets (9 rows x 6 columns), and VT6, the virtual tributary for DS–2, consists of 108 octets (9 rows x 12 columns). To make loading easier for payloads that consist of a mix of VTs, the SPE can be divided into seven equal compartments (9 rows x 12 columns) that contain four VT1.5s, or three VT2s, or two VT3s, or one VT6, or a compatible combination of signals. For payloads of only one size VT, the VTs are allowed to float in the SPE. **Figure 6.34** shows the principle of loading a mix of virtual tributaries into the SPE of an STS–1 frame.

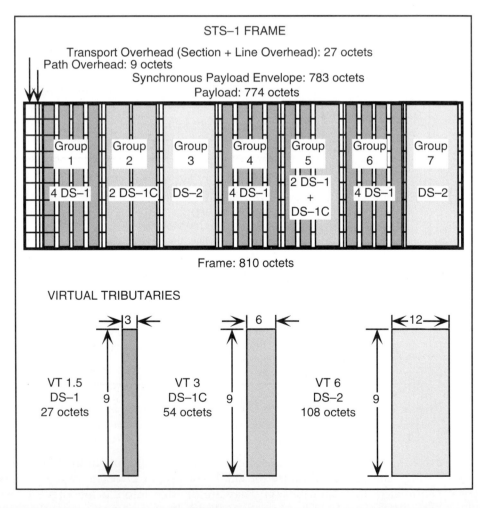

Figure 6.34 Virtual Tributaries and an Example of Their Arrangement in an STS–1 Frame

DS–3 signals are carried as STS–1 payloads. For SYNTRAN, it is a matter of mapping the signal (information octets + overhead bits) into the space available. For traditional DS–3, the technique must compensate for timing variations due to the asynchronous nature of the signal. This is achieved by providing an asynchronous stuffing bit in each SPE row (in the Z octet row). In addition, fixed stuffing bits are added to fill the STS–1 frame. The arrangement is shown in **Figure 6.35**.

(4) STS–N Frame

Of 125 μs duration, an STS–N frame consists of N x 6,480 bits arranged in 9 rows x 90N columns. STS–N signals are created by octet multiplexing lower-speed STS signals. For example, three STS–1 signals are octet multiplexed into an STS–3 signal, or four STS–3s are octet multiplexed into an STS–12 signal. **Figure 6.36** shows an outline of an STS–3 frame, and a multiplexing arrangement that will produce it. Before transmission at 155.52 Mbits/s, the entire frame is scrambled (see Section 3.1.7) to randomize long sequences of zeros.

Some input signals are larger than STS–1; suppose a signal is large enough to require an STS–N frame. Under this circumstance, the N STS–1 frames into which the input signal is mapped are defined as a *concatenated* structure, and the signal is designated STS–Nc. The frames are multiplexed and move through the network in the SPE of the STS–N frame as an entity. Octets H1 and H2 in the line overhead of each STS–1 frame are used to denote the concatenated condition. Because all the individual frames are part of the same signal and destined to take

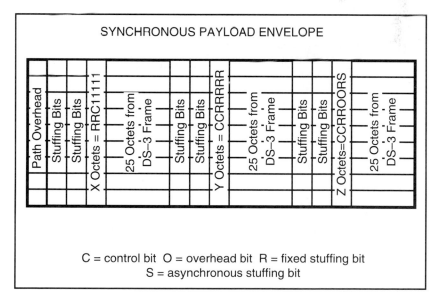

Figure 6.35 Synchronous Payload Envelope Containing an Asynchronous DS–3 Frame

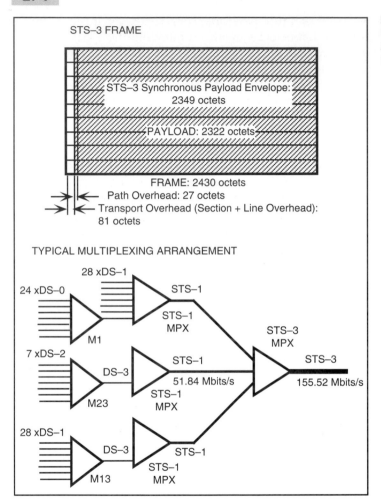

Figure 6.36 SONET STS–3 Frame and a Typical Multiplexing Arrangement to Fill It

the same path, path overhead information is confined to the first STS–1 frame of the group. Further discussion of SONET is found in Section 10.2.3(7).

6.4.2 SYNCHRONOUS DIGITAL HIERARCHY

To transport the broadband signals envisaged for BISDN (see Section 10.6.1), ITU-T has defined a synchronous digital hierarchy. Compatible with existing facilities, SDH handles ISDN basic and primary rate signals [i.e., 64 kbits/s (DS-0) and 1.544 Mbits/s (DS–1) or 2.048 Mbits/s (E-1)], and SONET STS–N signals. SDH provides transport at

- **Synchronous transport module level 1** (STM-1): a basic speed of 155.52 Mbits/s, i.e., 3 STS–1
- **Synchronous transport module level N** (STM-N): speeds that are multiples of STM-1—i.e., N 155.52 Mbits/s. STM-N frames are created by interleaving N STM-1 frames, octet-by-octet.

Table 6.1 compares levels and speeds in SDH and SONET.

(1) STM-1 Frame

The organization and structure of SDH are similar to SONET—unfortunately, some of the nomenclature is different. What is also different is that SDH is intended for use by microwave radio relays as well as optical fibers. SDH signals flow in fixed length frames of 125 µs duration. An STM-1 frame consists of 19,440 bits that are transported at a speed of 155.52 Mbits/s. Shown in **Figure 6.37**, it consists of 9 rows of 270 columns with one octet in each location (2,430 octets). In each frame, 9 rows x 9 columns (81 octets) are occupied by information related to the frame structure and information that facilitates frame transport from one node to the next (section overhead). The remaining 9 rows x 261 columns (2,349 octets) are the payload. Made up of a limited set of structures that have a specific size and fixed relationship to one another, it contains user information, and information that defines the network path for each payload structure (path overhead). By way of illustration, in Figure 6.37 the payload consists of three equal structures called administrative units (see below). Compatibility with other network signals is achieved by

RATE Mbits/s	SONET Electrical	Optical	SYNCHRONOUS DIGITAL HIERARCHY
51.84	STS-1	OC-1	
155.52	STS-3	OC-3	STM-1
622.08	STS-12	OC-12	STM-4
1244.16	STS-24	OC-24	STM-8
2488.32	STS-48	OC-48	STM-16
4976.64	STS-96	OC-96	STM-32
9953.28	STS-192	OC-192	STM-64

Table 6.1 Comparison of SONET and SDH Transport Levels

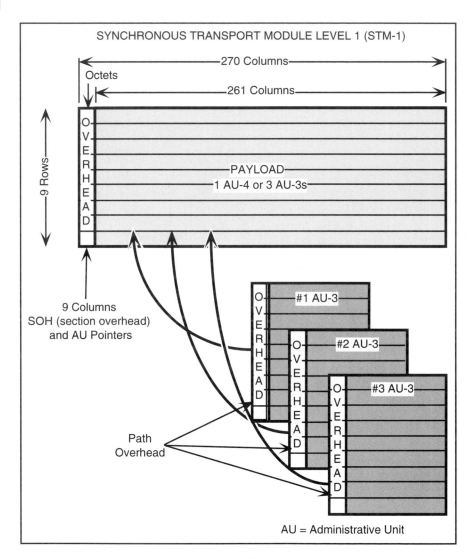

Figure 6.37 SDH STM–1 Frame and a Typical Payload of Three Administrative Units

- **DS–0**: transmitting frames every 125 μseconds
- **DS–1** (North America and Japan): the 192 payload bits in a DS–1 frame, and STM-related overhead bits (path overhead), are mapped into a 9 rows x 3 columns (27 octets) structure (24 octets for signals + 3 octets for overhead) that can be carried in the payload and delivered at a rate of 8,000/s
- **E–1** (Europe): an E-1 frame is mapped into 9 rows x 4 columns (36 octets, 32 octets for signals + 4 octets for overhead) that can be carried in the payload and delivered at a rate of 8,000/s

- **SONET**: the STS–1 SPE (783 octets) is exactly one-third of the STM-1 payload (2,349 octets), and the speed of STM–1 (155.52 Mbits/s) is three times the speed of STS–1 (51.84 Mbits/s).

(2) Creating STM-1

To create an STM-1 frame, a 125-μsecond segment of each *tributary* signal is mapped into a structure called a *container* whose size and designation (C-4, C-3, etc.) reflect the speed of the user's signal. For instance, if the tributary speed is 1.544 Mbits/s, the container (C-11) carries 24 octets, and if the speed is 44.736 Mbits/s, the container (C-3) carries 672 octets. As shown in **Figure 6.38**, the container is then joined to path overhead information to become a virtual container

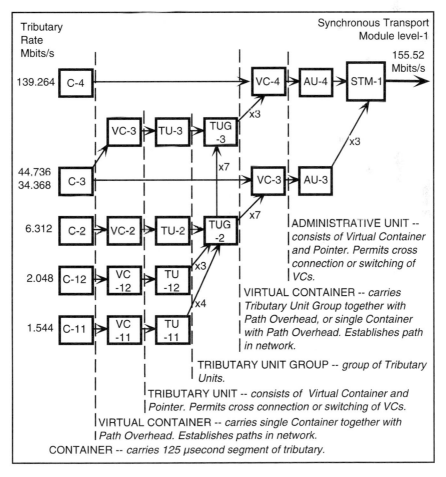

Figure 6.38 STM–1 Tributary Multiplexing Scheme
Tributaries are collected into larger groups before insertion in STM frame.

(VC). Next, to keep track of the start of the virtual container, pointer information is added, and the whole is identified as a tributary unit (TU). Continuing the multiplexing process, several TUs are collected together to form tributary unit groups (TUGs). In turn, several TUGs are loaded into another virtual container (at level 3 or level 4), more pointers are added to keep track of the segments, and administrative units (AUs) are formed. By ones (AU-4) or threes (AU-3), the AUs are formed into an STM-1 frame that is transported between nodes at 155.52 Mbits/s. Using the pointers to locate them, the segments of the payload are delivered to the receiving node. By operating with tributary rates between 1.544 Mbits/s and 139.262 Mbits/s, STM-1 can transport a mixed collection of signal streams in the same container.

REVIEW QUESTIONS FOR SECTION 6.4

1 What SONET features are missing in T–3 and T–4 systems?

2 Define STS–1 and STS–N. What are the corresponding optical signals?

3 Describe an STS–1 frame. Why is the repetition rate of the frames 8,000/s?

4 What functions are performed by the section, line, and path overhead? Explain the lower diagram of Figure 6.32.

5 Under what conditions does path overhead not start in column 4?

6 Define a tributary signal.

7 What is a virtual tributary? Define VT–1.5, VT–2, VT–3, and VT–6.

8 How does SONET accommodate the variable bit rates of DS–3 signals?

9 What is an STS–Nc frame? Why is it used?

10 Define STM–1 and STM–N frames.

11 How do the speeds of SONET and SDH relate?

12 How does SDH achieve compatibility with DS–0, DS–1, E–1, and SONET signals?

13 Describe how an STM–1 frame can be filled with containers carrying segments of different speed tributary signals.

6.5 RADIO SYSTEMS

In some ways, sharing a radio channel is the same as sharing a wire channel—frequency-division and time-division techniques work equally well. In other ways, it is quite different. Because radio allocations are limited, and radio waves travel in all directions, assigning a frequency or a time slot permanently to one station restricts the number of stations that can participate in a network. Accordingly, a normal way of working radio networks involves assigning communications resources to a station when it requests them. Called DAMA (demand-assigned multiple access), the technique is used with some communications satellite systems and some mobile cellular radio systems. An alternative technique is to employ CDMA (code-division multiple access). By using special orthogonal spreading codes, each station can transmit at will without interfering with other stations. With all DAMA techniques, the fact that the user must notify an entity of its need for a channel, and a channel must be assigned, means that there is a (substantial) setup time before communication can begin.

6.5.1 MULTIPLE ACCESS TECHNIQUES

Frequency-division multiple access (FDMA), time-division multiple access (TDMA), and code division multiple access (CDMA) techniques are used alone, or in combination, to create channels that can be used for radio communication.

- **FDMA**: a technique in which the available spectrum is frequency divided into channels. A physical channel is defined by carrier frequency and bandwidth available about the carrier. As needed, a central control facility assigns operating frequencies (and permissible bandwidths) to stations from among the frequencies available.
- **TDMA**: a technique that assigns the same carrier (and associated bandwidth) to stations in the network at different times (time slots). During its time slot, the full capability of the carrier is available to a station. A physical channel is defined by carrier frequency, associated bandwidth, and time slot. A master station defines a superframe with specific transmit times that define the slots available. Slots may be allocated to stations as they have messages to send, or are assigned permanently to a station. Each transmitter is synchronized to the start of the superframe and transmits only during its assigned slot.
- **FD/TDMA**: the available spectrum is frequency divided into channels, and each channel is time divided into time slots. During its time slot, a station sends and receives within the bandwidth associated with the assigned carrier frequency. A physical channel is defined by carrier frequency, associated bandwidth, and time slot. A master station defines a superframe for each frequency-defined channel. Transmitters are synchronized to the start of the superframe in their assigned channel. During its time slot, the full capacity of the channel is available to each station.
- **CDMA**: a spread spectrum technique (see Section 4.3) in which stations transmit at the same carrier frequency with approximately equal power and the same chip rate. To spread the signal spectrum, or to cause the frequency

to hop from one value to another in a random fashion, each station employs a chipping code that is orthogonal to, but synchronized with, the codes used by the other stations. The receiver employs the transmitter's unique code to detect the message. A physical channel is defined by the carrier frequency, associated bandwidth, and a chipping code. Each receiver sees the sum of the individual spread spectrum signals as uncorrelated noise. It can detect an individual signal if it has knowledge of the makeup of the unique spreading code associated with the sender's transmitter, and can synchronize with it. Cooperating stations transmit at the same carrier frequency with approximately equal power and the same chip rate using direct spreading or frequency hopping techniques. With CDMA, collisions are of no consequence, and the full bandwidth of the system is available to each station at all times. In practice, the spreading sequences will have a cross-correlation with the rest of around –20 db. For reliable performance in a multi-user environment, it is important to control the power level of competing signals to prevent higher power signals from capturing the receiver.

- **FD/CDMA**: the available spectrum is frequency divided into channels and stations employ a spread spectrum technique within their assigned frequency channels. A physical channel is defined by the carrier frequency, associated bandwidth, and a chipping code.

6.5.2 CELLULAR RADIO

Several active mobiles located in a single cell can be connected to the base station using FDMA, TDMA, FD/TDMA, CDMA, or FD/CDMA.

(1) FDMA

Using frequency-division techniques, a pair of carrier frequencies is assigned to a mobile that wishes to make a call. One frequency is used for mobile-to-base station traffic, and the other is used for base station-to-mobile traffic. When the call is completed, the frequencies become available for other mobiles. Because the discussion of operations in Section 5.4.2 assumes the use of FDMA, I will not repeat it here. Further discussion is found in Section 11.2.1.

(2) FD/TDMA

Using time-division multiple access techniques, a pair of carrier frequencies is assigned to n mobiles, and each mobile is assigned time slots on the carriers. In this way, the traffic from n mobiles is carried as a time-division multiplexed signal on the carrier pair. Each mobile transmitter sends bursts of information timed and synchronized so that they arrive at the cell site receiver at exactly the right time to follow the burst from the mobile ahead in the sequence and lead the burst from the mobile that follows. Accurate slot timing and carrier synchronization are essential. Fortunately, all participants in the mobile-to-base station stream receive signals from the same base station. Further discussion of the use of FD/TDMA is found in Sections 11.2.2 and 11.2.3.

(3) FD/CDMA

CDMA produces a wideband, noiselike signal, and the entire range of frequencies allocated to the system can be used in each cell. Discrimination between the mobiles in the same cell is achieved through the assignment of independent *spreading* codes (as described in Section 4.3.4). Discrimination between mobiles in one cell and mobiles in an abutting cell is achieved through the assignment of a different set of frequencies in the abutting cell. Under these circumstances, the number of mobiles that can be served is significantly greater than the number served with FDMA or TDMA. However, CDMA is not without complications. So that users close to the base station do not overpower users farther away, each mobile must regulate its transmitted power in inverse proportion to some power of the distance it is from the cell site.

Figure 6.39 shows the principle of CDMA in a direct spreading environment. Several transmitters operate with orthogonal chipping codes; at the receiver, each correlator responds only to a specific code. Performance depends critically on the receiver's ability to synchronize with and duplicate exactly, the spreading code. In

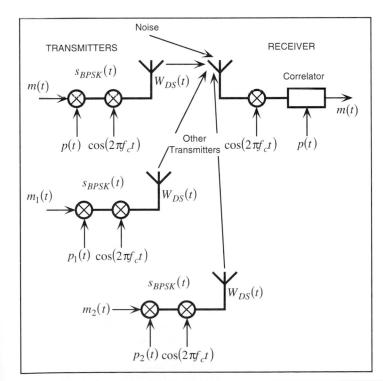

Figure 6.39 Principle of CDMA in a Direct Spreading Application

some applications, this is achieved with the assistance of timing signals from navigation satellites. Further discussion of the use of FD/CDMA is found in Section 11.2.4.

6.5.3 COMMUNICATION SATELLITES

In communication satellite systems, frequency division, time division, and code division are used to create channels that can be assigned on demand to users wishing to communicate. Moreover, a procedure for random access to channels has been developed. Known as *Aloha*, it can be characterized as sending when ready, and repeating if necessary; what might be called a *laissez-faire* approach to communicating.

(1) FD/DAMA System

In FD/DAMA, a central control facility assigns operating frequencies at the request of the network stations. The calling Earth station communicates to the control station requesting circuits for another station. Control selects a pair of frequencies and notifies both stations of the assignment. The principle is shown in **Figure 6.40.** It includes examples of fixed-assignment and demand-assignment

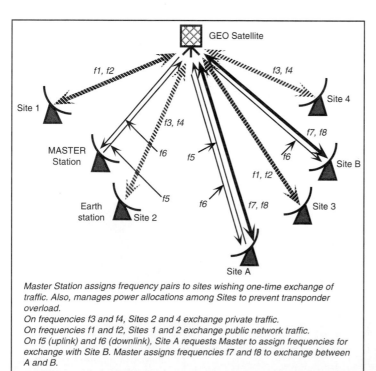

Master Station assigns frequency pairs to sites wishing one-time exchange of traffic. Also, manages power allocations among Sites to prevent transponder overload.
On frequencies f3 and f4, Sites 2 and 4 exchange private traffic.
On frequencies f1 and f2, Sites 1 and 2 exchange public network traffic.
On f5 (uplink) and f6 (downlink), Site A requests Master to assign frequencies for exchange with Site B. Master assigns frequencies f7 and f8 to exchange between A and B.

Figure 6.40 Principle of FD/DAMA Operation of Communication Satellite System

operations. Frequencies *f*1 (uplink) and *f*2 (downlink) are permanently assigned to Sites 1 and 3 for the exchange of public network traffic. Frequencies *f*3 and *f*4 are permanently assigned to Sites 2 and 4 for the exchange of private network traffic. Frequencies *f*5 and *f*6 are used to request channel assignments from the master station for occasional exchanges of traffic between Sites A and B. The master station assigns frequencies *f*7 and *f*8 to their exchange. Known as *single-channel per carrier* (SCPC) DAMA, this technique has been used with manual and computer-controlled stations to provide on-demand, voice, or data channel connections.

(2) TD/DAMA System

In TD/DAMA, time slots are assigned to each Earth station on a demand basis. During each time slot, the station has full and exclusive use of the entire transponder. The principle of a TD/DAMA system is shown in **Figure 6.41**. Messages are buffered at the Earth stations and sent to arrive at the satellite in a sequence that allows the transponder to serve each of the Earth stations in turn. In

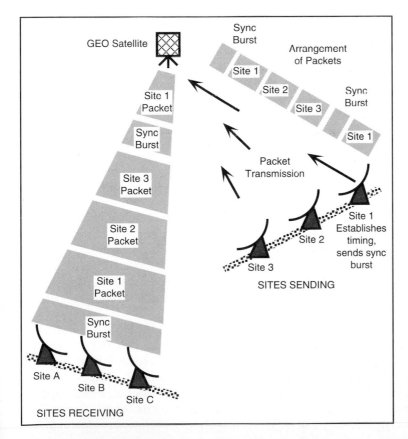

Figure 6.41 Principle of TD/DAMA Operation of Communication Satellite System

the example, bursts are shown transiting the link in an organized way. Because of the long path, several discrete message blocks may exist between stations and satellite.

A set of message blocks (one from each station) constitutes a frame. It starts with a synchronizing signal (synch burst) from the control station that includes a unique word for easy recognition. To determine when to send their message bursts, all other stations use this signal as a reference. Each message is opened by a preamble block that contains block and timing information, station identification, and other data, and may be closed by a block that performs error checking and other functions, and ends the sequence. A portion of the digital stream is reserved for channel requests and time-slot allocations. In the diagram, Site 1 functions as the master station that implements DAMA operation and provides network synchronization.

(3) Request Channel

Request and assignment channels are shared among all stations. The request channel may operate as a fixed allocation channel; each Earth station is assigned a portion of the channel. Alternatively, it may operate as a contention channel in which each Earth station sends its request when ready and competes for resources. With a fixed allocation strategy, the capacity of the request channel limits the absolute size of the network. With a random access strategy, the capacity of the request channel limits only the rate at which requests are made. The INMARSAT network that provides on-demand communications for ships at sea employs an Aloha (see below) random access request channel protocol.

(4) Random Access—Basic Aloha

Invented for a microwave radio system in the Hawaiian Islands, Aloha is a technique in which stations send data packets when they are ready. **Figure 6.42** shows the principle of satellite communication using this procedure. Should a collision occur at the transponder, the stations wait for a time and then send again. The procedure is repeated until the packet is received without errors. If packets are generated randomly, it can be shown that the maximum packet throughput is $S_{max} = 1/2et_p$, where t_p is the time required to send a packet. Maximum throughput occurs when the offered traffic is one-half the full-load capacity of the transponder. **Figure 6.43** shows the relationship between normalized throughput and the ratio of offered traffic to full-load traffic; clearly, a maximum occurs at 18% of the full-load packet rate.

If the offered traffic from all the stations is 50% of full load (i.e., one-half of the traffic that will just fill the transponder), approximately 36% of the packets will get through on the first attempt. If the offered traffic is 10% of full load, approximately 80% of the packets will get through on the first attempt. Thus, so long as the total offered traffic from all stations is not a very large fraction of the capacity of the transponder, many packets have a reasonable chance of finding the transponder unoccupied; they can be relayed without interference. However, if a

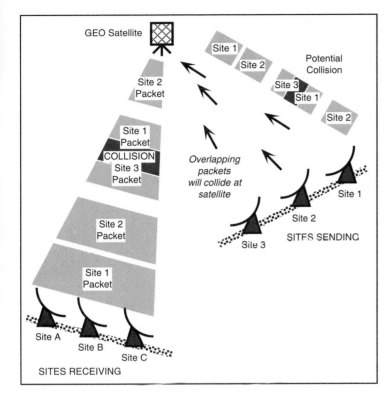

Figure 6.42 Principle of Satellite Communication Using Aloha

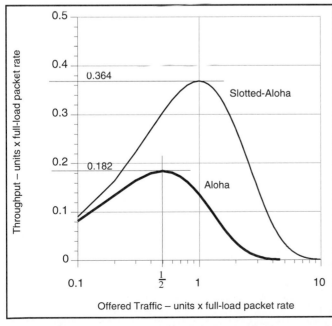

Figure 6.43 Relation Between Normalized Throughput and Offered Traffic for Aloha and Slotted-Aloha Operation

Using Aloha, 18% of the packets will get through without collisions if the offered traffic is one-half of the maximum channel capacity. Using slotted-Aloha, 36% of the packets will get through without collisions if the offered traffic is equal to the maximum channel capacity.

second signal arrives at the transponder while the first signal is still being received, a collision occurs and the receiving stations receive a garbled version of the colliding transmissions. When collisions occur, the sending stations wait for a random time (determined independently by timers in each station), and then transmit again.

(5) Slotted and Reservation Aloha

By modifying the process so that packets are sent to arrive at the transponder at the beginning of agreed-upon time slots, the throughput can be doubled. Called slotted-Aloha, the technique requires that all stations be synchronized in some fashion so that their packets only arrive at the transponder at a slot start time. Under this circumstance, the period of vulnerability to a collision is reduced to one packet time so that the maximum throughput is $1/et_p$ which is twice the value for simple Aloha. Figure 6.43 shows throughput as a function of input density for slotted-Aloha.

An extension of this technique allows stations to reserve time slots for their exclusive use. Called reservation-Aloha, the technique guarantees that messages will arrive one at a time at the transponder and will be relayed without collisions. Under these conditions, throughput is the full-load packet rate, and the average delay is the propagation time plus send time. Thus, throughput can be improved at the expense of imposing discipline on the senders.

(6) Code-Division Multiple Access

With approximately equal power and the same chip rate, all stations in the network transmit in a bandwidth that can be as wide as the passband of a transponder. Each station employs a code that is orthogonal to the codes used by the other stations. Each receiver sees the sum of the individual spread spectrum signals as uncorrelated noise. It can demodulate an individual signal if it has knowledge of the spreading code and the carrier frequency, and it can synchronize with the sender's code. With CDMA, collisions at the transponder are of no consequence, and the full transponder bandwidth is available to each Earth station at all times.

(7) Polling

The introduction of satellite links into a centralized computer network in which the FEP (front-end processor) attached to the host is the primary station, and all units in the network are secondary stations, significantly reduces throughput. To speed things up, *satellite delay compensation units* (SDCUs) can be added at all Earth stations. An exchange between the FEP and a remote DTE (data terminal equipment) is shown in **Figure 6.44**. At the Earth station serving the FEP, the SDCU acknowledges the polling command at the same time the polling packet is launched from the Earth station toward DTE1. By doing this, the SDCU frees the FEP to poll other stations in the network. The SCDU continues to respond to polls until data arrives from DTE1.

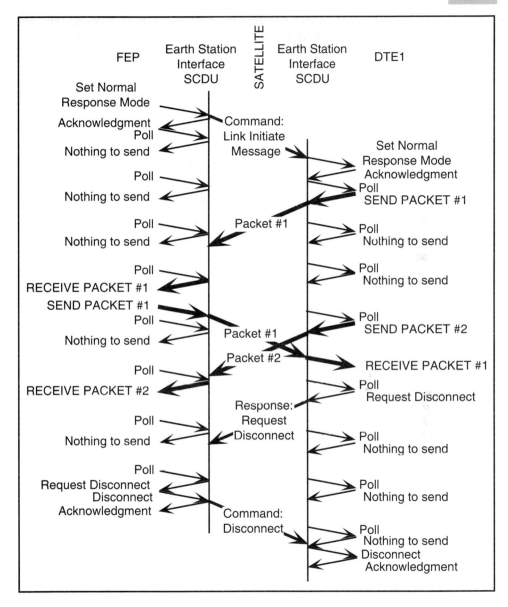

Figure 6.44 A Polling Sequence Through a Satellite Using SCDUs at the Earth Stations
Replies from the SCDUs allow the polling process to proceed without waiting for the satellite link delay.

If the primary has something to send to DTE1, the SDCU at the Earth station serving the FEP acknowledges receipt according to the ARQ (await-receiver-request) procedure employed while the packet stream is launched toward DTE1. At the Earth station serving DTE1, the SDCU polls DTE1 and receives responses from it. When DTE1 has something to send, packets are launched in the direction of the FEP and the SDCU serving DTE1 provides acknowledgments that keep the stream going. If the FEP detects an error, an error message is sent to DTE1, and it implements the ARQ procedure in use.

REVIEW QUESTIONS FOR SECTION 6.5

1 Using DAMA, when are resources assigned to a station?

2 What special attribute does CDMA have with respect to access to a radio channel?

3 Describe how the following techniques are used with cellular mobile radio: FDMA, TDMA, CDMA, FD/TDMA, and FD/CDMA.

4 Describe how the following techniques are used with communication satellites: FD/DAMA and TD/DAMA.

5 Describe two ways of operating a request channel.

6 Describe basic Aloha.

7 What are the advantages of slotted-Aloha and reservation-Aloha? What are their disadvantages?

8 Describe CDMA.

9 What can be done to improve the response of a communication satellite system that supports polling of secondary sites?

7

DATA
COMMUNICATION

Data communication involves the distribution or exchange of data messages between machines that process data. In this activity, the sequences of bits must be preserved while the signals that represent them are converted to forms compatible with the different phases of generation, transportation, and reception. Specialized equipment makes these transformations, some form of error control is likely to be invoked, and the processors follow common procedures. In this chapter, I describe the hardware functions required to implement communication between data processing machines. In Chapter 8, I describe the software functions required to implement communication between data processing machines.

Before describing hardware devices, some definitions are in order. Those of us whose professional careers span the 1960s are wont to talk of *terminals* and *hosts*; persons who began work in the 1980s speak of *clients* and *servers*. Are they different? Yes and no. They reflect the evolution of the data processing and computing environments—from ones that contained relatively unsophisticated video terminals connected to central computing resources that did all the processing—to powerful desktop computers that perform sophisticated processing on their own, and need volumes of data to turn into information. The following definitions may be helpful.

- **Terminal**: a device used to input and display data and information. It may have native computing and data processing capabilities

- **Client**: a terminal with significant computing capability
- **Host**: a mainframe computer that performs processing and computing tasks in support of terminals
- **Server**: a major data processing device that maintains databases and delivers data files to clients, on demand.

Is communication between terminal and host, and client and server, different? Again, the answer is yes and no. Communication between either combination is manifest by the exchange of streams of data bits, but, usually, clients require more data and can cope with faster datastreams.

7.1 CONNECTIONS

At the edges of the network, devices that perform special functions anchor data connections. They make the physical connection and process the datastream in ways that enhance the probability of data being received correctly.

7.1.1 DTEs AND DCEs

Even though they may in fact be implemented as a single device, it is customary to divide the equipment at the periphery of the network into two parts

- **Data terminal equipment** (DTE): a device that creates, sends, receives and interprets data messages
- **Data circuit terminating equipment** (DCE): a signal conversion device. DCEs condition (i.e., prepare) signals received from DTEs for transmission over communication connections, and restore signals received from the network so as to be compatible with receiving DTEs.

Figure 7.1 shows a range of DTE-to-DCE-to-network connections. The configurations are explained below.

(1) Data Terminal Equipment

DTEs are digital devices. Internally, their signals are likely to be simple unipolar pulses; externally they may use a more sophisticated digital signaling format. In the terminal/host world, DTEs are called *terminals*. For instance, they may be

- **Teletypewriters** (TTYs): devices that accept keyboard input, print hard copy output, and store and forward messages

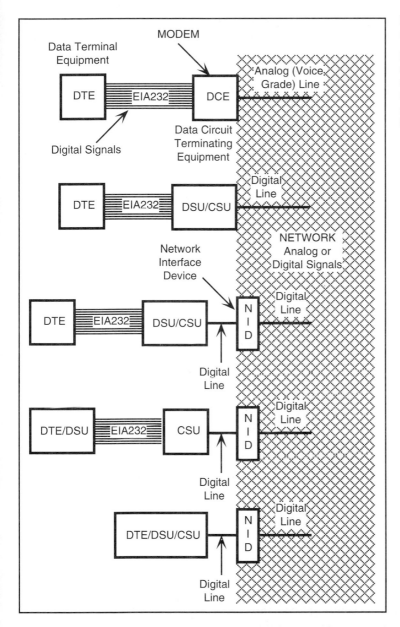

Figure 7.1 DTE to DCE to
Network Connections

For clarity, the pieces of
equipment are drawn as
separate entities. In fact,
several may exist on a
single card inserted in the
terminal.

- **Video display terminals** (VDTs): devices that accept keyboard input and
  display output on a video screen
- **Transaction terminals**: devices that capture and process on-line transactions.
  Examples are automatic teller machines and supermarket checkout terminals with bar-code readers.

- **Intelligent terminals**: devices programmed to perform some of the tasks usually assigned to a host machine—such as screen formatting and data editing, or more complex assignments—so as to reduce dependence on the central processor
- **Smart terminals**: devices that contain memory and logic to perform specific tasks—such as facsimile machines or order entry devices
- **Dumb terminals**: devices that depend on the host for all but the most routine functions.

However, in the client/server world, the designation DTE includes a broader category of equipment including

- **Workstations** or **personal computers**: devices that contain significant processing power in their own right, but need to access other processors
- **Front-end processors**: equipment that performs all of the communication-related functions at a host site. The front-end processor (FEP) relieves the much faster host of activities associated with the relatively slow processes required for data communication.

(2) Data Circuit Terminating Equipment

A DCE is a signal conversion device that assists DTEs to send or receive data messages over data circuits.

- If the connections are analog (voice-grade), the DCE is called a *modem*. When sending, it converts the digital signals received from the DTE to analog signals that match the bandwidth of the line. When receiving, the DCE converts the analog signals to digital signals and passes them on to the DTE.
- If the connections are digital connections, the DCE consists of two parts:
  - *digital service unit* (DSU). Over EIA 232 and similar cables [see Section 7.1.2(1)], a DSU receives unipolar digital signals from the DTE at 2.4, 4.8, 9.6, 19.2, 56, or 64 kbits/s, and 1.544 Mbits/s. The DSU converts them to bipolar signals. When sending, a DSU performs special coding to make the signals compatible with the operation of the transport system—such as zeros substitution codes for subrate data and 56-kbits/s applications [see Section 6.2.1(2)]. When receiving, a DSU removes any special codes inserted by the sending DSU and converts the bipolar signals to unipolar signals compatible with the DTE.
  - *channel service unit* (CSU). A CSU provides line conditioning, loopback (for testing), limited diagnostic capabilities, and, if needed, span powering for repeaters. When sending, it converts bipolar signals to AMI, 2B1Q, or other codes; when receiving, it converts AMI, 2B1Q, or other coded signals to bipolar. When operating at 1.544 Mbits/s, a CSU performs 8 zeros

substitution coding, and supports special framing requirements—such as superframe and extended superframe [see Section 6.2.1(3)].

Figure 7.1 shows DSU/CSU installations. DSUs and CSUs may be combined in the same piece of equipment, DSUs may be incorporated in DTEs, and DSUs and CSUs may be incorporated in DTEs. Furthermore, because FCC defines CSUs as customer-provided equipment (CPE), carriers may install a network interface device (NID) that performs carrier-controlled testing, loopback and powering functions.

7.1.2 INTERCONNECTIONS

Standard cables are used to connect DTEs to DCEs, and the DTE/DCE combination to the first network node.

(1) EIA232 Interface DTE ↔ DCE

Commonly, a DTE is connected to a DCE by a cable that conforms to standard EIA232 (also known by its previous designation of RS 232).[1] Figure 7.1 shows the arrangement. EIA232 describes a multiwire cable that terminates in 25-pin connectors. The standard limits the cable to asynchronous or synchronous operation at speeds up to 19.2 kbits/s. At 19.2 kbits/s, the standard limits cable length to 50 feet. The EIA232 circuits linking DTE and DCE carry signals that initiate, maintain and terminate communication between the two. **Table 7.1** provides information on the signals that flow over the cable. Transmitted in parallel between the terminating devices

- Data from the sending DTE (DTE1) appear on pin 2 and are transmitted over circuit AB to the sending DCE (DCE1)
- At DCE1, the datastream is conditioned to suit the circuit being used and passed through the network to the receiving DCE (DCE2)
- At DCE 2, the signal is converted to a digital datastream, placed on pin 3, and transmitted over circuit BA to the receiving DTE (DTE2).

Before transmission can start, the system must be *ready*. Marks (1s) present on circuits CA (request to send), CB (clear to send), CC (DCE ready), and CD (DTE ready) fulfill this condition. Four versions of the EIA232 standard have been issued. The latest, EIA232D, is displacing EIA232C.

Similar standard interfaces have been adopted by ITU–T. Known as V.24 and V.28, they are functionally similar to EIA232. In addition, ITU–T's V.35 supports speeds to 4.8 kbits/s, and X.21 provides connections to public data networks.

---

[1] EIA stands for Electronic Industries Association; RS stands for recommended standard.

Table 7.1 EIA 232 Interface Circuit Definitions

PIN#	CIRCUIT	NAME	DIRECTION	COMMENTS
		Ground Connections		
1	AA	Protective Ground		Machine ground
7	AB	Signal Ground		Ground reference for all circuits
		Data Signals		
2	BA	Transmit Data	DTE to DCE	Data generated by DTE
		Circuit BA may only be used if a mark condition exists on CA (Request to send), CB (Clear to send), CC (DCE ready), and CD (DTE ready).		
3	BB	Receive Data	DCE to DTE	Data received by DTE
		Control Signals		
4	CA	Request to Send	DTE to DCE	DTE has message to transmit
5	CB	Clear to Send	DCE to DTE	DCE is ready to accept and transmit DTE data
6	CC	DCE Ready	DCE to DTE	DCE is ready to operate
20	CD	DTE Ready	DTE to DCE	DTE is ready to operate
22	CE	Ring Indicator	DCE to DTE	DCE is receiving a ringing signal
8	CF	Signal Detect	DCE to DTE	DCE is receiving a carrier signal
21	CG	Signal Quality Detector	DCE to DTE	OFF indicates may be errors in received data
23	CH/CI	Data Signaling Rate Selector/Indicator	CH to DCE; CI to DTE	Selects higher of two signaling rates
		Timing Signals		
24	DA	DTE Transmit Clock	DTE to DCE	To help detect center
15	DB	DCE Transmit Clock	DCE to DTE	of signaling element on BA
17	DD	DCE Receive Clock	DCE to DTE	To help detect center of signaling element on BB
		Special Signals		
14	SBA	Secondary Transmit Data	DTE to DCE	)
16	SBB	Secondary Receive Data	DCE to DTE	)
19	SCA	Secondary Request to Send	DTE to DCE	) Special Operations
13	SCB	Secondary Clear to Send	DCE to DTE	)
12	SCF	Secondary Signal Detect	DCE to DTE	)

While the standard clearly limits the performance of the cable, in real applications it is used over a much greater range of lengths and speeds.

(2) Subscriber Lines DTE/DCE ↔ Network

Depending on the application, the connection between the peripheral DTE/DCE and the network can be implemented by any one of the customer access lines described in Section 6.3.

(3) Specialized Higher-Speed Interconnections

For higher-speed applications, several specialized interconnections have been developed. Some of them are

- **Enterprise systems connection** (ESCON): an optical-fiber connection operating up to 40 kilometers at 17 Mbits/s. Developed by IBM, and supported by SNA (see Section 13.4), ESCON extends the mainframe, high-speed, I/O channel (see Section 13.4.1) to remote peripherals.

- **High-performance parallel interface** (HIPPI): introduced by Los Alamos National Laboratory, and standardized by ANSI, HIPPI provides a connection operating up to 120 feet at 800 Mbits/s on a single cable, and 1,600 Mbits/s on two cables. HIPPI includes flow control and error correction capabilities. Over optical fiber, HIPPI operates up to 10 kilometers.

- **High-speed serial interface** (HSSI): supports data transfer at speeds up to 52 Mbits/s. Standardized by ANSI, HSSI describes the physical and electrical interface at the DTE. It can be used with T-3 multiplexers [see Section 6.2.3(4)].

- **Fiber channel** (FC): designed to operate up to 10 kilometers at speeds up to 800 Mbits/s on optical fiber, FC operates over electrical media up to 100 meters (shielded twisted pair or coaxial cable). Standardized by ANSI, and intended to replace HIPPI (see above), FC targets the interconnection of clusters of workstations, and the connection of workstations to supercomputers. When used to interconnect several workstations, FC is connected in star to a nonblocking switching device (see Section 9.6.1). For transmission, FC employs an 8B/10B code (8 bits are represented by 10-bit sequences) that detects all errors caused by an odd number of bits in error, and many other error patterns. Frames contain up to 2,112 payload bytes, a 24-byte header and a 4-byte cyclic redundancy check [see Section 7.3.2(2)] trailer. Provision is made for multiplexing, flow control [see Section 7.2.1(3)], and compatibility with a range of protocols (see Sections 8.3 and 8.4).

- **Firewire**: a fast electronic bus designed to interconnect computer and multimedia devices at speeds up to 400 Mbits/s. Standardized by IEEE as 1394 and IEC as 1883, Firewire is a serial bus that operates over a cable containing two shielded twisted pairs and two power wires. Used to connect devices in daisy-chain fashion, it configures itself automatically, and up to 63 devices can be added in series.

7.1.3 Modem Signals

The pervasive use of the World Wide Web, electronic mail, facsimile, and similar text- or graphics-based person-to-person services requires signaling methods that send data signals over the voice bandwidth of the public switched telephone network.

(1) Signaling Rate

The signaling rate of a transmission system is the number of independent symbols sent per second. Its unit is the *baud*—i.e., one symbol per second is a signaling rate of 1 baud. For binary symbols (1s and 0s) the circuit speed in bits per second is equal to the signaling rate in bauds. In the early 20th century, Harry Nyquist (a Dutch mathematician) showed there is a limit to the rate at which independent symbols can be transmitted.

- **Nyquist's limit**: the maximum signaling rate over a channel with a passband B Hz is $2B$ baud.

Nyquist's work assumes conditions impossible to duplicate on real bearers. As a result, the practical limits to signaling rates are quite a bit less than the Nyquist relationship predicts. For instance, for a channel with a 4-kHz passband, Nyquist's formula predicts a maximum of 8,000 baud; for most voice-grade telephone lines, a practical signaling rate is around 2,400 baud.

On voice-grade circuits, then, does this result mean that circuit speed can be no higher than 2,400 bits/s? Certainly, this will be the case if each symbol represents a 1 or a 0 (i.e., we employ a scheme in which each *symbol* represents 1*bit*). To achieve data rates higher than 2,400 bits/s, we must employ schemes that make each symbol represent several bits. This is not as difficult as it may sound.

(2) Overcoming the Signaling Rate Limitation

If we have four unique symbols (S1, S2, S3, and S4), they can represent all possible combinations of 2 bits. Thus:

$$S1 = 00; S2 = 01; S3 = 10; \text{ and } S4 = 11$$

If we have 16 unique symbols, they can represent all possible combinations of 4 bits. Thus:

$$S1 = 0000; S2 = 0001; S3 = 0010; S4 = 0011; S5 = 0100; S6 = 0101;$$
$$S7 = 0110; S8 = 0111; S9 = 1000; S10 = 1001; S11 = 1010; S12 = 1011;$$
$$S13 = 1100; S14 = 1101; S15 = 1110; \text{ and } S16 = 1111.$$

If we have 2^n unique symbols, they can represent all possible combinations of n bits.

As an example, consider the datastream

$\Leftarrow$0111111001011010000000000000000001000100001111110

- Using two symbols, S1 = 1 and S2 = 0 (a signaling efficiency of 1 bit per baud), and a signaling rate of 2,400 baud, we can transmit 2,400 bits/s
- Using four symbols (a signaling efficiency of 2 bits per baud), we can represent the datastream as

$\Leftarrow$S2S4S4S3S2S2S3S3S1S1S1S1S1S1S1S1S3S1S3S1S2S4S4S3

and, if the signaling rate is 2,400 baud, transmit 4,800 bits/s
- Using 16 symbols (a signaling efficiency of 4 bits per baud), we can represent the datastream as

$\Leftarrow$S8S15S6S11S1S1S1S1S1S9S9S8S15

and, if the signaling rate is 2,400 baud, transmit 9,600 bits/s.

In all three examples, the signaling rate is the same—2,400 baud (i.e., 2,400 symbols/s). The increase in circuit speed is obtained solely at the expense of complicating the symbol structure.

Information theory estimates that the theoretical maximum speed for an analog telephone channel is around 30,000 bits/s (depending on channel signal-to-noise ratio). Today, analog modems operate up to 33,600 bits/s.

(3) Creating Unique Symbols

Suppose 9,600 bits/s are sent over a 2,400-baud line. The number of signal states required to implement a full symbol set is $2^4 = 16$. They are created from 16 sinusoidal signals of the same frequency that are characterized by a particular amplitude and phase angle. For instance, we may choose to create 16 unique signals from two amplitude values (1 and 2) and eight phase angle values (0°, 45°, 90°, 135°, 180°, 225°, 270°, and 315°) applied to an 1,800-Hz signal. The 16 signal states are plotted in polar form in **Figure 7.2**. This diagram is known as a signal *constellation*. Four-bit sequences can be assigned to the signal states in any order. Of course, once assigned, they must be made known to both sender and receiver so that they will associate the correct 4 bits with the same symbol.

Sixteen unique signal states can be created in other ways. For instance, we can use

- 4 amplitude values (1, 2, 3, and 4) and 4 phase angle values (0°, 90°, 180° and 270°)
- 16 amplitude values and 1 phase angle
- 1 amplitude and 16 phase angle values
- Etc.

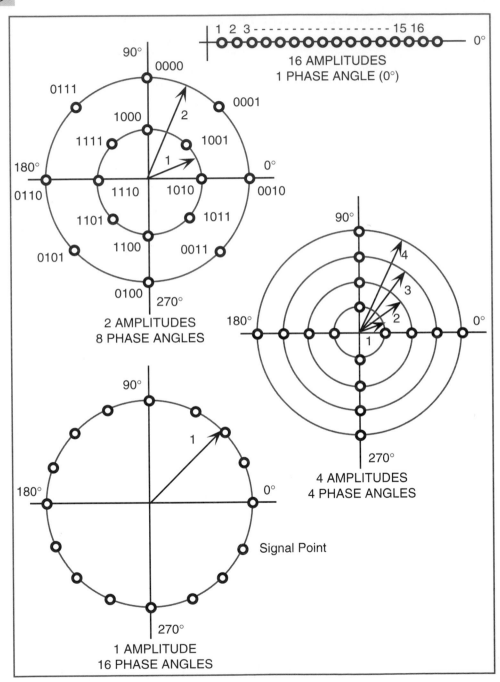

Figure 7.2 Four Signal Constellations That Create 4-Bit Symbols
The options range from 16 amplitude levels to 16 separate phase angles.

Constellations for each of these combinations of amplitudes and phase angles are included in Figure 7.2.

A unique symbol is formed from each of the modulated waves by selecting a segment of the signal. With an 1,800-Hz sinusoidal wave as carrier, and a 2,400-baud signaling requirement, each symbol will consist of a 270° segment of the signal. Using them, we can produce a stream of analog symbols that changes 2,400 times per second (2,400 baud), fits within the bandwidth of a voice-grade line, and transmits 9,600 bits/s. For the 48-bit datastream introduced above, **Figure 7.3** shows the physical signal that carries the message on the voice-grade line. Created from the 16 signals formed from the combinations of four amplitudes and four phase angles drawn in Figure 7.3, the symbols are concatenated according to the 4-bit symbol stream derived from the 48-bit datastream.

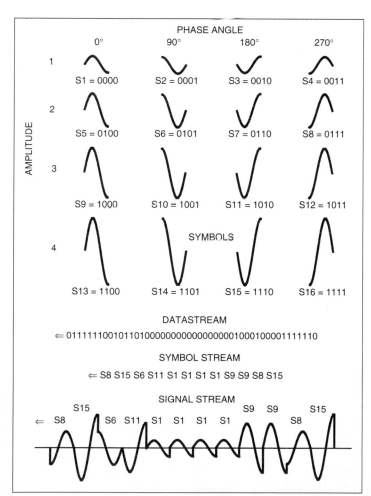

Figure 7.3 Creation of a 16–QAM Signal Stream

Individual symbols, each of which represents 4 bits, are assembled in a sequence determined by the datastream.

None of the streams described above are likely to be used in real-world systems. Instead, so that the margin for error can be about the same for all symbols, the signal points are distributed in a regular fashion over the signal space. **Figure 7.4**, shows a regular constellation—16 combinations formed from 3 amplitudes and 12 phase angles.

(4) Error Control and Data Compression

Forward error correction can be achieved by adding to the signal space

- **Trellis code modulation** (TCM): modulation that adds an extra bit to the n-bit symbol sequence to create twice as many signal points in the constellation as are needed to represent the data. Thus, with a basic 16-signal point code, TCM requires $2^5 = 32$ signal states, not $2^4 = 16$. The last two bits of the symbol set and the TCM-bit define eight subsets of the constellation with four symbols each. The first two bits of the symbol set identify which member of the subset (i.e., which signal state) is used. Doubling the number of signal points in this manner ensures that contiguous symbols fall far enough apart to be distinguished readily. In the demodulation process, the trellis decoder compares the received symbol with all possible valid symbols, and selects the best fit. As a result, errors are reduced, and TCM modems are able to operate at higher speeds than their non-TCM counterparts (for the same error rate).

Other techniques provide both data compression and error control. To perform error detection and correction, the datastream is segmented and assembled in frames that are tested by cyclic redundancy checks [see Section 7.3.2(2)], and the errored frames are repeated. Three techniques are

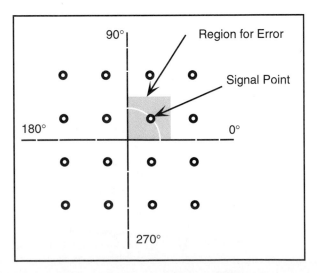

Figure 7.4 Signal Constellation for a Practical 16–QAM Modem

The signal points are arranged to provide an equal region for error for each point.

- **Microcom networking protocol** (MNP): by assembling the data in frames, performing cyclic redundancy checks, and transmitting the frame synchronously, MNP permits the receiving modem to verify the integrity of the received stream. When an error is detected, the receiving modem requests retransmission of the frame. Versions MNP-1 through 4 provide increasingly complex procedures for error detection and correction. MNP-5 uses an adaptive algorithm to compress the data by approximately 2:1. MNP-7 uses entropy coding [see Section 3.3.4(3)] to achieve compressions of approximately 3:1. MNP-configured products exist in several versions (known as Classes) that focus on different operating environments.

- **V.42**: performs error correction at the DCE using asynchronous to synchronous conversion and framing according to LAP–M [Link Access Procedure–Modem, a variant of HDLC (see Section 8.3.4)]. As an alternative, V.42 may use MNP-4 for error correction.

- **V.42 bis**: at the DCE, performs data compression of nominally 4:1 using BTLZ. Developed by Jacob Ziv and Abraham Lempel and updated and expanded by British Telecom, BTLZ is a powerful compression algorithm.

7.1.4 ANALOG MODEMS

Several ITU–T recommendations describe the characteristics of modems that employ echo cancellers to achieve duplex operation over two-wire circuits. They are

- **V.32**: a family of two-wire, duplex modems operating at data rates up to 9,600 bits/s on the public telephone network

- **V.32 bis**: a duplex modem operating at data rates up to 14.4 kbits/s on the public telephone network

- **V.34**: a duplex modem operating at data rates up to 28.8 kbits/s on the public telephone network.

(1) Modem Types

A wide range of analog modems is in use; divided roughly according to speed, they employ a variety of the techniques described above.

- **300 bits/s modems**: most employ FSK. Two frequencies are assigned to one direction, and two different frequencies are assigned to the other. In all, four carrier frequencies are used over a two-wire, duplex connection. AT&T 103/113 modems in North America and ITU–T V.21 modems in Europe, employ this procedure.

- **Multiple-speed modems**: may employ FSK or PSK. AT&T 212A modems employ BFSK at 300 bits/s and BPSK at 1,200 bits/s. AT&T 202 modems operate at 1,200 and 1,800 bits/s with FSK. ITU–T V.22 modems use BPSK at 600 bits/s and QPSK at 1,200 bits/s. ITU–T V.23 modems operate at 600 or 1,200 bits/s with FSK.

- **2,400 bits/s modems**: most employ PSK. AT&T 201 modems and ITU–T V.26 modems operate at 2,400 bits/s with BPSK.

- **4,800 bits/s modems**: most employ 8–PSK. AT&T 208 modems and ITU–T V.27 modems operate at 4,800 bits/s with 8–PSK.

- **9,600 bits/s modems**: AT&T 209 modems use 16–QAM to operate at 9,600 bits/s. ITU–T V.29 modems operate at 9,600 bits/s using a technique in which the datastream is divided into segments of 4 bits. The first bit determines the amplitude of the symbol signal and the next 3 bits determine the phase of the symbol signal (i.e., two amplitudes and eight phase angles). ITU–T V.32 modems employ TCM (32–QAM) and echo cancellation to operate at 9,600 bits/s over a two-wire circuit (i.e., two-wire, duplex).

- **TCM modems**: these modems operate at 2,400 baud, two-wire, duplex, at speeds of 9,600 bits/s (V.32) and 14,400 bits/s (V.32bis). They incorporate trellis coding and echocancellation. Other modems operate over private lines to achieve 14,400 bits/s (V.33), 16,800 bits/s, and 19,200 bits/s. In addition, if data compression is used, effective throughput rates of 40,000 to 50,000 bits/s are reported. The data rate achieved depends on the susceptibility of the datastream to the compression technique employed.

- **TCM modems with precoding and equalization**: using adaptive-bandwidth, adaptive-rate modulation, operation at 24,000 bits/s has been achieved over four–wire, private lines, and 19,200 bits/s over two–wire, dial-up circuits.

- **Packetized-ensemble-protocol multicarrier modems**: perhaps the ultimate in adaptive-bandwidth, adaptive-rate modulation modems, one embodiment employs a comb of 512 carrier tones within the voice band to send symbols in parallel. To start, the originating modem transmits all 512 tones; the receiver evaluates the signals and notifies the transmitter of tones that cannot be used because of noise, or for other reasons. With this information, the originating modem selects a signaling rate, breaks the datastream into packets that contain a fixed number of frame check sequence symbols and as many message symbols as there are acceptable carrier tones, and transmits a packet at a time, in parallel, on the available carriers. If the receiver detects an error, it generates a request for retransmission of the packet. Rates up to 19.2 kbits/s can be achieved with these modems.

- **33.6 kbits/s modems**: employing error control (ITU-V.42, MNP Classes 2–4), data compression (ITU-V.42, MNP Class 5), and echo cancellation, V.34 modems are capable of duplex operation at 33.6 kbits/s. Standardized by ITU–T, they are known as *best efforts* modems. The speed attained is a matter for negotiation when the connection is set up. Because of the variability of switched telephone lines, actual throughput is likely to be less than 33.6 kbits/s. V.34 uses the maximum possible bandwidth permitted by the channel and employs QAM [see Section 4.1.1(3)]. The symbol rate is specified as 2,400, 2,743, 2,800, 3,000, 3,200, and 3,429 baud. The quadrature carriers range from 1,600 to 1,959 Hz, depending on the symbol rate.

- **56 kbits/s modems**: a class of modems (ITU–T V.90) that provides analog operation upstream at ≈ 33.6 kbits/s and digital operation downstream at nominally 56 kbits/s. **Figure 7.5** shows the principle of connecting to a digital server across a digital network (Internet). The ISP is connected to the server by digital channels operating at ≥ 56 kbits/s. The user is connected to the ISP by an analog local loop. The ISP provides downstream data at ≤ 56 kbits/s and receives upstream signals at ≤ 33.6 kbits/s. The exact speeds of the data over the loop depend on the condition of the loop and the noise environment. In many situations the downstream speed will be between 45 and 56 kbits/s, and the upstream speed will be between 28.8 and 33.6 kbits/s.

(2) Modem Compatibility

There is a wide range of operating parameters among modems. Besides operating speeds, operating frequencies and signal constellations, the performance depends on operating mode (i.e., half-duplex or duplex), facility type (i.e., dial-up or private line), asynchronous or synchronous operation, type of error control (if any), type of data compression (if any), etc. Although modern modems adhere to ITU–T standards, the wide variety of in-place modems makes compatibility an issue to be considered whenever voiceband data circuits are involved.

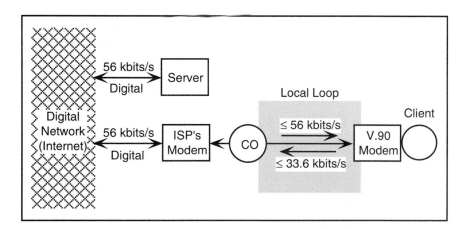

Figure 7.5 Use of a V.90 Modem to Connect to a Digital Server

The speed of the user's downstream channel is approximately 56 kbits/s. The speed of the user's upstream channel can be up to 33.6 kbits/s.

(3) Modem StartUp

Training is an important part of the setup of a data circuit. At the start of communication, to ensure the modems are operating at the best speed, and with mutually compatible error control, data compression, and perhaps other techniques, signals are sent between them. Negotiating compatibility between capabilities, stabilizing, and calibrating them can be complex. Consequently, the pair may exchange negotiating and training signals for several seconds before they are ready to support the exchange of user's data.

(4) Cable Modems

High-speed access channels for home-based servers (with heavy upstream and downstream traffic) and work-at-home (with moderate upstream and heavy downstream traffic) can be provided over CATV networks. In Section 6.3.4, the use of FDM to provide an upstream channel was described. With a suitable cable modem, a combination of downstream and upstream channels can be configured to operate at speeds up to 30 Mbits/s (with throughput of 24 Mbits/s). In the downstream direction, a 6-MHz channel with 64-QAM modulation can support a signaling rate of 30 Mbits/s and a payload of 24 Mbits/s. In the upstream direction, up to twelve 1.8-MHz channels each provide 2.6-Mbits/s signaling rate and a payload of 1.9 Mbits/s.

REVIEW QUESTIONS FOR SECTION 7.1

1   Define DTE and DCE.

2   List some types of DTEs in the terminal/host world.

3   List some types of DTEs in the client/server world.

4   When is a DCE called a modem? What functions does it perform?

5   When does a DCE consist of a DSU and a CSU? What functions do they perform?

6   Describe a popular cable used to interconnect DTE and DCE.

7   What signals must be present before operation can begin over an EIA232 cable?

8   Describe ESCON.

9   Describe HIPPI.

10   Describe HSSI.

11   Describe Fiber Channel.

12   Describe Firewire.

13   Define baud.

14   What is Nyquist's Limit?

15   Does Nyquist's Limit mean that circuit speed can be no higher than 2,400 bits/s?

16   Describe four ways of creating 16 unique symbols.

17   What is trellis code modulation? What advantages does it have? What is its disadvantage?

18   Describe MNP. What is it used for?

19   Describe ITU–T standards V.42 and V.42bis.

20   Describe ITU–T standards V.32, V.32bis, and V.34.

21   Comment on the following statement: Although modern modems adhere to ITU–T standards, the wide variety of in-place modems makes compatibility an issue to be considered whenever voiceband data circuits are involved.

22   Why is modem training necessary?

23   Describe the performance of a cable modem.

7.2 COMMUNICATION PROCEDURES

In voice communication, the basic rules were worked out thousands of years ago. For data communication, there is little history, and the rules must be all encompassing so that nothing is left to chance. *If it can happen, it will happen* is the maxim that must guide those who would design procedures (protocols) for the machine-centered environment. In this section I discuss the protocols needed to send data over a data link.

7.2.1 RANGE OF PROTOCOL FUNCTIONS

Communications protocols must include rules and procedures to cope with all normal and abnormal operating situations. Some of them are

- **Startup**: how does communication begin? Does the sender gain access to the line by being polled or by competing for it? Can any station initiate communication, or is one station in control? Once interrupted, how is communication reestablished?

- **Message format**: how are data organized and in what sequence are they sent? How are long messages handled? How are short messages handled?

- **Identification and framing**: how does the receiver separate the bit stream into characters? How does the receiver distinguish between control information and data characters? How does the receiver separate header information from text information, and text information from trailer information? How are frames identified?

- **Line control**: in half-duplex operation, under what circumstances is the line turned around?

- **Error control**: what process will be used for the detection and correction of errors? How, and when, does the receiver inform the sender no errors have been detected in the frame(s) received, or, conversely, errors have been detected?

- **Termination**: under normal conditions, how is communication terminated? Under abnormal conditions, how is communication terminated?

In addition, to be user friendly, communication procedures should have the following attributes

- **Transparency**: they must function in the presence of all bit sequences that occur in the text portion of the frame. Specifically, if data bytes and control octets share the same bit patterns, there must be a way to distinguish between a bit sequence that is data, and the same bit sequence that is used for a control purpose.

- **Independence**: their performance should not be influenced by the coding employed by the users.

- **Efficiency**: the number of bits added to the frame for control purposes must be small so as to devote most of the transmission time to sending information.

(1) Half-Duplex Line Discipline

In half-duplex transmission, data flow in one direction at a time.

- Usually, one station is the **primary station**; it determines when secondary stations shall send, and when they shall receive.

- Sometimes, information is sent frame by frame with the sending station pausing between frames for the receiver to acknowledge receipt of an error-free frame, or to request retransmission of the frame. This is **stop-and-wait ARQ** error correction (see Section 7.4.1).

- Frequently, the sending station sends a predetermined number of frames before pausing for acknowledgment. If a frame contains an error, the receiver discards it, ignores any subsequent frames sent, and, when able to do so, requests resending of all frames from the one found in error. This is **go-back-n ARQ** error correction (see Section 7.4.1).

- Because a limited number of bits are allocated to **frame numbers**, the receiver must acknowledge error-free receipt of frames at least once within each numbering cycle. If acknowledgment is not made, the sender must stop sending until it is.

(2) Duplex Line Discipline

In duplex transmission, data flow in both directions at the same time, and the stations may be receiving and transmitting simultaneously.

- For **equal status** stations, either station may initiate communication. To begin communication the station with something to send must inquire whether the other station is ready to receive the message.
- For **unequal** stations, one station will be designated primary and the other(s) will be designated secondary station(s). The primary station manages message flow (see polling, Section 7.2.2).
- **Information** is transmitted as a sequence of frames. Sending is interrupted to repeat frames in which the receiver has detected an error. Go-back-n ARQ and selective repeat ARQ error correction are used (see Section 7.4.1).
- Because a limited number of bits are allocated to **frame numbers**, the receiver must acknowledge error-free receipt of frames at least once within each numbering cycle. If acknowledgment is not made, the sender must stop sending until it is.

7.2.2 POLLING

For stations of equal status connected on a point-to-point (or multipoint-to-multipoint) basis, the stations poll each other.

- **Peer polling**: the stations poll one another, in sequence, continually. If the polled station has something to send, it sends it before polling the next station; if it has nothing ready, it polls the next station in the sequence. When all stations have been polled by another station, the process is repeated. To ensure equal access for all stations, and to provide continuing operation even when some stations have failed, it is essential that one station monitors the total process and is supplied with procedures to restart it if, for any reason, it fails.

For unequal stations connected on a point-to-point (or point-to-multipoint basis), the primary station manages communications by polling the secondary station(s)

- **Primary polling**: in turn, the primary station asks each of the secondary stations if they have anything to send. If a secondary station has a frame (or frames) to send, when polled, it transmits it (or them) to the primary; if the secondary station has nothing ready to send, it indicates nothing to send, and the primary station polls the next station. Under normal operation, the primary station interrogates the secondary stations continually. When the primary station has a frame (or frames) to send, it ceases to poll the secondary stations and sends its frames. Often, the primary station is a front-end processor that manages the communication process for a host.

7.2.3 FLOW CONTROL AND CONGESTION

Traffic may bunch up and overload the paths and nodes in a network. This occurs when network resources are in demand and processing speeds lag behind the speed of the incoming traffic. Two basic control methods can be distinguished

- **Credit-based flow control**: on the basis of the traffic it is capable of handling, the receiver monitors traffic on each incoming line and assigns *credits* to the upstream nodes. They may send message units up to their credit limit, but no more until they receive a new credit allocation. In effect, credit-based control stores traffic elements at upstream nodes in the network until they can be received without interference. Whether congestion is short term or long term, processing overhead is significant for this procedure.
- **Rate-based flow control**: on the basis of the amount of traffic it is receiving, the receiver controls the rate at which input nodes accept traffic intended for it. As a result, messages may be queued at the edges of the network, and at the receiving node, but not in intervening nodes. In effect, rate-based control stores traffic at source nodes until the addressed node is ready to receive them. For conditions in which congestion is long term but infrequent, rate-based control requires little overhead. However, if congestion is short term and frequent, rate-based control requires significant overhead.

In effect, to prevent congestion, rate-based flow control relies on reducing the traffic in the network, while credit-based flow control distributes excess traffic to storages in the network nodes until it can be handled. Even so, the implementation of flow control to manage congestion knows no simple solution. Several schemes are described in later chapters.

REVIEW QUESTIONS FOR SECTION 7.2

1 List some of the normal and abnormal operating conditions a communication protocol must accommodate.
2 What is meant by transparency with respect to communication protocols?
3 What is meant by independence with respect to communication protocols?
4 What is meant by efficiency with respect to communication protocols?
5 Describe the discipline involved in transmission over a half-duplex line.

6 Describe the discipline involved in transmission over a duplex line.

7 Describe peer polling.

8 Describe primary polling.

9 Distinguish between rate-based and credit-based flow control.

7.3 ERROR DETECTION

If the recipient is to have confidence in the integrity of the messages received, the transmission process must include means for error control. Arguably, error control—the detection and correction of errors—is the most important function performed by terminal equipment.

- **Error control**: a cooperative activity between sender and receiver in which the sender adds information to the character or frame to assist the receiver to determine whether an error has occurred. If it has, the sender and receiver work together to correct it.

A concept of this activity is shown in **Figure 7.6**. Before sending, the sender and receiver must agree on the information that will be added, how it will be used, and how an error will be corrected. In this section, I discuss methods for error detection.

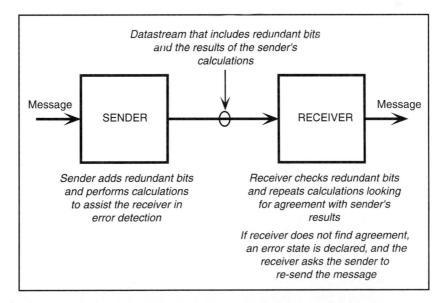

Figure 7.6 Error Detection Process

7.3.1 PARITY CHECKING

One method of error detection adds parity bits to individual characters or to other parts of the message.

(1) Vertical and Longitudinal Checking

Parity checking adds a bit (0 or 1) to a sequence of bits in the user's data to make the total number of 1s in the sequence even or odd.

- **Even parity check**: the parity bit makes (or leaves) the number of 1s even
- **Odd parity check**: the parity bit makes (or leaves) the number of 1s odd.

Parity bits may be added to individual characters, in which case the process is known as *vertical redundancy checking* (VRC), or to bit sequences formed from bits assigned to specific positions in a block of characters, in which case the process is known as *longitudinal redundancy checking* (LRC). When operating with start (0_s) and stop (1_s) bits, they frame the entire set of character bits and parity bit—i.e.,

$$\leftarrow 0_s\, xxxxxxxx(0/1)_p 1_s 1_s \quad.$$

Adding a parity bit to ASCII characters sent singly reduces the transmitted coding efficiency to 70%. Thus, to send 7 information bits requires 1 start bit + 7 character bits + 1 parity bit + 1 stop bit—i.e., a total of 10 bits. If two stop bits are employed, the efficiency drops to 64%. Employing parity bits with ASCII in a standard 136-octet frame (5 header, 128 text, and 3 trailer) produces an efficiency of 82%.

Parity is a relatively weak error detection technique; for one thing, it can detect the presence only of an odd number of errors. Nevertheless, parity checking can be productive. Consider the case of a very long datastream that is made up of 8 bit words in which 7 are message bits and 1 is a parity check bit. Let transmission errors occur at an average rate of 1 bit in error in 1,000 bits. If they occur randomly and are independent of one another, the probability of errors occurring is given by the binomial distribution. For an 8-bit word, the total probability of a word being *in error* is 0.008028—i.e., words with single, double, triple, etc., errors. The probability of a word in error being *detected* is 0.008—i.e., words with single, triple, etc., errors. Hence the ratio of errored words detected to errored words not detected is close to 300:1.

For a range of probabilities of error from 0.1 to 1 x 10⁻⁶, **Figure 7.7** shows the ratio of errored words detected to errored words not detected. While parity may be a weak form of error detection, it can be effective when transmission errors are random and independent. The efficiency of parity checking increases as the probability of bit error decreases.

(2) VRC and LRC Combined

Longitudinal redundancy checking adds parity bits to sequences formed from bits assigned to specific positions in a sequence of characters. The process

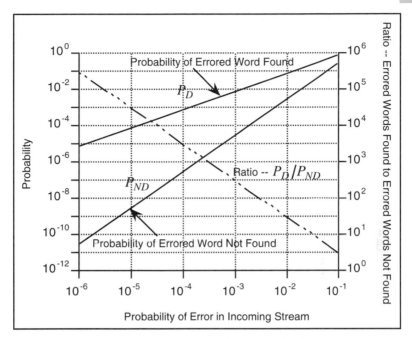

Figure 7.7 Probability of Error Detection by Parity Checking
Errors are assumed equally likely to be 1s or 0s that are randomly distributed throughout the datastream.

produces an additional character, known as the *block check character* (BCC), that is transmitted in the trailer of the frame. As with VRC, the use of LRC only detects the presence of odd numbers of errors.

Figure 7.8 shows vertical and longitudinal redundancy checks on a sequence of five ASCII characters. Bits added to the characters provide even parity checking. Bits added to the rows provide even parity checking of each row of bits; they form the block check character. At the receiver, the VRC bits and the BCC are recalculated. If agreement is found with the values sent, the sequence of data is assumed to have been received without odd transmission error(s).

In certain circumstances, the combination of VRC and LRC can locate the exact position of an error. The bottom half of Figure 7.9 shows the determination of the location of a specific bit error. However, the technique fails to uncover many multiple errors. For instance, the second example at the bottom of Figure 7.8 shows a combination of four errors that will not be detected.

7.3.2 TECHNIQUES THAT EMPLOY CALCULATIONS

By treating the entire datastream, or segments of the datastream, as continuous binary numbers, error detection can be based on modulo–2 arithmetic.

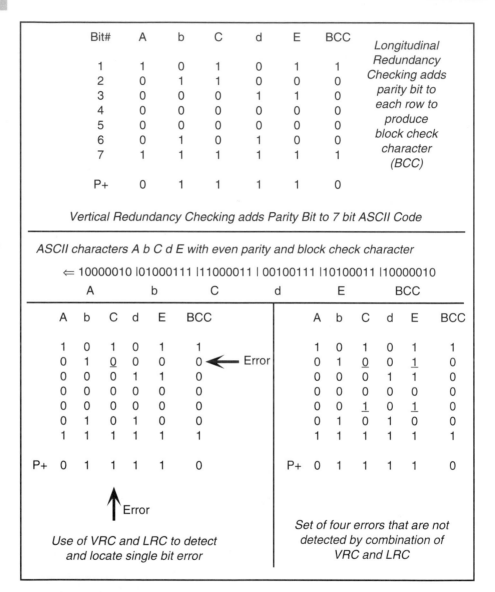

Figure 7.8 Vertical and Longitudinal Redundancy Checking

(1) Checksum

A technique the transmitter can use to assist the receiver in detecting the presence of errors employs addition. Called a *checksum*, the process treats the characters as binary numbers, adds them together, and expresses their sum as a 16-bit binary number to produce a 2-octet checksum. The transmitter performs the addition, and attaches the checksum to the datastream as it is sent to the receiver. The receiver performs the same addition on the message bits in the received datastream. If the difference between the receiver's checksum and the transmitted checksum is zero, it is likely the datastream has been received without error.

(2) Cyclic Redundancy Checking

Another technique treats the entire datastream, or a segment of the datastream, as a continuous binary number. Called *cyclic redundancy checking* (CRC), the process is illustrated in **Figure 7.9**. Given a k-bit message M_k, the sender divides it by a $n + 1$ bit prime number G_{n+1} to produce an n-bit remainder F_n [known as the *frame check sequence* (FCS)]. A prime number divisor is necessary to ensure that the remainders are unique. G_{n+1} is called the *generator function*; it is $n + 1$ bits long and is known to both the sender and receiver. The value of the FCS calculated by the sender is placed in the trailer of the frame. In the incoming frame, the receiver divides the k-bit message by G_{n+1} to produce the frame check sequence F'_n. If $F'_n = F_n$ it is almost certain that M_k has been received without error.

For convenience, the generating function can be expressed as a polynomial in a dummy variable (x, say) with coefficients that may be 1 or 0. Thus, the sequence 10100011 can be written

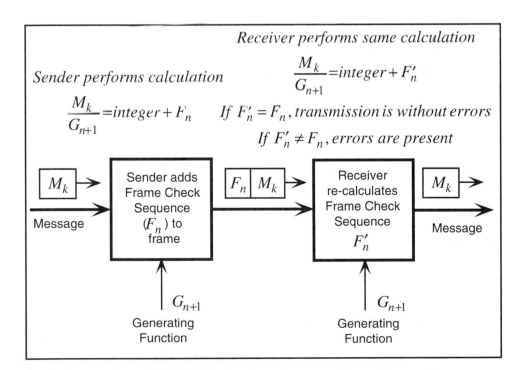

Figure 7.9 Principle of Cyclic Redundancy Checking

$$1x^7 + 0x^6 + 1x^5 + 0x^4 + 0x^3 + 0x^2 + 1x^1 + 1 = x^7 + x^5 + x + 1$$

For common use, several generating functions have been standardized. They are

- **CRC-10**: $x^{10} + x^9 + x^5 + x^4 + x + 1$. Generates a 10-bit FCS for streams of 8-bit characters.
- **CRC-12**: $x^{12} + x^{11} + x^3 + x^2 + x + 1$. Generates a 12-bit FCS for streams of 6-bit characters.
- **CRC-16**: $x^{16} + x^{15} + x^2 + 1$. Generates a 16-bit FCS for streams of 8-bit characters.
- **CRC-CCITT**: $x^{16} + x^{12} + x^5 + 1$. Generates 16-bit FCS for streams of 8-bit characters. •
- **CRC-32**: $x^{32} + x^{26} + x^{23} + x^{22} + x^{16} + x^{12} + x^{11} + x^{10} + x^8 + x^7 + x^5 + x^4 + x^2 + x + 1$. Generates 32-bit FCS for streams of 8-bit characters.

Each function is a binary prime number; they produce FCSs that are unique for all practical datastreams. They detect all single-bit errors, all double-bit errors, any odd number of errors, any burst errors for which the number of bits in error is less than the number of bits in the FCS, and most of the errors caused by larger bursts. Cyclic redundancy checking is by far the strongest of the common error checking procedures. For randomly distributed errors, some estimates place the likelihood of CRC-16 not detecting an error at 1 in 10^{14} bits. For a system transmitting continuously at 1.544 Mbits/s, this amounts to approximately one undetected error every two years.

7.3.3 CODING

Error detection can be accomplished through the use of relatively complex coding schemes. Because they contribute to both error detection and error correction, they are described in Section 7.4.

REVIEW QUESTIONS FOR SECTION 7.3

1. What is meant by error control?
2. Describe parity checking.
3. What is vertical redundancy checking?
4. What is longitudinal redundancy checking?
5. Why is parity checking considered to be a weak error detection tool? When is parity effective?

6   What is a block check character?

7   Describe how VRC and LRC can be combined to identify a specific bit in error.

8   Describe an error pattern that VRC and LRC cannot detect.

9   What is a checksum?

10   Describe cyclic redundancy checking.

7.4 ERROR CORRECTION

Once detected, an error must be corrected. Two basic approaches to error correction are

- **Automatic-repeat-request** (ARQ): upon request from the receiver, the transmitter resends portions of the exchange in which errors have been detected
- **Forward error correction** (FEC): employs special codes that allow the receiver to detect and correct a limited number of errors without referring to the transmitter.

7.4.1 ARQ TECHNIQUES

Three basic ARQ techniques are

- **Stop-and-wait**:
  - The sender sends a frame and waits for acknowledgment from the receiver
  - If no error is detected, the receiver sends a positive acknowledgment (ACK), which may distinguish between even-and odd-numbered frames (ACK0 and ACK1)
  - The sender responds with the next frame
  - If an error is detected, the receiver returns a negative acknowledgment (NAK)
  - The sender repeats the frame.
- **Go-back-n**:
  - The sender sends a sequence of frames and receives an acknowledgment from the receiver
  - On detecting an error in the sequence it is receiving, the receiver discards the corrupted frame and ignores all further frames in the sequence

    – The receiver notifies the sender of the number of the first frame in error

    – On receipt of this information, the sender begins resending a sequence starting with that frame.

Because replacement of a corrupted frame depends on both receiver and sender sharing the same frame number, the sender must stop sending after an agreed-upon number of frames (7 and 127 are common values). When the receiver acknowledges that all frames in the block have been received without detecting an error, the sender can reset the frame counter and begin to send another block.

- **Selective-repeat**: used on duplex connections only. On the return channel, the receiver returns negative acknowledgments for the individual frames found to have errors. The sender repeats the frames for which NAKs are received. This procedure puts a repeated frame out of sequence so that, for multiframe messages, the receiver must reorder the frames. The sender must not send any more frames than there are numbers to identify them before receiving an acknowledgment from the receiver.

(1) Throughput

To the network designer, the rate at which each link transports data is of vital importance. However, to the sender, the internal network speeds are much less important than the time it takes to deliver a message. Along with message information must go overhead to alert the receiver that a message is coming, to carry the address, to manage the data link, and to check for, and correct, errors.

If the frames are n bits long, they contain k information bits (user's data), and the bit rate is r bits/s

- **Throughput** (S): the effective rate at which user's data bits are delivered is

$$S = \frac{\text{number of user's data bits delivered in } m \text{ frames}}{\text{time to deliver all bits in } m \text{ frames}} \text{ bits/s}$$

- **Normalized throughput** (S'): the ratio of the time to deliver only the user's data bits (i.e., mk/r) to the time to deliver all bits in a sequence of m frames is

$$S' = S/r = \frac{\text{time to deliver only user's data bits in } m \text{ frames at bit-rate employed}}{\text{time to deliver all bits in} m \text{ frames}}$$

where
    – *all bits* includes the number of user's data bits (mk), the number of overhead bits in m frames [$m(n-k)$, where n is the number of bits in a frame], any additional bits included in signaling required by the protocol in use, and additional bits used to correct any errors detected

– *time to deliver all bits* includes the time to deliver *m* frames, propagation time(s), decision time(s), time for whatever additional signaling the protocol requires, and time for error correction, if needed.

Normalized throughput is a measure of protocol efficiency.

(2) Protocol Efficiency

Figure 7.10 shows examples of stop-and-wait procedures. In Figure 7.11A, over a duplex connection, the sender transmits frames at a rate of *r* bits/s, and the receiver acknowledges them at the same rate. In Figure 7.11B, frame F1 is found to be in error, the receiver sends a NAK, and the sender repeats F1. In Figure 7.11C,

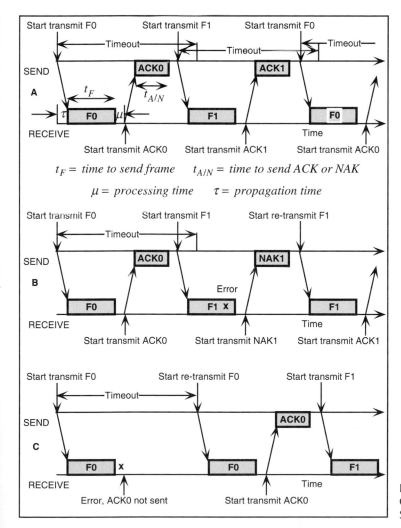

Figure 7.10 Data Communication Using Stop-and-Wait Procedures

the receiver fails to acknowledge frame F0. Because no acknowledgment has been received by timeout, the sender repeats frame F0. This time it is acknowledged.

The sender's frames are n bits long and contain k information bits. The propagation time—i.e., the time for signal energy to cross the distance between sender and receiver—is designated τ seconds. The time to transmit each frame is designated t_f seconds, the time for the receiver to determine the condition of the received frame (the processing time) is μ seconds, and the time to transmit ACK (i.e., received without error) or NAK (i.e., received, but error detected) is $t_{A/N}$ seconds. To simplify things, we assume the acknowledgment frames to be $n - k$ bits and the time for the sender to determine the next step in the procedure after receiving an acknowledgment is μ seconds. Then, in an error-free environment, the time to send user's bits is kt_F/n, and the time to send all bits is $2\tau = t_F + 2\mu + t_{A/N}$. Thus:

$$\eta_{ARQ} = \frac{k}{(2n - k) + 2r(\tau + \mu)}$$

and, if a protocol is employed that sends a block of m frames before pausing for acknowledgment and resetting the frame numbers:

$$\eta_{ARQ(m)} = \frac{mk}{\{(m+1)n - k\} + 2r(\tau + \mu)}$$

By way of illustration, if

- The sender sends frames of 1,088 bits (1,024 information bits, 64 overhead bits) to the receiver
- The receiver responds with acknowledgment frames of 64 bits
- Propagation time and decision time can be neglected (i.e., $\tau = \mu = 0$)

then the protocol efficiency is 0.89 when $m = 1$, 0.93 when $m = 7$, and 0.94 when $m = 127$. In an error-free environment, the only factor degrading performance is the overhead to signal the sender when 1, 7, or 127 frames have been received. These values of η_{ARQ} then, are the best that can be achieved.

Also, if

- The sender sends frames of 1,088 bits (1,024 information bits, 64 overhead bits) to the receiver
- The receiver responds with acknowledgment frames of 64 bits
- Error-free transmission occurs over a 1,000-mile optical fiber facility for which the propagation speed is 1.26×10^5 miles/s (see Section 5.2.1) so that the propagation time is 0.008 second
- The processing time to begin to send an acknowledgment or to decide to send the next frame is negligible (i.e., $\mu = 0$)

then **Figure 7.11** shows how the protocol efficiency varies with the number of frames in a block. The curves are asymptotic to the upper bound calculated above. In an error-free environment, the greater the number of frames in a block, the greater the protocol efficiency.

(3) Throughput of Stop-and-Wait ARQ with Errors

In stop-and-wait ARQ, the sender sends a frame and waits for acknowledgment from the receiver ($m = 1$). If no error is detected, the receiver sends a positive acknowledgment. The sender responds with the next frame. If an error is detected, the receiver returns a negative acknowledgment and the sender repeats the frame.

If the errors are randomly distributed, and if each error requires that a frame be repeated, the top diagram in **Figure 7.12** shows normalized throughput for $k =$ 1,024 bits, $n = 1,088$ bits, $\tau = 0.0008$ second, $\mu = 0.005$ second, $r = 4.8$ kbits/s, 56 kbits/s, and 1.544 Mbits/s, and $1 \times 10^{-8} \leq P_s \leq 1 \times 10^{-2}$. An important point is the assumption that errors are uniformly distributed and each error causes a frame to be repeated. Thus, on average, with frames of around 1,000 bits, data transfer ceases at levels of probability of error around 1×10^{-3}. The model does not take into account the probability of multiple errors in one frame.

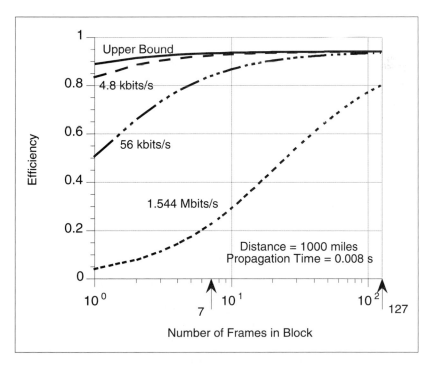

Figure 7.11 Effect of Distance and Bit Rate on Protocol Efficiency

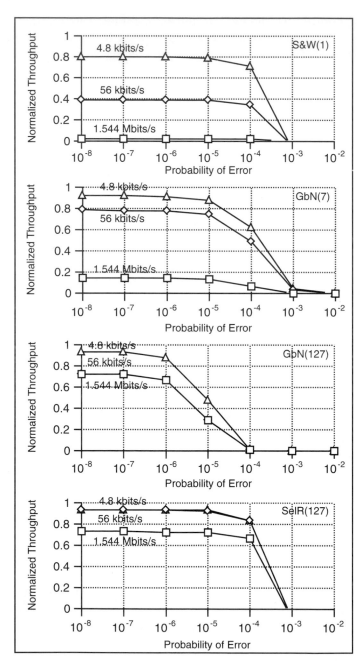

Figure 7.12 Normalized Throughput as a Function of Bit Rate and ARQ Procedure

(4) Throughput of Go-Back-N ARQ with Errors

In go-back-n ARQ, the sender sends frames in a sequence and receives acknowledgments from the receiver. On detecting an error, the receiver discards the corrupted frame and ignores any further frames. The receiver notifies the sender of the number of the first frame in error. On receipt of this information, the sender begins resending the data sequence starting from that frame.

If the errors are randomly distributed, and, on average, an error in a sequence of m frames causes $m/2$ frames to be repeated, the middle diagrams in Figure 7.12 show normalized throughput for $k = 1{,}024$ bits, $n = 1{,}088$ bits, $m = 7$ or 127 frames, $\tau = 0.0008$ second, $\mu = 0.005$ second, $r = 4.8$ kbits/s, 56 kbits/s, and 1.544 Mbits/s, and $1 \times 10^{-8} \le P_s \le 1 \times 10^{-2}$. Because errored sequences occur more frequently, the performance of GbN(127) lags GbN(7) in the mid-range of probabilities of error. However, for low values of error probability, performance at 1.544 Mbits/s improves by a factor of 4 when $m = 27$.

(5) Throughput of Selective Repeat ARQ with Errors

In selective-repeat ARQ, the sender repeats only those frames for which negative acknowledgments are received from the receiver. Without disturbing the flow of frames from the sender, NAKs are sent by the receiver over the return channel. The sender repeats the frames as they can be inserted in the message flow. The sender must clear all requests for repeat before beginning a new sequence of frames.

If the errors are randomly distributed, and each error causes a frame to be repeated, the bottom diagram in Figure 7.12 shows normalized throughput for $k = 1{,}024$ bits, $n = 1{,}088$ bits, $m = 127$ frames, $\tau = 0.008$ second, $\mu = 0.005$ second, $r = 4.8$ kbits/s, 56 kbits/s, and 1.544 Mbits/s, and $1 \times 10^{-8} \le P_s \le 1 \times 10^{-2}$. With selective repeat, the fact that NAKs are sent on a return channel without interrupting forward data flow reduces overhead delays significantly. As is to be expected, throughput performance improves rapidly as error probability decreases. However, corrected frames must be stuffed into the datastream, leading to frames out of sequence. In a multiframe message, the receiver must reconstruct the correct sequence before the file can be used. Further, the sender and receiver must clear all corrupted frames before proceeding to the next 7- or 127-frame sequence.

(6) Comparison of ARQ Techniques

Figure 7.12 can be used to compare the performance of different ARQ techniques on the basis of normalized throughput for bit rates of 4.8 kbits/s, 56 kbits/s, and 1.544 Mbits/s and error probabilities from 1×10^{-8} to 1×10^{-2}. For convenience, the discussion is divided into *low* levels of error probabilities(i.e., 1×10^{-8} to 1×10^{-6}); *moderate* levels of error probabilities (i.e., 1×10^{-6} to 1×10^{-4}), and *high* levels of error probabilities (i.e., 1×10^{-4} to 1×10^{-2}).

- **4.8 kbits/s**:
  - *low* levels of error probabilities. SelR(127) is the most efficient protocol; it is followed closely by GbN(127) and GbN(7)
  - *moderate* levels of error probabilities. The performance of GbN(127) drops off rapidly and S&W(1) becomes a competitor in the range 1×10^{-5} to 1×10^{-4}
  - *high* levels of error probabilities. All procedures perform poorly.
- **56 kbits/s**:
  - *low* levels of error probabilities. SelR(127) is the most efficient protocol; it is followed closely by GbN(127)
  - *moderate* levels of error probabilities. In this range, the performances of S&W(1), GbN(127), and GbN(7) drop off; SelR(127), is without a close competitor
  - *high* levels of error probabilities. All procedures perform poorly.
- **1.544 Mbits/s**:
  - *all* levels of error probabilities. Throughput performance is diminished due to the fraction of time required for propagation and processing delays as the frame time decreases with increased bit rate
  - *low* levels of error probabilities. SelR(127) is the most efficient protocol; it is followed closely by GbN(127)
  - *moderate* levels of error probabilities. SelR(127) is supreme
  - *high* levels of error probabilities. All protocols perform very poorly.

Obviously, the choice of a protocol is very dependent on the parameters of the operating environment. If the receiver can handle frames out of order and the connection is duplex, SelR(127) performs well under moderate and low probability of error conditions. If the receiver cannot handle frames out of order, or the connection is half-duplex, GbN(7 or 127) is an appropriate choice. In a high probability of error environment, if the receiver cannot handle frames out of order, low-speed S&W(1) may be the only ARQ technique that can be used. In the following section, we describe a way to achieve error correction that improves on ARQ techniques at high levels of error probabilities and permits operation at high speeds.

7.4.2 FORWARD ERROR CORRECTION

Provided the number of errors is less than a value determined by the coding, the receiver can detect and correct errors without reference to the sender. This convenience is bought at the expense of adding bits.

(1) Repetition Code

If there are only two messages to send, I can use an 8-bit repetition code to represent them. To send message A, I send a block of 8-1s, and to send message B, I send a block of 8-0s. If the receiver knows that the message is A or B and no other, and it is provided with the ability to determine the logical distance[2] between each incoming message and the two known messages, this strategy will allow the receiver to correct for up to 3 bits in error.

- Suppose the received word A′ is in error by 1 bit. The logical distance between A′ and A is 1 and the logical distance between A′ and B is 7; thus, A′ is likely to be A.
- Suppose A′ is in error by 3 bits. The logical distance from A is 3 and the logical distance from B is 5; thus A′ is likely to be A.
- Suppose A′ is in error by 4 bits. The logical distance from A is 4 and the logical distance from B is 4; thus A′ is equally likely to be A or B.

Continuing the sequence to higher levels of error makes A′ more likely to be B than A. For this particular case, then, the limit of correction is 3 bits in error.

By way of illustration, if

- In a very long datastream, transmission errors occur at an average rate of 1 bit in error in 10 bits
- The errors are randomly distributed and independent of one another
- An 8-bit repetition code is used to overcome the effect of the high error rate

then coding will correct up to three errors. Hence, the remaining probability of error is due to the occurrence of four, or more errors in a word. Because the transmission errors occur randomly and independently, the probability of k errors occurring in an n-bit word is given by the binomial distribution. When $k = 0.1$, and $n = 8$, the probability of words occurring with four through eight errors is 0.0076, so that the code has a probability of error of 0.0076. Without the coding, the probability of errors is 0.1. Of course, this performance is achieved with a coding efficiency of only 12.5%. If the bit error rate is reduced to 1 bit in error in 100 bits, an 8-bit repetition code has a probability of error of better than 1 in 10^6.

---

[2] The logical distance between two binary sequences is determined by finding the number of bits that must be changed to make one sequence into the other. This measure is also known as the Hamming distance.

CHAPTER 7

(2) Practical Codes

Codes used to provide forward error correction are more sophisticated than our example.[3] They are divided into two types

- **Linear block codes**: in linear block coding, a block of k information bits is presented to the encoder; the encoder responds with a unique code block (codeword) of n bits ($n > k$). A set of codewords is selected so that the logical distance between each codeword and its neighbors is approximately the same.

- **Convolutional codes**: convolutional codes are generated by continuously performing logic operations on a moving, limited sequence of bits (m bits, say) contained in the message stream. For each bit in the message stream, the encoder produces a fixed number (two or more) of bits whose values depend on the value of the present message bit and the values of the preceding $m - 1$ bits.

The 8-bit repetition code introduced above is an example of a block code in which 1 information bit is represented by a block of 8 bits ($k = 1$ and $n = 8$). It corrects any combination of three, or fewer, errors.

Parity checking is a special case of a block code; it is a single-error detecting code—it cannot correct errors. Exhaustive computer searching of possible implementations has determined optimal encoding and decoding strategies. If the number of errors is larger than the correction span, the receiver must resort to an ARQ technique to achieve retransmission of the corrupted message.

(3) Linear Block Codes

By adding redundant bits to information bits in a disciplined way, linear block codes build on the principle of parity checking. Pioneered by R. W. Hamming,[4] a 20th-century American mathematician, they allow us to reduce the information bit error rate while maintaining a fixed transmission rate. If an efficient message source generates M equally likely messages from a k bit code so that $M = 2^k$, and r redundant bits are added to each codeword, the messages are expanded into codewords of length n bits, where $n = k + r$. With this expansion, there are $2^n - 2^k$ redundant codewords. Block codes of this sort are known as *systematic* and called (n,k) codes. Further, they are said to have a *rate* of k/n. This quantity reflects the fact that the information bit rate is k/n times the bit rate of the (n,k) word. If the speed at which infor-

---

[3] For a comprehensive discussion of coding and codes, see Herbert Taub and Donald L. Schilling, *Principles of Communication Systems*, 2nd ed., (New York: McGraw-Hill, 1986), 529–78.

[4] R. W. Hamming, "Error detecting and correcting codes," *Bell System Technical Journal*, 29 (1950), 147–160.

mation is to be delivered is not to be changed, the *n*-bit word must be sent at a higher speed. In fact, the bit rate of the (*n*,*k*) word must be *n/k* times the bit rate of the uncoded word.

If the logical distance between two codewords is represented by d_{ij}, the value will vary with the combination *i,j*. Called the *Hamming* distance, the minimum distance d_{min} between words sets a limit to the effectiveness of the code. If there are ε errors in a received codeword, we shall be able to detect that the word is errored provided

$$\varepsilon \le d_{min} - 1$$

Also, if

$$\varepsilon \le \frac{d_{min} - 1}{2} \text{ , } d_{min} \text{ odd; and } \varepsilon \le \frac{d_{min}}{2} - 1 \text{ , } d_{min} \text{ even}$$

we can correct the errors by substituting the closest valid codeword.

In the 8-bit repetition code introduced above, we have only two valid codewords. The distance between them is 8; because there are only two, this is also the Hamming distance—i.e., $d_{min} = 8$. The maximum number of errors that can exist in a word detected to be in error is $\varepsilon = d_{min} - 1 = 7$. Anything less than eight errors produces an unknown codeword. The maximum number of errors in a word that can be corrected is $\varepsilon = d_{min} / 2 - 1 = 3$. With three or less errors, the received word is closer to the word it is supposed to be than to the word it is not supposed to be.

(4) Cyclic Codes

A subclass of linear block codes, *cyclic* codes have a structure that gives them an efficient code rate and makes them relatively easy to implement. Their construction is best illustrated by an example. Consider the codeword 01001101. By moving the bits in order one position to the right we can form a set of eight cyclic codewords—viz:

$$
\begin{aligned}
C1 &= 0\,1\,0\,0\,1\,1\,0\,1 \\
C2 &= 1\,0\,1\,0\,0\,1\,1\,0 \\
C3 &= 0\,1\,0\,1\,0\,0\,1\,1 \\
C4 &= 1\,0\,1\,0\,1\,0\,0\,1 \\
C5 &= 1\,1\,0\,1\,0\,1\,0\,0 \\
C6 &= 0\,1\,1\,0\,1\,0\,1\,0 \\
C7 &= 0\,0\,1\,1\,0\,1\,0\,1 \\
C8 &= 1\,0\,0\,1\,1\,0\,1\,0
\end{aligned}
$$

If the process is pursued, the set of codewords is repeated. By inspection, the logical distance between contiguous codewords, i.e., C2–C1, C3–C2, C4–C3, etc., is 6. Also, the distance between a codeword and all codewords other than those that are contiguous is 4. Thus, C3–C1 = 4, C4–C1 = 4, C5–C1 = 4, etc. Hence $d_{min} = 4$ so that up to three errors can be detected and one error can be corrected. Cyclic codes in common use have d_{min} = 3, 5, and 7, and correct one, two, or three errors. Called cyclic Hamming; Bose, Chaudhuri, and Hocquenghem;[5] and Golay,[6] after their inventors, the codes increase overhead significantly. Throughput is 71%, or less.

(5) Burst Error Correction

The parity bits (r) included in block codes allow correction of a limited number of errors. They are most effective if the average bit error rate is small and the errors are evenly distributed within the codeword sequence. However, in many situations, the errors are closely clustered together—they occur in *bursts*—and affect entire codewords and short sequences of codewords. Two techniques that are effective in overcoming error bursts are

- **Block interleaving**: information bits are arranged in sequential rows to form columns of bits to which forward error correction parity bits are added [this action is analogous to longitudinal parity checking in Section 7.3.1(1)]. The sequence of information bits is transmitted; it is followed by the sequence of parity bits. During transmission, a limited error burst will affect a sequence of bits in one or more rows. At the receiver, if the errors in a column are within the capabilities of the coding, the errors are corrected and the information bits can be read out in the original sequence. The principle of interleaving is shown in **Figure 7.13**.

- **Reed-Solomon (RS) code**: a block code that employs groups of bits, not single bits. Known as *symbols*, the RS code has k' information symbols, r' parity symbols, and codewords of length $n' = k' + r'$ symbols. The number of symbols in a codeword is arranged to be $n' = 2^m - 1$ where m is the number of bits in a symbol (the symbol length). Thus, when $m = 8$ there are 255 symbols in each codeword (2,040 bits). The code is able to correct errors in $r'/2$ symbols; thus, if we wish to correct errors up to bursts that affect 16 consecutive symbols, the number of parity symbols must be 32. In this case, $n' = 255 = 223 + 32$—i.e., there are 233 information symbols and 32 parity symbols in every 255-symbol codeword of 2,040 bits. The code rate (k'/n') is 223/255 or 0.8747. Reed-Solomon codes are not efficient codes for correcting random errors.

When the datastream is subject to both random errors and error bursts, it is possible to cascade coding techniques so that one handles random errors and another handles error bursts. This is called *concatenated* coding.

---

[5] R. C. Bose and D. K. Ray-Chaudhuri, "On a Class of Error Correcting Binary Group Codes," *Information and Control*, 3 (1960), 68–70.

[6] M. J. E. Golay, "Notes on Digital Coding," *Proceedings of the IRE*, 37 (1949), 657.

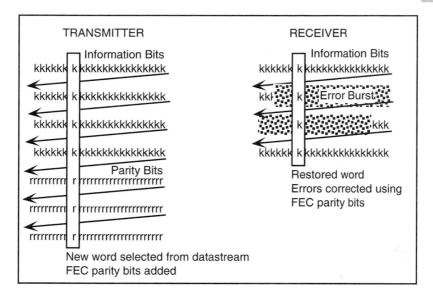

Figure 7.13 Principle of Burst Error Correction Using Block Interleaving

The codewords are bit interleaved so that a burst of interference is likely to destroy only some of the bits in each word. In many situations, there will be enough bits to reconstruct the most likely codeword.

(6) Convolutional Codes

Convolutional codes employ relatively simple encoders whose span may extend over many bits; in contrast, the decoders are quite complex. Fundamentally different from block codes, information bits are not collected in a discrete sequence to which check bits are added to form a block or codeword. Instead, a structured mapping technique is used to convert the sequence of information input bits (k) to encoder output bits. Decoding requires the receiver to undo the mapping and determine the most likely known codeword for an errored word.

Decoding is a matter of reversing the logical operations that produced the encoded bits (x_j, x'_j, x''_j) to determine the most likely message bit (m_j). In effect, the decoder estimates the path through the logical tree that was followed in the encoding process. Because complexity increases exponentially with the number of components, a common thread in decoder strategies is to reduce the number of paths to be searched. Three techniques have received attention

- **Sequential decoding:**[7] searches for the most likely logical path by examining one path at a time. The algorithm evaluates probabilities from node to node and abandons likely incorrect paths as the search proceeds.

---

[7] R. M. Fano, "A Heuristic Discussion of Probabilistic Coding," *IEEE Transactions on Information Theory*, IT-9 (1963), 64–74.

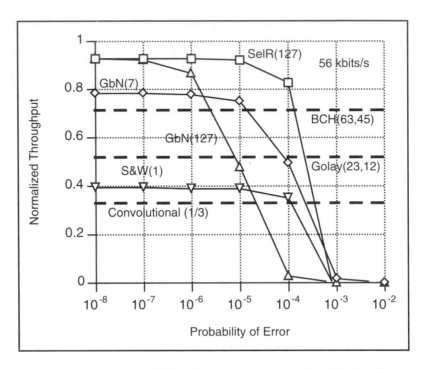

Figure 7.14 Comparison of Throughput as a Function of Probability of Error Using ARQ or FEC Techniques

- **Viterbi algorithm:**[8] implements maximum likelihood decoding. Employs an algorithm that limits searching the possible logical paths to a manageable number by creating a group of minimum (surviving) paths.
- **Feedback decoding**: searches the possible logical paths through a limited number of nodes. Selects the most likely and discards the remainder, then searches the section that fans out from this path, and so on.

Of the three methods, the Viterbi algorithm is the most complex. However, because it is robust and is optimum for many applications, it is the most used.

(7) Throughput Using Forward Error Correction

If we assume that a forward error correcting code of rate R ($= k/n$) corrects all errors in the datastream, then $R = S'$. In **Figure 7.14**, I plot throughput for BCH (63, 45) and Golay (23, 12) codes, and a rate 0.33 convolution encoder on ARQ curves

---

[8] A. J. Viterbi, "Error Bounds for Convolutional Codes and an Asymptotically Optimum Decoding Algorithm," *IEEE Transactions on Information Theory*, IT-13 (1967), 260–69.

from Figure 7.12. For this particular set of circumstances an overwhelming conclusion is that, when $P_\varepsilon > 1 \times 10^{-4}$, FEC should be considered.

REVIEW QUESTIONS FOR SECTION 7.4

1   What are the two basic approaches to error correction?

2   List and explain three basic ARQ techniques.

3   Define throughput and normalized throughput.

4   Define protocol efficiency.

5   At 1.544 Mbits/s, for low and medium probabilities of error, which ARQ technique is most efficient?

6   At 56 kbits/s, for low and medium probabilities of error, which ARQ technique is most efficient?

7   At 4.8 kbits/s, for low and medium probabilities of error, which ARQ technique is most efficient?

8   If I use a 4-bit repetition code to represent two messages, for how many bits in error can the receiver correct?

9   If I use a 16-bit repetition code to represent two messages, for how many bits in error can the receiver correct?

10   What two types of code are used for forward error correction?

11   What is a linear block code?

12   What is the Hamming distance?

13   What is a cyclic code?

14   Explain how block codes can be used for the correction of burst errors.

15   What is a convolutional code?

16   List three strategies for decoding convolutional codes.

17   Above what level of error should forward error correction be considered?

8

OPEN SYSTEMS ARCHITECTURE

Architecture for implementing data communication between cooperating systems has been standardized by the International Standards Organization (ISO). Called *open systems architecture* (OSA), it is embodied in the Open Systems Interconnection Reference Model (OSI model, also referred to as OSIRM). The model is a functional blueprint for data communication among a global community of cooperating, diversified computer systems.

8.1 MAKING A DATA CALL

Suppose an application process running on System 1 is compiling the bill of materials for a new product designed in several locations of a corporation on a variety of computers. When the application calls for data on a certain subassembly, the data are found to be absent from the product source file. The activity to locate and to obtain this data might proceed as follows:

- The application running on System 1 generates a request for the missing data (D).

To emphasize it is expressed in terms that are derived from the organization of System 1 (i.e., internal naming conventions, storage location numbers, etc.), I call the request R(D)S1.

- R(D)S1 is passed to the operating system in System 1 (OS1).

By means of the subassembly number code, a data directory, or other means, OS1 determines that the source of D is System 2. (System 2 is a host remote from System 1.)

- OS1 converts R(D)S1 to R(D)S2.

Using information describing System 2 (which must be available in System 1), OS1 converts the request from System 1 understandable terms [R(D)S1] to System 2 understandable terms [R(D)S2] for presentation to System 2. This conversion is essential if System 2 is to know what it is that System 1 wants.

- OS1 prepares to establish communication with System 2.

A session procedure is initiated that, with its counterpart in System 2, will supervise communication between System 1 and System 2 for the time required to request, find, and transfer D.

- From among the available resources, a network is selected that connects System 1 and System 2.

The need to send data between System 1 and System 2 causes System 1 to consult a routing table, or other information, to select a communication network (e.g., public data network, public switched telephone network, or private network) over which to transport R(D)S2. Network selection defines the transmission mode to be employed (packet or circuit, for instance).

- For the chosen network, the address of System 2 is determined.

This action requires access to a directory that identifies the specific connection to be used. This information defines any requirements that are special to the network (protocol version, block length, etc.).

- The front-end processor (FEP1) selects a data link that connects to System 2 through the chosen network.

Using the network information provided, the FEP selects a suitable data link from those it has access to, and provides the resources to support message flow.

- FEP1 and System 1 allocate and prepare to manage buffer memory.

Because the speed of the communication circuit is significantly less than the internal speed of System 1, the combination of FEP1 and System 1 allocates buffer memory. It is used to make the necessary speed conversions in FEP1 and in the host, and to exercise flow control in both directions.

- FEP1 establishes a communication connection to System 2.

Finally, the front-end processor attached to System 1 establishes a physical connection to the front-end processor (FEP2) attached to System 2.

- FEP1 requests permission to send.

FEP1 sends a frame to FEP2 that requests permission to communicate with System 2.

- FEP2 refers the request to OS2 (or to some other entity that can grant access).

FEP2 receives the frame, checks for errors, and refers the request to OS2. If OS2 determines that communication with System 1 is permitted, it initiates a session process that will link with that started in System 1 to manage the communication session. Also, OS2 authorizes FEP2 to reply affirmatively to System 1.

- FEP2 and OS2 allocate and prepare to manage buffer memory
- FEP2 replies to FEP1 with *ready to receive*
- FEP1 sends R(D)S2 to FEP2
- FEP2 checks for transmission errors, strips the text from the frame, and passes R(D)S2 to OS2
- OS2 validates R(D)S2.

If OS2 determines that System 1 can receive D, OS2 instructs the application on System 2 to assemble the data D.

- The application creates A(D)S2 and sends it to OS2.

The file server copies the data D to memory, adds any additional data required for the application program on System 1 to use it, and releases control to OS2.

- OS2 identifies the data as the answer to R(D)S2 and conditions it for the chosen network.

OS2 identifies the answer for System 1, causes the telecommunication software in System 2 to divide A(D)S2 into appropriate data blocks, and authorizes FEP2 to send them to System 1.

- FEP2 encapsulates the data blocks in frames and sends them in sequence to System 1
- FEP1 receives the sequence of frames, checks for errors, strips off protocol bits, transfers control of error-free data blocks to OS1, and replies to FEP2 with frames acknowledging error-free receipt or requesting retransmission of corrupted blocks
- FEP2 completes transmission of data blocks to FEP1
- FEP1 receives the final frame, confirms receipt of all frames to FEP2, and sends the final data block to OS1
- OS1 reassembles data blocks into A(D)S1, confirms all data blocks are present, and sends A(D)S1 to the application
- The combination of OS1 and OS2 terminate the communication session.

The hardware and software associated with System 1 and System 2 terminate the communication session, and the application running on System 1 proceeds with compiling the bill of materials for the new product.

What we have described is the bare skeleton of the procedure—to say the least, the steps for effecting communication between cooperating systems to achieve file transfer are complex, and they become even more so if the communication path spans several networks. Moreover, it is likely that not one, but many, communications will be going on at the same time in each system; there must be means to prevent one message from being confused with another. Too, each mainframe will be executing many jobs simultaneously, some that require communication support, and some that do not; they must be carefully managed to ensure the integrity of each task.

REVIEW QUESTIONS FOR SECTION 8.1

1 Complete the following statement: OSIRM is a functional blueprint for…

2 For System 1 to communicate with System 2, what information must System 1 have?

3 What does a session procedure do?

4 Comment on the following statement: Network selection defines the transmission mode to be employed.

5 Why is it necessary for buffer memory to be available before communication can start?

6 Why must OS2 approve OS1's request to communicate with OS2?

7 What determines the size of the data blocks sent from System 2 to System 1?

8.2 OSI MODEL

The OSI reference model has been created to formalize the activities listed in Section 8.1 and to ensure that all conditions that affect communication between cooperating systems are considered. **Figure 8.1** shows the model, and **Table 8.1** summarizes the activities of each layer.

The model does not assume specific system architecture, nor does it require a particular system implementation. Thus, it is able to accommodate evolving

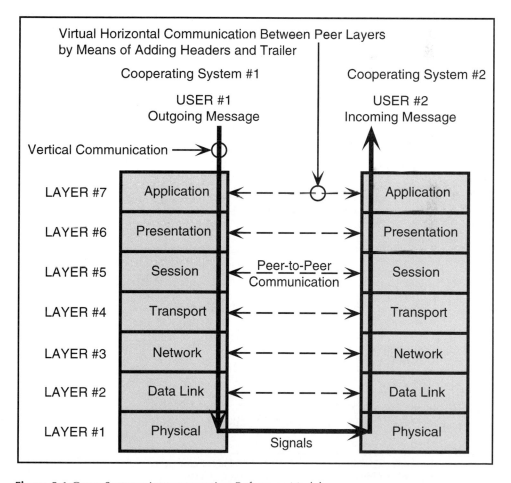

Figure 8.1 Open Systems Interconnection Reference Model

Peer-level messages are exchanged between layers by passing information down one stack and up the other.

Table 8.1 Summary of Activities in OSIRM Layers

	Action When Sending	Action When Receiving	Peer-to-Peer Communication
Layer 7 APPL	Organizes information from application agent into block of user data. Adds AH with identifier unique to this communication and these participants. Passes user data to Presentation.	Removes AH, converts user data block to information. Passes information through application agent to addressee.	Agreement on identifier unique to this communication and these participants.
Layer 6 PRES	Translates user data into recipient's code, or an agreed-on intermediate code. Performs compression and/or encryption, if used. Adds PH with encoding, compression, and encryption information. Passes data to Session.	Removes PH, deciphers and/or decompresses if appropriate, converts to recipient's code if intermediate code used. Passes user data to Application.	Agreement on code to be used for transmission; also schemes for compression and encryption, if used.
Layer 5 SESN	Manages line discipline, defines beginning and ending of message, provides checkpoints for failure recovery. Adds SH with markers.	Manages line discipline. Removes SH, notes any markers, holds until all data are received. Passes PPDU to Presentation.	Agreement on line discipline and markers.
Layer 4 TRSPT	May break user data into segments. Adds TH with sequence numbers, and FCS to form TPDU. Copies TPDU, retains until acknowledged.	Removes TH, uses FCS to verify TPDU. Acknowledges error-free TPDU, discards and requests resend of corrupted TPDU. If message was broken up, waits for all TPDU, reassembles message. Passes SPDU to Session.	Agreement on communication parameters such as size of TPDU, etc.
Layer 3 NTWK	May break TPDUs into smaller segments to match size limits of chosen network. Adds NH with sequence numbers and destination address to form NPDUs, or packets.	Removes NH, verifies destination address and sequence number. If TPDU was broken up, waits for all packets, reassembles TPDU. Passes TPDU to Transport.	Agreement to initiate, maintain, and terminate network level connection.
Layer 2 DLNK	Adds DH and DT to each packet to form frame. DH includes sequence number for DLC protocol. DT includes FCS. Copies frame, retains until acknowledged.	Removes DH and DT, uses FCS to determine if error has occurred in frame. Passes packet to Network.	Agreement on DLC protocol parameters, including error control procedures.
Layer 1 PHYS	Transmits each frame as a sequence of bits.	Receives each frame and reconstructs datastream.	

AH = Application Header. PH = Presentation Header. SH = Session Header.
TH = Transport Header. NH = Network Header. TPDU = Transport PDU.

technology and changing user requirements. It consists of seven layers that serve to decompose the complexity of achieving information flow between cooperating machines into consecutive steps that are substantially independent of one another. This independence is achieved by allowing the superior layer to depend on the functions (services) performed by the next lower layer without regard to the way in which the level is implemented. Thus, equipment from different manufacturers can work together, and new equipment can be introduced as technology changes.

Passing a message between Application 1 in System 1 and Application 2 in System 2, is achieved by a sequence of activities that descends through the seven layers of System 1 and ascends through the seven layers of System 2. To ensure success, it is important that each of the layers performs a well-defined set of functions, and that similar functions are assigned to the same layer.

8.2.1 COMMUNICATION TECHNIQUES

Communication between cooperating end-systems is accomplished by layer entities that perform the functions defined by the layer standard. The layer standard consists of two parts:

- **Service definition**: describes the functions the layer performs and what services it provides (to the next upper layer)
- **Protocol specification**: describes the procedures used within the layer, and between peer entities, to execute the functions defined by the service definition.

(1) Communication Primitives

Shown in **Figure 8.2**, a layer consists of the two corresponding layer entities in the cooperating end-systems. Each layer entity is implemented to send, receive, and act on messages exchanged with its peer. In the upper part of Figure 8.2, the upper layer entities communicate through *service access points* (SAPs) in the layer next down in the stacks (the SAP is defined by an address or identifier).

Communication between peer entities uses four primitives to which the generating entity attaches parameters that describe their scope. Their use is illustrated by the following exchanges

- **Request**: by making the request to its service layer (SLE1), upper layer entity (ULE1) requests a function be performed by its peer entity (ULE2). SLE1 passes the request down protocol stack 1 and up protocol stack 2 to its peer service layer SLE2
- **Indication**: SLE2 tells its upper layer entity (ULE2) that ULE1 has requested the function be performed

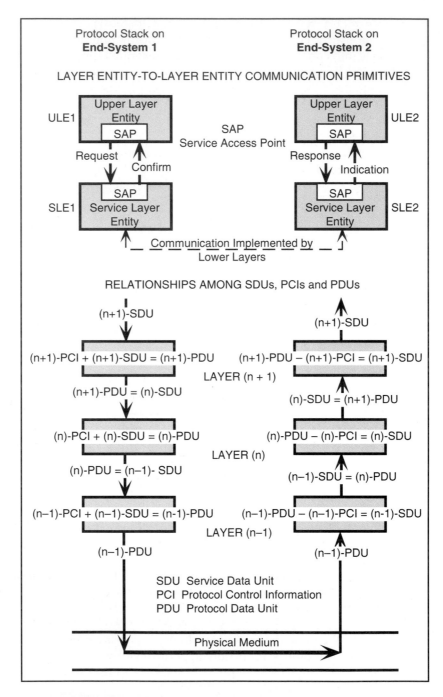

Figure 8.2 Components of Layer-to-Layer and Peer-to-Peer Communications
OSIRM employs four primitives in the communication process.

- **Response**: ULE2 informs SLE2 that it has/has not invoked the function requested, and SLE2 passes the response back to SLE1

- **Confirm**: SLE1 informsULE1 that ULE2 has/has not invoked the requested function. (A negative response may cause ULE1 to request the same function with a different set of parameters—or request another function. Negotiations continue between the entities until they agree on something each can implement.)

As examples, the function requested can be *connect to...*, in which case the layers involved are network and data link—or *encrypt* (parts of) the message, in which case the layers involved are presentation and session.

(2) Communication Components

The lower part of Figure 8.2 shows the communication components that are employed. They are

- **Service data unit** (SDU): consists of user data and control information created in the upper layers of the protocol stack

- **Protocol control information** (PCI): information exchanged by peer entities to perform certain tasks or settle on a format

- **Protocol data unit** (PDU): the combination of SDU and PCI.

Instructions are exchanged between peer entities by creating headers and combining them with others from the different layers, passing the composite message over the physical connection, and dropping off the appropriate header at the proper layer as the data sequence ascends the other stack.

As shown in Figure 8.2, to the SDU arriving at layer $(n + 1)$ in the sending stack [i.e., $(n + 1)$-SDU], the protocol control information [i.e., $(n + 1)$-PCI] that layer $(n + 1)$ wishes to transfer to its peer is added to form $(n + 1)$-PDU. This becomes the service data unit that is passed down the stack to layer (n). At layer (n), the protocol data information [i.e., (n)-PCI] that layer (n) wishes to transfer to its peer is added to form (n)-PDU. This becomes the service data unit that is passed down the stack to layer $(n–1)$. And so on.

In layer $(n–1)$ of the receiving stack, $(n - 1)$-PCI is stripped off to leave $(n - 1)$-SDU. $(n - 1)$-PCI contains data sent by layer $(n - 1)$ in the sending stack for use by layer $(n - 1)$ in the receiving stack. $(n - 1)$-SDU is sent up the stack to become (n)-PDU in layer (n). Proceeding further, (n)-PCI is stripped off in layer (n) and $(n + 1)$-PCI is stripped off in layer $(n +1)$. Layers (n) and $(n + 1)$ use them in the receiving stack.

(3) Encapsulation and Decapsulation

As we see in Section 8.2.2, in descending the protocol stack in the sending system, at each layer, protocol control information (PCI) is attached to the service data unit (SDU) as a header, and, in the case of the data link layer, as a header and trailer. The procedure is known as *encapsulation*, and the headers and trailer encapsulate the user data. In ascending the protocol stack of the receiving system, at each layer, PCI is stripped from the PDU. This procedure is known as *decapsulation*, and the user data is said to be *decapsulated*.

Finally, to understand how communication between user applications, and between peer entities, is implemented in the model, realize that

- All **information** exchanged between the end-systems enters or leaves the model through the two application layers
- **Headers** and **trailers** (the protocol control information) passing between peer entities flow from the entity in one end-system down the stack of layers to the physical layer, across to the other end-system, and up the stack of layers in the other end-system to the peer entity
- **Signals** are exchanged between the systems only over the connection between the physical layers.

8.2.2 DESCRIPTION OF LAYERS

Figure 8.3 shows more details of the layer structure of the OSI model and the encapsulation of user data. The activity is described below.

(1) Application Process

The application process—i.e., that part of an end-system that processes information for an application task that requires communication in an OSI environment—consists of two parts

- **Application agent**: exists between the user and the operating system of the processing resource to provide access to specific processor capabilities
- **Application entity**: exists within the application layer. As required, provides services to the application agent by exchanging protocol data units (PDUs) with peer entities (application entities in other machines).

They cooperate to transfer the data required to initiate communication with another end-system. As these data descend the stack associated with System 1, headers are added by the application, presentation, session, transport, and network entities, and a header and a trailer are added by the data link layer entity, to produce the data link PDU. Converted to a datastream, it is transferred to System 2 over the physical connection between them. As the information ascends the stack associated with System 2, the data link entity strips off the data link header and trailer—they constitute a message from the peer entity in System 1. The data link entity acts on the information they contain. In turn, the network, transport, session, presenta-

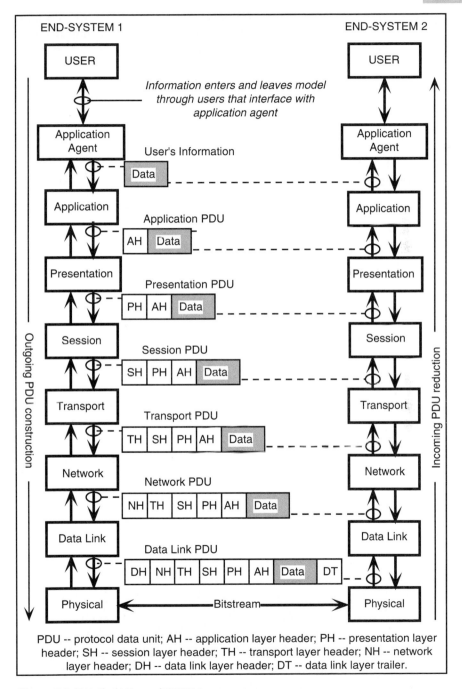

PDU -- protocol data unit; AH -- application layer header; PH -- presentation layer header; SH -- session layer header; TH -- transport layer header; NH -- network layer header; DH -- data link layer header; DT -- data link layer trailer.

Figure 8.3 Detailed View of OSIRM
Headers are added or removed at each level. The data link layer employs a trailer, also.

tion, and application entities strip off headers and act on the information they contain. Finally, the presentation entity delivers the information from the application running in System 1 to the operating system running in System 2.

(2) Application (or User) Layer

The application layer differs from the other layers of the OSIRM. Its users are not entities of the next higher layer but programs that provide services to the users of the end-system. No other layer of the model is accessed directly by a user.

Because the application layer interfaces with a wide variety of user programs, it contains many user-specific services that are hard to standardize. Some *application service elements* (ASEs) that have been standardized by ISO are

- File transfer, access, and management (FTAM)
- Electronic messaging (X.400)
- Job transfer and manipulation (JTM)
- Virtual terminal (VT)
- Directory (X.500).

They cover many of the reasons for engaging in machine and machine communications. To assist the interworking of open systems by providing the information they need to communicate, Recommendation X.500 describes a directory. Intended to provide *location transparency*, it holds information concerning many things (people, distribution lists, applications entities, etc.) and may be consulted by a user to obtain addressing information (telephone numbers, electronic mail address, postal address, facsimile number, service access point information, etc.). The actions of the application layer are

- When **sending**, the application layer
  - organizes information received from the application agent into a user data block
  - adds a header (application header, AH) that identifies this communication between specific participants
  - passes the application protocol data unit (APDU) on to the presentation layer.
- When **receiving**, the application layer
  - removes the application header from the application PDU
  - converts the data block to user information
  - passes it to the application agent for delivery to the user application identified by the header.
- **Peer-to-peer** communication is required to agree on the unique identifier for the communication.

(3) Presentation Layer

The presentation layer translates the information to be exchanged into terms that are understood by the end-systems. If the same (or compatible) equipment supports the users, this conversion may be trivial. However, if the systems are not compatible, conversion may be much more difficult. To make the translation capability required by each system less onerous, a common intermediate form may be used. The actions of the presentation layer are

- When **sending**, the presentation layer
  - performs translation services (and may perform data compression and encryption)
  - adds a header (presentation header, PH) that identifies the specific encoding, compression, and encryption employed
  - passes the presentation PDU on to the session layer.
- When **receiving**, the presentation layer
  - removes the presentation header from the presentation PDU
  - performs decoding, decompressing, and decrypting
  - passes the application PDU to the application layer.
- **Peer-to-peer** communication is required to agree on encoding, compression, and encryption.

The ISO connection-oriented presentation layer is defined in ISO 8822 (Service Definition) and ISO 8823 (Protocol Specification).

(4) Session Layer

The session layer directs the establishment, maintenance, and termination of the connection. It controls data transfer and manages the insertion of synchronization points into the information flow. These points structure the data exchange into a series of dialogue units so that the exchange can restart if service is interrupted. The actions of the session layer are

- When **sending**, the session layer
  - manages line discipline
  - tracks correspondence of requests and responses
  - defines the beginning and ending of the exchange
  - adds a header (session header, SH) that identifies any specific markers employed
  - passes the session PDU on to the transport layer.
- When **receiving**, the session layer
  - removes the session header from the session PDU

– notes any specific markers

– passes the presentation PDU to the presentation layer.

- **Peer-to-peer** communication is required to agree on line discipline and the use of markers.

The ISO connection-oriented session layer is defined in ISO 8326 (Service Definition) and ISO 8327 (Protocol Specification).

(5) Transport Layer

The transport layer ensures the integrity of end-to-end communication independent of the performance and number of networks involved in the connection between end-systems. It serves as interface between the three lower layers that are concerned with providing transparent connections between users, and the three upper layers that ensure that the data are delivered in correct and understandable form. The transport layer selects the networks that will be employed, provides network addresses for the parties, and may multiplex messages from several sources for transmission on a common physical connection. The transport layer is responsible for detecting the loss or duplication of message segments, and for requesting resends as needed. In conjunction with the data link layer (that supervises message flow at the frame level), it controls the flow of message segments so that overloads (in the host, or in the communication network) are prevented. The actions of the transport layer are

- When **sending**, the transport layer
  – may break the user data block into segments
  – adds a header (transport header, TH) that contains the sequence number
  – calculates a frame check sequence (FCS) that checks the entire segment
  – copies the transport PDU
  – passes it on to the network layer.
- When **receiving**, the transport layer
  – removes the transport header from the transport PDU
  – verifies the FCS
  – acknowledges an error-free transport PDU or discards and requests resend
  – passes the session PDU to the session layer.
- **Peer-to-peer** communication is required to agree on the maximum size of the transport PDU, and the network(s) that will be used for this communication.

Typical transport layer protocols are Internet's TCP (see Section 12.6.1) and OSI's TP0, TP1, TP2, TP3, and TP4 (see Section 8..4.2).

(6) Network Layer

The network layer provides communications services to the transport layer. To do this, it makes use of the underlying data link layer services to establish, maintain, and terminate transmission across the chosen network(s). It determines how to route the segments within the network chosen by the transport layer, selects the circuit on which they shall be sent, and breaks segments into data blocks of a size suited to the chosen circuit. The actions of the network layer are

- When **sending**, the network layer
    - adds a header (network header, NH) to each block that provides destination addresses (in a format understood by the chosen network) and sequence number
    - passes the network PDU to the data link layer.
- When **receiving**, the network layer
    - removes the network header (NH)
    - verifies destination address and sequence number
    - reassembles the transport PDU
    - passes it on to the transport layer.
- **Peer-to-peer** communication is required to initiate, maintain and terminate the network-level connection.

Typical network level protocols are Internet's IP (see Section 8.4.1); ISO's 8348 and 8648 (see Section 8.4.1); and ITU–Ts X.25 Packet Layer Protocol (see Section 9.2.1).

(7) Data Link Layer

The data link layer transfers data over a single communication link without intermediate nodes. The link can be point to point, point to multipoint, broadcast, switched, or nonswitched. The data link layer protocol prepares the data in a specified format (SDLC, Bisync, etc., see Sections 8.3.2, 8.3.3, and 8.3.4) for transmission over the selected circuit that may use a variety of media including twisted pairs, coaxial cables, optical fibers, and communications satellites or other radio links. The actions of the data link layer are

- When **sending**, the data link layer
    - adds a header (DH) and a trailer (DT) to the network PDU to form the data link PDU. The header includes the flag, class of frame identifier, sequence number for use in the control byte, and other information. The trailer includes an FCS to check the frame, and a flag
    - copies the frame
    - passes it on to the physical layer.

- When **receiving**, the data link layer
  - reconstructs the data link PDU from the logical bitstream received from the physical layer
  - removes both header and trailer from the data link PDU
  - verifies FCS and other layer information
  - passes the network PDU on to the network layer
  - requests resend, if necessary.
- **Peer-to-peer** communication is required to agree on data link protocol parameters and error correction procedures.

In conjunction with the transport layer (which supervises message flow at the segment level), the data link layer controls the flow of frames so that overloads (in the host or in the communication network) are prevented. Data link protocols are discussed in Section 8.3.

(8) Physical Layer

The physical layer provides the connection over which bitstreams flow between the users. It defines the physical (electrical and mechanical) standards and signaling required to make and break the data communication path and to send and receive the data signals over twisted pairs, coaxial cables, optical fibers, and communications satellites or other radio links. The meaning of the bitstream is of no significance to the physical layer.

- When **sending**, the physical layer
  - converts the logical datastream to a suitable electrical signal, including signal conditioning (i.e., pulse shaping, zero stuffing, scrambling, etc.)
  - transmits a sequence of electrical symbols that represents the frames received from the data link layer.
- When **receiving**, the physical layer
  - receives a sequence of electrical symbols
  - deconditions the signal (i.e., unstuffs zeros, unscrambles, etc.)
  - passes a logical sequence to the data link layer.
- **Peer-to-peer** communication consists of the signals that represent the total frame passed between System 1 and System 2.

Typical physical layer protocols are

- EIA 232/449 [see Section 7.1.2(1)]
- ITU–T X.21, which uses balanced signaling
- ISDN I.430, which provides basic rate access (see Section 10.4.1)

- ISDN I.431 that provides primary rate access (see Section 10.4.1)
- A range of procedures for different media components used to implement IEEE 802 local-area networks (see Section 12.1.2).

8.2.3 REALITY OF OSIRM

The OSI model is the result of many years of work by standards committees. It is an exhaustive solution to the problem of establishing data communication between any two machines, located anywhere, and contains more options than any individual communication situation requires.

In many cases, data communication is a matter of exchanging messages between many terminals and a few hosts, or among a few hosts, for purposes defined by a single enterprise. Often, the community of interest exists within an organization, and equipment is bought that communicates effectively. The result is there are many more quasi-homogeneous, more or less compatible, systems than heterogeneous systems. In addressing the heterogeneous case, the OSI reference model provides a general framework for the discussion (and comparison) of actual protocol stacks.

Despite the best efforts of influential organizations to promote the use of the OSI model, the world did not stand still while OSIRM was developed. Most of today's networks employ protocol stacks from ARPAnet's TCP/IP or IBM's SNA. While they both have things in common with OSIRM, they have many more differences—and they are different from one another.

In Figure 8.3, I traced the development of the data link protocol data unit (DPDU) that becomes the signal exchanged by the physical layers of cooperating machines. In the following sections I describe data link, network, and transport layer protocols—the layers that have to do with communication.

REVIEW QUESTIONS FOR SECTION 8.2

1 Describe the assumptions made in creating the OSI Reference Model.
2 Comment on the following statement: Equipment from different manufacturers can work together, and new equipment can be introduced as technology changes.
3 Describe the two parts of a layer standard.
4 List and describe the four primitives used for communication between peer layers.
5 List and describe the components used to achieve communication between layers.

6 Describe how instructions are exchanged between peer entities.

7 Define encapsulation and decapsulation.

8 How does information enter OSIRM?

9 How is information passed between the end-systems?

10 Where are signals exchanged?

11 Distinguish between an application agent and an application entity.

12 Describe the operation of the application layer.

13 Describe the operation of the presentation layer.

14 Describe the operation of the session layer.

15 Describe the operation of the transport layer.

16 Describe the operation of the network layer.

17 Describe the operation of the data link layer.

18 Describe the operation of the physical layer.

19 Comment on the following statement: The OSI model is an exhaustive solution to establishing data communication between any two machines, located anywhere, and contains more options than any individual communication situation requires.

20 Comment on the following statement: Most of today's networks employ TCP/IP or SNA.

8.3 DATA LINK PROTOCOLS

In this section I describe the headers and trailers associated with common data link layer protocols.

- **Data link control** (DLC) **protocol**: a set of rules that governs the exchange of messages over a data link.

DLC protocols are divided into two classes, as follows

- **Asynchronous operation**
  - start-stop DLC protocol.
- **Synchronous operation**
  - character-oriented DLC protocol (e.g., BISYNC)
  - byte-count-oriented DLC protocol (e.g., DDCMP)
  - bit-oriented DLC protocol (e.g., SDLC).

8.3.1 START-STOP DLC PROTOCOLS

Start-stop DLC protocols are used by terminals, and other input-output devices, over relatively slow speed links that employ asynchronous transmission. They manage a stream of characters framed by start and stop bits. The framed characters are transmitted one at a time and are independent of one another.

(1) USART

In contemporary systems, an integrated circuit chip called a *universal synchronous/asynchronous receiver transmitter* (USART) is employed to implement start-stop protocols. As a transmitter, USART

- Adds a parity bit (if so instructed)
- Frames the character sequence with start and stop bits
- Generates unipolar signals.

As a receiver, USART

- Extracts timing information from the start bit
- Converts the incoming signal to a bit sequence
- Checks parity
- Strips off the start, stop and parity bits.

In the idle state, the data link is held at a voltage level that corresponds to 1. Transmission begins by dropping the line level to zero to send a start bit (0). Bits that represent the character being sent follow the start bit. At the end of the character sequence, a stop bit returns the line to the level representing 1. It remains at that level until the next character is transmitted. **Figure 8.4** shows this sequence.

(2) Synchronization

At the receiver, bit synchronization is achieved by developing a timing mark at the center of each start pulse. Because the characters consist of only a few bits, sampling the ensuing voltage levels at nominal pulse-width intervals from this mark permits the receiver to test the incoming signal at approximately the center of each bit. Because operation is asynchronous, synchronization at byte (octet) and frame levels is not possible.

(3) Error Control

Error control relies on parity checking. If parity is not confirmed, the USART is likely to flag the condition. In some cases, the received bit sequence is echoed back to the sending terminal and displayed. If it is the character the user intended to send, it confirms that the correct character was received. If it is not the character the user intended to send, the user corrects it. In many cases, the

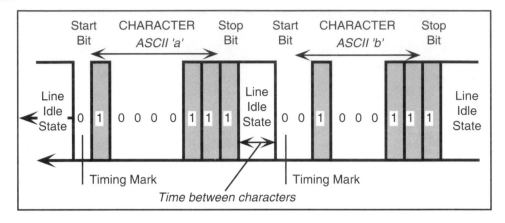

Figure 8.4 Line Levels with USART
The line idle state is 1. The start bit drops the level to 0.

user is not conscious of echoing and assumes that the appearance of a character different from the one typed was caused by improper keyboard input. Unless echoing is employed, the sender has no indication of whether the characters have been received. Start-stop DLC protocols are the simplest of communications procedures.

8.3.2 CHARACTER-ORIENTED DLC PROTOCOLS

Character-oriented DLC protocols employ special characters that are reserved in the code set to facilitate communication. A common data link control protocol is known as BISYNC (binary synchronous data link control). Developed by IBM, it is the oldest data link control protocol in common use; as a consequence there are many user-specific variations in its implementation.

(1) Binary Synchronous DLC Protocol

BISYNC supports ASCII and EBCDIC codes; it uses specific control characters identified by mnemonics (see Table 3.2). Thus

- **ACK0** and **ACK1** (positive acknowledgment): no error detected in data block. To provide a measure of security, ACK0 is used for even-numbered data blocks, and ACK1 is used for odd-numbered blocks.
- **DLE** (data link escape): changes the meaning of following character(s)
- **ENQ** (enquiry): requests a response from another station
- **EOT** (end of transmission): concludes a transmission that may have contained several data blocks

- **ETB** (end of text data block): end of a text data block—requires receiver to acknowledge receipt
- **ETX** (end of text): terminates text portion of frame
- **ITB** (end of intermediate text data block): does not require receiver to acknowledge receipt; block check character (BCC) follows ITB
- **NAK** (negative acknowledgment): error detected in data block
- **SOH** (start of header)
- **STX** (start of text)
- **SYN** (synchronizing character): begins frame—usually sent in pairs.

(2) BISYNC Frame

A BISYNC frame has the form

$$\leftarrow SYN\ SYN\ SOH\ Header\ STX\ Text\ ETX\ BCC$$

- It employs **synchronous transmission**—i.e., no start or stop bits and transmission in response to an external event
- The **frame** begins with a repeated *SYN* character
- **Control characters** are used as needed to identify portions of the frame
- A **block check character** is used to check the frame from *SOH* to *ETX*. *BCC*s are used only with information frames (i.e., those that contain text)
- On **point-to-point connections**
  - no addressing is required
  - if stations have equal status, the station with a frame to send asks the other station for permission to send
  - the header is optional and is specified by the users.
- On **multipoint connections**
  - the header includes the address of the secondary station taking part in the communication
  - if the secondary station is sending, its address tells the primary station from which secondary station the frame comes
  - if the primary station is sending, it is the address of the secondary station for which the frame being sent is intended
  - if several frames are included in the exchange, the header is likely to include the frame number.

Table 8.2 portrays an exchange that employs stop-and-wait ARQ on a half-duplex, multipoint connection between primary station A and secondary stations B, C, and D. When polled by A, station B has nothing to send, and responds with *EOT*. Station C has a two-frame message to send. When polled by A, C sends the first

Table 8.2 BiSync on Half-Duplex Multipoint Circuit with Stop-and-Wait ARQ

A polls B. *ENQ 'B' SYN SYN* ⇒ The control character *ENQ* asks if B has any information to send. After sending, the line is turned around, A becomes the receiver, and B becomes the sender.
B responds to A. ⇐ *SYN SYN 'B' EOT* *'B'* tells A the reply is from station B. The control character *EOT* indicates no information to send. After sending, the line is turned around, B reverts to receiver, and A becomes the sender again.
A polls C. *ENQ 'C' SYN SYN*⇒ After sending, the line is turned around, A becomes the receiver, and C becomes the sender.
C responds to A. ⇐ *SYN SYN SOH 'C' '1' STX Text1 ETB BCC* C sends frame 1 (*'1'* in the header indicates the frames are numbered). It includes text block 1 (*Text1*); the use of *ETB* (end of text block) indicates more blocks to come. After sending, the line is turned around, C becomes the receiver, and A becomes the sender.
A finds the VRCs and LRCs it calculates are not in agreement with the values sent. An error exists. A destroys the frame and sends *NAK* to C. *NAK 'C' SYN SYN* ⇒ After sending, the line is turned around, A becomes the receiver, and C becomes the sender.
C repeats frame 1. ⇐ *SYN SYN SOH 'C' '1' STX Text1 ETB BCC* After sending, the line is turned around, C becomes the receiver, and A becomes the sender.
Having received the frame and detected no error this time, A sends a positive acknowledgment to C. *ACK1 'C' SYN SYN* ⇒ After sending, the line is turned around, A becomes the receiver, and C becomes sender.
C sends frame 2 which includes text block 2. ⇐ *SYN SYN SOH 'C' '2' STX Text2 EOT BCC* *EOT* indicates there are no more blocks to follow and sending is complete. After sending, the line is turned around, C becomes the receiver, and A becomes the sender.
A acknowledges error-free receipt of frame 2. *ACK0 'C' SYN SYN* ⇒ At this point, A retains the role of sender.
A polls the next station (*'D'*). *ENQ 'D' SYN SYN* ⇒ After sending, the line is turned around, A becomes the receiver, and D becomes the sender, and so on.

frame. A finds it to contain error(s), destroys it, and requests C to repeat the frame. On being sent the second time, it is good, and A sends a positive acknowledgment to C. C then sends the second frame and signifies there are no more frames to come. A confirms receipt of frame 2 and polls D. If D has nothing to send, A polls B again, and so on.

Even though there are no messages to send, the circuit is in constant activity as A polls each station in turn, repeatedly. When A has a message to send, it breaks into the polling to send it. When sending is completed, A returns to polling.

(3) BISYNC Transparent Operation

Defining certain characters as control characters means that sequences in the text portion of the frame may provoke a control action. For instance, in EBCDIC, 00000011⌋ represents *ETX*—this is also the binary code for 3. Thus, if the text contains $b3$ (binary number 3), the receiver will take it to mean *ETX*, expect that the next octet(s) are error checking information, and that this is the last frame of the series.

(4) Use of DLE

In the text, to preserve the integrity of bytes whose bit patterns duplicate control characters, the escape character (*DLE*) is employed. It is inserted before all control characters in the text portion of the message; thus, *DLE STX, DLE ITB, DLE ETB*, etc. In this way a bit pattern in the text portion of the message that happens to duplicate any control character except *DLE* will be treated as a character intended to be part of the text. In EBCDIC, *DLE* is $00010000 \Rightarrow$—i.e., the bit pattern for the binary number 16. If nothing else is done, the occurrence of $b16$ in the text portion of a message will cause the receiver to expect the number following to be a control character. If the number following is $b3$ ($= ETX$), the receiver will understand that the text portion of the message has ended. If the character that follows $b16$ is not the same as a control character, the receiver will understand it is dealing with an unknown and assume an error has been made. In either event, the text will not be processed properly.

(5) Use of DLE DLE

A technique employed to prevent this confusion and to obtain completely transparent operation is as follows. After inserting *DLE* in front of the *STX* character, the transmitter scans the text for occurrence of the *DLE* bit pattern. Upon detection of $b16$ (the *DLE* bit pattern), the transmitter inserts *DLE* ahead of it. At the receiver, receipt of *DLE STX* signifies the beginning of the text of a message that is being sent in transparent mode. Hereafter, until *DLE ETB, DLE ETX*, or *DLE EOT* is received, the receiver will interpret the sequence *DLE DLE* as the binary number 16. In this way, control characters are interpreted correctly, and the text is protected. As an example, if a frame contains a portion of text that includes the binary numbers 3 and 16, the transmission will be

$\Leftarrow$ *SYN SYN SOH Header DLE STX b29 b36 DLE b16 b3 b29 DLE ETX BCC*

DLE is used ahead of control characters that define the text portion of the frame, and ahead of *b16*—the binary number that mimics *DLE*—to ensure that it is interpreted as the binary number 16. Because it is not preceded by *DLE*, the next number *b3* is treated as *b3* not *ETX*. *DLE ETX* closes the text segment.

(6) BISYNC Synchronization

Usually, two synchronizing characters are sent at the beginning of the frame. Two are needed to ensure that the receiver can establish character synchronization reliably. Some systems employ more *SYN* characters. In addition, *SYN* characters may be transmitted when the line is idle to maintain character synchronization between the end equipment. Also, if the text block is long, they may be sent in the middle of the text to ensure continued synchronization throughout reception of the frame—e.g.:

$$\Leftarrow SYN\ SYN\ SOH\ Header\ STX\ Text1\ ITB\ BCC\ SYN\ SYN$$
$$STX\ Text2\ ETB\ BCC$$

The first *BCC* checks the frame up to its position in the frame; the second *BCC* checks the remainder of the frame.

8.3.3 BYTE-COUNT-ORIENTED DLC PROTOCOLS

In byte-count-oriented DLC protocols, the number of bytes (or octets) in the text field is carried in a byte-count field in the header. Set numbers of bits are present in all fields except the text field. The leading example of such a protocol is DDCMP, a protocol developed by Digital Equipment Corporation.

(1) DDCMP Frame

DDCMP (Digital Data Communications Message Protocol) employs a header that contains exactly 10 octets, a text portion that contains a stated number of octets, and a trailer that consists of a 2-octet error detection sequence (FCS). The fields in a DDCMP frame are shown in **Figure 8.5**.

- Two *SYN*s are used to establish **character synchronization**
- The **Class field** contains one of three control characters:
    - *SOH* if the frame is an information frame
    - *ENQ* if the frame is a supervisory frame
    - *DLE* if the frame is a maintenance frame.
- **Supervisory frames** do not contain a text field
- The **Text Byte Count field** contains the number of bytes that are present in the text field. With 14 bits in the field there can be 2^{14}—i.e., a maximum of 16,383 bytes in the text

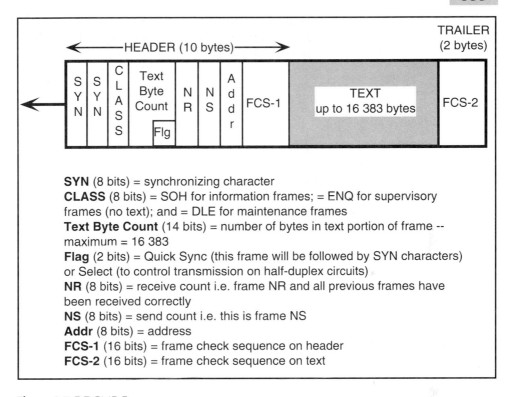

Figure 8.5 DDCMP Frame

- The **Flag field** contains 2 bits:
  - *Q*-bit (Quick Sync) tells the receiver whether *SYN* characters will follow this frame
  - *S*-bit (Select) tells the receiver whether the line can be turned around (in half-duplex operation).
- The **Receive Count field** notes the number of the last frame received without error. A count of x implies that frame x, and all lower-numbered frames have been received without detecting errors. Thus, one receive count can acknowledge a number of frames
- The **Send Count** is the number of the frame being transmitted. Each station maintains its own send count
- *FCS-1* is the frame check sequence for the **header**
- *FCS-2* is the frame check sequence for the **text** field.

Supervisory frames are used to start communication and to initialize frame counts. Also, if only one station is sending information frames, supervisory frames are used for acknowledgments.

(2) DDCMP Transparency

Because the exact number of bits or bytes is known for each field, there can be no confusion between control information and text information so long as the receiver keeps accurate count. Thus, special arrangements to ensure transparency are not needed in DDCMP.

(3) DDCMP Synchronization

Two *SYN* characters are sent at the beginning of the frame. Two are needed to ensure that the receiver can establish character synchronization reliably. No other *SYN* characters are employed.

8.3.4 BIT-ORIENTED DLC PROTOCOLS

Bit-oriented DLC protocols make use of a special character, the flag character, to mark the beginning and ending of the frame. Between these markers, the header and the trailer fields are of predetermined lengths. Whatever data lie between them are the text field. There is no need for other special characters to separate the frame into its major divisions. The original bit-oriented protocol was IBM's

- **SDLC** (Synchronous Data Link Control Protocol).

Introduced in 1972, SDLC was modified and standardized by ITU–T and ISO as

- **HDLC** (High Level Data Link Control Protocol).

Since then, other bit-oriented protocols have been standardized.

- **ADDCP** (Advanced Data Communications Control Procedure)—an ANSI standard
- **LAP-B** (Link Access Procedure—Balanced)—a component of ITU–T's X.25 standard
- **LAP-D** (Link Access Procedure—D Channel)—for ISDN D-channel data transport
- **LAP-E** or **LAP-F** (Link Access Procedure—Extended or Frame Relay)—a version of LAP-D used in frame relay applications.
- **LAP-M** (Link Access Procedure—Modem)—a version of LAP-D used with modem mediated links.

Different in the detailed meaning of specific control field bits, all of these protocols share a common structure. In the order that they are transmitted, they consist of the following fields: flag, address, control, text, frame check sequence, and flag. I shall use SDLC to illustrate the use of bit-oriented protocols.

(1) SDLC Frame

An SDLC frame is shown in **Figure 8.6.**

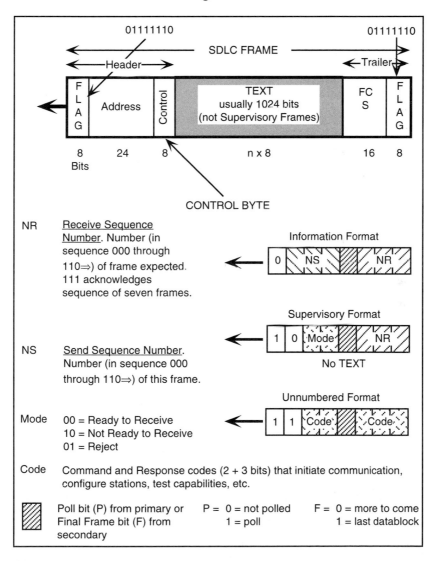

Figure 8.6 SDLC Frame

- It employs **synchronous transmission**
- The **flag** is the sequence 01111110. It defines the beginning and ending of the frame
- Usually, the **address field** is 3 octets long
- Usually, the **text** portion of the frame is limited to 1,024 bits (128 characters)
- The **frame check sequence** (the cyclic redundancy check remainder) follows immediately on the text
- **0** as the first bit in the control byte identifies an **information frame**—i.e., one that contains text. The state of the number 5 bit in the control byte indicates whether information transfer is complete with the present frame (F = 1) or whether there is more to come (F = 0)
- ⇐**10** as the first 2 bits in the control byte identify a **supervisory frame**. It initiates and manages communication. It does not contain text. The state of the number 5 bit in the control byte indicates whether a reply is required (P = 1) or not required (P = 0)
- **11** as the first 2 bits in the control byte identifies an **unnumbered frame** (also called a nonsequenced frame). An unnumbered frame sets the operating mode and initializes transmission. The state of the number 5 bit in the control byte indicates whether the access method employs polling (P = 1) or contention (P = 0). It may contain data in the text field
- **NS** (send sequence number) and **NR** (receive sequence number) are 3-bit numbers used to track individual frames. NS is the number of this frame in the sending sequence (the sequence is unique to the sender). NR is the number of the frame next expected in the receiving sequence. If it is one more than the last frame sent (i.e., NR = NS + 1), it indicates that all frames received so far are without detectable errors. If it is the number of the last frame transmitted, or the number of an earlier frame (i.e., NR $\leq$ NS), it indicates that error has been detected in frame NR, and that the receiver is still expecting a good NR. Using go-back-n, it will be the frame from which the sequence must be restarted. Using selective repeat, it will be the number of the frame that must be repeated. NR = 111 acknowledges the error-free receipt of all frames within a sequence of seven. In this context, the frames are numbered:

Frame 1 = 000	2 = 001 ⇒	3 = 010 ⇒	4 = 011 ⇒
5 = 100 ⇒	6 = 101 ⇒	7 = 110 ⇒	Frames 1 through 7 = 111

An exchange between primary station A and secondary station B that employs SDLC and go-back-n error correction on a half-duplex, point-to-point connection, over which a maximum of four frames is sent before pausing for acknowledgment is given in **Table 8.3**. A polls B. B has a five-frame sequence to send. In frame 3, A detects an error and destroys frames 3 and 4. After four frames are sent, B queries

A. A responds that it is still waiting for frame 3. B repeats frames 3 and 4 and sends frame 5. A sends an information frame to B that acknowledges correct receipt of all five frames and sends the first block of a two-block message. A sends the second block, and B responds with a positive acknowledgment for the two frames. With go-back-n ARQ, B's five text block message to A, and A's two block message to B, were received in sequence.

An exchange that employs the same scenario as Table 8.3 but with selective repeat ARQ is given in **Table 8.4**. A polls B. B begins to send the five-frame message. When A receives frame 3 and detects an error, it destroys frame 3 and immediately requests B to repeat frame 3. After sending frame 4, B repeats frame 3 and then sends frame 5. At the same time, A acknowledges frames through 4 and begins to send a message to B. B responds with an all frames acknowledgment to A. Through the use of selective-repeat ARQ, B's five-text block message to A was received in the sequence 1, 2, 4, 3, and 5. Station A must reorganize the frames to reconstruct the message.

(2) Expanding the Frame Size

The address, control, text, and frame check sequence fields of the frame can be expanded to provide greater operational flexibility. Thus

- **Address**: the basic address module is an octet. It may be extended to any number of octets (SDLC uses three). To do this, bit 1 in the first and suc-

Table 8.3 SDLC on Half-Duplex Point-to-Point Circuit with Go-Back-N

A sends a supervisory frame (*01* ⇒) to B. *Flag* I *FCS* I *000 1 00 01* I *'B'* I *Flag* ⇒ The mode bits (*00*) indicate A is ready to receive. The poll bit is set—a response is required. After sending, the line is turned around, A ceases to be sender and becomes receiver, and B becomes sender.
B responds with information frame 1 (⇐ *0 000*) that contains text 1. ⇐ *Flag* I *'B'* I *0 000 0 000* I *Text1* I *FCS* I *Flag* The final bit is not set—there is more to come. NR = *000* indicates that B is expecting the first frame from A—i.e., A has sent no information frames to B.
B continues with information frame 2 (⇐ *0 100*). ⇐ *Flag* I *'B'* I *0 100 0 000* I *Text2* I *FCS* I *Flag* The final bit is not set—there is more to come. NR continues to indicate no information frames have passed from A to B.
B continues with information frame 3 (⇐ *0 010*). ⇐ *Flag* I *'B'* I *0 010 0 000* I *Text3* I *FCS* I *Flag* The final bit is not set—there is more to come. NR continues to indicate no information frames have passed from A to B. However, A detects error in frame 3 and destroys it.

Table 8.3 SDLC on Half-Duplex Point-to-Point Circuit with Go-Back-N (continued)

B continues with information frame 4 (⇐ *0 110*). ⇐ *Flag* I *'B'* I *0 110 0 000* I *Text4* I *FCS* I *Flag* The final bit is not set – there is more to come. NR continues to indicate no information frames have passed from A to B. However, because of error in frame 3, A destroys frame 4.
B sends a supervisory frame (⇐ *10*) that indicates B is ready to receive A's acknowledgment of the first four frames. ⇐ *Flag* I *'B'* I *10 00 0 000* I *FCS* I *Flag* After sending, the line is turned around, B ceases to be the sender and becomes the receiver, and A becomes sender.
A sends a supervisory frame to B (*01* ⇒). *Flag* I *FCS* I *010 1 00 01* I *'B'* I *Flag* ⇒ The mode bits indicate A is ready to receive. The poll bit is set—a response is required. NR is set to 3 (*010*), indicating that A is expecting frame 3—i.e., an error was detected in frame 3 and A has discarded frames 3 and 4. After sending, the line is turned around, A ceases to be sender and becomes receiver, B becomes sender.
B resends information frame 3. ⇐ *Flag* I *'B'* I *0 010 0 000* I *Text3* I *FCS* I *Flag*
B resends information frame 4. ⇐ *Flag* I *'B'* I *0 110 0 000* I *Text4* I *FCS* I *Flag*
B sends information frame 5 (⇐ *001*). ⇐ *Flag* I *'B'* I *0 001 1 000* I *Text5* I *FCS* I *Flag* The final bit is set—the message is complete. NR continues to indicate no information frames have passed from A to B. After sending, the line is turned around, B ceases to be sender and becomes receiver, A becomes sender.
A sends an information frame to B. *Flag* I *FCS* I *Text1* I *101 0 000 0* I *'B'* I *Flag* ⇒ NS = *000*—this is the first information frame A has sent to B. The final bit is not set—there is more to come. NR is set to *101* indicating that A is expecting frame 6 from B thereby confirming that frames 1 through 5 have been received without errors being detected.
A sends a second information frame to B. Flag I FCS I *Text2* I *111 1 001 0* I 'B' I Flag ⇒ NS = *001* – this is the second information frame sent from A to B. The final bit is set – there is no more to come. NR is set to *111* indicating all frames from B have been received and no errors have been detected. After sending, the line is turned around, A becomes receiver and waits for B's acknowledgement.
B responds with a supervisory frame. ⇐ *Flag* I *'B'* I *10 00 0 010* I *FCS* I *Flag* The mode bits indicate B is ready to receive. NR is set to *010* indicating that B is expecting frame 3 from A thereby confirming that frames 1 and 2 have been received without errors being detected. After sending, the line is turned around, B reverts to receiver status while A reverts to sender status.

Table 8.4 SDLC on Duplex, Point-to-Point Circuit with Selective Repeat ARQ

A sends a supervisory frame to B. *Flag / FCS / 000 1 00 01 / 'B' / Flag* ⇒ The mode bits indicate A is ready to receive. The poll bit is set—a response is required. NR is set to *000*—i.e., A is expecting frame 1. Since this is a duplex connection, A and B can send or receive at any time.
B responds with information frame 1 (*000*) that contains text 1. ⇐ *Flag / 'B' / 0 000 0 000 / Text1 / FCS / Flag* The final bit is not set—there is more to come. NR = *000*—i.e., B is expecting frame 1 from A.
B sends information frame 2 (⇐ *100*). ⇐ *Flag / 'B' / 0 100 0 000 / Text2 / FCS / Flag* The final bit is not set—there is more to come. NR continues set to *000* indicating B is still expecting A's first frame.
B continues with information frame 3 (*010*). ⇐ *Flag / 'B' / 0 110 0 000 / Text3 / FCS / Flag* The final bit is not set—there is more to come. NR continues to indicate no information frames have passed from A to B.
B continues with information frame 4 (⇐ *110*). ⇐ *Flag / 'B' / 0 110 0 000 / Text4 / FCS / Flag* The final bit is not set—there is more to come. NR continues to indicate no information frames have ~~pas~~sed from A to B.
While B is sending information frame 4 to A, A destroys frame 3 and sends a supervisory frame to B. *Flag / FCS / 010 1 00 01 / 'B' / Flag* ⇒ The mode bits indicate A is ready to receive. The poll bit is set—a response is required. NR is set to 3 (*010*) indicating that A has found an error in frame 3.
B resends information frame 3. ⇐ *Flag / 'B' / 0 010 0 000 / Text3 / FCS / Flag*
A sends an information frame to B. *Flag / FCS / Text1 / 100 0 000 0 / 'B' / Flag* ⇒ NS = *000*—this is the first information frame A has sent to B. The final bit is not set—there is more to come. NR is set to *100* ⇒—i.e., A is expecting information frame 5. This indicates that information frames through 4 were received without detecting an error.
B sends information frame 5 (⇐ *001*). ⇐ *Flag / 'B' / 0 001 1 100 / Text5 / FCS / Flag* The final bit is set—the message is complete. NR = ⇐ *100*—i.e., B is expecting information frame 2; also acknowledging error-free receipt of information frame 1 from A.
A sends a second information frame to B. *Flag / FCS / Text2 / 111 1 001 0 / 'B' / Flag* ⇒ NS = *001* ⇒, the final bit is set—there is no more to come. NR is set to *111* indicating all frames from B have been received and no errors have been detected.
B sends a supervisory frame. ⇐ *Flag / 'B' / 10 00 0 111 / FCS / Flag* The mode bits indicate B is ready to receive. NR is set to *111* indicating all frames from A were received without detecting an error.

ceeding octets (except the last) is set to 0. In the last octet of the address, bit 1 is set to 1

- **Control**: in the basic control octet, NS (send sequence number) and NR (receive sequence number) are limited to 3 bits. In an extended mode format, the control field is expanded to 2 octets and NS and NR are set at 7 bits each. Thus, the number of frames that can be sent before acknowledgment must be made is increased from 7 to 127 (assuming 111 and 11111111 are reserved for acknowledgment of entire sequences of frames)

- **Information**: usually, in SDLC the information field is limited to 1,024 bits (128 characters). However, because the header format is fixed, and the trailer consists of the FCS that occupies the last 2 octets before the terminating flag, the information frame can be as long as the users require. What is more, the text need not have a regular byte structure, nor need it contain an integral number of bytes. This is important when transmitting data that have been compressed using entropy coding and similar techniques

- **Frame check sequence**: each frame contains an FCS in the last 2 octets before the terminating flag. For most purposes, a 16-bit sequence is adequate. However, if greater certainty of error detection is required, a 32-bit FCS may be employed.

(3) SDLC Communication Modes

Originally developed to function in an environment dominated by a single mainframe, SDLC designates one station (usually the mainframe or the front-end processor attached to the host) as the primary station and all other stations as secondary stations. Frames that pass from primary to secondary are called *commands*, and frames that pass from a secondary station to the primary are called *responses*. Only one primary station may be associated with a link at any one time. These conventions lead to the *unbalanced point-to-point link* and the *unbalanced multipoint link* shown in **Figure 8.7**. In the normal response mode, a primary station manages the exchange; secondary stations respond when polled. On a duplex connection, if there is only one secondary station, it may initiate transmission without explicit permission from the primary station to produce an *asynchronous unbalanced response mode*.

With the expansion of communications to include national and international networks, a *balanced point-to-point link* structure has been added. Shown in Figure 8.7, it consists of two stations that combine primary and secondary status. Over a duplex connection between stations with equal communication responsibilities, either station may initiate transmission. When sending a command, the station acts as a primary. When receiving a command, or sending a response, the station acts as a secondary. This is known as an *asynchronous balanced response mode*.

(4) Classes of Communication Procedures

Bit-oriented DLC protocols are subject to many options. Before reliable communication operations can be established in a specific environment, all stations

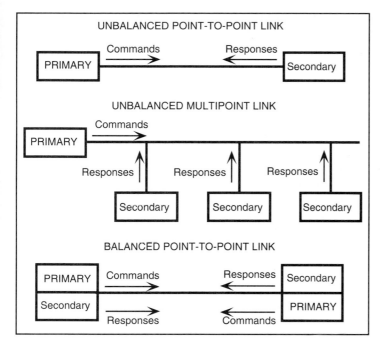

Figure 8.7 SDLC Communication Modes

The balanced, point-to-point link connects hosts with equal status.

must be set up to implement the same class of communication procedures. For instance, it will not do for some stations to be expecting 32-bit FCSs and others to be expecting 16-bit FCSs. (For HDLC, international standards define three basic classes of procedure and 14 options that can be invoked.)

(5) SDLC Transparency

In SDLC, the only special character is the flag, 01111110. Between the beginning flag and the ending flag, whenever a sequence of five 1s is detected the transmitter inserts a 0. When a sequence of five 1s is detected by the receiver, the receiver deletes the following 0. In this way, bit patterns that mimic the flag do not provide false information to the receiver. Generally, this technique is described as *zero stuffing*. As an example, consider the following datastream

⇐ <u>01111110</u>010111011<u>1111111111</u>00000100<u>1111110</u>11100101<u>111110</u>

The first 8 bits and the final 8 bits are underlined—they are the beginning and ending flags (01111110). In between, there is a section of the datastream that mimics the flag (also underlined). Before transmission, between the beginning and ending flags, the transmitter inserts a 0 (denoted Ø for clarity) after sequences of five 1s. This makes the transmitted datastream

⇐ 01111110010111011111Ø11111Ø110000010011111Ø1011100101111110

The section of code in the datastream that mimicked a flag no longer does so. At the receiver, the inserted Øs are removed to produce the original signal. Note that zero stuffing takes place without regard to the octet structure.

(6) SDLC Synchronization

The flag character is used to synchronize the receiver; no synchronizing characters are inserted in the data.

REVIEW QUESTIONS FOR SECTION 8.3

1 What is a data link control protocol?
2 Describe the two classes into which DLCs are divided. Give examples.
3 What functions are performed by a USART?
4 How is bit synchronization achieved in a USART?
5 How is error control performed in a USART?
6 What are character-oriented DLC protocols?
7 List (by mnemonics) and identify BISYNC control characters.
8 Sketch and describe the frame format employed in BISYNC.
9 With BISYNC, what special conditions apply to point-to-point connections?
10 With BISYNC, what special connections apply to point-to-multipoint connections?
11 Explain Table 8.2.
12 In BISYNC, explain how text characters are prevented from causing a control action.
13 In BISYNC, how is bit synchronization achieved?
14 What are byte-count-oriented DLC protocols?
15 Describe the frame format of DDCMP.
16 How is transparency achieved in DDCMP?
17 How is synchronization achieved in DDCMP?
18 What are bit-oriented protocols?
19 List some examples of bit-oriented protocols.
20 Describe an SDLC frame. Explain the uses made of the control byte.
21 Explain Table 8.3.
22 Explain Table 8.4.
23 In SDLC, how can the frame size be expanded? Describe the areas of expansion.

24 In SDLC, explain the meaning of primary and secondary stations, and commands and responses.

25 In SDLC, what is the purpose of a balanced point-to-point link?

26 In SDLC, how is transparency achieved?

27 In SDLC, how is bit synchronization achieved?

8.4 TRANSPORT- AND NETWORK-LAYER PROTOCOLS

ISO has made some attempt to define network- and transport-layer protocols.

8.4.1 ISO NETWORK-LAYER PROTOCOL

For connection-oriented operation, ISO has not developed a complete network layer protocol. The principal difficulty is the dependence of routing and flow control functions on implementation. However, standards do exist that describe the organization of the network layer (ISO 8648), the services the network layer provides to the transport layer (ISO 8348), and network layer addressing. ISO 8648 divides the network layer into three sublayers

- **Subnetwork independent convergence functions** (SNICF): concerned with internetwork routing and relay functions, and global network protocol. The SNICF sublayer contains functions that implement internetwork protocols; they do not depend on specific networks.

- **Subnetwork dependent convergence functions** (SNDCF): concerned with correcting network anomalies and deficiencies. The SNDCF sublayer is included to accommodate situations in which some networks do not provide all the services assumed by SNICF.

- **Subnetwork access functions** (SNAF): concerned with data transfer within individual networks. The SNAF sublayer defines how network-layer entities make use of the functions networks provide.

ISO 8648 describes a connection-oriented process. Divided into three phases—connection setup, data transfer, and disconnection—the standard describes services that establish the connection, transfer data, recover from interruptions, confirm receipt, expedite transfer, and release the connection.

ISO has defined a connectionless network-layer protocol (CLNP, ISO 8473). It is a *best-efforts* datagram service in which data units are routed across the network by nodal decisions (distributed routing) without establishing a connection. Delivery is not guaranteed, and when the datagrams are delivered, they may not be in the sequence in which they were sent.

8.4.2 ISO Transport Layer Protocol

ISO has defined five transport-layer protocols for use with three different levels (A, B, and C) of network performance.

- The high standard of **type A networks** eliminates the need for failure recovery services, and services to handle loss of data and other delivery problems.
- In contrast, the poor standard of **type C networks** makes it necessary to detect and recover from network failures, and to detect and correct for loss of data, out-of-sequence data, duplicated data, misdirected messages, and other delivery problems.
- **Type B networks** lie somewhere between these extremes.

The transport-layer standards define the following connection-oriented headers

- **TP0**: assumes a type A network with error control, sequencing, and flow control. It provides minimal connection establishment, data transfer, and connection termination facilities. TP0 was developed for Teletex applications [see Section 2.3.1(2)].
- **TP1**: assumes a type B network. It provides basic error recovery services including recovery from resets, clears, and restarts.
- **TP2**: assumes a type A network. It provides the ability to multiplex several transport-layer connections over a network layer circuit, and control packet flow in the transport layer.
- **TP3**: assumes a type B network. It adds the error recovery services of TP1 to the multiplexing and flow control services of TP2.
- **TP4**: assumes a type C network. It provides comprehensive error detection and recovery services. Five headers are defined for *transport-connection management* (connection request, connection confirm, disconnect request, disconnect confirm and reject), and five for *data transfer* (data, acknowledgment, expedited data, expedited acknowledgment, and TPDU error). These functions are similar to those performed by TCP (see Section 12.6), but TP4 and TCP are not compatible.

In **Figure 8.8**, I show the formats of connection request, connection confirm, data, and acknowledgment headers. Most of the field names are self-explanatory. Ones that require explanation are

- **Class** (CLS): protocol class (i.e., 0, 1, 2, 3, or 4)
- **Credit** (CDT): credit allocation for flow control—i.e., the number of TPDUs that can be transmitted from the last acknowledgment

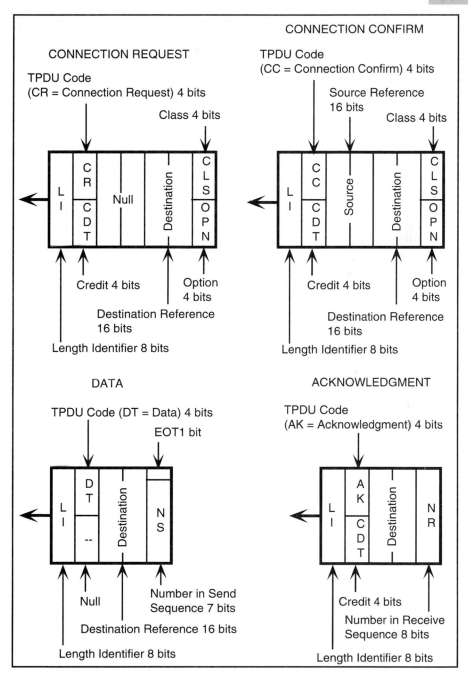

Figure 8.8 Some ISO Transport Layer Headers

- **Option** (OPT): specifies flow control fields. In normal operation, 7 bits are allocated to sequence numbers (i.e., possible sequence of 128 TPDUs), and 4 bits are reserved for credit (i.e., up to 16 TPDUs may be sent from last acknowledgment). In extended operation, 31 bits are allocated to sequence numbers (i.e., possible sequence of over 2 billion TPDUs), and 16 bits are reserved for credit (i.e., up to 65536 TPDUs may be sent from last acknowledgment).

- **EOT**: if the session-layer data unit has been segmented, the bit is set to 1 in the TPDU containing the last segment.

ISO has defined a connectionless transport-layer protocol (CLTP, ISO 8602). It is a *best-effort* datagram service in which data units are transported between sender and receiver on the basis of destination address without establishing a connection. Delivery is not guaranteed, and when the datagrams are delivered, they may not be in the sequence they were sent.

REVIEW QUESTIONS FOR SECTION 8.4

1 List and describe the three sublayers of ISOs network-layer protocol. Why has it not been developed further?

2 Describe the three types of networks for which ISO has defined transport-layer protocols.

3 Describe the five connection-oriented headers associated with ISO's transport layer protocols.

8.5 INTERCONNECTING NETWORKS

It is apparent that networks differ greatly in access methods, frame sizes, frame formats, sequence numbering techniques, addressing conventions, and other items. Yet data that originate at a station served by one network may have to be delivered to a station served by another kind of network. Further, the connection information that originates in one network, must be processable in the network in which it terminates and in any intermediate networks that it traverses. Also, the user's data must be understandable in the network in which it terminates.

Differences between networks can be organized on the basis of the OSI model layer in which they occur. Thus, differences in

- **Electrical** and **mechanical properties** of the communication medium (e.g., cable vs. twisted pair, connectors, signaling methods, etc.) are differences in the implementation of the *physical* layers of the two networks

- **Types of frame information** (e.g., synchronizing patterns and flags, control bits, sequence numbering, error control, flow control, etc.) are differences in the implementation of the *data link* layers
- **Segment sizes** and **routing techniques** (e.g., centralized vs. distributed) are differences in the implementation of the *network* layers
- Message **flow control** and **integrity** are differences in the implementation of the *transport* layers
- **Line discipline**, **failure recovery** strategies, and communication **management** techniques are differences in the implementation of the *session* layers
- **Codes** used and **encryption** and **compression** techniques employed are differences in the implementation of the *presentation* layers.

The capabilities of the device used to effect interconnection will depend on the highest layer in which differences exist. Thus:

- If the differences are limited to the implementation of the physical layers, **repeaters**—devices that link two connections together and amplify and regenerate signals—are used
- If the differences reach no higher than the implementation of the data link layers, **bridges**—devices that operate with medium access control (MAC) sublayer addresses and connect networks of the same general type at the data link level—are used
- If the differences reach no higher than the network layers, **routers**—devices that connect at the network level and operate with network layer addresses to route frames between users—are employed. Frames are routed using
  - *Source routing* also called *centralized routing*: The sender places information in the PDU that specifies the addresses of the nodes that form the route to be taken by the frame including the addresses through any intermediate networks.
  - *Nonsource routing* also called *distributed routing*: The sender specifies the final destination of the frame. Each intermediate node determines the route the PDU shall follow to its destination from a routing table maintained by the node. If the destination address is not included in this table, the PDU is discarded.

  Routers may perform flow control, change the size of PDUs, and perform other functions, as required
- If the differences involve the implementation of layers above the network layer, **gateways**—devices that link networks that are not of the same type and connect at whatever layer is required—are used.

Generically, the four devices are called *relays*—they connect two networks and move information from one to the other—and, more formally, they can be called *internetworking units* (IWUs). Some of the distinctions between these devices have been lost as protocol stacks that vary from the OSI model are implemented, and the terms gateway and router are used interchangeably in many situations.

The use of repeaters, bridges, routers, and gateways to connect networks is illustrated in **Figure 8.9**. The application to each specific combination of networks must be worked out in exact detail. Not only must normal operations be supported, but also abnormal operations, including recovery from failures, must be assured.

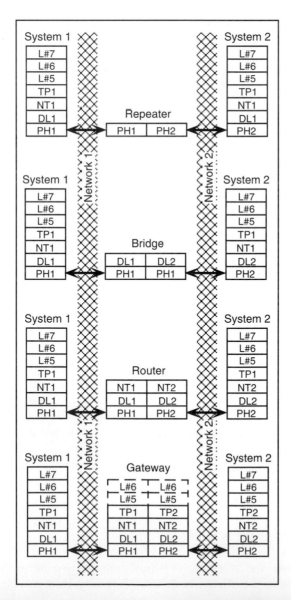

Figure 8.9 Illustrating the Use of Repeaters, Bridges, Routers, and Gateways

The device used to interconnect networks depends on the highest level in which a difference exists between them.

REVIEW QUESTIONS FOR SECTION 8.5

1 If differences occur in the electrical and/or mechanical properties of the communications medium that connects two networks, which level of OSIRM is involved?

2 If differences occur in the types of frame used in two networks, at which level of OSIRM is a connection made?

3 If differences occur in the segment sizes and routing techniques used in two networks, at which level of OSIRM is a connection made?

4 If differences occur in the flow control and error-checking techniques used in two networks, at which level of OSIRM is a connection made?

5 If differences occur in the line discipline and failure recovery strategies used in two networks, at which level of OSIRM is a connection made?

6 If differences occur in the codes used, and the encryption and compression techniques employed in two networks, at which level of OSIRM is a connection made?

7 Between what levels of OSIRM does a repeater connect?

8 Between what levels of OSIRM does a bridge connect?

9 Between what levels of OSIRM does a router connect?

10 Between what levels of OSIRM does a gateway connect?

8.6 NETWORK MANAGEMENT

In the same way that other resources are managed to optimize their contributions to the enterprise (public or private), communication networks must be managed. As they become more complex and sophisticated, the management task becomes more difficult, and the penalty for poor performance becomes greater. In this section, I describe a framework for network management that has been developed by ISO. It provides a common set of concepts for managers, users, and owners. Part of the management task is to ensure message and facility security, and information privacy. These topics are discussed in Chapter 13.

8.6.1 WHAT IS NETWORK MANAGEMENT?

- **Network management**: the planning, organizing, monitoring, and controlling of activities and resources within a communication network to provide a level of service that is acceptable to the users at a cost that is acceptable to the owner.

The activities that are managed are those associated with the use of the network to facilitate communication at a distance among what is probably a large number of subscribers. The resources that are managed are the combination of persons, hardware, and software that have been put in place by the owner to support the needs of the subscribers.

As in other management assignments, network management includes the selection, training, motivation, direction, and evaluation of persons. In addition, network managers must be willing to market network resources to potential users, and to assist them in solving their telecommunication problems. A manager of shared resources cannot neglect to treat users as special persons—*customers*.

Except to note that the quality of the persons brought to bear is critical to the success of any operation, I will discuss neither the human resources aspects of network management, nor the marketing of network services. The principles involved in both of these activities are not markedly different from those that govern them in other areas of commerce. What I will address are the specialized, network-related aspects of network management; they involve tasks such as

- **Monitoring** communications activities so as to detect and correct congestion, errors, or other problems
- **Planning and coordinating** equipment additions or changes to add new services or to improve the level of existing services, including the need for training or the acquisition of additional persons
- **Administering** user-oriented information (station numbers, calling restrictions, security procedures, etc.)
- **Imposing** a communications architecture and maintaining accurate records describing nodes, links, and protocols
- **Operating** a network simulator, or other tools, to test opportunities for expansion or upgrade of equipment, and to examine the effects of new tariffs and other changes in the environment
- **Analyzing** costs and performance to determine the cost-effectiveness of continuing to operate a private network.

It is possible to build a software description of the physical network that can be used to

- **Track** the status of all network components
- **Predict** the effects of changes
- **Provide** a decision support tool that will handle the complexities of traffic analysis, alternate routing, and other network activities, automatically.

In large networks, the status of links and nodes can be reported to displays in the management center; in smaller networks, real-time reporting may be too expensive, and performance records may be entered off-line and examined manually.

8.6.2 OSI MANAGEMENT SYSTEM

In the 1980s:

- Continuing growth in the applications of telecommunications-based information systems
- The need to communicate with other companies
- The prevalence of takeovers and mergers that forced dissimilar networks together

focused attention on the need to implement truly open system architecture—an architecture that will permit cooperating equipment to work together, and to be managed as an integrated entity, no matter who supplies it. The reader will recognize this statement as containing the objective of the open systems interconnection reference model. Using this concept, individual equipment manufacturers can produce products that will respond to operations expressed in standard form with notifications that can be understood by the managing process.

(1) Open Systems

With the support of many organizations and governments in developed countries, the International Standards Organization has extended the concept of OSI to include integrated management of heterogeneous networks. By defining the environment in an abstract way and concentrating on the interfaces between equipment (hardware and software), the ISO standards (X.700 series) have extended and expanded the OSI model to cope with the problems of network management.

(2) Managed Objects

Using the concepts of object-oriented programming, ISO defines a number of building blocks to describe the management process. Chief among them is the idea of a managed object

- **Managed object**: a network entity that is included in the management activity. A managed object may be a switch, a PBX, a router, a multiplexer, a workstation, or other hardware device, or a buffer management routine, a routing algorithm, an access scheme, or other software package.

The *management process* is concerned with the state of these objects as they exist in the network being managed. It is not interested in the internal procedures, but only in the external performance of the objects. A managed object is described by four parameters

- **Attributes**: describe the current state and condition of operation of a managed object. Attributes consist of a type and one or more value(s). Thus, *status* is an integer type attribute that may take one of the values *disabled*, *enabled*, *active*, or *busy*.

- **Operations**: actions that are performed on the managed object, such as
  - *Create* a managed object
  - *Delete* a managed object
  - *Perform an action* on a managed object.

  Actions are also performed on specific attributes of the managed object. Thus:
  - *Get* value of an attribute
  - *Add* value to an attribute
  - *Remove* value from an attribute
  - *Replace* value of an attribute
  - *Set* value of an attribute.
- **Notifications**: reports provided by the managed object about events that occur within its domain
- **Behavior**: actions exhibited by the managed object in response to operations performed on it, or to internal stimuli (e.g., notifications). Also, includes lack of response due to constraints placed on the behavior of the managed object (e.g., report on conditions only when critical).

Managed objects are classified by how they fit into the OSI model—e.g.,

- **(N)-layer-managed object**: object specifically contained within an individual layer (e.g., data link protocol)
- **System-managed object:** object pertaining to more than a single layer (e.g., router or gateway)

and can be identified as a class

- **Managed object class**: objects with similar characteristics (i.e., attributes, operations, and notifications).

(3) Other Building Blocks

Per se, network management is performed with the aid of a network management program—an application layer process. In the ISO environment, it is divided into two parts

- **Managing process**: the part of the application process concerned with overall management tasks
- **Agent process**: the part of the application process concerned with performing functions on managed objects at the request of the managing process.

In addition, agent processes and managing processes share a common database through which they communicate with one another

- **Management information base**: contains both long-term and short-term information concerning managed objects.

The top part of **Figure 8.10** shows the organization of these entities in a network management system. The network consists of hardware and software entities that are designated managed objects. Managing agents communicate with them by means of operations, and they respond by changing behavior and/or providing notifications. The managing process attends to assimilating and analyzing information, initiating state changes as required, maintaining logs and records, and providing information in understandable form to persons in the control center.

The management information base serves as the record repository for the network. It contains descriptions of all network elements and their status over the last period.

(4) Managers and Agents

An important function of the agent process is to delete trivial notifications so that the managing process can focus on important events such as faults, congestion, and the like. This is achieved through the use of filters, the employment of scoping, and event forward discriminators. They are shown in the bottom part of Figure 8.10. Their functions are

- **Filter**: uses Boolean operators to chain test conditions that determine which events in the network are to be reported
- **Scoping**: identifies the subtree of managed objects to which a particular filter is applied
- **Event forward discriminator**: directs reports to the appropriate destination.

Implemented in software at various points in the network equipment, these tools allow the managing process to manage by exception, and the network manager to investigate the utilization of any portion of the network.

(5) Architecture of ISO Network Management System

The system management application entity resides in the application layer of the OSI model. It consists of a collection of cooperating application service elements (ASEs). **Figure 8.11** shows the arrangement.

- **System Management Application Service Element** (SMASE): creates and uses the protocol data units that are exchanged during the management process. It employs communications services supplied by specific applications service elements (ASEs) or by CMISE.

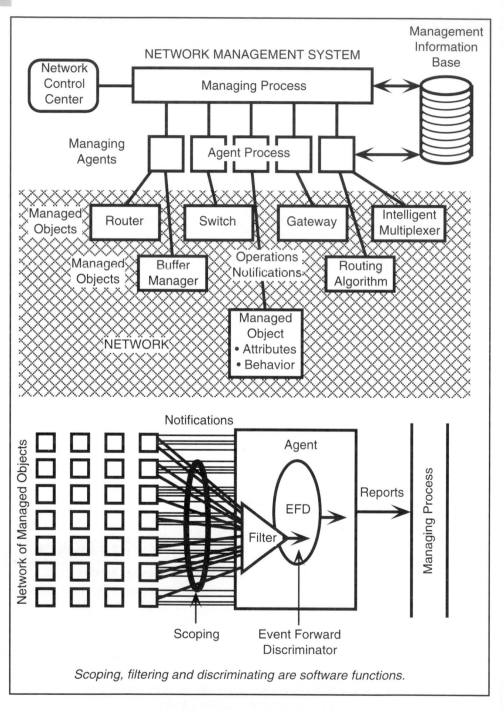

Figure 8.10 OSI Management Model

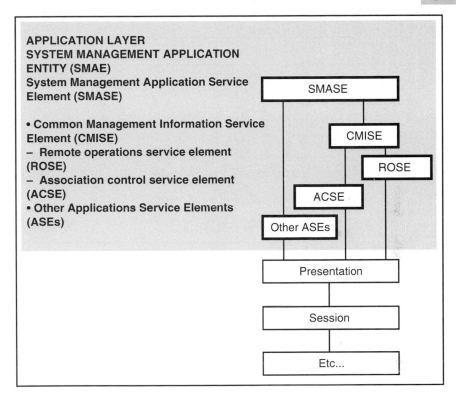

Figure 8.11 Systems Management Application Entities

SMAE exists in the application layer. It is supported by several application service elements.

- **Common Management Information Service Element** (CMISE): defines a basic structure for use in all management areas. Uses a connection-oriented transport protocol to ensure reliable data transfer. Provides management notification services that can be used to report any event about a managed object, and management operation services that define the operations required to perform actions on a managed object. In addition, provides a framework for common management procedures that can be invoked from remote locations. CMISE employs scoping and filtering; it supports the services listed in **Table 8.5**.

- **Remote Operations Service Element** (ROSE): general software support system employed in a number of OSI applications (e.g., X.400—Message Handling System, and FTAM—File Transfer and Access Management) that supports distributed, interactive applications processing between remote objects. In synchronous mode, ROSE requires the receiver to reply before the originator can invoke another operation. In asynchronous mode, ROSE permits the originator to continue invoking operations without waiting for a response.

Table 8.5 Services Supported by CMISE

Service	Description
get	Requests agent to get value(s) of specific attribute of managed object
set	Requests agent to set specific attribute of managed object to particular value
action	Requests agent to perform specific action on managed object
delete	Requests agent not to report specific attribute(s) of managed object
cancel	Requests agent to cancel previous request
event	Agent reports occurrence of an event in a managed object

- **Association Control Service Element** (ACSE): general software support system that works between application processes that are independent of application-specific needs. Employed in a number of OSI applications (e.g., X.400–Message Handling System, and FTAM) it supports the establishment, maintenance, and termination of a cooperative relationship between application entities.

8.6.3 MANAGEMENT ACTIVITIES

ISO has defined five areas of specialized, technical responsibility.

(1) Areas of Activity

Known as specific management function areas (SMFAs), the areas are

- **Configuration management**: maintaining, adding, and updating records of hardware and software components; defining their functions and usability; initializing the network; and ensuring that the network is described correctly and completely at all times. Also, assign, maintain, and update records of users, including billing responsibilities and communication privileges. A critical feature of configuration management is the ability to cope with, and control, change.
- **Fault management**: provision, monitoring, and response to facilities that detect and isolate abnormal conditions in the network so as to restore the network to normal functioning and correct the aberrant condition(s).
- **Security management**: control access to, and use of, the network according to rules established by the network owner.

- **Accounting management:** identify costs, collect and process billing information, prepare budgets, etc.
- **Performance management**: evaluate the performance of components, and the network as a whole, with regard to availability, response time, accuracy, throughput, and utilization.

(2) System Management Functions

For each of these areas, ISO working groups have identified a specific set of functions, and the procedures and services needed to implement them. Known as system management functions (SMFs), there is overlap between them, and many SMFs serve more than one SMFA. Briefly, some of them are

- **Object management function**: describes the generation of reports on object creation, deletion, name changes, and changes in attribute values
- **State management function**: describes the generation of reports on the current state, and changes in state, of managed objects
- **Relationship management function**: defines relationships between managed objects
- **Alarm reporting management function**: facilitates reporting of problems by managed objects
- **Event reporting management function**: describes the selection of events that are to be reported to the managing process in a given period
- **Confidence and diagnostic testing**: supports means to determine if a managed object is able to perform its designated functions
- **Log control function**: defines rules for construction of network operations management logs
- **Security alarm reporting function**: defines the rules for the generation of security alarms
- **Security audit trail function**: defines rules for construction of network audit trail management logs
- **Objects** and **attributes for access control**: defines levels of access controls
- **Accounting meter function**: defines rules for constructing usage and accounting data for managed objects
- **Workload monitoring function**: provides a model for monitoring the performance and measuring the workload of a managed object.

(3) Performance Management

On a day-to-day basis, the most important activity to be managed is network performance. Constant monitoring will provide measurements of the magnitudes and trends of performance aberrations. Study and analysis (perhaps aided by decision support software) of the quantitative data will provide insight into problems and faults that develop. Both will aid in the selection of a strategy for recovery or tuning the network. Network performance information falls into two categories

- **Service-oriented**: items such as availability, response time, and accuracy—parameters that are important to the users
- **Efficiency-oriented**: items such as throughput and utilization—parameters that are important to the network owner.

It is the network manager's job to balance one with the other so as to maintain good relations with the users and the owner. Among other things, network performance is measured by

- **Throughput**: the effective rate of transfer of messages across the network
- **Response time**: the time between generating and satisfying a request. Associated with this measure are those times that have to do with setting up, using, and taking down a circuit—namely, connection establishment delay, transit delay, and connection release delay
- **Waiting time**: the time spent in queues waiting for service, the time spent in processing and forwarding messages, and the time between arrival events (at a queuing point, for instance)
- **Workload**: number of managed objects idle, number of managed objects active, number of managed objects busy, and number of managed objects rejecting service requests because of overload
- **Statistical analysis**: of event records to determine quantities such as utilization, availability, error rate, mean-time to failure, mean-time to repair, etc.

Network performance management requires continuous data gathering and analysis, the ability to make judicious allocations of resources to overcome potential problems, and the opportunity to hear from customers about their problems and concerns.

8.6.4 FAULT MANAGEMENT

Once a fault occurs, its effect must be contained, and network performance restored as soon as possible. Fault management includes the ability to detect faults, to perform diagnostic tests on a managed object to define the fault, and to perform operations on the managed object to correct the fault, or to take the object out of service for physical maintenance. In addition, fault management includes the creation of a fault log that can be analyzed by the managing process and used as the basis for preventive maintenance activities. Among others, fault management makes use of the following SMFs: event reporting, confidence and diagnostic testing, log control, and alarm reporting.

8.6.5 CONFIGURATION MANAGEMENT

In today's dynamic network environment, resources can be enabled and disabled on command and redistributed as required to follow changing demands for services. Under these circumstances, configuration management must report on hardware and software changes in a timely way, and maintain an up-to-the-minute picture of the network. By operating on the management information base, configuration management programs maintain naming conventions, identify new managed objects, set initial values for attributes, supervise relationships between (among) managed objects, change operational characteristics of managed objects, report on changes in the state of managed objects, and delete managed objects. In describing the states of the objects in the network, configuration management distinguishes between administrative states and operational states. Administrative states employ the following descriptors

- **Unlocked**: the managed object can be used
- **Locked**: the managed object cannot be used
- **Shutting down**: the managed object will honor current users, but will not accept new users.

Operational states employ the following descriptors

- **Enabled**: the managed object is operable, available, and not in use
- **Disabled**: the managed object is not available
- **Active**: the managed object is available for use and can accept new users
- **Busy**: the resource is available but cannot accept new users.

Combinations of these states are permitted, and used. **Table 8.6** shows their meanings.

8.6.6 SECURITY MANAGEMENT

Pervasive telecommunication facilities connected into networks are creating opportunities for ill-disposed persons to compromise transactions. In addition, there is always the possibility of a genuine mistake leading to an unexpected outcome. Security management is directed toward the detection of these occurrences. I take up the subject in Section 13.6.

8.6.7 ACCOUNTING MANAGEMENT

Accounting management refers to the collection of data that can be used to charge for communications capabilities provided to the users on the basis that, in aggre-

Table 8.6 Administrative and Operational States of Managed Objects

ADMIN STATE	OPERATIONAL STATE			
	Disabled	Enabled	Active	Busy
Unlocked	• MO in service but disabled by some problem. State may change to locked.	• MO in service and can be used when required.	• MO in service, in use and has more capacity.	• MO in service, in use, but has no more capacity.
Shutting Down	• MO with problem is being taken out of service.	• working MO is being taken out of service.	• MO in service, in use and has more capacity, is being taken out of service.	• MO handling traffic but can handle no more, is being taken out of service.
Locked	• MO out-of-service; disabled by some problem.	• MO out-of-service, can be used; probably in standby.	Not possible	Not possible

MO = managed object

gate, it should cover the cost of doing business and provide a reasonable return to the owners.[1]

(1) Communications Charges

In a public network, a call-record is produced at the time of placing a call. It is used to charge for the service on the basis of

- Time consumed and distance transmitted
- Bandwidth employed

---

[1] Although slightly dated, the reader will find a wealth of information in Nathan J. Muller, *Minimum Risk Strategy for Acquiring Communications Equipment and Service* (Norwood, MA: Artech House, 1989).

- Number of PDUs (or octets, characters, or bits) transmitted
- Any combination of these and other parameters.

The carrier has adjusted the charge to cover the cost of capital equipment, running expenses, taxes, interest, and profit for the corporation. Because it relies on internal bookkeeping, in a private network, the cost of communications is what the accounting department says it is, or what the general manager stipulates. In truth, the network manager has only two controllable accounts—salary and outside expenses. Nevertheless, many organizations make a good-faith effort to determine the true cost of their communications.

(2) Annual Charge

The annual cost associated with purchasing, owning, maintaining, and using up plant is known as the *annual charge*. For items of plant that are owned by the private network operator, summing annual charges gives the total annual charge for them. Because they have been purchased on behalf of the network operator, these items must be managed throughout their economic and accounting lifetimes. They cannot be discarded without considering the point at which they are in their depreciation cycle and their salvage value. In the case of equipment that is retired early, a displacement charge (consisting of first cost minus accumulated depreciation minus salvage value) is incurred.

Prudent management suggests that plant that is owned should be generally useful to the corporation for at least the depreciation period.

(3) Lease Cost

For items of plant that are leased by the private network operator, the annual cost includes the annual charges incurred by the lessor together with an additional charge for lessor's profit. Within the terms agreed to in the lease, these items can be replaced at the initiative of the network operator. Often, early termination results in a penalty charge to compensate the lessor for loss of revenue that was intended to cover the initial investment.

Prudent management suggests that plant that is leased should be generally useful to the corporation that leases it for at least the initial lease period.

(4) Rental Cost

For services that are used on demand over other companies' facilities, the service provider will charge the network operator a fee. For the use of public networks, the fees may be based on time and distance and may be subject to the control of state and federal agencies. They will have been set to cover all the relevant costs of the provider. Available on demand, the services are paid for as used. In addition, it is likely that there will be a basic monthly charge that represents a fee for connection. It is paid so long as the user wants to have the services available for immediate use.

Prudent management suggests that services should be rented until demand is great enough to lease or buy plant that will provide them, or so long as rental costs are less than the cost of providing the service in other ways. It is the latter situation that is exploited by providers of virtual networks.

(5) Options

For private networks, some general options for allocating telecommunications costs are the following

- **Usage records** may provide the basis for charging the cost of the facilities to individual users or departments
- The cost of communications may be included in **general overhead** and distributed to company departments on the basis of payroll, sales, or some other operational quantity
- Communications may be provided by a separate **cost center**, with payments for services negotiated between the network manager and individual using departments.

All three arrangements (and several others) have been implemented in firms in the United States—and none is perfect.

(6) Equipment That Is Owned

For general-use equipment (PBXs, telephones, etc.) owned by the corporation and generally useful for at least the depreciation period, a strong case can be made to treat the expense as a cost of doing business. Expenses collected at the corporate level can be allocated to each operating unit. Doing so simplifies accounting for each piece of equipment and lends credibility to a corporate telecommunications unit. However, the procedure tends to mask inefficiencies in the telecommunications operation, and operating units can become frustrated by what they perceive as a lack of accountability of the telecommunications unit, particularly during periods of fiscal pressure.

(7) Equipment That Is Leased

For general-use equipment that is leased, the cost can be bundled with the cost of other plant and allocated to each business unit. Leased equipment that is used by one or more units for a specific purpose can be charged directly to the using units. In this way, there is the possibility of a critical review of its contribution to the business—at least every budget cycle.

(8) Services That Are Rented

In providing these services, the supplier is likely to offer individual billing details that can be used to allocate exact costs to each user. Through their use, individual responsibility for the costs incurred is established, and there is no doubt who must pay. Extending the principle to virtual networks supplied by public carriers, it is possible to receive billing details that permit the allocation of charges to relatively small organizational units and have the charges match the use made of the facilities. However, because these records are an invitation to build an internal bureaucracy to check them, there is a point at which such detail becomes nonproductive.

(9) Effect of the Number of Switching or Multiplexing Locations

In a network that serves several locations, switches, routers, and/or multiplexers will be employed to minimize the total network cost. By connecting the backbone circuits to local networks, they make local connections available to anyone on the network. By concentrating traffic on tie lines between the switches, low-cost long-distance connections make it possible for remote users to employ local area connections at a cost lower than the cost of individual WATS facilities. Generally, as the number of nodes increases

- The **investment** in switches and multiplexers increases
- The **cost of access lines** decreases as the average connection to the serving switch or multiplexer gets shorter
- The **cost per hour of calling** may rise as the average loading on each link diminishes
- The internode **trunk mileage** (backbone network) costs increase.

The result is a complex tradeoff between switch and multiplexer investments and transmission expenses. Minimization of the total cost can lead to several, quasi-optimum network topologies. The general nature of these relationships is shown in **Figure 8.12**. To discover what to do in any practical situation a computer-based study is required.

8.6.8 TELECOMMUNICATIONS MANAGEMENT NETWORK

The ISO Network Management System described above formalized the activities needed to ensure reliable data communication in a heterogeneous computing environment. Building on OSIRM, system management is achieved through the exchange of messages between application-level processes. Most importantly, the network management system established a common structure for information exchanges between managing and managed entities. However, as networks grew in size and complexity and automated maintenance and administration became necessary, it was apparent that a model with substantial telecommunications underpinnings was needed. Spurred on by industry bodies, ITU–T has created the architecture[2] of a system for the management of telecommunications networks

---

[2] ITU–T Recommendation M3010, *Principles for a Telecommunications Network*, 1996.

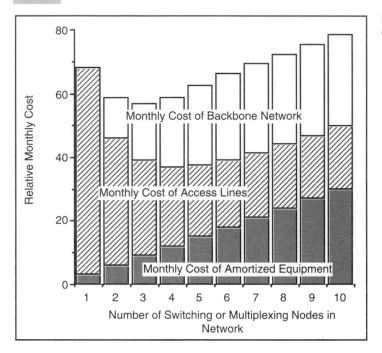

Figure 8.12 Cost Components of Various Size Networks

and services. Known as *Telecommunications Management Network* (TMN), it employs the OSI concepts of *manager* and *agent* and application service elements (ASEs). The functional, information, and physical communicating entities of TMN are described in this section.

(1) TMN Architecture

Figure 8.13 shows the entities in a simple TMN. They communicate with one another over the interfaces shown. Of the five types of function blocks, one, NEF (network element function block), represents the managed system. Three other blocks, OSF (operations system function block), MF (mediation function block), and WSF (workstation function block), represent the managing system. The fifth block, QAF (Q adaptor function block), serves as a translator between TMN and already existing (legacy) systems that employ non-TMN management systems. Data communication between function blocks occurs over the DCF (data communications function). Operators use WSF to supervise the network. In common with the OSI model, the managed system and the managing system exchange information (expressed in terms of the attributes and behavior of managed objects) through the management information base (MIB).

At a minimum, each function block contains a TMN physical block that implements the function. In addition, it may contain other physical blocks. The physical blocks are

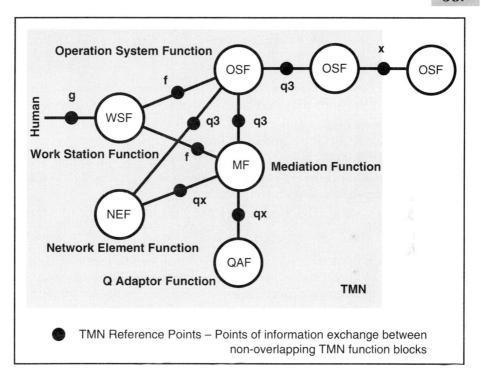

Figure 8.13 Functional Entities and Interfaces in Telecommunications Network Architecture

- **OS**: operations system
- **MD**: mediation device
- **NE**: network element
- **WS**: workstation
- **QA**: Q adaptor.

(2) TMN Interfaces

Communication between function blocks occurs through TMN reference points whose physical embodiments are called TMN interfaces. Thus

- **q3**: interface between OS, and MD, NE and other OS blocks
- **qx**: interface between MD, and NE and QA blocks
- **f**: interface between WS, and OS and MD blocks
- **x**: interface between OS in one TMN and OS in another TMN
- **g**: interface between operator and WS block.

(3) TMN Logical Layered Architecture

The operations system is responsible for four activities. Defined as layers in the model, they are

- **Element management layer** (EML): managing a subset of NEs (individually or as a group), and mediating between NEs and the network
- **Network management layer** (NML): managing all NEs and interconnecting links (DCF), and assuring end-to-end integrity
- **Service management layer** (SML): managing providing services to customers, including creating service orders, processing complaints, billing, and measuring quality of service
- **Business management layer** (BML): managing finance, budgets, goals, and plans.

When expressed in this form, the functional architecture is called *logical-layered architecture*. There is a greater degree of freedom from vendor specifications at the network management layer than at the element management layer. In addition, there is a greater degree of freedom from technical implementation at the services management layer than at the network management layer.

(4) Managers, Agents, and Manager/Agents

TMN defines two roles for entities when engaging in information exchange

- **Manager**: seeks information on managed entities from agent, evaluates status of managed entities, and issues directives to agent to change state of managed entities
- **Agent**: provides information on managed entity to manager, receives and implements directives from manager.

Each TMN function block can perform as manager, agent, or combine the functions as a manager/agent. In the latter role, the components of the manager/agent communicate through the appropriate MIB. Within reasonable limits, the actions of manager and agent are identical to those performed by similar entities in the ISO management model.

(5) Status of TMN

Originally intended to manage the operation, administration, maintenance, and provisioning (OAM&P) of network facilities and services in an open, multi-vendor environment, TMN has been in the ITU–T standards development process since the early 1980s. Much of the work has been concerned with definitions of interface protocols and information formats (managed objects). Because very little of today's equipment is TMN-compliant, applying the principles of TMN to today's (legacy) networks requires special engineering. Replacing legacy interfaces with TMN interfaces can call for many specialized Q adaptor devices. In some administrations, TMN interfaces are being implemented to manage certain performance parameters (e.g., quality of service) while leaving legacy interfaces in place for other parameters (e.g., OAM&P).

As has happened before, the multivendor environment will not wait for slow-moving standards efforts. A consortium of operators, vendors, and users has been formed to develop TINA (telecommunications information networking architecture). Known as TINA–C, their work is described in Section 12.10.2(1). What is more, the evolution and growth of Internet has produced a simplified architecture for network management. It is described in Section 12.10.1.

REVIEW QUESTIONS FOR SECTION 8.6

1 Define network management.

2 Describe the network-related aspects of network management.

3 In the 1980s, what conditions focused attention on the need to implement truly open system architecture?

4 What is a managed object? List and explain four parameters that describe it.

5 What is an (N)-layer managed object?

6 What is a system-managed object?

7 Distinguish between a managing process and an agent process.

8 For what purposes is a management information base used?

9 How do managing agents communicate with managed objects? What does the managing process do? Explain the top part of Figure 8.10.

10 Explain how managing agents delete trivial notifications. Explain the bottom part of Figure 8.10.

11 Explain the structure and functions of the system management application entity. Use Figure 8.11.

12 List and explain the five areas of specialized, technical responsibility defined by ISO.

13 What are an SMF and an SMFA?

14 Enumerate some SMFs.

15 What are the two categories of network performance information?

16 List and explain five factors used to measure network performance.

17 Comment on the following statement: Once a fault occurs, its effect must be contained and network performance restored as soon as possible.

18 List, and explain, the three administrative states used to describe the status of a managed object.

19 List, and explain, the four operational states used to describe the status of a managed object.

20 Describe Table 8.5.

21 Comment on the following statement: In aggregate communication charges should cover the cost of doing business and provide a reasonable return to the owners.

22 Comment on the following statement: Because it relies on internal book-keeping, in a private network, the cost of communications is what the accounting department says it is, or what the general manager stipulates.

23 What is an annual charge?

24 Comment on the following statement: Prudent management suggests that plant that is owned should be generally useful to the corporation for at least the depreciation period.

25 Comment on the following statement: Prudent management suggests that plant that is leased should be generally useful to the corporation that leases it for at least the initial lease period.

26 Comment on the following statement: Prudent management suggests that services should be rented until demand is great enough to lease or buy plant that will provide them, or so long as rental costs are less than the cost of providing the service in other ways.

27 Use Figure 8.12 to describe how the number of nodes it contains affects the cost of a private network.

28 Describe the relationship between OSI network management model and TMN.

29 Describe the objectives of the five entities in Figure 8.13.

30 Identify the physical blocks in the functional entities of Figure 8.13.

31 Define the five types of interfaces in a TMN.

32 Describe the four management layers in the logical-layered architecture model.

33 Justify the following statement: There is a greater degree of freedom from vendor specifications at the network management layer than at the element management layer.

34 Justify the following statement: There is a greater degree of freedom from technical implementation at the services management layer than at the network management layer.

35 Are the roles of managers and agents the same in OSI management network and TMN?

36 Why is TINA under development?

TRANSFER MODES

It is not practical for users to have permanent connections to all those with whom they may wish to communicate. Instead, temporary connections are formed from the facilities of communication networks. Each network routes traffic along links and across nodes to satisfy the user's immediate communication objectives. The work of the nodes is performed by a spectrum of transfer machines (called switches or routers). They range from those that route variable bit-rate and bursty signals over paths of opportunity with variable delays (e.g., packet switches) to those that route fixed bit-rate signals over reserved paths with fixed delays (e.g., circuit switches). The links employ specialized transport mechanisms to suit the nodes employed.

9.1 TRANSFER MODE

The combination of node and transport mechanism defines a transfer mode.

- **Transfer Mode**: the manner of transfer of information and overhead bits between (among) users. Characterized by the switching/routing technique employed and the transmission technique that supports it. The transfer mode may be

   - *asynchronous*. Synchrony does not exist throughout the transfer facilities. Operation does not require that the clocks in the transmitter and receiver are synchronized

    – *synchronous*. Synchrony exists throughout the transfer facilities. Operation requires that the clocks in the transmitter and receiver are synchronized.

A typical *synchronous transfer mode* is the combination of digital circuit switch and T-carrier (and/or SONET) multiplexing used in telephone voice networks. All facilities are synchronized to coordinate activities and minimize delays in the system. A typical *asynchronous transfer mode* is the combination of packet switch and statistical frame multiplexing used in packet data networks. Facilities are not globally synchronized, and traffic is bursty. There will be fluctuations in delivery delay, and segments may not arrive in the sequence they were sent.

    To lend some confusion to this chapter, the name Asynchronous Transfer Mode (ATM) is used to describe the technology employed to transfer voice, data, and video information, and overhead, among users of wideband ISDN (see Section 10.6). It defines a specific combination of switch (ATM switch) and transmission format and speeds (STM–n).

REVIEW QUESTIONS FOR SECTION 9.1

    1 What do you understand by the term transfer mode?

    2 Distinguish between asynchronous and synchronous transfer modes. Give an example of each.

9.2 PACKET RELAY

In the late 1960s, the Advanced Research Projects Agency (ARPA) of the U.S. Department of Defense began to construct a packet-switching network. Called ARPAnet, it was the precursor of many of today's commercial X.25 ventures and gave rise to a host of networks that linked military, defense industry, and academic research facilities around the world. Based on these developments, specialized common carriers offered data communication over packet-switched networks. Known as value-added networks (because they provide error control and other *valuable* services), they provide connection-oriented and connectionless, point-to-point and point-to-multipoint data services.

9.2.1 X.25 PROTOCOLS

ITU–T Recommendation X.25 describes the user network interface (UNI) employed with packet networks. Usually, the packet is relatively short so that a user's message must be subdivided before it is sent.

(1) Architecture

The division of user's data into packets, and encapsulation in an LAP-B frame, is shown in **Figure 9.1**. LAP-B is the vehicle for communication between the originating DTE and the DCE at the packet network boundary.

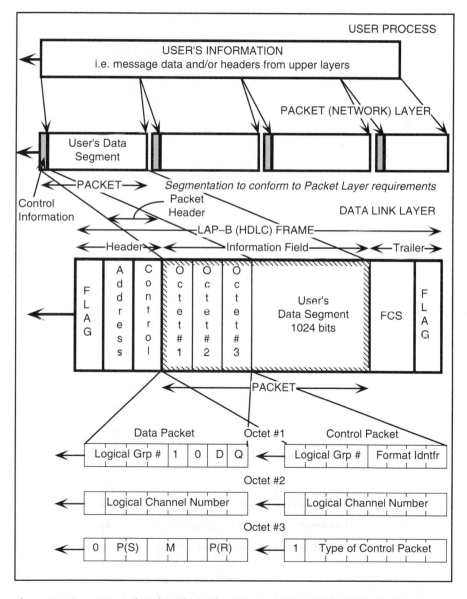

Figure 9.1 Formation of Packets from User Data and Encapsulation in an LAP-B Frame

Each packet may be encapsulated in several different frames during its journey through the network and must carry along sufficient information to ensure it will traverse the network successfully. This information is contained in the 3 octets of control information that transit the route with the user's data segment.

The send and receive sequence numbers – N(S) and N(R) in the control octet of the HDLC frame, and P(S) and P(R) in octet 3 of the packet—are different. N(S) and N(R) counts are assigned consecutively to all frames crossing the link irrespective of to which call the packets they contain are assigned. P(S) and P(R), on the other hand, refer to the sequence of packets that comprises a specific call.

Between the originating DTE and the DCE that serve as entry to the packet network, X.25 defines three layers that are analogous to the network, data link, and physical layers of the OSI reference model. As shown in **Figure 9.2**, at the DTE

- In the **packet layer**, or **X.25-3 layer** (network layer), packets are formed by adding control information to user information segments

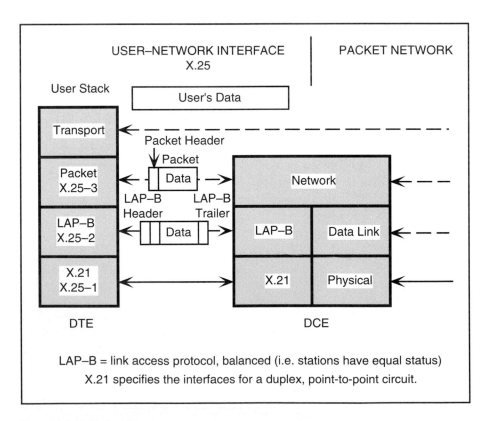

Figure 9.2 X.25 Architecture
ITU–T X.25 defines the user network interface for a packet switched network.

- In the **X.25-2 layer** (data link layer), the packet is encapsulated in an HDLC frame that implements LAP-B (a bit-oriented protocol that operates in asynchronous balanced mode—i.e., either station may send without waiting for permission from the other, and all information frames are treated as commands)
- In the **X.25-1 layer** (physical layer), the frame is transmitted synchronously to the network DCE.

At the DCE

- The **physical layer** reconstructs the datastream
- The **data link layer** interprets the header and trailer in accordance with LAP-B rules and passes the packet to the network layer
- The **network layer** hands off the packet to the proprietary protocols that make whatever modifications are necessary for it to enter the packet network.

(2) Virtual Circuits

As each packet in a call transits the network, resources are allocated on a link-by-link basis. Other packets use these resources at other times. Octets 1 and 2 of the packet header contain a 4-bit logical group number, and an eight-bit logical channel number; together, they provide a 12-bit identifier. If the all-zeros possibility is excluded, this identifier can discriminate among 4,095 channels. At the start of each session, to identify the stream of packets in the coming exchange of data, the originating DTE and the terminating DCE assign logical channel numbers. In the same way that the telephone numbers of the calling and called parties identify a telephone circuit, the originating and terminating logical channel numbers identify a virtual circuit. It is up to the network administration resource to coordinate the network portion of the journey to satisfy the requirements of the communicating DTEs.

9.2.2 PACKET NETWORKS

X.25 packet-switched networks multiplex packets of data across links between store-and-forward nodes to share the resources among all network users. Users employ X.25 protocols to connect to the network. Within the network, the nodes route packets in accordance with information carried in individual packets, information stored in the node, and the state of the network. Between networks (a transport layer concern), ITU–T's Recommendation X.75 applies.

(1) What Is an X.25 Network?

The realms of application of the standards governing connections to, and connections between, packet-switched networks are shown in **Figure 9.3**. At the top of the network is a front-end processor (FEP/Host) combination that connects to the network using X.25 procedures. At the bottom of the network, a synchronous DTE employs a similar technique.

Asynchronous DTEs are incapable of handling synchronous X.25 protocols. To provide them with the ability to use packet-switched networks, a special device, known as a packet assembler and disassembler (PAD), is required. As the

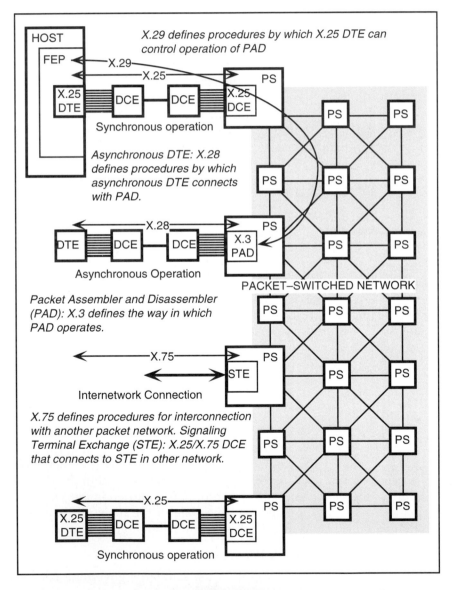

Figure 9.3 Realms of Application of ITU–T Standards Governing Connections to and Connections Between Packet Switched Networks

name implies, it assembles packets from characters received from asynchronous DTEs and disassembles packets into characters for transmission to asynchronous DTEs. ITU–T Recommendation X.3 defines the PAD. Usually it is implemented in software resident at the boundary of the network. ITU–T Recommendation X.28 defines the procedures whereby the asynchronous DTE connects to the PAD, and X.29 defines the procedures that permit a packet-mode device to control the operation of the PAD.

At top center of Figure 9.3 is an asynchronous DTE that connects to a PAD through the use of X.28 procedures. At bottom center is a signaling terminal exchange (STE), a DCE that implements X.75 procedures to connect to another packet network.

(2) Packet Flow

Packets migrate across the network in different ways depending on the routing technique employed. In **Figure 9.4** we show three examples. DTE A sends groups of packets to terminals B, C, and D.

- **Packets 1, 2, 3,** and **4** are sent to DTE D using a *connectionless* service and *distributed* routing. Starting off in sequence, the packets are sent over different links at the direction of the individual nodes based on current knowledge of the traffic loads in the networks. They arrive at DTE D in the order 2, 3, 1, and 4. It will be up to the receiving DTE to reorder them, if necessary.

- **Packets 5, 6,** and **7** are sent over a *permanent virtual circuit* to STE C for transmission to another network. They arrive at C in sequence.

- **Packets 8, 9,** and **10** are sent to DTE B using a *connection-oriented* service and *centralized* routing. A call request packet is used to establish a virtual circuit (primary and alternate route). When packets 8 and 9 have traversed the primary route, we assume a failure occurs that requires packet 10 to traverse the alternate route.

In a packet network, the integrity of each packet is checked at a node before the packet is passed on to the next node. This means that the entire packet must be received before it can be routed toward its destination, or destroyed, and retransmission requested. Thus, at each node, packets are delayed for at least one packet time. Further slowing the throughput is the fact that long messages are segmented into limited-length packets that contain additional overhead information to facilitate control, identification, sequencing, and delivery. In all, some 30 processing steps may be required to check the integrity of the packet before the node can begin to pass a packet to the next node on its path through the network. The presence of so many steps reflects the fact that conventional packet networks were designed to link dumb terminals to mainframes over relatively unreliable transmission facilities and that much checking is required to achieve acceptable network performance. Because these procedures are time-consuming, X.25-based networks provide slow-speed transport service [see Section 9.2.2(6)].

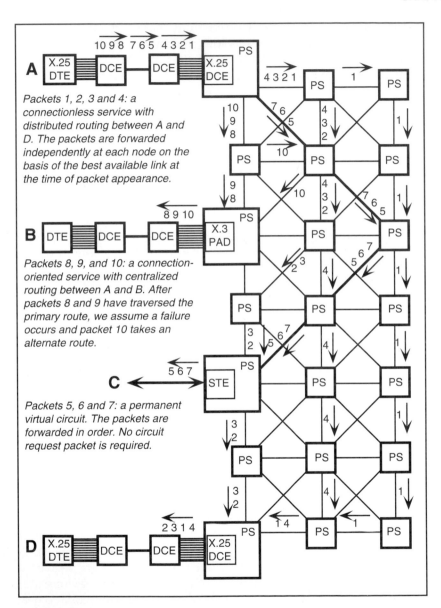

Figure 9.4 Three-Packet Routing Techniques

(3) Flow Control

A packet is a small segment of information. If 128 bytes is the limit of the information field, it can contain approximately 20 average words—and that may be less than two lines of text. Strings of packets, then, are common, and procedures are needed to ensure that they are not sent so rapidly as to block the receiving network links, DCE, or the receiving DCE/DTE combination. One technique employs *sliding window* flow control.

- **Sliding window flow control**: the transmitter transmits up to W packets and must then wait for an acknowledgment from the receiving DTE. The value of W is $1 \leq W < 8$ for data packets in which P(R) is a 3-bit number, and for packets that employ an extended format $1 \leq W < 128$—i.e., P(R) is a 7-bit number.

If the sequence of W packets in the window is received correctly, the receiving DTE will acknowledge them by updating the value of P(R). This permits the sending DTE to transmit another W packets. The value of W is negotiated between sending and receiving DTEs.

At any time, the receiving DCE may send a Receive Not Ready (RNR) indication to the sending DTE; also, the receiving DTE may send RNR or Reject (REJ) indications to the sending DCE. Thus, the terminating equipment has several options to exercise control over the flow of packets between the end DTEs. The value of the D-bit in the format identifier field of octet 1 in the header of an information packet (see Figure 9.1) distinguishes between local (i.e., DTEn ↔ DCEn) and end-to-end (i.e., DTE2 ↔ DTE1) flow control.

(4) Making a Packet Call

Virtual calls are identified and managed on the basis of logical channel numbers that are assigned as required.

- **Logical channel**: a communication channel identified by a unique binary number. The channel is created when the identifier is assigned, and destroyed when the number is withdrawn.

Over a physical link, any number of logical channels can be created. They occupy the channel one at a time in statistical time-division manner. An example of the procedure to set up, establish, and clear a packet call is shown in **Figure 9.5**.

To initiate a virtual call to DTE2, DTE1 assigns an *outgoing* logical channel number that defines the virtual connection from DTE1 to DCE1 and sends a *Call Request* packet to DCE1. The packet identifies DTE1 and DTE2, and contains the outgoing logical channel number. At DCE1, the packet enters the network. On the basis of its destination, the Call Request packet is routed in a distributed fashion to DCE2. DCE2 assigns an *incoming* logical channel number that defines the virtual connection between DCE2 and DTE2 and sends an *Incoming Call* packet to DTE2.

Assuming DTE2 is authorized and ready to communicate with DTE1, DTE2 responds to DCE2 with a *Call Accepted* packet. It contains the same incoming logical channel number as DCE2 assigned to the Incoming Call packet so that DCE2 shall know to which incoming call DTE2 is responding. At DCE2, the Call Accept-

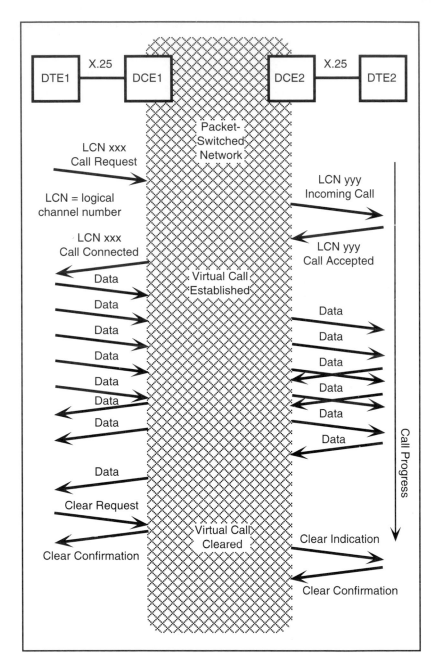

Figure 9.5 Making a Packet Call

The call begins with a request for a call over a specific logical channel between DTE1 and DCE1. It ends with a clear confirmation message from DCE1 to DTE1 over the same logical channel.

ed packet enters the network and is routed to DCE1. DCE1 assigns the original outgoing logical channel number to a *Call Connected* packet and sends it in an HDLC frame to DTE1.

DTE1 may now proceed to send packets to DCE1 for DTE2. They are routed through the packet-switched network on the basis that they arrive at DCE1 with the outgoing logical channel number already assigned by DTE1, and exit from DCE2 with the incoming logical channel number already assigned by DCE2. All of the packets in the call follow the same series of links through the network so that they arrive at their destination in sequence.

To clear the virtual call, DTE1 sends a *Clear Request* packet to DCE1. DCE1 replies with a *Clear Confirmation* packet and sends the Clear Request to DCE2. DCE2 sends a *Clear Indication* packet to DTE2; it answers with a *Clear Confirmation* packet. The logical channel numbers are now available for assignment to other calls.

For short messages, such as those encountered in automatic teller machine transactions, the call connect and disconnect procedures described above can be circumvented by using a *Fast Select Call Request* packet in which up to 128 bytes of user's data can be included. The network delivers the packet in the usual way as an incoming call request. The receiver responds to the data with a *Fast Select Clear Indication* packet that contains up to 128 bytes of information in reply. It is delivered to the original sender as a Clear Indication packet.

(5) Effect of Access Speed on Throughput

Suppose 4,096 bits of user's data are sent in four equal packets from DTE 1 to DTE 2 over 3 nodes of a packet network with a network speed of R_n bits/s. If

- The **UNI** is X.25
- The **data speed** across the UNI is R_{UNI} bits/s
- **Packets** are delayed by one packet time at each node or device. (I assume checking and routing are performed in no additional time as soon as the packet is collected at the node. One packet time = total number of bits in packet divided by the line speed in bits/s)
- No **errors** are detected
- **Propagation delays** may be neglected,

then the network model and time diagram are as shown in **Figure 9.6**.

The 4,096 user's data bits are divided into exactly 4 x 1,024 bits, a 3-octet packet header is added to each segment, and the whole is encapsulated in an HDLC frame with a 3-octet HDLC header and a 3-octet HDLC trailer. Thus, the total number of bits in each packet is 1,096 and the packet efficiency is 0.93. From Figure 9.6, to send the *first* packet from DTE 1 to DTE 2 requires $2T + 4t$ seconds. (This is the time from when the first bit begins its journey from DTE 1 until the last bit arrives at DTE 2.) If packets 2, 3, and 4 are sent immediately after packet 1 (this

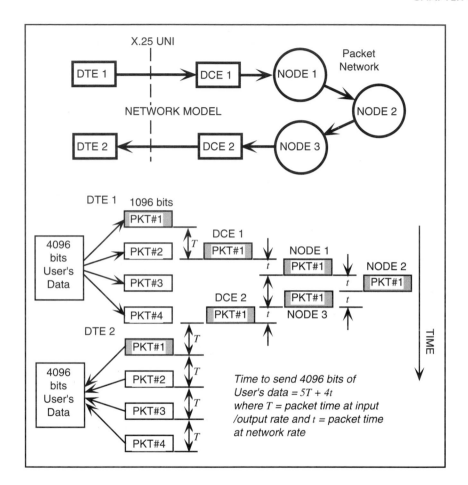

Figure 9.6 Network Model and Time Diagram

is known as *pipelining*), the total time to send 4,096 user's data bits is $5T + 4t$ seconds. Using these principles, **Figure 9.7** shows the general case of throughput (expressed as normalized to the network data rate) plotted against the ratio of the network data rate to the UNI data rate for 1, 3, 7, and 100 nodes.

Figure 9.7 leads to the following conclusions:

- Because of the waiting time at each node, as the number of nodes in the network increases, the throughput decreases
- As access speed (R_{UNI}) increases, the time spent in the network becomes a greater fraction of the total transfer time, and throughput becomes more dependent on packet-processing time

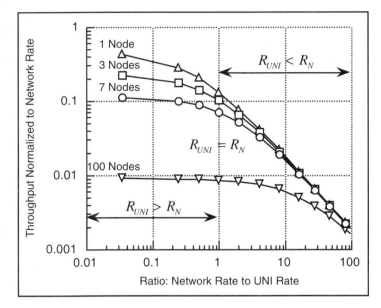

Figure 9.7 Throughput of Packet Network Normalized to Network Rate for Networks Consisting of 1 Node to 100 Nodes

Throughput of each network is plotted as a function of the ratio of network rate to UNI rate. As the network rate becomes greater than the UNI rate, throughput normalized to the network rate diminishes rapidly.

- When the data rate across the UNI is less than the data rate in the network, throughput decreases as the ratio of network data rate to UNI data rate increases

- When the UNI data rate is more than the network data rate, throughput increases as the ratio of the network data rate to the UNI data rate decreases

- The more nodes involved in transporting the message, the less important is the data rate across the UNI.

(6) Effect of Packet Network Delays on Throughput

The situation used to draw Figure 9.7 assumes the only delay at each node is the time required to collect packets prior to sending them on their way. No time is allowed for error detection, consulting the routing table, or waiting for traffic already in the queue to clear the link. **Figure 9.8** shows the relative effect of average delays of one, three, and five packet times at each node for a range of networks up to 10 nodes. The UNI and network data rates are 1.544 Mbits/s, and the message is the four packets shown in Figure 9.6. It can be seen that network delay significantly degrades the throughput of a packet network.

In contrast, if operation is based on permanent virtual circuits, and routing is assumed to begin as soon as the packet begins to be received by the node, the average nodal delay is close to zero packet times. Throughput performance is represented by the horizontal line drawn in Figure 9.8. This is the principle of frame relay (see Section 9.3).

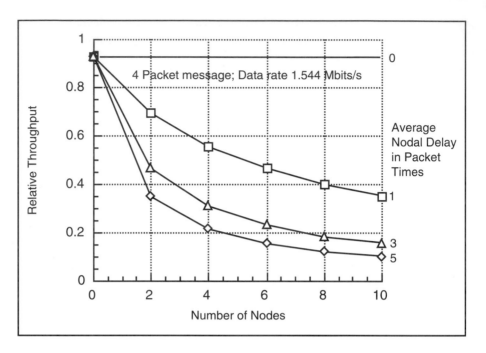

Figure 9.8 Relative Throughput of Packet Network as a Function of the Number of Nodes and the Average Delay at Each Node

9.2.3 PACKET SWITCH

A packet-switched network has been described as a network of queues. Each packet switch consists of controllers, databases, and queuing elements designed to make the receipt, holding, and forwarding of packets as rapid and as reliable as possible.

- **Packet switch**: a network node that receives packets, performs error and congestion control, may store the packets briefly, and sends them on toward their destination over routes determined by information carried in the packet, by information stored at the node, and by the state of the network.

In **Figure 9.9**, we show the principle of a packet-switching node. Each port consists of buffers in which packets are queued to await processing according to data they carry and data stored in the switch. The network map provides the node with information on the state of the network; routing information is contained in a database; and port status is available to all controllers. In the switch, the controllers execute the logical procedures assigned to layers 2 and 3 of X.25.

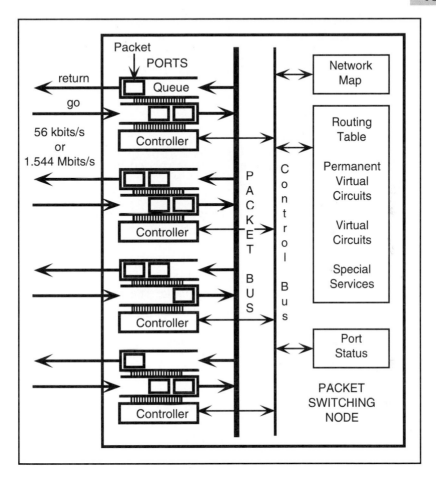

Figure 9.9 Principle of Packet-Switching Node
Each port buffers incoming packets and routes them according to the information
they contain.

(1) Data Link Layer

X.25 protocol includes LAP–B, the balanced link access procedure for ISDN
(see Section 10.3.3) at layer 2. Over the links to the switch, LAP–B

- Recognizes flags (to define frame limits)
- Executes bit-stuffing (to achieve bit transparency)
- On the transmit side, generates frame check sequences [FCSs, see Section 7.3.2(2)]
- On the receive side, confirms FCSs

- On the receive side, acknowledges receipt of frames and requests retransmission to correct errored frames (go-back-n or selective repeat ARQ)
- On the transmit side, repeats unacknowledged frames
- Performs flow control
- Using logical channel numbers, multiplexes packets over the links.

(2) Network Layer

The decision to forward a packet over one link or another (a layer 3 action) involves the processing of information carried in the 3 octets that comprise the packet header (see Figure 9.1).

- If it is a *call request packet*, forwarding will depend on its destination and the best link available to facilitate migration toward this destination at the time the packet is received.

The number of the link selected and the logical channel number carried by the packet will be stored in the routing table. The presence of a call request packet implies a connection-oriented service is being used and establishes a virtual circuit.

- If it is an *information packet* carrying a logical channel number that *is* stored in the routing table, it is part of a connection-oriented service over a virtual circuit.

It will be routed over the same link as the call request packet that preceded it.

- If it is an *information packet* carrying a logical channel number *not* stored in the directory, it is part of a connectionless service.

It will be forwarded over the best available link without regard to the way any other packet has been forwarded.

- If it is an *information packet* that carries a logical channel number that is associated with a permanent virtual circuit, it will be afforded priority treatment over the assigned link.

These examples allow us to distinguish two methods of packet routing

- **Distributed routing**: on the basis of information about traffic conditions and equipment status (network map, port status), the node decides which path the packet shall take to its destination.

Each packet is routed without regard to packets that may have preceded it. The node's decision is based solely on the situation existing at the time of packet arrival.

- **Centralized routing**: a primary (and perhaps an alternate) path is (are) dedicated to a pair of stations at the time of need.

For the sequence of packets that constitutes a block of user information, each packet uses the same primary (or alternate) en-route resources. Information on all possible routes that can be assigned is stored in a database (routing information). Routes established in this manner are known as *virtual circuits*.
     In addition, circuits comparable to private lines are employed,

- **Permanent virtual circuit**: a virtual connection that is permanently assigned between two stations.

The sending DTE and the receiving DCE use fixed logical channel numbers. Because it is a permanent (virtual) connection, no connect or disconnect packets are required to set up the circuit, and information packets may be transmitted at will by the originating DTE.

(3) Performance Measures

Packet error rate, packet loss rate, and packet insertion rate are three measures that can be used to determine the level of performance of a packet switch, or of an entire network. They are defined as follows:

- **Packet error rate (PER)**: in a given time period,[1] the number of errored packets received divided by the number of packets sent—i.e,

$$PER = \frac{\text{Number of errored packets received in given period}}{\text{Number of packets sent in same period}}$$

Errors in the header may cause packets to be misrouted (and not delivered), and congestion may cause packets to be dropped (destroyed). For these situations

- **Packet loss rate (PLR)**: in a given time period, for a specific sender-receiver pair, the number of packets sent, but not received, divided by the number of packets sent—i.e.,

$$PLR = \frac{\text{Number of packets lost in a given period}}{\text{Number of packets sent in the same period}}$$

Errors in the header may cause packets intended for another destination to be misrouted and received. For this situation:

---

[1] The time period must be long enough to acquire an accurate average.

- **Packet insertion rate (PIR)**: in a given time period, for a specific sender-receiver pair, the number of packets received that were not sent by the sender, divided by the number of packets sent to the receiver in the same time period, i.e.:

$$PIR = \frac{\text{Number of packets inserted in a given period}}{\text{Number of packets sent in the same period}}$$

9.2.4 IMPROVING THE SPEED OF OPERATIONS

When packet networks were developed, the quality of the available transmission links was poor. Consequently, at every node in a classical packet-switching network, much time is spent checking the parameters of the received packet for errors introduced by individual links and, when necessary, retransmitting the packet over the offending link. As a result, packet networks are slow and operation is complex. The protocol stack for communication over two links connected by a single node is shown in **Figure 9.10A**. Full error control is performed at the link (X.25–2, layer 2) level.

With the introduction of digital services (ISDN) and the subsequent upgrading of transmission facilities (particularly through the use of optical fibers) it has been possible to relax some of these requirements to simplify and speed up operations. In one approach, the user's frames are not divided into packets, the data link layer protocol is changed to a limited set of capabilities known as LAP–D core (link access procedure for ISDN signaling channel), and it is applied in two steps.

- **LAP–D core**: supports limited error control on a link by link basis. It recognizes flags (to define frame limits), executes bit stuffing (to achieve bit transparency), generates or confirms frame check sequences, destroys errored frames, and, using logical channel numbers, multiplexes frames over the links

- **LAP–D remainder**: performs error control on an end-to-end basis. Acknowledges receipt of frames, requests retransmission of destroyed frames, repeats unacknowledged frames, and performs flow control by limiting sending rate when requested by receiver.

Known as *frame relay*, the protocol stack for communication over two links connected by a single node is shown in Figure 9.10B.

Still higher speeds are achieved by eliminating the layer 2 functions from the node and moving all checking and control to the edges of the network. Known as *cell relay*, the standard packet is replaced by 55-byte cells. The protocol stack for communication over two links connected by a single node is shown in Figure 9.10C. In the next two sections, frame relay and cell relay are described.

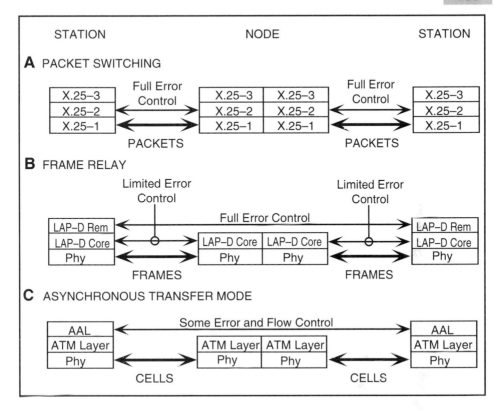

Figure 9.10 Protocol Stacks for Packet-Switched Network, Frame Relay Network, and Cell Relay Network

Frame relay and cell relay are techniques that take advantage of the improved performance of modern transmission facilities to eliminate much of the node-by-node checking of packet relay.

REVIEW QUESTIONS FOR SECTION 9.2

1. Why are some networks known as value-added networks?
2. What does ITU–T Recommendation X.25 describe?
3. Describe Figure 9.1.
4. Distinguish between the send and receive sequence numbers in the control octet of the HDLC frame, and those in octet 3 of the packet.
5. When sending, what function does the packet layer perform? What does it do when receiving?
6. When sending, what functions does the X.25–2 layer perform? What does it do when receiving?

7 When sending, what function does the X.25–1 layer perform? What does it do when receiving?

8 For a packet call, what do the originating and terminating logical channel numbers define?

9 Explain Figure 9.3.

10 Explain Figure 9.4.

11 Comment on the following statement: X.25-based networks provide slow-speed transport service.

12 What is sliding window flow control?

13 Explain Figure 9.5.

14 Explain Figure 9.7.

15 Describe the principle of a packet switch.

16 On what does the forwarding of a call request packet depend?

17 On what does the forwarding of an information packet depend?

18 What is distributed routing?

19 What is centralized routing?

20 What is a virtual circuit?

21 What is a permanent virtual circuit?

22 State three performance measures for the performance of a packet network.

23 How have better transmission facilities made it possible to improve the speed of operations?

9.3 FRAME RELAY

Desktop computing has led to the installation of a growing number of local-area networks (LANs) and to the need to interconnect them—around the business campus, or across the country. Like intra-LAN traffic, inter-LAN traffic is bursty, so that X.25-based packet-switching networks would appear to be well suited to the transport of such messages. However, LAN traffic is fast, and packet networks are slow. In addition, LAN bursts may be large (perhaps many thousands of bytes) and require several packets (with individual overheads). When intelligent end-user systems are linked over reliable, high-speed, digital transmission facilities, a frame relay network provides users with the high-speed connections for transferring data messages and lengthy files between fixed points.

- **Frame Relay**: a connection-oriented, data link layer packet-switching technology that transfers data frames from a source to a destination at speeds up to DS–1/E-1 (1.544/2.048 Mbits/s).

9.3.1 User Network Interface

Frame relay transfers data in variable-length frames over permanent virtual circuits (PVCs). For this reason, the user network interface supports data link and physical layers only

- **Frame relay user network interface** (FRI): a broadband network access arrangement that transfers information in variable length frames (262 to 8,189 octets). Error detection (but not correction) is performed, and corrupted frames are destroyed (without requesting retransmission). Normally, the interface is limited to DS–1/E–1 speeds.

The user network interface employs a set of *core* functions derived from LAP-D (ISDN D-channel data link protocol). Called LAP-D core, it includes

- Frame delimiting and alignment (using beginning and ending flags)
- Inspection to ensure that the frame consists of an integer number of octets that is less than the maximum permitted
- Transparency (using zero-bit stuffing)
- Frame multiplexing and demultiplexing (using the address field)
- Detection of transmission errors using cyclic redundancy checking CRC).

On an end-to-end basis, LAP–D remainder performs

- **Flow control**: through the use of a control bit (backward explicit congestion notification, BECN, see page 413), the receiver requests the sender to limit the sending rate. A node will destroy frames, if necessary.
- **Error control**: requests retransmittal of missing frames (including those destroyed because of detected errors).

Contained in ITU–T standard Q.922, and ANSI–T1.618, **Figure 9.11** shows frame relay frame formats. The fields are

- **Flag**: 01111110.
- **Address field**: a 2-octet routing label that may be expanded to 3 or 4 octets. It contains routing information and control bits as follows:
  - *data link connection identifier* (DLCI). 10, 16, or 23 bits that identify the logical connection between end-point and nodal processor. They are used to multiplex virtual circuits on each link. Some bits are used for control or management purposes. A DLCI is reserved for *consolidated link layer management* (CLLM), an activity in which the frame relay node notifies the end users about operational problems, status, etc.

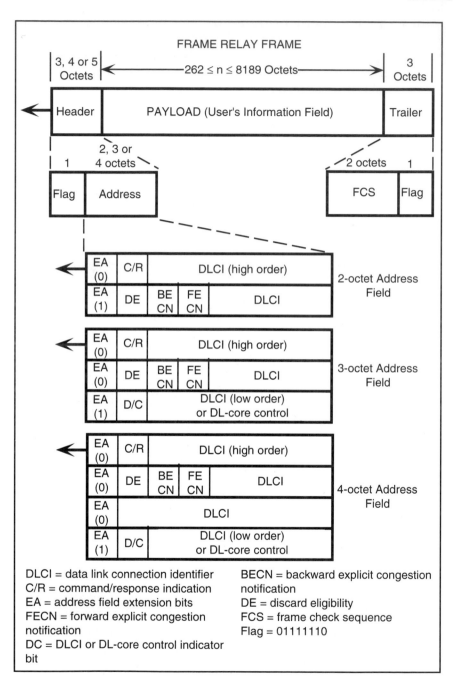

Figure 9.11 Frame Relay Frame

- *command/response* (C/R). Identifies the frame as a command or a response. This information is used to correlate messages.
- *forward explicit congestion notification* (FECN). This bit notifies the receiver that the path through the network is congested, and that following frames may be delayed.
- *backward explicit congestion notification* (BECN). This bit notifies the sender that the path through the network is congested, and the sender should institute congestion avoidance procedures. It is set in a frame passed backward from node to sender. If congestion continues to increase, the network will begin to discard frames (see DE bit, below)
- *extended address* (EA). Bits that indicate the length of the address field: 01 = 2 octets (DLCI = 10 bits); 001 = 3 octets (DLCI = 16 bits); 0001 = 4 octets (DLCI = 23 bits)
- *discard eligibility* (DE). This bit is set by the sender to indicate that the frame can be dropped before others (not so marked) in case of congestion
- DLCI or DL-core bits indication (DC). Indicates whether the last 6 bits of DLCI are to be interpreted as DL-core bits
- *DL-core bits*. 6 bits containing administrative information.

- **Information field**: an integral number of octets between 262 and 8,189. Contains user's information at the data link level (i.e., LLC + MAC, SDLC, or X.25-2). When the number of octets exceeds 4,096, the 16-bit FCS is likely to be ineffective.
- **FCS** (frame check sequence): 16-bit remainder from cyclic redundancy check (CRC) performed by entity-generating frame relay frame.

As soon as the header has been received at the network node and the DLCI validated, the frame is started on the way to the next node. Within the network, the entire frame may be transmitted at once, or it may be broken up into smaller segments (cells). The reasons for this maneuver are discussed in Section 9.4.2(3).

Predominantly, frame relay service provides speedy data packet delivery between nodes connected by permanent virtual circuits. However, there are occasions when it would be advantageous to have switched service (i.e., a quasi-permanent virtual circuit can be established at time of need). ITU–T, ANSI–T1, and Frame Relay Forum have published procedures for user-to-network, and network-to-network signaling that permit the use of switched virtual circuits (SVCs). In addition, Frame Relay Forum and other standards bodies have defined a set of interconnection standards so carriers can hand off frames to one another. The standards define procedures for carriers to exchange status messages across the node-network interface (NNI) and include the set up of SVCs.

9.3.2 FRAME RELAY OPERATION

To cope with the exigencies of communication, frame relay relies on intelligent end-user systems connected by quality transmission facilities. Because it does not perform detailed checking, frame relay does not guarantee faultless delivery of data. It,

- Detects, but does not correct, transmission, format, and operational errors
- May discard frames because they contain errors, or to clear congestion. When an invalid frame is detected (for any reason), the node discards the frame.
- Does not acknowledge frames or request retransmission of frames. It is left to the receiving end-user system to invoke an end-to-end, error control strategy to restore the frame.

Despite these caveats, frame relay is a technique of choice for data networks that interconnect LANs separated by substantial distances.

(1) Frame Relay UNI

Section 9.3.1 introduced the structure of the (data link level) frame PDU that is transferred across the user network interface (UNI) to gain network access. Just as X.25 is directed to the user and network connection, so frame relay (defined in ITU–T I.233 and ANSI T1.606) is a network access technique. Within the network—i.e., over the node-network interface (NNI) defined by the Frame Relay Forum in 1992, the procedures employed may be frame relay, cell relay, X.25, or ISDN.

Frame relay depends on the existence of permanent virtual circuits; the necessary route information (i.e., DCLIx → DCLIy) must have been loaded into the routing tables at the nodes before operation is begun. Flow control is effected by destroying frames marked as eligible for destruction (DE bit), destroying other frames if necessary, and communicating to contiguous nodes and senders (FECN and BECN bits). The receiving end-user must invoke end-to-end, error control strategies to restore the lost frames.

(2) Committed Information Rate and Other Measures

Defining acceptable network performance in a bursty data environment is difficult. A measure that is used is

- **Committed information rate** (CIR): the average rate, in bits/s, at which the network agrees to transfer data.

If the maximum number of bits (sometimes called the *committed burst size*) is B_C, and the time in which they will be transferred is T, $CIR = B_C/T$. Thus, for a network with a committed burst size of 1,024 kbits, and a CIR of 256 kbits/s, the time to transfer is 4 seconds. For the data to be transferred in 1 second, the CIR must be increased to 1,024 kbits/s.

If data are submitted at an average rate that is greater than the CIR, the user may lose some data. Data in excess of CIR can be carried only if the capacity (bandwidth) is available. For frames that exceed the CIR, the network sets the DE (discard eligibility) bit; they will be destroyed first if congestion occurs.

Two other parameters have been defined.

- **Excess information rate** (EIR): the number of bits sent in a certain time (bits/s) minus the CIR.
- **Residual error rate** (RER): the total number of frames sent minus the number of good frames received divided by the total number of frames sent.

9.3.3 FRAME RELAY APPLICATION

Frame relay is an important technique for the rapid transport of long streams of LAN data between specific nodes in private networks. Used as a hub, a frame relay nodal processor can reduce the cost of a private network significantly.

Figure 9.12 shows a group of LANs fully connected by routers and private lines. The speed of each line must be as great as the speed of the data bursts between the locations it connects. When any distance separates the locations, this can be expensive.

Figure 9.13 shows the same group of LANs connected in a hub arrangement through frame relay configured routers and a frame relay nodal processor. The routers encapsulate the LAN frames in frame relay frames that include data link connection identifiers (DLCIs). The frame relay nodal processor (FRNP) interprets this information and forwards the frames over the circuits they define to the routers serving the destinations. There, the routers strip off the frame relay headers and trailers and restore the frames to formats compatible with the receiving LANs. At the bottom of Figure 9.13, the protocol stacks for this operation are shown. The transport layers of the users' stacks perform error control, and the network layers of the routers' stacks perform routing.

As you might expect, there are considerable cost savings between the configurations of Figures 9.12 and 9.13. To illustrate the magnitude, suppose

- All transmissions occur at DS–1
- In Figure 9.12, the total number of miles of T–1 is 18,000, and the monthly charge is $800 per 100 miles
- In Figure 9.13, the frame relay nodal processor is colocated with one router, and T–1 facilities connect to the other five routers. The total length is 4,500 miles, and the monthly charge is $900 per 100 miles.

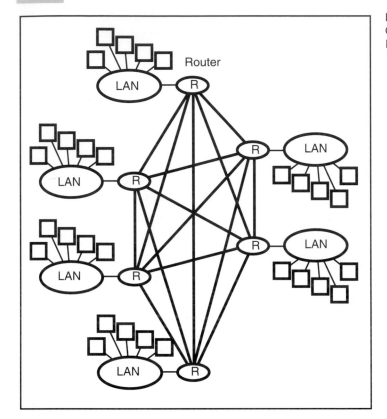

Figure 9.12 LANs Fully Connected by Routers and Private Lines

- Routers are connected to IXC POPs by T–1 circuits at an average monthly charge of $250, and the monthly charge for POP access is $300
- Converting a standard router to a frame relay configuration costs $1,000, and a frame relay processor costs $50,000.

The network in Figure 9.12 consists of 30 loop circuits, 30 POP access connections and 18,000 miles of T–1 circuits. Hence the monthly charge is

$$\$7,500 + \$9,000 + \$144,000 = \$160,500$$

Also, the network in Figure 9.13 consists of 5 loop circuits, 10 POP access connections, and 4,500 miles of T–1 circuits. Hence, the monthly charge for transmission circuits is

$$\$1,250 + \$3,000 + \$40,500 = \$44,750$$

In addition, there are capital costs—costs to upgrade the routers and the cost of the FRNP—that amount to

$$\$6,000 + \$50,000 = \$56,000.$$

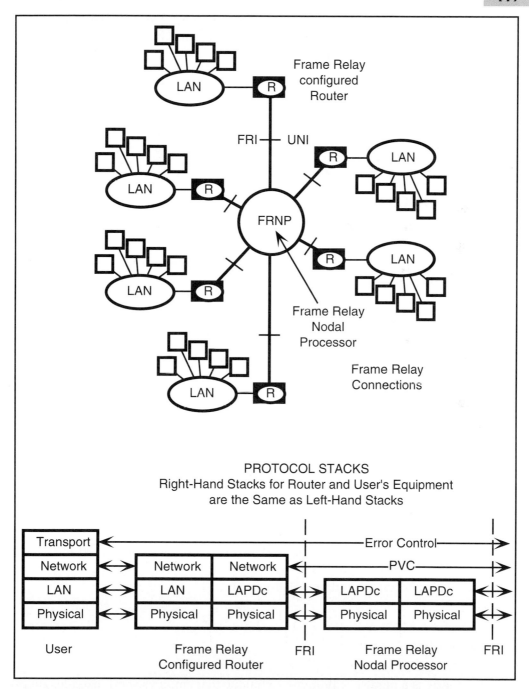

Figure 9.13 LANs Connected by Frame Relay Configured Routers to Frame Relay Nodal Processor in Hub Configuration

Amortizing (straight line for simplicity) this amount over 60 months, the monthly equipment expense is approximately $1,000. Thus, the total monthly charge for the FRPN network is $45,750.

For the numbers chosen, the monthly charges for the FRPN network are approximately 30% of the mesh-connected router network. Obviously, real situations require real tariffs and real distances. Nevertheless, this simple example shows the kind of dollar savings that are possible with some frame relay installations. This is particularly true when the transmission distances are large, an FRNP can be used as a hub, and the alternative is a fully connected mesh network of T–1 lines.

REVIEW QUESTIONS FOR SECTION 9.3

1 Define the frame relay user network interface.

2 Explain the following statement: the frame interface supports data link and physical layers only.

3 What is LAP-D core?

4 What is the difference between LAP-D and LAP-D core?

5 In the address field, describe the use of DCLI.

6 In the address field, describe the use of FECN and BECN.

7 In the address field, describe the use of DE.

8 What is the size of the information field?

9 What is implied by the following statement: frame relay does not guarantee faultless delivery of data.

10 Describe the set of core functions employed by the frame relay UNI.

11 Comment on the following statement: Frame relay depends on the existence of permanent virtual circuits.

12 How is flow control achieved?

13 Define committed information rate.

14 Comment on the following statement: For frames that exceed the CIR, the network sets the DE bit.

15 Describe the operation of Figure 9.13.

9.4 CELL RELAY

Cell relay is a high-speed routing technique that employs short, fixed-length cells that are statistically multiplexed over virtual connections. Cell relay is implemented by specialized technology. The service and technology are defined as follows

- **Cell relay service** (CRS): a bearer service that can transport voice, video, and data messages simultaneously.
  - *voice.* Carried as a constant bit rate (CBR) stream with low delay and cell loss requirements.
  - *video.* Carried as a CBR stream or a real-time variable bit rate (VBR) stream (as explained in Section 10.6.2.). The bit rate cannot exceed the peak cell rate (PCR) negotiated with the network.
  - *data.* Carried as a VBR stream that uses the available bit rate (ABR), or a stream for which the bit rate is unspecified (UBR). With UBR, the sender transmits as fast as it can (up to its PCR), but the network does not guarantee service quality.
- **Asynchronous transfer mode** (ATM):[2] a packet-switching technology that uses short, fixed length cells (packets) to implement CRS. ATM supports the transport of
  - isochronous streams
  - connectionless data packets
  - connection-oriented data packets.

9.4.1 ARCHITECTURE

In describing the architecture of ATM, I focus on the standards sponsored by ITU–T.

(1) Cell Composition

Figure 9.14 shows the arrangement of a cell. It consists of 48 octets of payload and 5 octets of header information. The 53-octet cell size is a compromise between the 32-octet payload supported by voice users, and the 64-octet payload supported by data users. The headers used at the user network interface and network node interface are different.

- **ATM UNI** header: consists of
  - 4-bit *generic flow control* (GFC) field that is used to assist in controlling the flow of traffic at the UNI
  - 24-bit *connection identifier* that explicitly associates each cell with a specific virtual channel on a physical link. Consisting of two parts, a 16-bit virtual channel identifier (VCI) and an 8-bit virtual path identifier (VPI), the connection identifier is used in multiplexing, switching, and routing cells

---

[2] Martin de Prycker, *Asynchronous Transfer Mode: Solution for Broadband ISDN*, 2nd ed., (New York: Ellis Horwood, 1993).

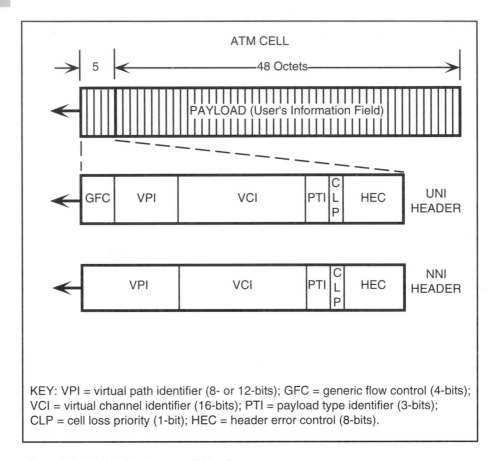

Figure 9.14 Cell Relay Frame and Headers

The UNI header contains a 4-bit flow control field and an 8-bit virtual path identifier. The NNI header contains no flow control information and a 12-bit virtual path identifier.

through the network. A connection identifier is assigned when the connection is established and is retained for the duration of the connection.

– 3-bit *payload type identifier* (PTI) that indicates whether the cell contains upper-layer header information or user data.

– 1-bit *cell loss priority* (CLP) field used to identify cells that can be discarded if the network has to discard cells to alleviate congestion

– 8-bit *header error control* (HEC) that is used for error detection and correction of the header

• **ATM NNI header**: similar to UNI except that the GFC field is replaced by four additional VPI bits to make the VPI field 12 bits at the NNI.

Figure 9.15 shows the protocol layers for CRS bearer service. They are

- **Physical layer**: the physical layer defines the transport of digital signals over multiplexed connections in a synchronous digital network. The bearer is likely to be optical or digital microwave radio (SONET, or SDH, see Section 6.4).

- **ATM layer**: multiplexes and demultiplexes cells belonging to different connections identified by unique virtual channel identifiers (VCIs) and/or virtual path identifiers (VPIs). Translates VCI and VPI at switches and cross-connection equipment. Passes cell *payload* (48 octets) to and from the ATM adaptation layer without error control or similar processing. Performs flow control at UNI. Implements connection-oriented and connectionless services.

- **ATM adaptation layer** (AAL): at the sender, AAL converts messages into sequences of cells for use by the ATM layer; at the receiver, AAL converts sequences of cells to messages for use by the user layers. More specifically

    - when sending, AAL receives different classes of information from higher layers in service data units (SDUs) and adapts them to occupy fixed-length, ATM cells

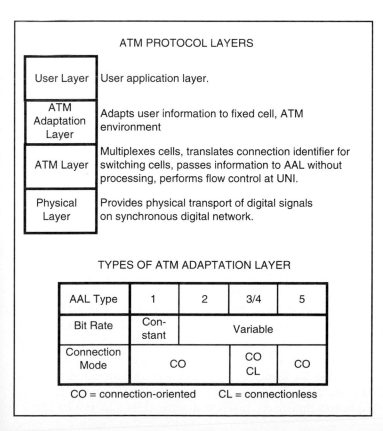

Figure 9.15 ATM Protocol Layers

The adaptation layer accommodates different styles of data to fixed-length cells.

– when receiving, AAL receives data in 48-octet payloads from ATM layer and reassembles it to fit user's service data unit requirements.

AAL-specific information passed to peer AALs (information field length and sequence numbers, for instance) is contained in the payload, not the cell header. AAL is divided into two sublayers:

– *segmentation and reassembly* (SAR) *sublayer*. Receives 47-octet blocks from convergence sublayer, adds 1-octet headers to form SAR-PDUs, and delivers 48-octet SAR-PDUs to ATM layer; or receives 48-octet SAR-PDUs from ATM layer, strips 1-octet headers, and delivers 47-octet blocks to CS

– *convergence sublayer* (CS). Performs a variety of functions depending on the service, such as separation of fill (dummy octets) from user octets, correcting cell delay variations, clock recovery, monitor lost cells, etc.

CS/SAR is discussed in more detail in Section 9.4.1(3).

(2) Types of AAL

Four classes of services are supported by AAL

- **Class A**: a time relationship exists between sender and receiver, the bit rate is constant, and the service is connection-oriented (e.g., voice or fixed-rate video)

- **Class B**: a time relation exists between sender and receiver, the bit rate is variable, and the service is connection-oriented (e.g., variable bit rate audio and video)

- **Class C**: no time relation exists between sender and receiver, the bit rate is variable, and the service is connection-oriented (e.g., TCP data)

- **Class D**: no time relation exists between sender and receiver, the bit rate is variable, and the service is connectionless (e.g., datagram).

Five types of AAL layers have been defined; however, as a practical matter, types 3 and 4 have been combined so that AAL types 1, 2, 3/4, and 5 are in use. Shown in Figure 9.15, they are described as follows

- **AAL type 1**: supports constant bit-rate, connection-oriented services that have specific timing and delay requirements (voice and video, for instance). Provides segmentation and reassembly, may detect lost or errored information, and recovers from simple errors. Error indications such as loss of timing, flow problems, and corrupted user information are passed to the appropriate entity.

- **AAL type 2**: supports connection-oriented, variable-bit-rate services (compressed video and audio, for instance). Provides segmentation and reassembly, and detection and recovery from cell loss or wrong delivery. The variable bit rate complicates timing recovery, and some cells may not be filled.

- **AAL type 3/4**: an all-purpose layer type that supports connection-oriented and connectionless variable-bit-rate data services. Two operating modes are defined
  - *message mode*. Each service data unit (SDU) is transported in one interface data unit (IDU). Employs cyclic redundancy checking and sequence numbers. Lost or corrupted units may be repeated. Flow control is optional.
  - *streaming mode*. Variable-length SDUs are transported in several IDUs that may be separated in time.
- **AAL type 5**: supports connection-oriented, variable-bit-rate, bursty data services, on a best-effort basis. Performs error detection, but does not pursue error recovery. AAL type 5 is essentially a connection-oriented-only type 3/4 layer. AAL 5 is also known as the *simple and efficient* layer (SEAL).

(3) Convergence, Segmentation, and Reassembly

By breaking them up into small cells at the sending point and reassembling them at the receiving point, convergence, segmentation, and reassembly (CS/SAR) are activities that facilitate the transport of large data packages. **Figure 9.16** shows the principle.

The information to be transferred is contained in the service data unit (SDU) that can be up to 65,535 octets long. In the header (and possibly a trailer) is information germane to the application. The SDU is divided into 44-octet segments that become frames in the convergence sublayer (CS). The CS adds a header and trailer to form CS PDUs. They are encapsulated in 48-octet frames in the segmentation and reassembly sublayer (SAR). The SAR adds a header and a trailer. From the SAR, the frames are passed to the ATM layer, where they are joined to 5-octet headers to become the 53 octet cells of Figure 9.16.

The headers and trailers added in the type 3/4 ATM Adaptation Layer contain the following information

- **Convergence sublayer**: in the header
  - *common part indicator*. 1 octet used to set units for subsequent fields in the header and trailer (e.g., buffer allocation size)
  - *beginning tag*. 1 octet that is duplicated in the trailer. Incremental frame by frame, *Btag* and *ETag* (intrailer) identify all CSPDUs associated with a session. At the receiver, they are used to correlate them.
  - *buffer allocation size*. 2 octets that inform the receiver of the maximum size of the buffer required to handle the payload. *BASize* is greater in streaming mode than in message mode

  In the trailer
  - *alignment*. 1 octet of 0s used to provide bit alignment for the trailer
  - *ending tag*. 1 octet used with *Btag* in the header

424

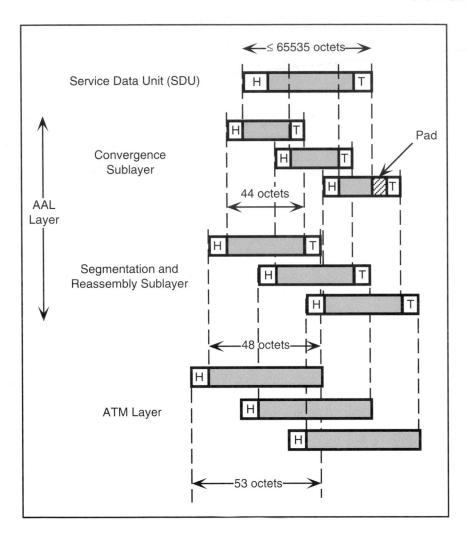

Figure 9.16 Principle of Segmentation of Service Data Unit to 53-Octet Cells and Reassembly to ≤65535-Octet SDU

- *length*. 2 octets that provide the length of the payload. Some frames are not filled with payload data and must be buffered to size (see Figure 9.16).
- **Segmentation and reassembly sublayer**: in the header
  - *segment type*. 2 bits that indicate whether segment is first, continuation, or last in the string, or a single segment
  - *sequence number*. 4 bits that are incremented segment by segment. Used to identify lost or inserted segments

- *multiplexing identifier (MID)*. 10 bits that allow multiplexing several user connections on a single network connection. For instance, *MID* is used to manage the interconnection of several terminals through a LAN gateway to servers on an ATM network.

In the trailer

- *length indicator*. 6 bits that provide the length of the payload
- *frame check sequence*. 10-bit remainder from cyclic redundancy check calculation. Used to detect errors in segment.

Headers and trailers associated with other types of AALs differ in detail.

(4) Virtual Paths and Virtual Circuits

As in the networks already described in this chapter, over a cell relay network, stations communicate using virtual connections. Once established, these connections are known as *virtual circuits*. To divide them into manageable groups, VCs are assigned to *virtual paths* (VPs) that connect the same endpoints. Each VP consists of physical resources (links) that are shared among the VCs it supports, and the quality of service the VCs receive depends exclusively on the activity over this particular path. Thus, congestion (or other poor performance) on one path will not affect other paths. By dividing the VCs into groups, the VP concept decomposes the global allocation of switch resources into a set of smaller decisions.

When a request for a new connection is received, the switch (traffic controller) attempts to place it on an existing VP where resources are available, and the call will have no effect on existing operating circuits. If this cannot be done, among other things, the controller may elect to place the call on the path and accept service degradation on the calls in progress, add resources to the path, seek another existing path, establish a new path, or refuse the call.

(5) Call Setup

Signaling is achieved over a separate, permanently assigned network. Call setup (and termination) information is sent over a high-speed signaling connection to the controller at the serving node. Each station is connected to one controller, and the nodes communicate with one another over other high-speed connections.

9.4.2 Cell Relay Applications

Figure 9.17 shows the provision of cell relay bearer service (i.e., delivery of messages) over a synchronous transport network. Defined by the ATM Forum in 1992 and 1993, physical transport is implemented at STM–1 or STS–3 (155.2 Mbits/s, see Section 6.4). Later ATM Forum specifications have addressed operation at DS–3 (44.736 Mbits/s), STS–1 (51.84 Mbits/s), and other speeds.

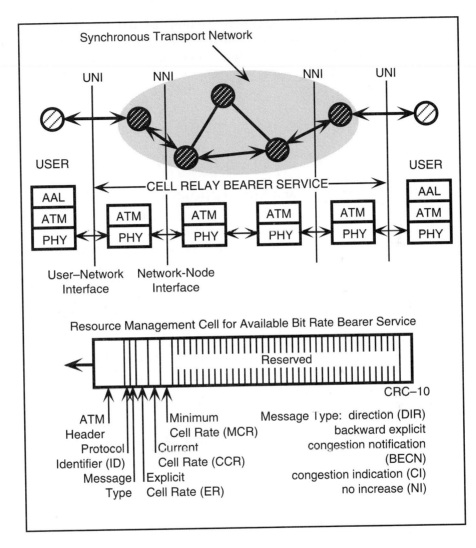

Figure 9.17 Protocol Stacks for Cell Relay Bearer Service

(1) Cell Relay Bearer Service

When cell relay carries voice and video messages, the network delay must be short and constant. This is achieved by making operations connection-oriented (i.e., a virtual circuit must be set up before communication begins) so that the cells pursue the same path through the network, and sufficient resources are assured for the message transfer. If the resources cannot be made available, the call is refused. An important aspect of assessing their adequacy is the queueing delays

being encountered at the switching fabrics in the path. The length and variability of these delays will affect the length and constancy of the circuit delay and impact on the ability of ATM to deliver time-paced signals correctly.

When cell relay carries data, operations may be connection-oriented for the purpose of establishing a virtual connection, or connectionless. Delay through the network is not a major issue, and the order in which cells arrive at their destination may, or may not, be an issue. It can be satisfied by the choice of connection-oriented or connectionless services.

(2) Available Bit Rate Service

To transfer cells as quickly as possible, a sender may try to use all of the bandwidth (bit rate) that is not allocated to other traffic. To do so without loss of data, the source must adjust its sending bit rate to match conditions as they fluctuate within the network due to the ebb and flow of traffic. If it does not do this, it runs the risk of losing packets because of buffer exhaust or other congestion mechanisms.

To control the source bit rate when using ABR service, the ATM Forum has devised a scheme in which *resource management* (RM) cells are introduced periodically into the sender's stream. Shown in the bottom half of Figure 9.17, RM cells are sent from sender to receiver (*forward* RM cells), and then turned around to return to the sender (*backward* RM cells). Along the way, they provide rate information to the nodal processors and may pick up congestion notifications. When an RM cell reaches the receiver, it (the receiver) changes the direction bit ready to return the cell to the source. If the destination is congested and cannot support the explicit cell rate, it (the destination) sets the congestion indication bit and reduces the explicit cell rate (ER) value to a rate it can support. On the return of the RM cell to the source, the sending rate is adjusted accordingly. If the RM cell returns to the source without the congestion indicator (CI) bit set, the sender can increase the sending rate and set a higher ER. As conditions fluctuate, the sender adjusts its rate in harmony with the feedback the RM cells provide.

(3) Combining Frame Relay and Cell Relay

The frame relay arrangement shown in Figure 9.13 is characteristic of private networks in which there is a limited number of destinations. The arrangement in **Figure 9.18** employs frame relay nodal processors to feed ATM networks. Frame relay is used to gain access to cell relay switches for the transport of data. Advantage is gained from the change in speed—many DS–1 (1.544 Mbits/s) streams can be multiplexed in the STS–3 or STS–12 (155 or 622 Mbits/s) streams of cell relay. A complicating factor is that frames must be converted to 55 octet cells, moved through the cell relay portion of the network, and then reassembled into frames. (A maximum-length frame relay PDU becomes approximately 200 cells.) To achieve transparent operation (i.e., the FR user is not aware of the use of CR in part of the network), **Figure 9.19** shows some of the relationships that must be maintained. They are

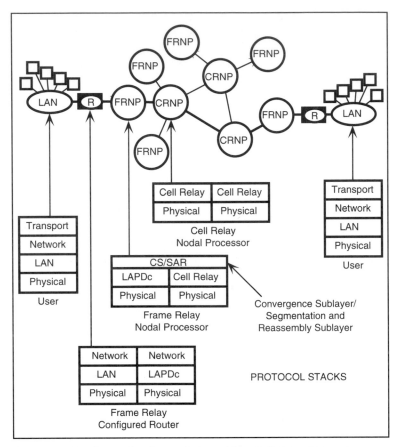

Figure 9.18 Implementing Frame Relay with a Cell Relay Network

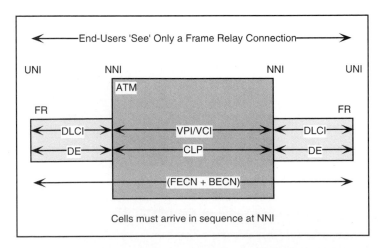

Figure 9.19 Showing Quantities That Must Be Preserved for the Transition from Frame Relay to Cell Relay to Frame Relay to Be Transparent to Users

- The DLCI of frame relay must be mapped to the VPI/VCI of cell relay and mapped back to the proper DLCI
- The DE bit must be mapped to the cell relay CLP and mapped back to DE
- For flow control in the frame relay portion of the network, FECN and BECN messages must be preserved
- The cells must arrive in sequence at the NNI to reconstruct the FR PDU.

In the FRNP protocol stacks of Figure 9.18, the UNI (FRI) is the same as in Figure 9.13. However, at the NNI, to accomplish the *frame* ⟷ *cell* transformations, the stacks include the convergence, and segmentation and reassembly, sublayers of the ATM adaptation layer. Cells are created and transmitted as soon as they can be formed from the incoming frame PDU. In the cell relay portion of the transmission path, cells are routed using the virtual identifiers (VPI/VCIs) assigned to the logical path defined by the DLCI. At the FRNP that serves the receiver, the segments are reassembled and the frame PDU is passed to the receiver. The FECN and BECN bits in the address field of the frame PDU implement end-to-end congestion control.

Figure 9.20 shows more detail of the protocol data units produced in the arrangement of Figure 9.18. At the sending NNI, a header and trailer are attached to the frame relay PDU (FRPDU) to create the convergence sublayer PDU (CSPDU). Assuming the use of AAL 5, the CSPDU is divided into 48-octet payload segments to which 6-octet trailers are attached to become segmentation and reassembly PDUs (SARPDUs). The SARPDUs are then divided into 48-octet payloads to which 5-octet headers are attached to become cells. These are the cells that traverse the ATM network.

At the destination NNI, the sequence of 48-octet payloads in the cells is reassembled into 54-octet SARPDUs. In turn, they are reassembled into the CS PDU. Finally, the header and trailer are stripped off the CS PDU to produce the original FRPDU. Throughout this process, the relationships drawn in Figure 9.19 must be maintained.

9.4.3 ATM NODE

During its development, ATM has been known as *fast packet switching* and *asynchronous time division* (ATD). The technique is *asynchronous* because the sending clock and the receiving clock are not synchronized; differences between them are compensated by the insertion or removal of unassigned cells. The short cell reduces delay and jitter, and the fixed length permits synchronous operation of the switch fabric. Before information is accepted from a source, a virtual circuit must be established to the receiver. If resources are not available, the request for service is refused.

(1) ATM Switch Components

Figure 9.21 shows the major components of an ATM switch. The switching matrix, or switching *fabric* as it is called, operates synchronously. The ports receive

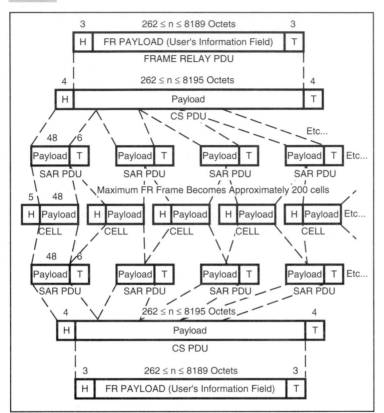

Figure 9.20 Segmentation and Reassembly of Frame Relay PDU in Passage Over Cell Relay Network

The maximum size FR frame becomes 200 cells approximately. The speed across the CR network is approximately 100 times FR.

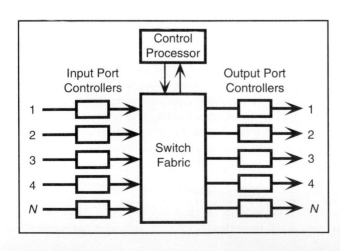

Figure 9.21 Principal Components of an ATM Switch

The switching fabric operates synchronously. The ports receive and send 53-octet cells at 155 or 622 Mbits/s.

or send bits at speeds of 155 or 622 Mbits/s (STS–3 and STS–12, or STM–1 and STM–4). The components perform the following tasks:

- **Input port controllers**: manage the input streams; among other things, they
    - buffer incoming cells
    - align cells for switching
    - on the basis of information in the headers, identify output ports and establish a path across the fabric for the cells, or attach self-routing labels to guide the cells across the switching fabric.
- **Output port controllers**: manage the output streams; among other things, they
    - strip off self-routing labels
    - may buffer cells awaiting transmission
    - align cells and transmit them to the next node.
- **Switching fabric**: routes incoming cells to the correct output ports. The structure of the switch fabric varies with the size of the switch.
- **Control processor**: directs the operation of the switch. To achieve very high speed operation, as many of the logical and control operations as possible are performed in hardware. The control processor contains a network database (with virtual circuit and other information), and performs system maintenance and administration tasks.

(2) Switching Fabrics

For *small* ATM switches –the sorts used in private networks—a common time-divided structure is employed as the switching fabric. Since every incoming cell occupies the fabric for a short time during which it is available to all outgoing ports, the structure can be used for point-to-point and point-to-multipoint (multicasting) operation. Two popular structures are

- **Time-divided bus**: all incoming cells flow over the shared bus. Its speed determines the maximum speed of the switch, and access to the bus is granted to each input port, in turn.

With this type of fabric, operation has been reported at speeds over 20 million cells per second (approximately 1 Gbit/s). The concept is shown in **Figure 9.22** and is similar to the packet switch shown in Figure 9.9.

- **Multiplexing and storage**: each cycle of incoming cells occupies the common storage for a slot time. They may be read out in any order to the output ports.

With this type of fabric, operation has been reported at speeds over 10 million cells per second (approximately 0.5 Gbit/s). The concept is shown in Figure 9.22. In principle, this technique is similar to the timeslot interchanger shown in Figure 9.29.

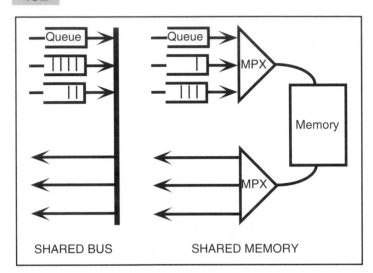

Figure 9.22 Small ATM Switches With Common Time-Divided Switching Fabric

Small ATM switches can be implemented with a shared bus or a shared memory.

For *large* ATM switches—the sort likely to be deployed in public networks—the switching fabric consists of elementary switching units (switching elements) that are connected in a specific manner. The network is implemented with solid-state switching elements to provide one, or several, nonblocking paths between ports. At regular intervals (known as *slot* times) the switch fabric transfers one cell from as many of the input controllers as have received cells, to the appropriate output controllers. Paths across the network are selected when calls are set up, or the cells are *self-routing*. In this technique, the input port controller adds a tag to each arriving cell that identifies the port to which it must be delivered. The switching elements in the fabric route the cells by inspecting their tags. If a number of cells destined for the same output port arrive simultaneously at different input ports, the output processor must be fast enough to handle them all in one slot time, or some must be delayed or discarded. Processing speeds up to 80 Gbits/s appear to be realizable.

Several structures are of interest for the large switch application. Dividing them into single-path and multiple-path fabrics, I describe some of them briefly.

- **Single-path switching fabrics**: a unique path is available between each input/output pair
    - *fully interconnected mesh*. Each input port is permanently connected to all output ports. Because it increases in complexity in proportion to the product of the number of input and output ports (i.e., N^2), this network design is not economical for very large switches. However, fully interconnected mesh structures can be used in moderate-size, nonblocking, self-routing switch networks, and in switching elements that form other fabrics.

– *crossbar* or *matrix*. The connection between individual input/output pairs is made by closing a gate to connect the input horizontal to the output vertical. The arrangement is similar to the matrix of the highway interchanger shown in Figure 9.30. Because it increases in complexity as the product of the number of input and output ports (i.e., N^2), this network is not economical for very large switches. However, crossbar and matrix structures can be used in moderate-size, nonblocking, self-routing switch networks, and in switching elements that form other fabrics.

– *Banyan*. A family of self-routing networks composed of 2 x 2 switching elements. With a complexity that increases as $N \log N$, it is better suited to large networks—but it is a blocking network. If switching elements larger than 2 x 2 are employed, performance is improved, but blocking remains a problem. However, if the incoming cells are sorted in the order of their destination ports and presented on adjacent network inputs, the network is nonblocking, provided that multiple inputs are not destined for the same output at the same time.

– *Batcher–Banyan*. A network that sorts incoming cells on the basis of their output port sequence (Batcher), connected by a *perfect* shuffle interconnection network to a self-routing network of 2 x 2 switching elements (Banyan). The addition of the Batcher network significantly increases the complexity of the whole. A Batcher-Banyan network is shown in **Figure 9.23**. To overcome the problem of multiple inputs seeking the same output port at the same time, all but one of the interfering inputs must be buffered, otherwise delayed, or destroyed.

- **Multiple-path switching fabrics**: several paths are available between each input/output pair

    – *augmented Banyan*. Adding switching stages to a simple Banyan network produces a multiple-path, self-routing fabric

    – *Benes*. A particular case of an augmented Banyan network

    – *Clos*. A nonblocking arrangement of switching elements and links that has long been used in telephone switches. It may be the technique of choice for very large networks.

    – *parallel switching fabrics*. Single-path or multiple-path switching fabrics can be arranged in parallel to create multiple-path networks. They have the advantage of redundancy so that the failure of a plane is inconvenient, but not catastrophic. Also, as described below, they can be connected so that more than one cell can be delivered to an output port at the same time.

(3) Performance Requirements

Several ATM switch designs are being pursued by major research and development organizations. In common, they are seeking to fulfill the following requirements:

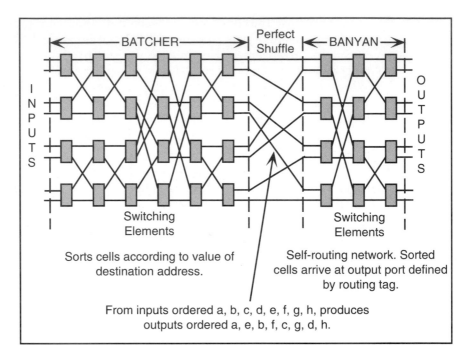

From inputs ordered a, b, c, d, e, f, g, h, produces
outputs ordered a, e, b, f, c, g, d, h.

Figure 9.23 Batcher-Banyan Switching Fabric

- **Information rates**: varying from as little as 100 bits/s to more than 100 Mbits/s, inputs may be synchronous or asynchronous.
- **Multiple receivers**: to implement electronic mail or video distribution services (for example), one-to-many, one-to-all, and many-to-many connections are required. Implementing these multicasting and broadcasting services requires a *copy* capability within the switching fabric.
- **Blocking**: before a message is sent, resources must be available to complete the transfer. If they are not, the call attempt will be refused (blocked). If the switching fabric is nonblocking, temporary congestion can be accommodated by queues at the input and output. If it is blocking, temporary congestion may require internal queues.
- **Delay**: when queues are present, network transit time depends on the cell arrival-time distribution, the number of cells waiting for service, and the service time. The difference in cell delay between light traffic conditions and congested conditions can be large and will affect channel throughput significantly. Under such conditions, priority handling may be necessary for synchronous traffic.
- **Cell loss**: when traffic to a port exceeds the capability of whatever queues are available, cells are lost. Also, when traffic is misrouted—i.e., delivered to

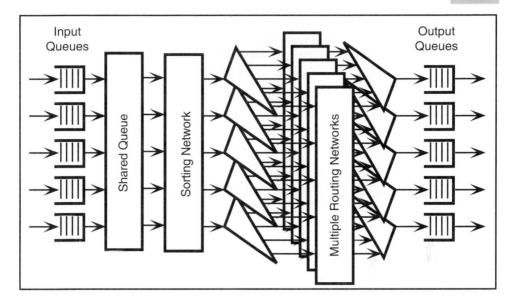

Figure 9.24 Principle of Large Self-Routing ATM Switch

the wrong port—the cells are lost. Service integrity requires that cell loss be ≤ 1 cell in 10^8 cells.

(4) ATM Switch

Figure 9.24 is a high-level diagram of an ATM switching fabric that incorporates a sorting network, parallel routing networks, and several queues. Queues of cells may be formed

- At each input port (to buffer incoming cells)
- At the input to the sorting stage (to buffer arrays of cells that will be routed each slot time)
- At the outputs of the routing stages (to accommodate several cells seeking the same output port)
- In the recirculation buffer (to accommodate those cells seeking the same outputs that cannot find room in the output buffers).

The sorting network is likely to be a Batcher network, and the routing networks will be self-routing and based on a Banyan or augmented Banyan network. Not all of the buffers shown will be implemented in every design. Those cells that cannot be buffered or delayed are destroyed.

REVIEW QUESTIONS FOR SECTION 9.4

1 Define cell relay. Explain the arrangement of a cell.

2 Describe the header of the ATM UNI.

3 What is the difference between the headers employed by ATM UNI and ATM NNI?

4 What are the protocol layers used in ATM bearer service? What do they do?

5 Explain the four classes of service supported by AAL.

6 Describe the four types of AAL layers in use.

7 Use Figure 9.16 to explain the principles of convergence and segmentation and reassembly.

8 Describe the entries in the convergence sublayer of the type 3/4 ATM adaptation layer.

9 Describe the entries in the segmentation and reassembly sublayer of the type 3/4 ATM adaptation layer.

10 Distinguish between ATM bearer service for voice only, and ATM bearer service for data only.

11 Describe Figure 9.18. What are the advantages of using FR with CR?

12 What are the disadvantages of using FR with CR?

13 Explain Figure 9.20.

14 Define an ATM node. Why is the term asynchronous included in ATM?

15 Describe the major components of an ATM switch.

16 What structures can be used as switching fabrics for small ATM switches?

17 What structures can be used as switching fabrics for large ATM switches.

18 Explain Figure 9.24.

9.5 SWITCHED MULTIMEGABIT DATA SERVICE

Switched Multimegabit Data Service (SMDS) is a high-speed, public, packet-switched, connectionless data service that employs cell relay. SMDS supports higher-access speeds and greater throughputs than frame relay networks. It is being implemented as a public service across the United States.

9.5.1 SMDS UNI

Based on IEEE 802.6 Metropolitan Area Network (MAN) standards (see Section 12.2.3), SMDS is implemented through an interface that supports data link and physical layers.

- **SMDS user network interface**: a broadband network access arrangement that supports switched, high-speed, connectionless data services.

Through the use of a three-level, SMDS *interface protocol* (SIP), SMDS transfers information in variable-length frames up to 9,188 octets over connections that employ cell relay techniques. Each PDU is routed independently to its destination using a 10-digit address based on the North American Numbering Plan. Error detection (but not correction) is performed, and corrupted frames are discarded (without requesting retransmission). Connectionless service shifts the onus for communication integrity to a higher layer. SNI supports DS–1, DS–3, and SONET speeds.

SIP is a three-level protocol that addresses, frames, and transmits PDUs as cells. **Figure 9.25** shows the protocol stacks, PDUs, and frame formats associated with the use of SIP. Up to 9,188 octets of user's data is collected in an SDU (called a Service Data Unit or an SMDS Data Unit) and passed from the upper layers of the router stack to Level 3 of SIP. There, it is combined with a 36-octet header and a 4-octet trailer to become the Level 3 PDU. At Level 2 of SIP, the PDU is segmented in 44-octet payloads and joined to other headers and trailers to become Level 2 PDUs. This action is accomplished on the fly—as soon as enough octets are present, the first PDU is formed and sent on its way, to be followed immediately by the remaining PDUs. They form a stream that passes to Level 1 of SIP, where it is converted into signals suited to cell relay.

(1) SIP Level 3 PDU

The fields in the header and trailer of the SIP Level 3 PDU are

- **Reserved** (Rsvd): carried as all zeros. Reserved for possible use later.
- **Beginning tag** (Btag): binary number between 0 and 255. In conjunction with end tag, used to define first and last segments of Level 3 PDU.
- **BAsize**: buffer allocation size. Gives length in octets from next field (DA) to end of SDU.
- **Destination address** (DA): contains two subfields
  - *address type*. 4 bits set to 1100 when the address is a 60-bit individual address and to 1110 when the address is a 60-bit group address.
  - *address*. 10 binary-coded decimal digits that are the SMDS address of the destination equipment (similar to 10-digit telephone number). 4 bits set to 0001, and 16 bits set to zero, frame the address. It is this address that guides the stream of Level 2 PDUs through the network.
- **Source address** (SA): contains two subfields
  - *address type*. 4 bits set to 1100. (The source address is always an individual address.)
  - *address*. 10 binary-coded decimal digits that are the SMDS address of the source equipment. 4 bits set to 0001, and 16 bits set to zero, frame the address.

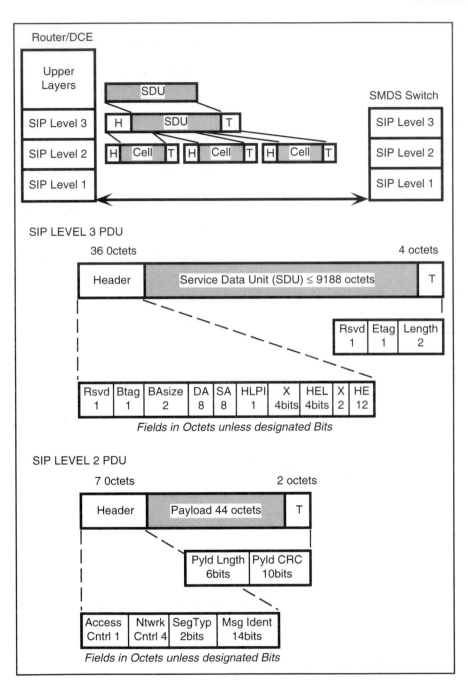

Figure 9.25 Protocol Stacks, PDUs, and Frame Formats for SMDS Interface Protocol

- **Higher-level protocol identifier** (HLPI): contains information on SIP/DQDB protocol formats for use by higher layers at destination
- **X**: these fields are for management purposes. They are carried across the network unchanged
- **Header extension length** (HEL): indicates the number of 4-octet groups in the HE field
- **Header extension** (HE): repeats key header values
- **Service data unit** (SDU): also called SMDS data unit. Variable-length field that contains user's data up to 9,188 octets. If needed, the field is padded by 1, 2, or 3 octets of 0s to bring the number of octets to a value divisible by four.
- **End tag** (Etag): binary number between 0 and 255. In conjunction with beginning tag, used to define first and last segments of level 3 PDU.
- **Length**: length indication field. Used to check that reassembly has been completed successfully.

(2) SIP Level 2 PDU

The fields in the header and trailer of the SIP level 2 PDU are

- **Access control**: bit 1 indicates whether the level 2 PDU contains information (1) or is empty (0). The remaining 7 bits are used by the SMDS switch and other devices.
- **Network control**: distinguishes between an empty PDU and one with user's data
- **Segment type**: 2 bits set to
  - *10*. Segment begins message
  - *01*. Segment ends message
  - *00*. Segment is continuation of message
  - *10*. Single-segment message
- **Message identifier**: identifies level 3 PDU to which segment belongs
- **Payload**: 44-octet field that contains segment of level 3 PDU
- **Payload length**: indicates how many octets in payload contain data (may not be 44 if the segment type is 01 or 11)
- **Payload CRC**: contains 10-bit FCS for error detection.

(3) SIP Level 1

The physical layer supports transmission over the SNI at DS–1, DS–3, and STS–3 speeds. Each access path is dedicated to a single user. Several devices (belonging to the same user) can access SMDS over this path.

(4) Example of Protocol Stacks

Figure 9.26 shows the arrangement of protocol stacks employed to transfer an IP datagram (see Section 12.6.1) through an SMDS-configured router to an SMDS network. Data originates in the upper layers of the DTE stack. It is encapsulated in an IP datagram that includes the IP destination address. The subnetwork access protocol (SNAP) adds a unique identifier (OUI) and a protocol iden-

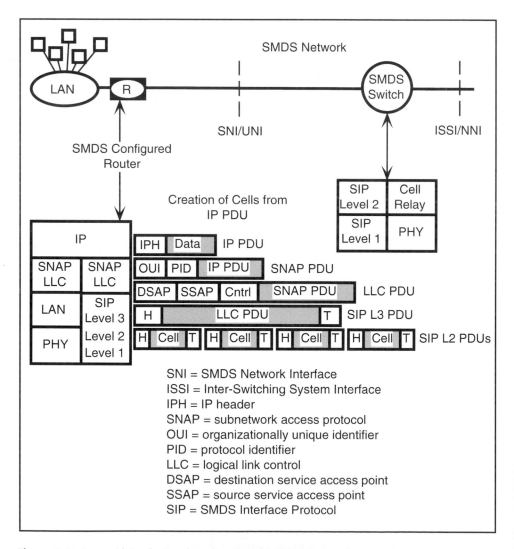

Figure 9.26 Protocol Stacks Employed to Transfer IP Datagram Over an IEEE 802 LAN Through an SMDS Router to an SMDS Network

tifier (PID). Among other things, the LLC level adds destination and source service access points, and SIP level 3 adds 10-digit destination and source network addresses for the sending and receiving routers. The entire PDU is segmented, or reassembled, at SIP level 2 and pipelined through the network by SIP level 1.

As well as SNI, two other interfaces are defined.

- **Interswitching system interface** (ISI): facilitates the interconnection of switches from different manufacturers in the same network.
- **Intercarrier interface** (ICI): facilitates the interconnection of intra-LATA and inter-LATA networks.

They permit the implementation of a countrywide, public SMDS network.

(5) Addresses

SMDS users can have significant flexibility in the way in which they use SMDS services. For instance, they can opt for

- **Logical private network**: restricts communication to a selected set of addresses
- **Source address screening**: lists addresses of sources from which data will/will not be accepted
- **Destination address screening**: lists addresses of destinations to which data may/may not be sent
- **Group addressing**: the same message may be sent to up to 128 destinations
- **Multiple addresses**: up to 16 addresses (associated with the same customer) can employ one SNI.

9.5.2 NETWORK PERFORMANCE

The carriers have established goals for network performance. Some of them are

- **End-to-end network delay**: for 9,228-octet PDUs and *2 DS–3 access lines*. 95% of PDUs are delayed $\leq$ 20 milliseconds
  - With *1 DS–3 and 1 DS–1 access line*, 95% of PDUs delayed $\leq$ 75 milliseconds
  - With *2 DS–1 access lines*, 95% of PDUs delayed 130 milliseconds
- **Lost PDUs**: $\leq$ 1 in 10^4 SIP level 3 PDUs lost
- **Duplicate PDUs**: $\leq$ 5 in 10^8 SIP level 3 PDUs duplicated
- **Misdelivered PDUs**: $\leq$ 5 in 10^8 SIP level 3 PDUs delivered to wrong address
- **Missequenced PDUs**: $\leq$1 in 10^9 SIP level 3 PDUs delivered out of sequence
- **Errored PDUs**: $\leq$ 5 in 10^{13} SIP level 3 PDUs delivered with errors.

REVIEW QUESTIONS FOR SECTION 9.5

1   What is SMDS?

2   Define SMDS UNI (SIP).

3   What is SIP? What does it do? What is the size of the SDU?

4   Describe the fields in the header and trailer of SIP Level 3 PDU.

5   Describe the fields in the header and trailer of SIP Level 2 PDU.

6   What transmission arrangements does SIP support?

7   Explain Figure 9.26.

8   Describe the range of addresses and addressing schemes available to SMDS users.

9   Describe some performance goals for SMDS.

9.6 SYNCHRONOUS TRANSFER MODE

In networks designed principally for voice services, the transfer of information and overhead between (among) users is accomplished by a combination of time-division circuit switches and time-division multiplexed transmission channels. The switches selectively establish and release time slot-based connections between digital transmission facilities to provide dedicated circuits for the exchange of messages between users. The time slot sequence is established before the information exchange begins, and, until the users terminate their session, the time slots are dedicated to their exclusive use. If electronic jitter is ignored, during each exchange, the delay between the parties is constant. It consists of propagation delay and a small number of octet delays due to multiplexing and speed changing in the network. However, before they began to communicate, there was a delay of several seconds while the calling party entered the called number and the telecommunication network hunted for, and established, the sequence of time slots that became the circuit.

9.6.1 TDM Switch Hardware Model

To understand the performance of a digital circuit switch, we need a model. **Figure 9.27** shows an arrangement that is typical of present day, Class 5, switching facilities that serve a customer community equipped with analog loops. The functions of the principal elements are as follows.

(1) Concentration and Expansion

At any specific time, the number of active loops is likely to be very much less than the total number of loops installed in an exchange area. Accordingly, the number of circuits in use at the same time is substantially less than the total number of lines supported by the office. A concentration stage is employed to connect the active loops to the input ports of the switching stages. As soon as the communication is completed, the idle loop is disconnected so that other active loops can be accommodated. In reverse, the concentration stage acts as an expansion stage to connect active lines from the switching stages to the loops that complete the calls.

Figure 9.27 shows the concept of a concentration and expansion stage in the environment of a digital end-office. The specific implementation involves a matrix in which connections can be made between horizontal and vertical members by ener-

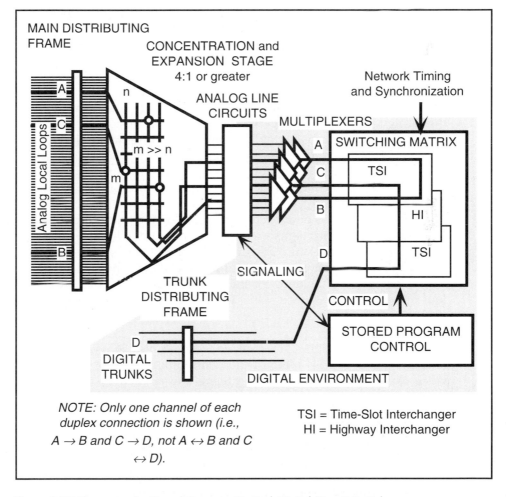

Figure 9.27 Elements of a Stored-Program Control Digital Circuit Switch

Most end-office switches are digital machines; most local loops are analog. On average, each loop is used 0.2% of the time. Thus, it is unnecessary to have the same number of switch ports as loops.

gizing cross points. The number of horizontal links (m) corresponds to the number of loop appearances at the distributing frame; the number of vertical links (n) corresponds to the number of channels that can be handled simultaneously by the switching matrix ($m << n$). In practice, to improve reliability and permit easy expansion of the system, the large matrix serving all $m \times n$ possible connections is subdivided into a set of identical, smaller matrices that perform a fraction of the task.

Traditionally, the concentration factor has been based on years of experience with average call lengths and calling rates in a voice-based environment. With the growth of data usage, particularly Internet use, this experience is less useful. Internet calls are likely to tie up a circuit for a long time (perhaps hours), reducing the opportunity to share the circuit among other customers. In many switching centers, to overcome the potential of a relatively few Internet users creating severe congestion for voice users, calls to Internet Service Providers (ISPs) are diverted from the concentrator stage to a separate switch that connects to the relevant IPOPs (ISP point of presence).

(2) Analog Line Circuit

To be processed by a digital end-office, analog signals must be converted to digital signals. An analog line circuit that performs this conversion, and the functions associated with powering, ringing, and testing the local loop, is shown in **Figure 9.28**. The subscriber line interface circuit (SLIC) is responsible for the two-to-four-wire conversion function of the hybrid, for signaling detection, and for battery feed. In the PCM codec, the analog voice signal is converted to a 64 kbits/s, companded, PCM, unipolar signal for transmission to the multiplexer. The reverse process is employed to produce analog voice from 64 kbits/s, companded, PCM, unipolar signals received from the demultiplexer. Since most of the terminations in an end-office are conventional analog loops, the cost of the line circuits becomes a significant fraction of the total cost of the switching system. This cost is increased by the need for them to withstand lightning surges and power line crosses, and to interface with a variety of station equipment.

(3) Multiplexer

Standard techniques are employed to multiplex 24 voice channels in a digital PCM stream at 1.544 Mbits/s and demultiplex the 1.544 Mbits/s stream to 24 voice channels. The multiplexed streams are applied to the switching matrix shown in Figure 9.27.

(4) Switching Matrix

In the switching matrix, connections are made so that a circuit is completed between the calling and called parties. The stored program control seeks out a suitable combination of time slots and multiplexed highways to perform this task and instructs the hardware elements to employ them for the duration of the communication session. Usually, the switching matrix consists of sections devoted to time slot interchanging and highway interchanging.

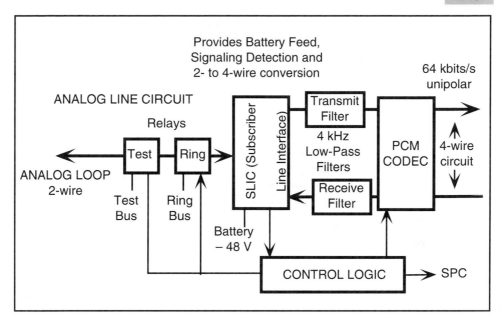

Figure 9.28 Analog Line Circuit

The analog line circuit conditions signals so that they can be exchanged between the digital switch and the analog loop.

(5) Time Slot Interchange Process

Time slot interchangers (TSIs)are known as time-switching stages and designated by the symbol T. The principle of time slot interchanging is shown in **Figure 9.29**. Incoming frames on a multiplexed highway are read into buffer memories; one memory handles odd-numbered frames, the other memory handles even-numbered frames. While one frame is read into one buffer, the preceding frame is being read out of the other buffer. As shown in Figure 9.3, frames are read in as a sequence of 24 octets—1 octet from each of 24 voice channels. Once the memory is filled, the octets are read out in a sequence defined by the stored program control (SPC). The result is an outgoing frame on the same multiplexed highway that contains the same octets as the incoming frame. However, the outgoing frame is delayed by one frame time, and the octets are in a different order. For small switching systems in which it is possible to multiplex all active lines on one highway, a single TSI stage is all that is required to interconnect the users.

(6) Highway Interchange Process

In addition to time slot interchangers, larger systems employ highway interchangers. Described as time-shared, space-division switching that employs a space array, highway interchangers are known as space-switching stages and des-

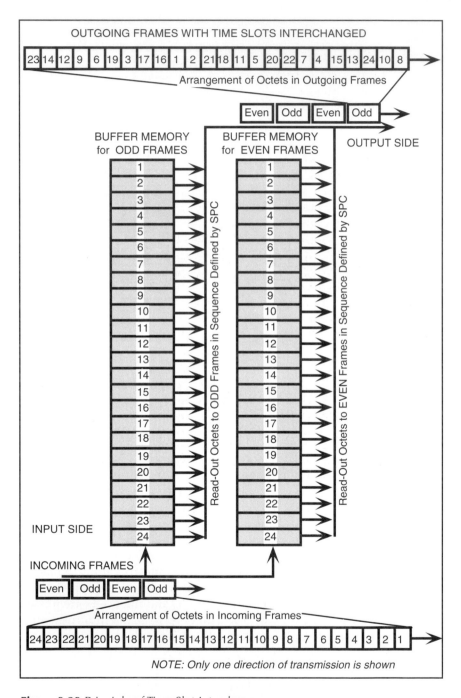

Figure 9.29 Principle of Time Slot Interchanger
By reorganizing the frame, time slots are changed without changing the signal highway.

ignated by the symbol S. The principle of highway interchanging is shown in **Figure 9.30**. Incoming highways are represented as the horizontal lines in a two-dimensional matrix; outgoing highways are the vertical lines. Incoming and outgoing highways are interconnected by gates that are activated by enabling signals (E). Thus, E(1,1) operates the gate that connects incoming highway (horizontal) 1 to outgoing highway (vertical) 1. The enable signal is a pulse of octet length that is synchronized with the incoming octets to switch an octet at a time on to different highways. Figure 9.30 shows the decomposition of an incoming frame of 24 octets on highway 1 into octets that are switched to different outgoing highways. Because there is no octet or frame buffering, there is no delay through the interchanger, and the octets leave the highway interchanger (HI) in the same order, and at the same time, as they entered; however, they are on different highways. The enabling signals are created in response to instructions from the SPC.

(7) Timing

For TSIs and HIs to operate successfully, the entire digital environment must perform in synchrony. It is vital that the processes change state at the exact instant that one octet ends and another octet begins. Achieving this condition requires that all operations be governed by a highly stable, digital clock and that adjustment procedures be included in those components in which delays occur.

(8) Time-Space-Time Switching Stages

To outfit practical, wire-center size switches, combinations of time and space switches are used. A Time-Space-Time array is shown at the top of **Figure 9.31**. In this architecture, the space stage interconnects the time stage serving the calling party with the time stage serving the called party. The time stages rearrange the order of octets from the channels that serve the customer loops to match slots available on the internal channels associated with the space stage. The stored program control does not associate the incoming and outgoing time slots with any particular space stage time slot, and any free slot may be chosen to create a connection. The path-hunting process is a matter of finding an available time slot at each end of the space array. In the circumstance that a pair cannot be found, blocking ensues.

(9) Space-Time-Space Switching Stages

A Space-Time-Space array is shown at the bottom of Figure 9.31. In this architecture, the time stage interconnects the space stage serving the calling party with the space stage serving the called party. Because the calling and called parties define the external timeslots (input of calling party and output to called party), the time stages must interchange them to create the connection. Flexibility is provided by the ability to choose among any of the TSIs that can perform the required slot interchange. In the circumstance that a TSI with the necessary capability cannot be found, blocking ensues. TST and STS networks can be designed with identical call-carrying capacities and blocking probabilities.

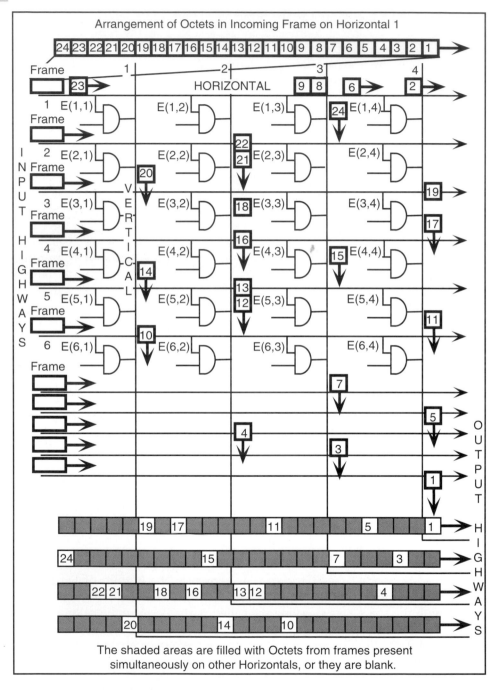

Figure 9.30 Principle of Highway Exchanger

By seizing time slots in different highways, highways are changed without changing time slots.

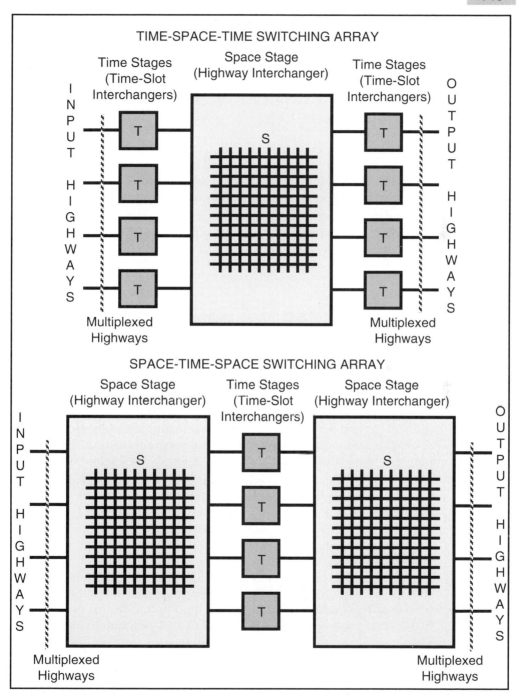

Figure 9.31 TST and STS Switch Architecture

9.6.2 TDM Switch Software Model

The software that drives switching machines must exhibit robustness, speed of operation, reliability, and extendibility that are several levels better than its data processing cousins. Without such properties, it will be impossible to handle the real-time requirements of a continuous stream of call requests and circuit allocations and accommodate undefined future services. Switching software can be divided into two parts

- **Generic program**: a general-purpose software package installed in all machines, it provides instructions that perform call processing, system maintenance, and system administration
- **Office database**: collection of data that is unique to a specific switching location. In the environment of a particular office, the generic program uses the information on attached equipment and customers to perform specific functions.

(1) Call Processing

Call processing software initiates, maintains, and terminates a message path between calling and called parties. It consists of six basic steps

- **Activity detection**: approximately every 100 milliseconds, scan programs test each customer line for signs of call origination, and each terminating trunk for signs of an incoming call. When the scan program detects a customer going off-hook, it stores the equipment number of the originating line (LENO, i.e., line equipment number—originating) in a service request buffer. At regular intervals, the executive control program reads the line equipment numbers requesting service and passes them to the origination program.
- **Number interpretation**: the origination program uses the equipment number to search the office database to identify the characteristics of the calling station (directory number, dial-pulse, or touchtone signaling, etc.). It creates a call record in temporary storage, locates an idle digit receiver that matches the signaling characteristics of the station, finds a path through the network between the calling loop and the digit receiver, and connects them. Completion of this sequence is followed by return of dial-tone to the calling station and initiation of the digit scan program that scans the digit receiver for dialed digit, and collects them in a digit buffer. At this point, the executive control program invokes the digit analysis program to identify and validate the called numbers. When satisfied that the number has been received correctly and that it is a valid directory number, the program enters it in the call record.
- **Ringing and answering**: the termination program uses the directory number to search the office database to obtain information on the line equipment

number of the loop serving the terminating station (LENT), and any associated routing instructions. Then the terminating line is checked to determine it is not busy, and network paths are found between LENO and LENT and between the ringing circuit and the terminating line. If the terminating loop is not busy, and a network path is available between the loops serving the calling and called parties, ringing voltage is applied to the terminating loop and ringback is applied to the originating loop.

- **Talking**: the scan program is activated to detect an answer (going off-hook) from the called station, or a disconnect (going on-hook) from the calling station. When an answer is detected, the executive control program disconnects ringing, establishes the talking path, notes the start of call in the call record, and releases the resources that have been used to establish the call.

- **Terminating**: the scan program maintains surveillance of the connection to detect a disconnect signal from one, or the other, party. At that time, disconnect programs return the resources to general availability.

- **Charging**: based on the call record, a charge is computed by the billing program (if appropriate).

(2) System Maintenance and Administration

Approximately 80% of the generic program is concerned with maintenance and administration functions. Among others, the functions are

- **System maintenance functions**: designed to maintain the correct operation of the switching system, they include
    - fault recognition and recovery procedures that detect improper hardware and software operation and enable the system to regain normal operation
    - congestion control and management techniques that enable the system to cope with traffic loads close to the ultimate capacity for which the system was designed
    - audit routines that appraise the use of memory and other resources and recover unused assets
    - recovery procedures that restore the system to normal operation when it encounters a major system-affecting problem. They include reloading the entire generic program and office database from permanent storage, as well as procedures that contain less drastic options.
- **System administration functions**: designed to allow the operating company to manage the switching system as a network resource, they include
    - collection of traffic engineering statistical information
    - office database change capability
    - verification of contents of system memory locations and office database entries.

(3) Office Database

The generic program is customized to each end-office environment by the office database. It contains two types of data

- **Customer data**: customer-related information such as
  - line equipment numbers for the loops that connect customers
  - directory number to line equipment number conversion
  - features available to each customer
  - identification of trunks available to remote offices
  - call-routing information
  - call-billing information.
- **Equipment data**: defines the hardware configuration of the end-office, including
  - number of equipment frames, and number of circuits within each frame
  - switching network configuration
  - addresses of frames, circuits, and switching network components
  - size of memory, and other information relevant to processor capabilities and performance.

The uniqueness of each end-office is reflected in the size and variety of the entries in the office database. In designing its structure, care is taken to preserve the independence of the generic program interface. This can be accomplished by requiring entry through a master head table that resides in a fixed location in memory, and by structuring the database as a hierarchical array of lists accessed from this table.

9.6.3 DIGITAL CROSS-CONNECT SYSTEMS

- **Digital cross-connect system** (DCS): a computer-based facility that reroutes channels contained in synchronous transmission systems without demultiplexing.

DCSs are not switches in the sense that they set up and tear down circuits on a call-by-call basis. Used in relatively large digital trunking situations, DCSs perform as the digital equivalents of analog distribution frames. Connections are loaded into the connection map memory to meet the average needs of the network, and entering new information from a keyboard changes connections. In some systems, several maps may be stored in memory so that the channel distribution can be changed easily in response to recurring situations.

Like a distribution frame, DCS permits reconfiguration of circuits; unlike a distribution frame, DCS operates within the multiplexed stream and performs the task of rerouting in an electronic matrix. Time slot by time slot, DCSs remap DS–0 channels among T–1 lines without demultiplexing. All channels must carry the same type of traffic and employ the same format. For instance, DS–1 lines operating with B8ZS, cannot connect with DS–1 lines using ZBTSI; DS–1 lines containing 56 kbits/s data channels cannot connect with DS–1 lines containing clear–64 channels, etc. DCSs are time slot manipulators, not format changers. For this reason, DCSs are not applied to asynchronous T–2 and T–3 systems. Finding a specific DS–1 signal that is bit-multiplexed in a traditional DS–3 signal requires demultiplexing the stream. However, DCS can be applied to synchronous DS–3 signals and to SONET streams.

Figure 9.32 shows a simple application of DCS. T–1 carrier systems containing inter-LATA traffic originating from two exchanges, home on a digital cross-connect system. On leaving the originating exchanges (AAA and BBB), the traffic

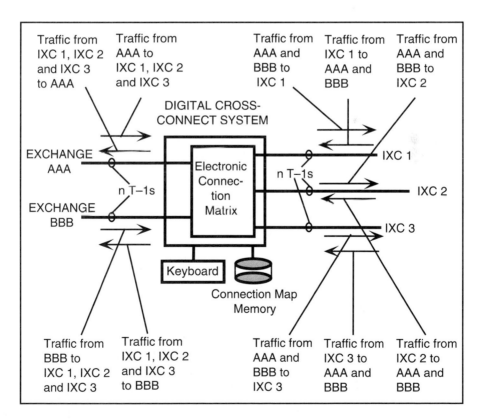

Figure 9.32 Principle of Digital Cross-Connect System

A DCS performs as the digital equivalent of an analog distributing frame. Entering information at the keyboard changes connections.

is segregated in time slot groups assigned to specific IXCs. At the DCS, traffic in the incoming time slots is moved to outgoing timeslots in the multiplexed lines that serve the intended IXC. Traffic from the IXCs is routed to the intended terminating exchange in a similar manner. As each IXC's long-term traffic patterns change, the connection map is modified to accommodate them.

9.6.4 SYNCHRONIZATION

The deployment of digital switching and transmission facilities throughout public and private networks requires that all the transmission paths connected to each of the digital switches deliver signals that are timed to satisfy the requirements of switch operations. They are sequenced so that information exchanges occur between the correct pairs of senders and receivers.

(1) Types of Synchronization

When digital signals are transmitted over synchronous systems, synchronization must exist at the frame, time slot and bit levels.

- **Frame synchronization** is necessary so that the sender and receiver can agree on the beginning and ending of a group of associated bits, and proceed to identify the time slots within the frame.

- **Time slot synchronization** is necessary so that the slots allocated to each sender/receiver pair can be identified correctly.

- **Bit synchronization** is necessary so that the receiver will reconstruct the same bit sequence as the sender transmitted.

(2) Bit Stuffing

Conceptually, synchronizing all the channels dependent on a single isolated switch is not a difficult a task. However, when thousands of communicating public and private network switches are involved, digital traffic is passed over transmission paths that vary significantly in length (and therefore propagation delay), and through equipment that may vary widely in performance (and therefore introduce jitter and other uncertainties). Despite these perturbations, activities at each node and associated equipment must occur in intervals that begin at the same instant. This *must* be so if frames are to be reorganized and rerouted in a reliable manner. To do this requires that all nodes have access to sources that can report time in exactly 125-microsecond intervals—the basic unit of time established by the sampling rate of digital voice. Further, all transmission facilities must have the ability to introduce *stuffing* bits that adjust the datastream to match the exact speeds required by the switching facilities, and adjust delays to compensate for different transmission paths. And these functions must be performed in real time so that no more delay is introduced than absolutely necessary. The

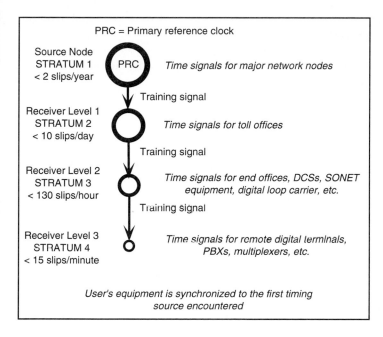

Figure 9.33 Timing Nodes for Network Synchronization

locations of stuffing bits introduced for these purposes in a DS–3 frame are shown in Figure 6.21.

(3) Hierarchical Synchronization

Practical network synchronizing techniques employ a hierarchical network approach.[3] The levels are represented in **Figure 9.33**. They are

- **Source nodes**: one or more network nodes serve as timing reference points. Known as source nodes, they are equipped with the most stable clocks, are in communication with receiver nodes equipped with less stable clocks, and, in large networks with more than one source node, each source node is connected to at least two other source nodes for coordination and security.
- **Level 1 receiver nodes**: the receiver nodes connected to the source nodes are equipped with less stable clocks. They employ the source nodes as timing references to synchronize (discipline) their clocks.
- **Level 2 receiver nodes**: these receiver nodes are connected to level 1 nodes. They employ level 1 nodes as timing references to discipline their clocks, and so on.

---

[3] Stephano Bregni, "A Historical Perspective on Telecommunications Network Synchronization," *IEEE Communications Magazine*, 36, 6(1998), 158–66.

The accuracy of these clocks is expressed in terms of their frequency stability, and the number of *slips* they permit.

- **Slip**: event that occurs when a timing error causes a time slot to be repeated or deleted.

With more precision, synchronization techniques are described as

- **Plesiochronous**: the signals are maintained arbitrarily close in frequency but are not sourced from the same timing reference. Over time, they will differ in phase. Timing inaccuracies produce slips that cause frames to be repeated or deleted.
- **Synchronous**: the signals are sourced from the same timing reference so that they all have the same frequency
- **Asynchronous**: the signals are sourced from independent timing references so that they are likely to differ in frequency and phase
- **Isochronous**: the timing information is embedded in the signals.

(4) AT&T Switched Network Synchronization

Through a network of 16 source nodes equipped with primary reference clocks (PRCs), AT&T provides timing and synchronizing services to its own digital network, to the digital networks of many local-exchange carriers, and to private digital networks. The clock hierarchy is

- **Stratum 1**: this group of 16, free-running PRCs consists of an ensemble of primary or secondary atomic standards that achieves a frequency accuracy of better than 1×10^{-11}, and permits no more than two slips per year. Fourteen of the clocks are located in the continental network; the others are located in Puerto Rico and Hawaii.
- **Stratum 2**: disciplined by the PRCs, these clocks are located in toll offices. They achieve a frequency accuracy of better than 1.6×10^{-8}, and permit no more than 10 slips per day.
- **Stratum 3**: disciplined by clocks in stratum 2, these clocks are located in local networks. They achieve a frequency accuracy of better than 4.6×10^{-6}, and permit no more than 130 slips per hour.
- **Stratum 4**: disciplined by clocks in stratum 3, these clocks are located in private networks. They achieve a frequency accuracy of better than 3.2×10^{-5}, and permit no more than 15 slips per minute.

On optical-fiber circuits, the major source of errors is the accuracy of the clocks that preserve the 125-microsecond rhythm of the network.

REVIEW QUESTIONS FOR SECTION 9.6

1   Comment on the following statement: In networks designed principally for voice services, the delay between the parties is constant. However before they began to communicate, there was a delay of several seconds.

2   Explain why concentration is used between the analog loops and the switch ports in a Class 5 switch.

3   Describe the function of an analog line circuit.

4   Explain the time-slot interchange process.

5   Explain the highway interchange process.

6   Why must the entire environment of a digital switch perform in synchrony?

7   Explain the use of a TST switching stage.

8   Explain the use of a STS switching stage.

9   Distinguish between the generic program part and the office database part of switching software.

10   List and describe the six basic steps in call processing.

11   Describe the system maintenance and administration functions performed by the generic program.

12   Describe the two types of data stored in the office database.

13   Define a digital cross-connect system.

14   Explain the following statements: DCSs are not switches. DCSs perform as the digital equivalents of analog distribution frames.

15   Explain Figure 9.32.

16   When digital signals are transmitted over synchronous systems, why must synchrony exist at the frame, time slot, and bit levels?

17   What is bit stuffing? Why is it used?

18   Explain the principle of hierarchical synchronization.

19   What is a slip?

20   Describe AT&T's network synchronization strategy.

10

TELEPHONE NETWORKS

As soon as the first installations demonstrated the feasibility of real-time communication between separated persons, the demand for telephone service created the telephone exchange. Staffed by human operators, it became a daunting assembly of plug boards and patch cables that were connected on-demand to provide talking paths between subscribers. Eventually, automatic switches were perfected, and the public switched telephone network (PSTN) was born. In this chapter, I describe the modern-day PSTN that is adapting to the role of providing pervasive, predominantly digital network facilities for voice, data, and video signals.

10.1 TELEPHONE INSTRUMENTS

Of course, the most frequently used terminal on a telephone network is a telephone. Starting as a cumbersome contraption of hand-cranked magneto, fixed microphone, and hand-held earphone, it has matured into an ultrareliable, integrated, electronic-driven device. The second most used terminal is the facsimile machine. Called *fax* for short, it has matured into a digital device capable of transferring graphics between terminals separated by continental distances. While both are familiar to the reader, how they perform their functions may not be. Accordingly, I will describe modern examples of both devices and briefly cover their operation.

10.1.1 DUAL-TONE MULTIFREQUENCY TELEPHONE

The elements of a dual-tone multifrequency (DTMF) telephone are shown in **Figure 10.1**. It consists of the following major components

- **Ringer**: or alerting device
- **Switch**: known as the *hook* switch, it closes when the handset is removed from its cradle
- **Touchpad**: generates bursts of two frequencies when a button is pressed to enter a digit

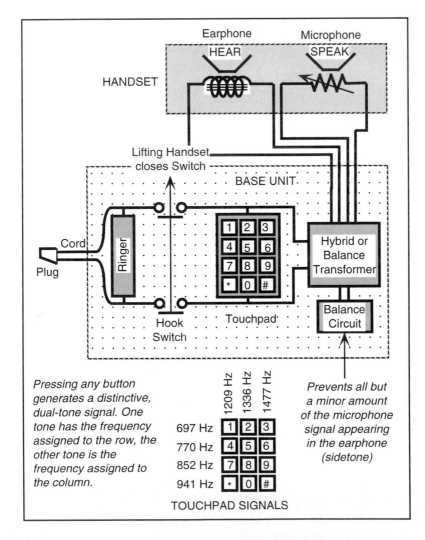

Pressing any button generates a distinctive, dual-tone signal. One tone has the frequency assigned to the row, the other tone is the frequency assigned to the column.

Prevents all but a minor amount of the microphone signal appearing in the earphone (sidetone)

TOUCHPAD SIGNALS

Figure 10.1 Elements of a DTMF Telephone

- **Microphone**: transducer contained in the handset into which the user speaks
- **Earphone**: transducer contained in the handset to which the user listens
- **Transformer**: known as the *hybrid* or balance transformer, it is contained in the base and separates the user's talking signal from the signal to which the user is listening.

Additional electrical components complete the circuit.

10.1.2 USING THE TELEPHONE

To make a telephone call, the telephones serving the *calling* and *called* parties must be connected together so that signals can flow between them. Establishing such a connection is a function of the telephone network. For the sake of simplicity, consider the case of persons whose telephones are connected to the same end-office so that their call is a *local* call. The major events are illustrated in **Figure 10.2**.

To initiate a call, the caller lifts the handset, thereby causing the hook switch to close and permitting current to flow from the end-office. This action is known as going *off-hook*. On detecting the current, equipment at the end-office connects the calling loop to a tone receiver and sends dial-tone to indicate the office is ready to receive the called number. On hearing dial-tone, the caller presses the buttons corresponding to the digits of the called number, an action that sends pairs of tones to the signaling receiver at the end-office. The digit information is encoded as a combination of two tones selected from seven tones in the range 697 to 1,477 Hz. Because the range of frequencies falls within the voiceband, it is *in-band* signaling, and the user can hear the digit tones while keying.

When the logic attached to the signaling receiver is satisfied that a complete number has been received, equipment in the end-office determines the status of the loop serving the called number. If the called loop is

- **Not in use**, a ringing voltage is applied to it. At the same time, a ringback signal is applied to the calling loop so that the caller is informed that the called number is being rung
- **In use on another call**, and the called party does not have *call-waiting* service, equipment at the end-office returns a *busy* signal to the caller to indicate that the call cannot be completed at the moment
- **In use on another call**, and the called party has *call-waiting* service, equipment at the end-office signals that a call is waiting. The called party has the option of interrupting the ongoing call to take the new one, or ignoring the alerting signal. After repeating the alert, if the called party does not respond, equipment at the end-office returns a *busy* signal to the caller.

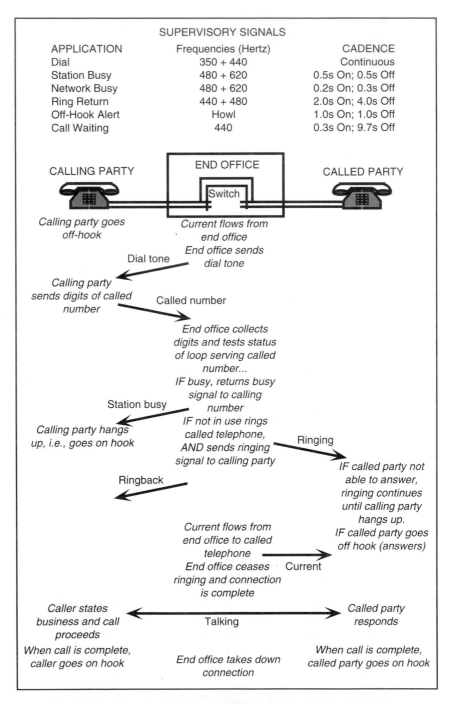

Figure 10.2 Major Events in Placing a Local Call

By applying a voltage to the pair of wires serving the called telephone, equipment at the end-office activates the bell, or alerting device. Lifting the handset of the ringing telephone, the person answering the call changes the state of the hook switch, thereby disconnecting the ringer, and connecting the microphone and earphone circuits to the line. The change in impedance that this produces indicates to the end-office that the call has been answered and that the connection is made. Accordingly, equipment at the end-office ceases to provide the ringing voltage to the called party and ringback signal to the calling party.

The voice of the answering party activates the microphone and sends a signal along the line to the caller. Because the microphone is connected at the center of the balance transformer, most of the signal it produces is not passed to the earphone. The small fraction of the signal that does get to it produces a sound known as *sidetone*. At the proper level, it provides a balance to the sound entering the user's other ear over the air path between ear and mouth. The result is a natural feel to the conversation. The voice signal from the caller is connected directly to the answerer's earphone through the transformer windings.

When the parties have completed their conversation, they hang up—i.e., they go *on-hook*. This action opens the hook switches on the calling and called telephones interrupting the currents flowing from the end-office and indicating disconnect. This signal triggers a series of events that return the loops to *idle* status ready to participate in other calls.

If only one party goes on-hook (hangs up), the end-office waits for approximately 15 seconds before applying dial-tone to the off-hook telephone. If the user of the off-hook telephone does not dial a number in a reasonable time, the end-office assumes the user forgot to hang up and places an off-hook alert signal on the loop. After a while, if the user still fails to hang up, the end-office ceases attempts to get the user's attention, and the loop is left open and silent.

Power-line crosses and lightning may subject the wire connections between the end-office and each telephone to high voltages. Because of this eventuality, the telephone set must be able to withstand a wide range of interfering voltages. The telephone company provides a measure of protection against lightning and other destructive surges by attaching arrestors and fuses at the network interface point.

10.1.3 Facsimile Machines

Many users of the telephone network employ facsimile machines to transfer images. In Section 3.3.5, we describe some compression schemes used in coding the image. **Figure 10.3** shows the essential components of a digital facsimile connection. At the sender, page-size documents are passed under a scanner where the images are converted to signals that are quantized, compressed, and transmitted by a modem as a modulated signal stream. At the receiver, the images are reconstructed by printing them dot by dot on a sheet of paper using the information contained in the signal stream, and the understanding that the reconstruction

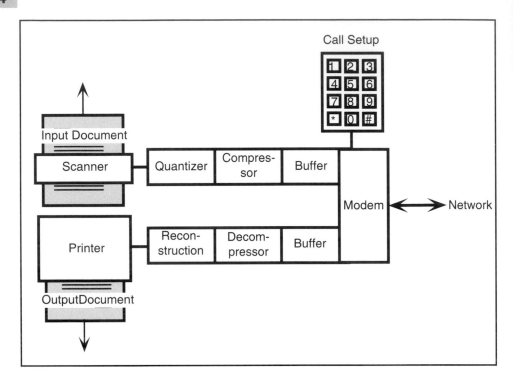

Figure 10.3 Elements of a Digital Facsimile Machine

process is the inverse of the signal processing that was employed by the sender. A letter-size document that is scanned with a horizontal and vertical resolution of 200 dots per inch (dpi) could generate 3.2 million bits. Transmitting this amount of data over a 56/64 kbits/s channel takes approximately 1 minute. To achieve faster service, data compression techniques are used. With modern techniques, speeds of up to 20 pages a minute are possible when the page is scanned at resolutions of 200 dpi x 200 dpi, and speeds of around 10 pages a minute are possible for 300 dpi x 300 dpi resolutions.

(1) ITU–T Recommendations

ITU-T defines four classes of facsimile machines. Some of their features are

- **Group 1**: scans documents at a resolution of approximately 100 lines per inch. Uses frequency modulation to transmit a page in 4 to 6 minutes. (Obsolete)
- **Group 2**: essentially the same as Group 1, but includes signal compression techniques to speed up delivery to 2 to 3 minutes. (Becoming obsolete)
- **Group 3**: scans documents at approximately 200 x 100, 200 x 200, or 200 x 400 dpi. Scanning generates digital signals that are subjected to variable-length cod-

ing (modified Huffman or modified Read coding, see Section 3.3.5) to reduce redundant information. Transmits a page (black and white) in approximately 20 seconds over modem-terminated telephone lines operating at 9,600 bits/s

- **Group 3 bis**: updates Group 3 to a speed of 14.4 kbits/s and a resolution of approximately 400 x 200 dpi. Transmits a page (black and white) in approximately 10 seconds.

ITU–T Recommendation T.4 defines the characteristics of a Group 3 facsimile terminal that transmits black and white, or color, documents. The procedures used are defined in Recommendation T.30. In Recommendations T.31 and T.32, ITU–T describes facsimile DTE and DCE that support physical, data link, and session functions, and in T.36, ITU–T describes techniques for the encryption of Group 3 facsimile transmissions.

- **Group 4**: a digital system that employs variable-length coding (advanced, modified Read coding, see Section 3.3.5) to reduce redundant information. Transmits a page (black and white) in approximately 6 seconds over a 64 kbits/s digital connection. The group is divided into three classes according to combinations of horizontal and vertical resolutions that range from approximately 200 dpi to 400 dpi.

ITU–T Recommendation T.563 defines the terminal characteristics for Group 4 facsimile equipment including optional color capability. In recommendation T.42, ITU–T describes a color representation method. ITU–T has defined simplex transmission speeds for facsimile machines as

- **V.17**: up to 9,600 bits/s (up to 14.4 kbits/s between fax-modems)
- **V.27 ter**: at 2,400 and 4,800 bits/s
- **V.29**: at 7,200 and 9,600 bits/s.

(2) Fax-Modems

Facsimile service can be implemented with a personal computer and a *fax-modem*. Combining the facsimile function and the modem function, they communicate with stand-alone machines or with a modem-equipped personal computer. Fax-modems are divided into three classes

- **Class 1**: an extension of the Group 3 standard for facsimile machines. The computer handles call set up and segmentation of the facsimile signal into data packets
- **Class 2**: the fax-modem is responsible for call setup and segmentation into data packets
- **Class 2.0**: in addition to call setup and segmentation of the facsimile signal into data packets, the fax-modem can also send and receive data files.

ITU–T V.17 includes one way transmission at 14.4 kbits/s between fax-modems.

(3) Fax Relay

In Section 3.4.7, several techniques for voice compression were discussed that depend on properties of the voice signal. Increasingly, they are being applied to voice facilities (e.g., 16- and 32-kbits/s ADPCM, digital channels) to improve the occupancy of long-distance circuits. Except for data rates $\leq 2,400$ bits/s, connecting facsimile machines or fax-modems over these circuits can produce severe distortion. For instance, ADPCM works because the values of voice samples change gracefully; the same is not true of facsimile signals. Accordingly, voice compressors are installed with a voice/fax activity detector. When voice is detected, the signal is encoded, compressed, framed, and dispatched to a digital transmission facility. When fax is detected, the signal is demodulated to create the original (14.4 kbits/s) datastream, framed, and forwarded to the digital transmission facility. Fax data received from the digital transmission facility are remodulated and sent to the analog facility. The technique is shown in **Figure 10.4**; it is called *fax relay*, or, more descriptively, demodulation/remodulation (or simply mod/remod).

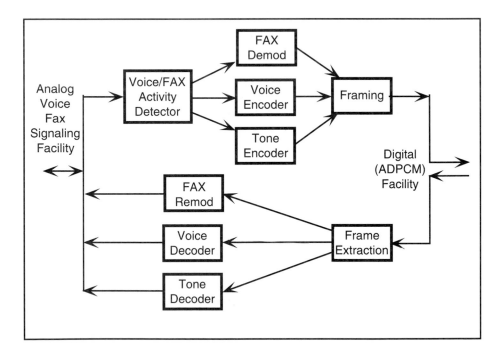

Figure 10.4 Principle of Fax Relay

To send fax over the same channel as compressed voice (ADPCM, for instance), the fax signal is received from the analog facility, demodulated to create the original 14.4-kbits/s signal, framed, and sent to the digital transmission facility. Fax data received from the digital facility are remodulated and sent to the analog facility.

REVIEW QUESTIONS FOR SECTION 10.1

1   Identify and describe the functions of the major components of a dual-tone multifrequency telephone.

2   Describe the sequence of events that causes the establishing of a connection between callers served by the same local exchange.

3   Identify and describe the functions of the major components of a digital facsimile machine.

4   List, and describe, the four classes of facsimile machine defined by ITU.

5   What is a fax-modem? Why is it used?

6   What is fax relay? Discuss Figure 10.4.

10.2 LOCAL AND LONG-DISTANCE NETWORKS

To users, telephone services appear to be provided by a seamless network that makes connections to just about anywhere. In fact, this network is composed of several networks that are operated by different organizations. While cooperating with one another in accepting, handling, and forwarding traffic, they may also compete for the same customers.

To describe the major features of the telephone network, it is convenient to divide providers into local exchange companies (LECs, or intra-LATA carriers) and interexchange companies (IXCs, or inter-LATA carriers) that operate local networks and long-distance networks, respectively [see Section 2.1.6(1)]. In this section, I discuss traditional local and long-distance networks in which transmission is achieved over wires, cables, and optical fibers. In Chapter 11, I discuss wireless networks.

10.2.1 SIGNALING

Described as the command and control component of a telecommunication network, signaling is the common thread that unites all the companies participating in the PSTN, and their customers. It is the name given to the exchanges of specialized messages that inform the users, and the facilities, of what actions to take to set up, maintain, terminate, and bill a call.

(1) Types of Signaling

For convenience, the action of signaling is divided into

- **Station signaling**: the exchange of signaling messages over local loops that connect the users' stations to the serving switch; station signaling is limited to an exchange area

- **Interoffice signaling**: the exchange of signaling messages between switches that serve the routes linking calling and called parties. Interoffice signaling occurs between the switches that are involved in completing a call within the LATA, or completing a call through IXC facilities and those of an LEC in another LATA.

Signaling messages are of four kinds

- **Alerting**: indicates a request for service (such as the caller going off-hook, or the terminating office ringing the called party)
- **Supervising**: provides call status information so that the users and/or the intermediate facilities can set up, maintain, and terminate the call (such as a station busy signal or a ringback signal)
- **Controlling**: provides tones, announcements, and other information related to the call (such as the tone that indicates ready to receive credit card number, and announcements such as *number changed to...*, or *number no longer in service*)
- **Addressing**: provides the called number, and other routing-associated information such as 1 + (area code) + (7-digit number).

(2) Telephone Numbering Plan

So that persons may telephone others around the world, ITU–T maintains a universal numbering plan in which customer stations are identified by subscriber codes, exchange codes, area codes, and country codes. The range of digits in each of these codes is shown in **Table 10.1**. However, national telephone authorities have modified it to accommodate local numbering plans, and no country employs all 13 digits. For instance, the country code

- For the United States is 1 (1 digit, not 3 digits)
- For Brazil is 55 (2 digits, not 3 digits) and the area code for Sao Paulo is 11 (2 digits, not 3 digits

Table 10.1 ITU–T Telephone Numbering Plan

Country Code	Exchange Code
XXX (NXX) NXX-XXXX	
Area Code	Subscriber Code
where, X = any digit and N = digits 2 through 9.	

- For Ireland is 353 (3 digits), the area code for Dublin is 1 (1 digit, not 3 digits) and the area code for Galway is 91 (2 digits, not 3 digits).

Some general principles apply to the codes.

- **Country code**: ITU–T divides the world into nine areas numbered 1 through 9. Countries have country codes beginning with the digit assigned to their area. Countries with few telephones are assigned two additional digits. Countries with more telephones are assigned one additional digit.

For Canada, the United States, and certain Caribbean countries—countries that participate in the North American Numbering Plan (NANP)—their country code is 1, the digit that identifies the geographical area North America. Breakdown by subarea in the NANP area is accomplished by three-digit area codes, three-digit exchange codes, and four-digit subscriber codes.

- **Area code**: North America is divided into subareas that are assigned three-digit area codes. With the restriction that the first digit is one of eight digits, 800 codes are possible. Some of them (e.g., 500, 800, 888, and 900) are assigned to special services
- **Exchange code**: identifies specific switching facilities located in a central office within the area defined by an area code. Many central offices handle more than one exchange code
- **Subscriber code**: identifies an individual station loop that connects the subscriber to the serving switching facilities.

A consequence of competition in local (i.e., intra-LATA) services has been the activation of many new area codes. They are required to ensure that current codes are left with enough numbers in reserve to accommodate new entrants.

In North America, long-distance (i.e., inter-LATA) calls are placed by dialing 1 + area code + exchange code + subscriber code. Dialing 1 as the first digit, indicates to the serving switch that an inter-LATA call is about to be placed and prepares it to access the long-distance network. The digit 1 is the network access code; it is not part of the number that identifies the called station, nor is it the NANP area country code; that is only used when placing a call from another area of the world.

(3) Signaling Arrangements

Signaling is described in several, nonexclusive ways as

- **Associated**: signaling messages are sent over the same route (transmission and switching facilities) as the information that is exchanged between the parties

- **Disassociated**: signaling messages are sent over a route different from the route taken by the information that is exchanged between the parties
- **Common-channel**: when signaling messages relating to many circuits are transported over a single channel by addressed messages
- **In-band**: when the associated signals are analog tones that exist within the bandwidth used for the information exchange
- **Out-of-band**: when the associated signals are analog tones that do not occupy the same bandwidth as the information exchange.

In the case of digital message signals, in-band and out-of-band are replaced by the terms

- **In-slot**: when the associated digital signals use a bit position contained within each digital message octet (this is known as bit-robbing)
- **Out-of-slot**: when the associated digital signals use a time slot assigned exclusively for signaling.

Some of the features of a signaling network are shown in **Figure 10.5**. In the United States, signaling between switching nodes employs disassociated, common channel signaling. The signaling data are encapsulated within a frame addressed to the terminating switch and sent to a signal transfer point (STP) that serves many switching centers. Here, the message is routed directly to the terminating switch, or directly to the STP that serves the terminating switch. The combination of signaling data links between switching centers and STP, and the interconnected STPs, forms a network. Called the *common channel signaling network* (CCSN), it is separate from the one that handles the user's information exchange. To increase the reliability of the signaling network, STPs are duplicated. Also, they are fully interconnected so that a signaling message need only pass through two pairs of STPs—the duplicated STPs serving the originating switch and the duplicated STPs serving the terminating switch.

The deployment of a common channel signaling network has led to the development of specialized services. Databases used for these purposes reside in network control points (NCPs) that are connected to signal transfer points. Signaling intended to use these capabilities is called direct signaling. More information on signaling networks is contained in Sections 10.3 and 10.4.

10.2.2 OUTSIDE PLANT

The result of many years of evolution in telephone cables and construction techniques, the signal transport network that serves the telecommunication needs of an established community contains a mixture of materials and installation methods. They reflect the longevity of telephone practices and the relatively slow speed

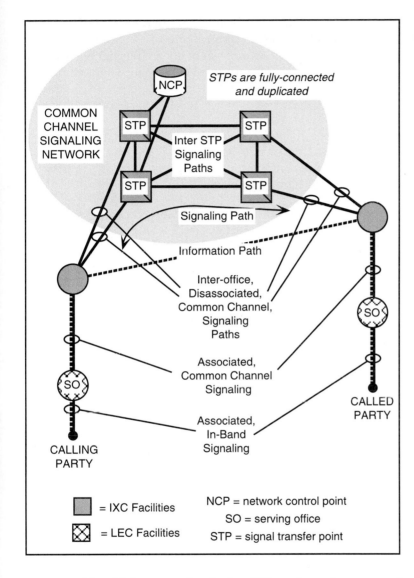

Figure 10.5 Features of Signaling Network

Between IXC facilities, a common channel signaling network is used. Between IXC and LEC facilities, the signaling path is likely to be associated with the information path. Between the using parties and the LEC facilities, signaling is likely to be in-band.

with which new technologies diffuse through a system in which it is important to recover the investment in in-place equipment. As noted in Section 6.3.1, all the facilities situated between the customer's premises and the distribution frame in the serving office are known as *outside plant* (OSP).

(1) Cables and Twisted Pairs

Close to the switch, the feeder cables contain many wire pairs. To conserve duct space, small diameter wires (26 AWG) are employed. In the outlying feeder cable sections, the diameter of individual wire pairs may be increased (e.g., to 22 and 19 AWG) so that the total resistance of the loop does not exceed 1,100 to 1,300 ohms. This is the value that allows sufficient current to flow from the serving office battery for ringing. Since there is no point to being able to signal if the parties cannot understand each other, inductance is added to long loops to balance the stray capacitance between the wires and reduce the voice-band attenuation of the connection. A loop must satisfy both resistance and transmission design parameters to provide satisfactory service to the user.

As digital loop carrier equipment (see Section 6.3.2) is installed, and individual physical loops between the carrier serving area and the customer are replaced by digital subscriber lines (see Section 6.3.3), the concepts of resistance design and transmission design become obsolete.

(2) Outside Plant Productivity

Customer loops and terminating telephones are the least productive telecommunication facilities. On average, residential facilities are in use about one-half hour each day, and business telephones may be used as much as one hour each working day. Further, while most pairs may be assigned to specific customer stations, there is a fraction of the plant that is idle waiting to be assigned to meet future customer requirements. Because outside plant represents up to one-third of the total capital investment in an exchange area, increasing the intensity of use of these assets is important to local exchange companies.

One strategy to increase the productivity of the OSP is to shorten loop lengths by moving switching closer to the customer. This creates a requirement for remote switching units, and for short-distance multiplexing to carry the traffic to the serving office. Other strategies introduce concentration and multiplexing at appropriate points where they can serve a community of customers. Known as *pair-gain* techniques, they replace a larger number of underutilized loops by a smaller number of loops that are in use more often. Pair-gain techniques are readily implemented in digital circuits and form the basis for modern digital end-offices in which a large base unit switch is supported by remote switching units and digital loop carrier equipment (see Section 10.2.3).

(3) Penetration of Digital Electronics and Optical Fibers

Fine wire close to the central office, bridged taps [see Section 5.1.2(2)], and loading coils have the primary objective of providing a local network that will support voice frequency communications at a reasonable cost. For all but low-speed applications, these techniques are harmful to the carriage of digital signals. To pass digital voice (DS–0), the loops must be *groomed*—i.e., loading coils and bridged taps must be removed.

As station growth takes place, an increasing amount of digital electronics is being placed in the analog loop plant to expand the number of channels without adding new bearers. As demand for digital channels picks up, optical fibers (see Section 5.2.2) and digital loop carrier equipment (see Section 6.3.2) are replacing feeder and distribution cables. They are the technologies of choice for new construction, particularly new construction for business customers.

(4) Loop Plant Topology

The traditional loop plant shown in Figure 6.24 can be described as a double star. The feeder cables centered on the serving central office (SCO) constitute one star, and the distribution cables centered on the remote ends of the feeder cables constitute the other. Because it contains no electronic devices, the arrangement is described as *passive*. The introduction of digital loop carrier (DLC) equipment has created an active double-star configuration in which the number of circuits is no longer dependent on the availability of individual twisted pairs. The deployment of optical fibers in the loop makes other arrangements possible. For instance, a bus or a ring can perform the function of the distribution plant, leading to star bus and star ring configurations, and the feeder plant may eventually become a second ring, leading to a ring-ring topology. These possibilities are illustrated in **Figure 10.6**.

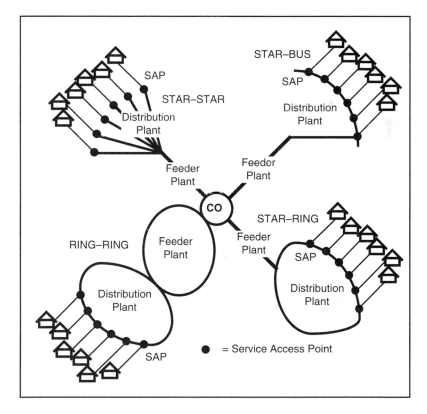

Figure 10.6 Evolution of Loop Plant

Present loops are organized in star-star configuration. With modern technology, the topology can advance through star-bus to star-ring and to ring-ring.

To date, most of the optical fibers in the local loop have been placed there in support of digital loop carrier applications. With this equipment deployed in increasing numbers, extending optical fibers to pedestals (this is known as fiber to the curb, FTTC), becomes possible and will make 64 kbits/s digital circuits (and multiples, i.e., n x 64) available to just about all who want them. However, demand for interactive, broadband services is insufficient to justify widespread deployment to the home (this is known as fiber to the home, FTTH).

In fact, the introduction of digital subscriber lines (see Section 6.3.3) that employ twisted pairs has probably set back, or halted, deployment of fibers much beyond the carrier serving area. It will occur when the cost of optical technology is at, or below, the cost of maintaining and replacing today's copper systems. As always, when there is competition between technologies, advances are made on both sides. The combination of very reliable integrated circuits and copper pairs will be a hard one to replace.

10.2.3 LOCAL NETWORKS

In simpler times, the caller picked up the earpiece, cranked the generator handle on the telephone instrument to call central, spoke with an attendant in the central office, and waited for the connection to the called party to be completed manually.

(1) Telephone Exchange and Exchange Area

As telephone facilities evolved, a switching machine was installed in the central office and became the focus of telephone activity in a given area. Called the telephone exchange, its customers defined an exchange area. They were connected to the exchange by loops—the dominant component of the outside plant.

To complete a network that spanned the United States, switches were arrayed in a hierarchy of five levels. A Class 5 switch served the telephone exchange. The other switching levels were Class 4–Toll Center switch, Class 3–Primary Center switch, Class 2–Sectional Center switch, and Class 1–Regional Center switch. With the separation of long-distance transport from local transport, the upper layers have been abandoned in favor of fully interconnected mesh networks of large digital switches [see Section 10.2.4(1)].

(2) End-office, Serving Office, and Wire Center

Digital switches and other electronic equipment have created a different telephone exchange. The changes are reflected in current names, such as

- **End-office**: a facility that contains the lowest node in the hierarchy of switches that comprise the network. It is the network interface with the caller—connecting the caller's loop to the called party's loop if they share the same end-office, or connecting the caller's loop to another switch if they do not

While the switch may serve a single exchange code, it is more likely to serve several. It is still called a Class 5 switch.

- **Serving off**ice: also known as serving central office. The end-office that serves a particular user

- **Wire center**: a physical location at which many cables come together to be served by a switching facility that spans several exchange codes. They may be wire pairs contained in feeder cables that carry analog voice from individual telephones, or optical fibers that carry digital, multiplexed traffic from remote switching units or pair-gain equipment in the exchange area—or from other network nodes.

Figure 10.7 provides a summary view of a modern end-office that serves an urban area with business and residential customers. The office contains a digital

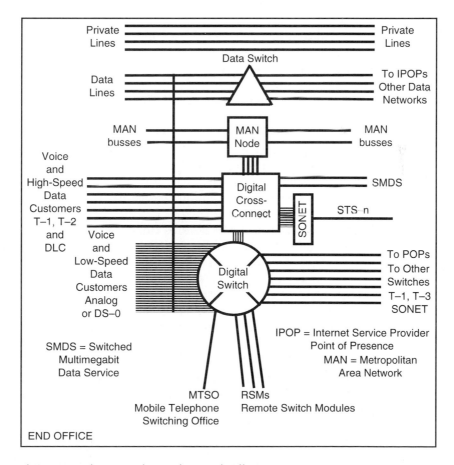

Figure 10.7 Elements of Digital Central Office

switching machine that supports remote switching modules, a digital cross-connect system, SONET connections, and other multiplexer interfaces. In addition, if there is sufficient high-speed data traffic, it may contain a separate data switch, and/or a MAN node (see Figure 12.12). The majority of circuits are provided over feeder cables to voice and low-speed data customers. Some customers use T–1 or T–2 digital loop carrier systems for voice or high-speed data, and other customers employ private lines that transit the office. The end-office connects to other end-offices, to a mobile telephone switching office (see Section 11.2.1) and to IXC POPs, for long-distance connections.

- **Point of presence** (POP): the facility interface between an LEC and an IXC. A POP must be established on each trunk path that connects a switch that belongs to a local carrier and a switch that belongs to a long-distance carrier. On its side of the POP, the LEC is responsible for service; on the other side, the IXC is responsible.

Digital, local switches are designed to perform in all of these situations. They contain different complements of hardware and software modules that customize their performance to the environment in which they are to operate.

(3) Inter- and Intra-LATA Traffic Arrangements

To collect inter-LATA traffic from several exchange areas, the LEC may employ a tandem switch. In this role, it is known as an access tandem (AT). Also, it may provide alternate connections between exchange areas in the LATA and be the principal repository of resources for the intelligent network (see Section 10.4). **Figure 10.8** shows some typical exchange areas and the interoffice connections that are required to serve the local and long-distance needs of the users. Briefly

- **Exchange area A**: a single end-office (that may include several exchange codes) serves the needs of users in exchange area A. Intra-LATA, interexchange traffic is sent directly to the end-offices in exchange areas B and C. Inter-LATA traffic directed to IXC (A) is trunked directly from the end-office to IXC (A)'s toll switch. Long-distance traffic intended for other IXCs is routed through an access tandem switch to the individual carrier's switches. At times of heavy demand, traffic intended for IXC (A) is carried through this switch

- **Exchange area B**: an end-office with a remote switching module serves the needs of users in exchange area B. Intra-LATA, inter-exchange traffic is sent directly to the end-offices in exchange areas A and C. Inter-LATA traffic intended for any IXC is trunked to the access tandem switch

- **Exchange area C**: an end-office with a remote switching module serves the needs of users in exchange area C. Intra-LATA, interexchange traffic is sent directly to the end-offices in exchange areas A and B. Inter-LATA traffic intended for any IXC is trunked to the access tandem switch

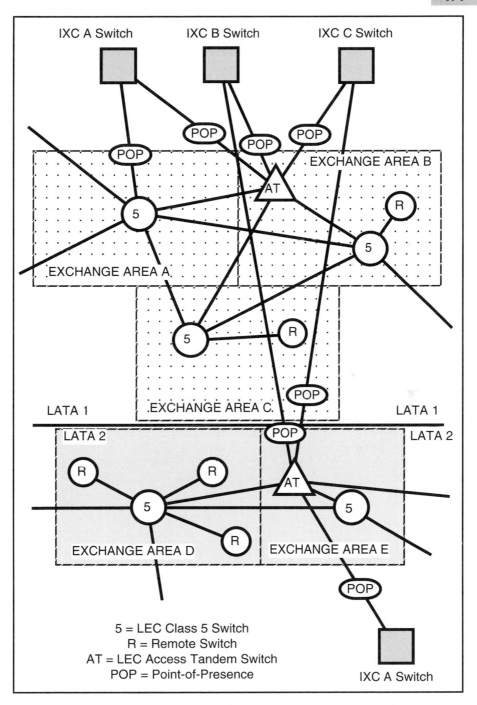

Figure 10.8 Inter- and Intra-LATA Traffic Arrangements

- **Exchange area D**: an end-office with three remote modules serves the needs of users in exchange area D. Intra-LATA, inter-exchange traffic is sent directly to the end-office in exchange area E. Inter-LATA traffic intended for any IXC is trunked to the access tandem switch in LATA 2. Traffic between exchange areas A, B, or C and exchange area D is carried through a toll switch operated by one of the IXCs

- **Exchange area E**: a single end-office serves the needs of users in exchange area E. Intra-LATA, interexchange traffic is sent directly to the end-office in exchange area D. Inter-LATA traffic intended for any IXC is trunked to the access tandem switch in LATA 2. Traffic between exchange Areas A, B, or C and exchange area E is carried through a toll switch operated by one of the IXCs.

Within networks, traffic is carried between nodes (switches) over multiplexed routes that employ digital techniques to transport DS–0, DS–1, DS–3, and SONET signals. The physical facilities employ twisted wire pairs in specially constructed cables, microwave radio, or optical-fiber lightguide bearers. Which is used depends on the number of channels and on the cost of competing installations, although optical fibers are clearly the preferred technology.

(4) Special Transmission Paths

To accommodate the needs of businesses, three specialized connections are available. They are

- **Tie lines**: nonswitched lines directly connecting two private network facilities (e.g., PBXs, data centers, etc.) for the exclusive, full-time use of those that lease them. Cost depends on mileage and performance required. They may be analog voice circuits or digital voice or data circuits with several circuit speeds. Sometimes, tie lines are called tie trunks

- **Foreign exchange lines**: tie lines that connect a private network facility (e.g., a PBX) to a public exchange in a distant city so that callers may have the benefit of local service in the distant calling area. Cost consists of the cost of the tie line between the cities, plus the cost of local service in the distant city, plus the cost of terminating the tie line in the distant city

- **WATS lines**: a service that offers bulk rates for station-to-station calls that are dialed directly. Calling categories are intrastate and interstate, and incoming or outgoing:
  - *outgoing WATS*. On a dedicated local access line, the subscriber may make calls to specific areas of the United States. The calling categories are intrastate and interstate. Interstate calling areas are divided into bands that span the continental United States, and serve Alaska, Hawaii, St. Thomas, St. John, and Canada. Costs depend on the originating area code, the band selected, and the number of calls made per month. Interstate

WATS tariffs are offered exclusively by inter-exchange carriers. Rate structures vary somewhat by company

– *incoming WATS*. On a dedicated local access line, the subscriber may receive calls from specific areas of the United States and certain other countries. The calling categories are intrastate, interstate, and international. Usually called *800 service*, the costs associated with all calls are borne by the subscriber.

Table 10.2 compares these classes of connections and includes direct distance dialing (DDD), the transmission path of last resort. It is used when these special connections are not available.

Table 10.2 Comparison of Some Special Service Lines

TIE LINE	Line directly connecting two private telephone facilities (e.g. PBXs) for the exclusive, full-time use of those that lease them.	*May be between any two points selected by user. Cost depends on mileage.*
FOREIGN EXCHANGE (FX) LINE	Tie line that connects a private telephone facility (e.g. a PBX) to a public exchange in a distant city so that callers may have the benefit of local service in the distant calling area.	*May be between user's facility and any local exchange in US. Cost consists of the cost of the tie line between the cities, plus the cost of local service in the distant city, plus the cost of terminating the tie line in the distant city.*
WIDE AREA TELEPHONE SERVICE (WATS) LINE	Connections that support wide-area telephone service -- a service that offers bulk rates for station-to-station calls that are dialed directly. On a dedicated local access line, the subscriber may make calls to, or receive calls from, specific areas of the United States.	*Calling categories are intrastate, and interstate. The service may be outgoing, or incoming (in this case it is known as 800 service). Interstate calling areas are divided into bands. Cost depends on the originating area code, the band employed, and call volume.*
DIRECT-DISTANCE-DIALING (DDD) LINE	Line available on-demand that connects user to any other user.	*Calling categories are local, long-distance and international. Cost depends on distance, time of day, and holding time.*

(5) Twisted-Pair Transmission Systems

Twenty-four and 48 PCM voice channels (i.e., T-1 carrier at 1.544 Mbits/s and T-1C carrier at 3.152 Mbits/s) are transmitted on wire pairs contained in multipair cables installed between local exchange facilities. Regenerators are spaced at intervals of approximately 6,000 feet. In older systems, in-slot signaling is used so that the channel speed for data is reduced to 56 kbits/s. In newer systems, common channel signaling, and B8ZS coding [see Section 6.2.1(3)] are employed to restore the channel speed to 64 kbits/s. These channels are known as *clear-64* channels.

(6) Digital Microwave Radio Transmission Systems

Introduced between New York City and Boston in 1947, microwave radio relay systems quickly expanded across the continent to provide long-distance routes to support direct distance dialing (DDD). The first systems employed frequency modulation (FM) and carried frequency-division multiplexed (FDM), single sideband, suppressed-carrier voice channels (see Section 6.1.2). Later, the capacity of existing systems was improved by converting the carrier modulation to single-sideband amplitude modulation (SSBAM).

With the introduction of robust integrated circuits, the cost per channel-mile of digital radio systems has dropped below FM/FDM and SSBAM/FDM radio systems for most routes. Coupled with the rapid adoption of digital switching, in-place analog microwave radio equipment is being replaced by digital radio—or optical fibers. To facilitate an orderly changeover, digital microwave radios employ the same frequencies and channel widths as analog microwave radios. Thus, in the 4, 6 and 11 GHz common carrier bands, channel spacings are 20, 29.65 and 40 MHz respectively. Commonly, 1344 voice circuits (2 x DS–3 = 90 Mbits/s) are carried in each channel. Digital radios that conform to SDH (synchronous digital hierarchy, see Section 8.3.3) transport 8 or 16 STM–1 streams at 6 GHz, and 24 STM–1 streams at 11 GHz (STM–1 = 155.52 Mbits/s). Some of the digital modulation schemes are described in Section 4.2.

(7) SONET

In Section 6.4.1, the transport speeds and framing structure of synchronous optical network were described. Originating and terminating signals are electrical, and members of the digital signal hierarchy (i.e., DS-0, DS-1, and DS-3). Intermediate transport is provided by optical carrier signals (i.e., OC–1, OC–3, etc.). To provide point-to-point, hub and ring connections, SONET makes use of multiplexers, digital cross connects, and other terminals.

Some basic configurations are shown in **Figure 10.9**. At the top are two examples of point-to-point connections. The first multiplexes DS-1 or DS-3 channels to make a broadband private line; the second shows a SONET digital loop carrier providing access service between the serving-area interface (SAI) and the serving office. Typically, a SONET DLC will operate as a multi-port device at OC-3 and will be capable of carrying up to 2,016 DS-0 signals. In the third example,

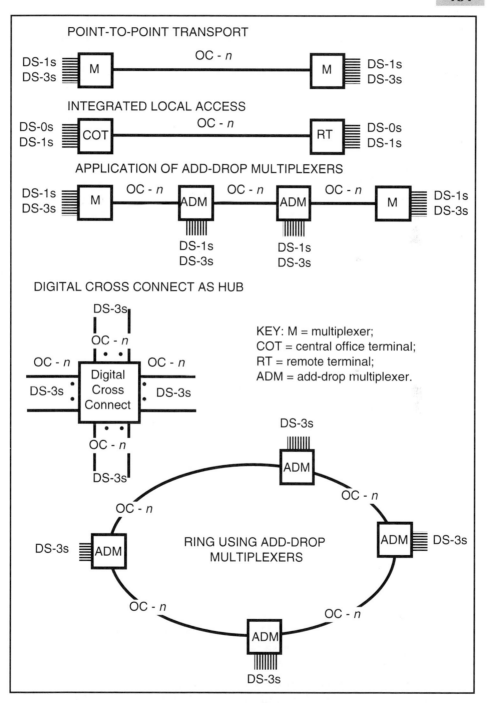

Figure 10.9 Basic Configurations of SONET Components

the system multiplexes DS-1s and, without demultiplexing the entire stream, drops and adds DS-1 channels at two points along the way. This application is made possible by the synchronous multiplexing SONET employs, and the inclusion of pointers in the line overhead that enable the add/drop multiplexers (ADMs) to keep track of the starts of individual envelopes. In addition, it enables the use of digital cross connects as hubs in large networks, and the construction of rings based on ADMs.

Because the individual fiber connections carry many channels, an important consideration for all these configurations is the provision of alternate routes that can be invoked rapidly in the case of a cable cut or node failure. Automatic protection switching (APS) can be used to switch any one of a number of working lines to a backup line. In addition, SONET permits bidirectional rings in which the second fiber acts as backup for the first. In the event of a failure in the primary line, the backup fiber is used to bridge the gap by sending the frames around the ring in the reverse direction until they reach the other side of the fault. This action has been described as *self-healing*.

SONETs may contain equipment that performs six specific functions. However, each function is not implemented by a separately defined network device; rather, the performance is obtained by deploying equipment in certain ways. The functions are

- **Add/drop multiplexer** (ADM): aggregates or splits SONET traffic at various speeds so as to provide access to SONET without demultiplexing the SONET signal stream. May be inserted in a ring to provide access to/from the ring for several terminal multiplexers. Generally, has two equal speed network connections

- **Terminal multiplexer** (TM): an end-point or terminating device that connects originating or terminating electrical traffic to SONET. Converts to/from non-SONET channels to SONET channels. Has only one network connection

- **Digital cross-connect** system (DCS): redistributes (and adds or drops) individual SONET channels (or virtual tributaries) among several STS-n links. Consolidates and segregates STS-1s and can be used to separate high-speed traffic from low-speed traffic so as to feed one to an ATM switch and the other to a TDM switch. Two types of SONET DCS are available:
  - *Broadband DCS*. B–DCSs cross connect between DS-3 interfaces, and OC-n SONET optical carriers, on the basis of the configuration software stored in the DCS
  - *Wideband DCS*. W–DCSs cross connect between DS-3 or VT-1.5 (DS-1) interfaces, and OC-n SONET optical carriers, on the basis of the configuration software stored in the DCS

- **Digital loop carrier** (DLC): used to link serving offices with carrier serving area (CSA) interface points (see Section 6.3.2). Typically, SONET DLCs concentrate DS-0 signals into OC-3 signals.

- **Matched node** (MN): pairs of MNs are used to interconnect SONET rings and provide alternate paths for recovery in case of link failure. SONET traffic is duplicated and sent over two paths between the rings. One set of MNs provides the active path; the other set is on standby in case of failure of the active connection. Pairs of DCCs or ADMs realize MNs.

- **Drop and repeat node** (D+R): SONET devices configured to split SONET traffic and copy (repeat) individual channels on two, or more, output links. Applications include the distribution of residential video and alternate routing.

All of these devices combine electronic and optical technology. While optical media are employed to achieve transmission of SONET signals, electronic techniques must be used to manipulate the signals.

Some possible applications are shown in **Figure 10.10**. In the local exchange area, interconnected ADMs can form SONET rings. Initially, they are likely to be used to provide new lines for business customers. As growth occurs in the neighborhood of expanding businesses, and new residential developments are built, the ring can be used to provision circuits for this growth. As demand rises, the optical carriers can be upgraded from OC–3 to OC–12, and to OC–48, without laying new fiber in the ring, although new fiber may be required to connect to additional TMs. It is the advent of SONET that has inspired the concepts shown in Figure 10.6.

10.2.4 LONG-DISTANCE NETWORKS

Over 100 companies compete to serve the long-distance communications needs of the public and commercial sectors of the United States. The largest long-distance network is owned by AT&T. It employs a network of some 130 switching machines, over 100 Network Control Point (NCP) databases, and 20 Signal Transfer Points (STPs).

(1) AT&T's Network

Figure 10.11 characterizes the scope of AT&T's Network. Virtually all the switching machines are fully interconnected. There are more than 7,500 point-to-point, multiplexed links in the network. Signaling employs Common Channel Signaling System 7 (SS7). In Section 9.6.4(4), we noted that AT&T maintains 16 Primary Reference Clocks (PRCs) that provide precise timing at critical locations in the network. Their output is used to discipline a hierarchical network of clocks with lesser accuracies that synchronize digital equipment at all levels of the network. A Timing Monitor System (TMS) verifies that this is done satisfactorily.

(2) Dynamic Nonhierarchical Routing

Calls are routed by a technique known as dynamic nonhierarchical routing (DNHR); they proceed from originating to terminating switches directly, or through a single intermediate switch so that over 130 routing options are available

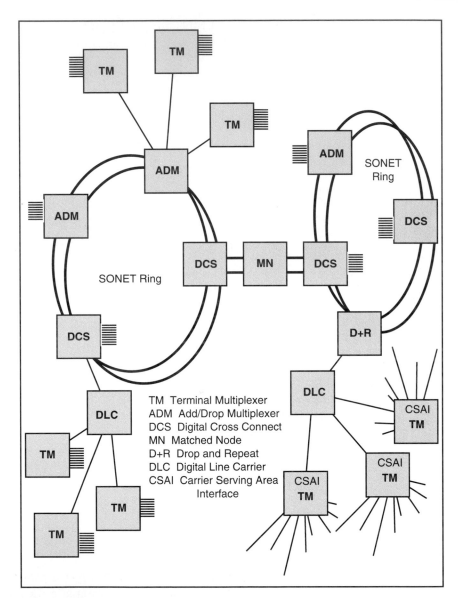

TM Terminal Multiplexer
ADM Add/Drop Multiplexer
DCS Digital Cross Connect
MN Matched Node
D+R Drop and Repeat
DLC Digital Line Carrier
CSAI Carrier Serving Area
Interface

Figure 10.10 Some Examples of SONET Connections

between each pair of switches. No hierarchy of switching points is involved—hence, the description *non*hierarchical routing.

Routing patterns vary by hour of day and day of week. At the bottom of Figure 10.11, we show examples of DNHR between switches located in San Diego and White Plains. The first route attempted is the direct San Diego to White Plains connection. The next attempt might be through the switch located in Albany. A

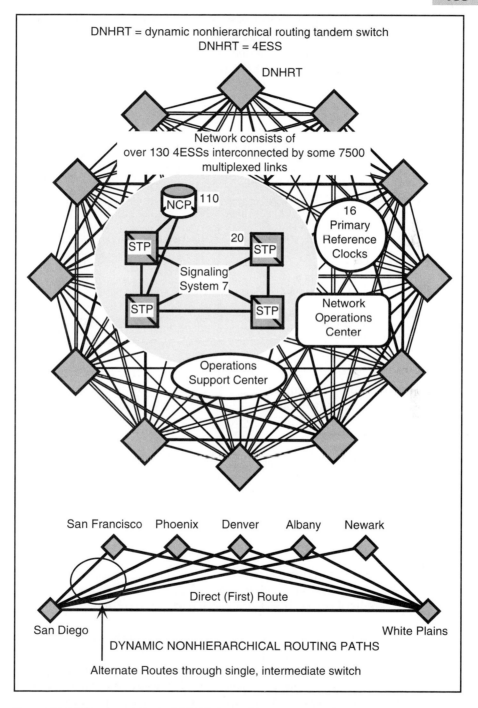

Figure 10.11 Representation of AT&T's Long-Distance Network

third route could be through the switch located in Phoenix—and so on. Which is used will depend on the overall state of the network (possible restrictions due to failure or maintenance problems), the time of day (is it busy hour in Albany or Denver?), and day of week (is it a workday, a weekend, or Mother's Day?). In no case does a call employ more than one intermediate switching point.

(3) Routing Controls

The possibility of selecting any one of a large number of routes to carry calls between switch pairs can result in an increase in the average number of circuits required to complete a call, and reduce the network capacity available for other calls. Under overload conditions, automatic expansive routing techniques may force the network into a congested, inefficient state and lead to a decline in overall call-carrying capacity. (If all calls are completed through an intermediate switch, twice as many network links are involved as when all calls are first routed.) To forestall this condition, protective routing techniques are employed. They assess the probability of completion of calls attempted during periods of high use and favor those that are judged likely to be completed. Specific examples of protective routing techniques are

- **Selective trunk reservation** (STR): activated by a reservation level placed on individual trunk groups, STR controls two link calls more stringently than single-link traffic. When the number of idle trunks in a particular group drops below a specified threshold, the idle trunks remaining are used preferentially to complete single-link (first-routed) calls. In cases of significant, widespread network congestion, alternate routed calls seeking to use the trunk group resources are rejected.
- **Hard to reach** (HTR): the ratio of the number of calls accepted by, and the total number of calls presented to, a specific switch, in a given time period is called the answer to bid ratio (ABR). It is computed by analyzing call completions. When the ABR of a switch drops below a given threshold, it is classified as hard to reach. On subsequent bids for connection, the originating switch is informed over SS7 of the HTR status of the destination switch. By building a list of such information, individual switches reduce the number of times they request connection through a congested switch.
- **Selective dynamic overl**oad (SDO): relieves congestion at an overloaded switch by sending an instruction to all switches connected to it to reduce the volume of their requests for service for a fixed time interval.

To make these strategies work requires a centralized network management authority with the tools needed to detect the onset of congestion, and the real-time ability to tune the decision thresholds used in protective routing schemes on the basis of the latest performance data. AT&T employs a centralized Network Operations Center for this purpose. With modern switching machines, it is possible to determine the optimum one- or two-path link on every call. Called *real-time dynamic traffic routing* (RTDTR), the routing calculation can be based on traffic congestion, events, or the state of the network (the busy-idle condition of available trunk groups, for instance).

(4) Other Long-Distance Carriers

The other major carriers employ the same principles in networks that they have built from optical fibers and digital switches. Smaller carriers may own specialized switches tailored to their target markets and lease transmission capacity from the larger companies.

REVIEW QUESTIONS FOR SECTION 10.2

1   Distinguish between station signaling and interoffice signaling.

2   List and describe four kinds of signaling messages.

3   Describe ITU's universal numbering plan. Why is it needed? Give examples of its use.

4   Distinguish among associated, disassociated, and common channel signaling.

5   Distinguish between in-band and out-of-band signaling. What are they called when the channel is digital?

6   Explain Figure 10.5.

7   Define outside plant.

8   For local loops, what is meant by resistance and transmission design?

9   Give an approximate average value for the length of time each day residential telephones are used. What is the corresponding value for business telephones?

10   Discuss Figure 10.6.

11   Comment on the following statement: the introduction of digital subscriber lines that employ twisted pairs has probably set back, or halted, deployment of fibers much beyond the carrier serving area.

12   What do you understand by end-office, serving office, and wire center?

13   What would you expect to find in a modern end-office that serves an urban area with business and residential customers?

14   Define point of presence.

15   Discuss Figure 10.8.

16   Distinguish among tie line, foreign exchange line, and WATS line.

17   Comment on the following statement: To facilitate an orderly changeover, digital microwave radios employ the same frequencies and channel widths as analog microwave radios.

18 What channel spacings are employed in the 4-, 6- and 11-GHz common carrier bands. How many digital voice circuits are carried in each channel?

19 Describe the basic SONET configurations shown in Figure 10.9.

20 In SONET, what do you understand by self-healing?

21 List and describe six specific functions that SONET equipment may perform.

22 Comment on Figure 10.11.

23 What is dynamic nonhierarchical routing?

24 Describe the routing controls used in DNHR.

10.3 TELEPHONE TRAFFIC

Too much traffic-handling capacity and the users are paying for idle facilities; too little traffic-handling capacity and users cannot get through. Accordingly, it is helpful to understand the principles involved in traffic engineering.

10.3.1 TRAFFIC MEASUREMENTS

The number and duration of calls are measures of telephone traffic. Three quantities are defined

- **Traffic density**: the number of calls in progress at a given moment
- **Traffic intensity**: the traffic density averaged over a one-hour period
- **Holding time**: the time between the moment when the circuit is first seized (the user goes off-hook) and the moment when the circuit is released (the user goes on-hook) and equipment becomes available for use by others.

(1) Traffic Units

A measure of the telephone traffic flow produced by a call is the length of time that the call occupies a telephone circuit (the holding time). Thus, if n calls are made, and, on average, each call occupies h seconds, the total traffic flow is nh call seconds. Two traffic units are in common use

- **Erlang**: 1 erlang is the traffic due to a call that lasts for 1 hour; it is also the traffic due to 60 calls that last 1 minute each, or any combination of calls for which the sum of the holding times (in seconds) is 3,600.
- **ccs** [100 (centum) call seconds]: 1 ccs is the traffic due to a call that lasts 100 seconds, or any combination of calls for which the sum of the holding times (in seconds) is 100.

Because there are 3,600 seconds in 1 hour, it follows that 36 ccs are equal to 1 erlang. In addition, the traffic in erlangs is the traffic density averaged over 1 hour, or the traffic intensity.

(2) Variations in Traffic Density

In many businesses, telephones are used intensively during the working day. Usually, for firms whose operations are in a single time zone, the traffic density peaks during the morning and afternoon hours. For firms that span several time zones, the activity on some links will exhibit other peaks as the business activities in different locations across the country overlap. For instance, when it is 9 A.M. in San Francisco, it is noon in New York; and, when it is 5 P.M. in New York, it is 2 P.M. in San Francisco. If lunch is from noon to 1 P.M. in both cities, New York-based employees and their San Francisco counterparts are likely to call each other between 1 P.M. and 3 P.M., and 4 P.M. and 5 P.M., New York time. There will be two traffic peaks in the afternoon on the facilities linking New York and San Francisco.

(3) Call-Blocking

Traffic varies from minute to minute, hour to hour, day to day, and possibly from season to season. What is more, each message varies both with respect to the time it consumes and its destination. During periods of heavy traffic, users may have to compete for the network resources needed to service their calls. If elements in the network are unable to handle all of the call attempts immediately, some of them will be blocked. The state is defined as

- **Blocking**: the traffic condition when a call attempt cannot be completed immediately. Network resources are fully occupied serving other calls.

Call-blocking events are likely to occur at the local switch (where common equipment is shared) or within the network (where transmission paths are shared).

(4) Classes of Traffic

For the calling process we can write

$$Calls\ attempted = Calls\ in\ progress + Calls\ blocked$$

Multiplying each term by the average holding time of the calls in progress gives

$$Offered\ traffic = Carried\ traffic + Blocked\ traffic$$

where

- **Carried traffic** is the number of calls in progress in the network, times the average holding time. It represents the traffic that is carried to its destination.
- **Offered traffic** is the number of calls attempted, times the average holding time of the carried traffic. If all attempted calls are completed when placed,

this traffic is the same as carried traffic. If some attempted calls are blocked, this traffic is a measure of the demand for service

- **Blocked traffic** is the number of calls blocked, times the average holding time of the carried traffic. It represents the amount of traffic that cannot be carried to its destination on the first attempt because the equipment is fully occupied with other calls.

What happens to blocked traffic depends on the configuration of the facilities employed. The possibilities are that the blocked traffic becomes

- **Lost traffic**: some blocked calls are likely to be abandoned by the callers. When they are to points outside the local (free) calling area, they represent lost revenue to the operator
- **Retried traffic**: other blocked calls may be retried at a later time at the discretion of the callers. (In this case, revenue is not lost.)
- **Delayed traffic**: in a private network, some (or all) blocked calls may be delayed until facilities are available to handle them.

10.3.2 TRAFFIC MODEL

Figure 10.12 shows a model of the calling process that includes carried, delayed, retried, and lost traffic. Some points to be made are

- **Number of sources**:
  - if the number is very large and the messages are very short, the input traffic will be random
  - if the number is finite, the input traffic will not be random.
- **Number of service channels**:
  - *full availability*. A source has access to all service channels
  - *limited availability*. A source can connect with only a fraction of the service channels.
- **Method of handling blocked calls**:
  - *blocked-calls-cleared* (BCC). On receiving a busy signal that indicates all facilities are occupied and the call cannot be handled, the caller hangs up so that the call is lost to the system. The caller may retry over the network at a later time. Alternatively, the caller may place the call over another network.
  - *blocked-calls-delayed* (BCD). The call is held in the system until it can be processed. In many systems, the caller may be alerted to the delay by an automated announcement, kept occupied with recorded music, and updated from time to time on the estimated waiting time remaining.

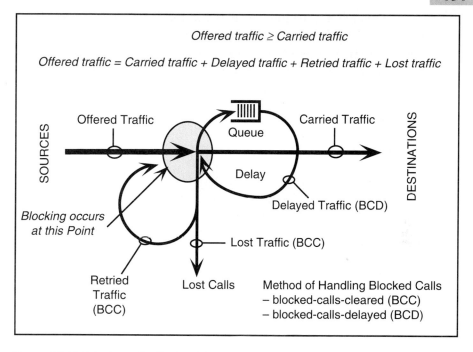

Figure 10.12 Telephone Traffic Model

The degradation in service caused by traffic competing for network elements can be measured by observing a large number of occurrences of several critical events that occur in placing and completing a call. Known as grade of service (GOS) criteria, the most important is the number of calls blocked. It is expressed in terms of

- **Blocking probability**: the probability of the first call attempt being rejected because of the unavailability of facilities.

A common value for blocking probability in a private network is P(blocking) ≤ 0.05—i.e. the probability of being blocked is less than or equal to 1 in 20 (5%) for each attempt. In public networks, P(blocking) is likely to be ≤ 0.01—i.e. the probability of being blocked is less than or equal to 1 in 100 (1%) for each attempt.

Other quantities can be measured to establish GOS; they include the times associated with the setup and teardown of calls.

- **Time to dial-tone**: the time between the moment the user goes off-hook, and the moment dial-tone is returned from the local switch. Common values of this parameter are μ (mean value) = 1 second, and 95% of all callers receive dial-tone within 3 seconds—i.e. P(0.95) = 3 seconds.

- **Connection time**: the time between the moment the last digit is entered and the calling circuit is completed
- **Call setup time**: the sum of the two times defined above, plus the time required entering the called number
- **Time to disconnect**: the time between the moment the user goes on-hook and the moment the circuit is available to others
- **Operating time**: the time to establish and to release a connection. It is the sum of call set-up time and time to disconnect.

These GOS parameters depend on the amount of common equipment at the originating and terminating switches. Performance is improved by adding more equipment so that each call attempt is less likely to be blocked and more likely to receive electronic attention as soon as the caller goes off-hook.

10.3.3 BUSY HOUR

The one-hour period in any day in which most use is made of the network (greatest traffic intensity) is known as the *busy* hour. For obvious reasons, it is important that a good grade of service is maintained during this period, and networks are engineered to cope with busy-hour traffic intensity. Consequently, at times other than the busy hour, the grade of service will be even better. However, because circumstances vary widely from day to day, there may be considerable variations in the daily busy-hour traffic intensities. To overcome this difficulty, we may choose to work with an artificial busy hour that represents the average of the busy hours observed over an extended cycle of business activity. Alternatively, we may use the busy hour that registers the highest traffic intensity during the period of observation.

(1) Which Busy Hour?
Busy hour is defined in three ways.

- **Daily busy hour**: on any particular day, the 60-minute period in which the traffic intensity on a group of telephone facilities is the highest
- **Average busy hour**: a 60-minute period in which the traffic intensity is the average of the traffic intensities observed in a series of daily busy hours
- **Peak busy hour**: that daily busy hour selected from a series of daily busy hours that exhibits the highest traffic intensity.

Choosing the peak busy hour as the basis on which to engineer network performance to achieve a given grade of service produces a GOS that is always equal to, or better than, the objective—but the network cost will be high. Choosing the average busy hour as the basis on which to engineer network performance to achieve a given grade of service produces a GOS that, on average, is equal to the objective—

at a lower network cost. Choosing the daily busy hour will produce a GOS that is equal to the GOS objective on the day the busy hour was observed, and on all other days with the same daily busy hour. On days on which the busy hour is different from the day observed, the GOS will be different from the GOS specified. Usually, as a practical matter, busy hour is taken to mean average busy hour.

(2) Example of Busy-Hour Traffic

In **Figure 10.13**, the number of calls in progress (measured by sampling a group of lines at 1-minute intervals) during a daily busy hour, is plotted. The

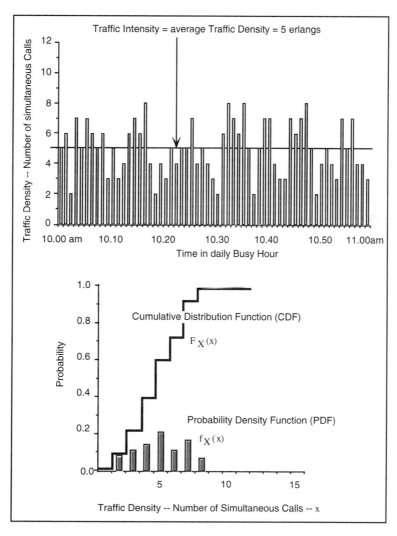

Figure 10.13 Example of Busy-Hour Traffic

number ranges between two and eight calls and the average number of calls in progress is five—i.e. the traffic intensity is 5 erlangs. We may envision this traffic intensity as equivalent to five circuits that are occupied throughout the busy hour.

While Figure 10.13 gives an impression of the traffic during busy hour, it is not easy to analyze. To do this we employ statistical techniques. The lower chart in Figure 10.13 shows the cumulative distribution function (CDF) and the probability density function (PDF) for this sample of traffic. The mean (or expectation) of these distributions is 5.12 erlangs, the variance is 2.23, and the variance-to-mean ratio (α) is 0.435. The distribution of holding times encountered in the sample daily busy hour is shown in **Figure 10.14**. With an average holding time of 1.5 minutes, the total number of calls placed in the busy hour is 200.

(3) Range of Traffic Distributions

Generally, traffic patterns can be characterized as smooth, peaked, or random.

- **Smooth traffic**: consider a telemarketing operation in which five callers place a continuous stream of calls that average 1.5 minutes each. In a typical hour, they generate 200 calls, and the traffic intensity of their operation is 5 erlangs. Because of the particular application, calling is regular and steady, and five lines are required to serve them. The mean traffic is 5 erlangs, the variance is zero, and the variance-to-mean ratio is zero. The traffic they generate is said to be *smooth*.

- **Peaked traffic**: next, consider a brokerage house where 50 investors are watching the variation in the prices of stocks. Suppose the market begins to seesaw so that they all want to place, buy, or sell orders simultaneously. To satisfy their demand, 50 lines will be required. If each conversation results in a holding time of 1.5 minutes, and the event occurs four times in 1 hour, at the end of the hour, the traffic intensity will be 5 erlangs, the value in the previous example, but the variance will be 225, and $\alpha = 45$. The traffic generated in this example is said to be *peaked* (some persons describe it as *rough*).

- **Random traffic**: finally, consider an office of several hundred persons who place 200 calls with an average holding time of 1.5 minutes in the busy hour. They will require more than five lines, but fewer than 50 lines. If their calling pattern is random, the traffic intensity will be 5 erlangs, the value in the previous examples, the variance will be 5, and $\alpha = 1$. (It can be handled with 11 lines with less than 1% blocking; see Figure 10.17.)

In these examples, the dispersion of the calls is significantly different. To put a mathematical definition to them, we employ the value of the ratio of the variance to the mean of these distributions

- **Smooth traffic**: $0 < \alpha < 1$
- **Random traffic**: $\alpha = 1$
- **Peaked traffic**: $1 < \alpha < \infty$.

Figure 10.14 Distribution of Holding Times for Busy-Hour Traffic

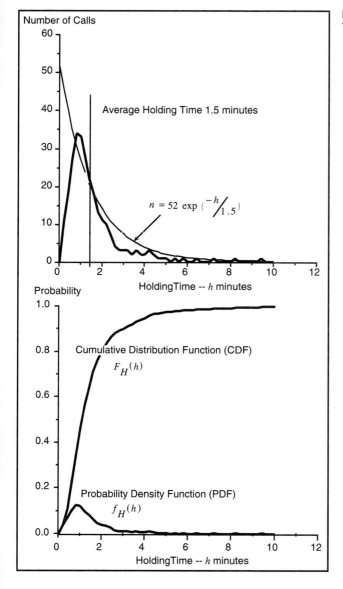

Diagrams of traffic intensity, and probability of number of simultaneous calls, for the situations described above are shown in **Figure 10.15**.

(4) Effect of Traffic Distribution on Blocking

For a traffic intensity of 5 erlangs, **Figure 10.16** illustrates the increase in the number of serving channels that is required to maintain a given level of blocking, as

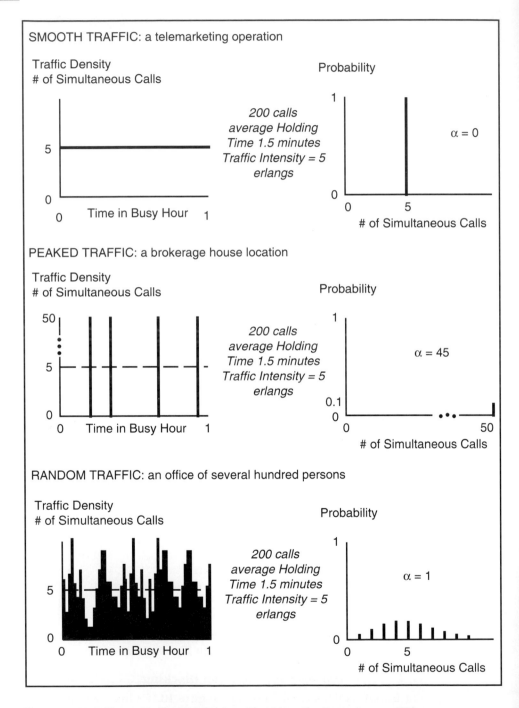

Figure 10.15 Different Traffic Distributions That Have the Same Average Value

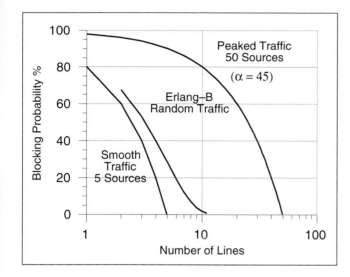

Figure 10.16 Effect of Traffic Distribution on Blocking Probability

the offered traffic becomes more peaked. The random traffic case is calculated using the Erlang-B formula—something that is discussed below [in Section 10.3.4(1)]. For the cases of smooth and peaked traffic, we calculate the levels of blocking produced by less than the full number of channels from the scenarios given above. Blocking of 20% is produced by 4 channels for very smooth traffic ($\alpha = 0$), by 6 channels for random traffic ($\alpha = 1$), and by 40 channels for very peaked traffic ($\alpha = 45$). Obviously, for any calculation of the number of channels required to serve a given situation, an accurate knowledge of the traffic distribution is essential.

10.3.4 TRAFFIC MANAGEMENT

Today's traffic engineering problems are solved by specialists using proprietary software who explore all possible alternatives and arrive at one or more near-optimum solutions. They employ analysis and/or simulation techniques and produce results that can guide business decisions. Generally, one of the strong points of their programs is that they can evaluate potential remedies, taking into account the spectrum of tariffs and arrangements that are available from an assortment of carriers. Faced with a real-life telecommunication problem, the responsible manager will do well to take advantage of these services. Understanding the scope and validity of their solutions may be easier if we examine a few simple problems without the aid of automation.

(1) Erlang's Formulas

In the early 20th century, the Danish mathematician A. K. Erlang developed basic traffic analysis tools. His formulas predict the performance of multi-channel communication systems under varying input loads (offered traffic). He assumed

- **Call requests** are Poisson-distributed; users may request a channel at any time
- **Service times** are exponentially distributed
- The **number of channels** is limited; all idle channels are available to every call request.

The Erlang B and Erlang C formulas[1] are applied to common trunking situations. Using tabulated values, *Erlang–B* (or Erlang Loss), and *Erlang–C* (or Erlang Delay) predict facility blocking for two policies—blocked-calls-cleared (BCC), and blocked-calls-delayed (BCD)—when a large number of sources generate relatively short messages at random. For more usual situations in which a small number of sources generate messages that are related to business events, the results should be used as a guide. Under these circumstances, computer simulations must be employed to obtain better information.

- **Erlang B Formula**: determines the probability of call blocking in a communication (trunking) system. Because the system provides no queueing capability, the blocked call is *lost* and must be *retried* at a later time. This strategy is known as *blocked-calls-cleared* (BCC) or *lost*-calls-cleared (LCC). In a steady state,
    - the probability of the next arriving call being blocked is

$$P[\text{blocking}] = \frac{A^C/C!}{\sum\limits_{k=0}^{C} A^k/k!}$$

    where A is the mean value of the offered traffic (in Erlangs) and C is the number of service channels in the trunking system
    - the probability of finding x calls in progress simultaneously is

$$P[x \text{ calls in progress}] = \frac{A^C/x!}{\sum\limits_{k=0}^{C} A^k/k!} = P[\text{blocking}] \, C!/x!$$

Under a BCC strategy, a certain number of calls will always be carried (C erlangs). Arriving calls that are blocked are immediately discarded and do not remain in the system. The probability of the next call being blocked is approximately 28% when the offered traffic is equal to the system capacity (C erlangs), and approximately 58% when the offered traffic is twice the system capacity (see Figure 10.17).

---

[1] For a modern derivation, see Theodore S. Rappaport, *Wireless Communications* (Upper Saddle River, NJ: Prentice Hall, 1996), 555–64.

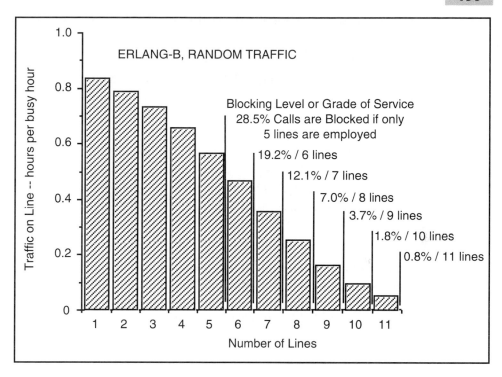

Figure 10.17 Occupancy of Lines Serving 5 Erlangs of Random Traffic
When 5 lines are available, a random traffic intensity of 5 Erlangs produces a blocking probability of 28.5%. If 10 lines are available, a random traffic intensity of 5 Erlangs produces a blocking probability of 1.8%.

- **Erlang C Formula**: determines the probability of call delay in a communication system that includes a queue to hold all call requests that cannot be assigned immediately to an idle channel. The strategy is known as *blocked call delayed* (BCD) or *lost call delayed* (LCD). In a steady state,
  - the probability of the next arriving call being delayed is

$$P[\text{call delayed}] = \frac{A^C}{A^C + C!(1 - A/C)\sum_{k=0}^{C-1} A^k/k!}$$

  - the probability that any call is delayed (in the queue) for more than t seconds is

$$P[\text{wait} > t] = P[\text{call delayed}]\exp\{-(C - A)t/H\}$$

  where H is the average duration of a call (holding time).

  – the average delay for all calls in the system is

$$D = P[\text{call delayed}]\,\{H/(C-A)\}$$

Under steady-state conditions, when the offered traffic is equal to the system capacity, the average delay for all calls becomes very large. With an 80% load, the probability of the average wait exceeding

- 0.3 holding time is 41%
- 1 holding time is 20%
- 3 holding times is 3%
- 10 holding times is 0.003%.

(2) Number of WATS Lines—Effect of Type of Traffic

A common telecommunications arrangement employs WATS to provide telephone connections [see Section 10.2.3(4)], and permits callers to overflow to the public network when all WATS lines are busy (i.e., BCC). The question arises—how many WATS lines are needed to minimize the total cost of telephone service?

Among other things, the answer depends on the traffic intensity, the areas called, the carriers used, and the current tariff structure. Suppose all calls are placed to an area located in a single WATS band, that there are many sources so that traffic may be considered to be random, and that the traffic intensity in the average busy hour is 5 erlangs. *Erlang–B* yields the chart shown in **Figure 10.17**. It depicts the way in which random traffic will occupy up to 11 lines that are selected in rotary fashion—i.e., each call attempt tries line 1, then line 2, etc., until it finds an unoccupied line that can be seized. With a traffic intensity of 5 erlangs from many sources, 28.5% of calls will overflow to the public network if five WATS lines are employed, and 19.2% of calls will overflow to the public network if six WATS lines are employed. If 10 WATS lines are employed, 1.8% of the calls will overflow to the public network.

As an alternative to the presentation of Figure 10.17, I include **Figure 10.18**. It provides a more general view of blocking levels as a function of the number of lines available for random offered traffic. For almost certain transmission (i.e., probability of blocking = 0.001, or one-tenth of 1%), 14 lines are required when the offered traffic is 5 erlangs.

But what if we have peaked traffic for which the variance-to-mean ratio is 4 and the traffic intensity is 5 erlangs? **Figure 10.19** shows a diagram for this case. In contrast to the random environment, 21.1% of the calls will overflow to the public network if 10 lines are employed, and 19 lines must be employed to reduce blocking to 2%. The message of this example is not the absolute number of lines that are likely to be required, but the fact that the nature of the traffic must be defined before the number of lines can be determined.

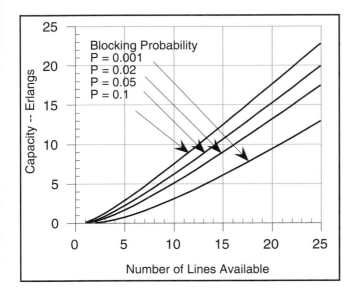

Figure 10.18 Traffic Capacity as a Function of Blocking Probability and Number of Lines Available

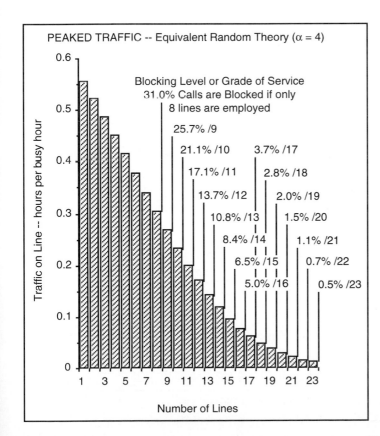

Figure 10.19 Occupancy of Lines Serving 5 Erlangs of Peaked Traffic

When 8 lines are available, a peaked traffic intensity of 5 Erlangs produces a blocking probability of 31.0%. If 20 lines are available, a peaked traffic intensity of 5 Erlangs produces a blocking probability of 1.5%.

Table 10.3 Variation of Traffic Density During Working Day

Hour in Work Day	1st	2nd	3rd	4th	5th	6th	7th	8th	9th	10th
Traffic Intensity—E	1	3	5	3	1	1	3	5	3	1

(3) Optimum Number of WATS Lines—BCC Policy

In a given BCC situation, the optimum number of WATS lines to use will be the number for which the cost of calls that overflow to the public network is less than the cost of an additional WATS line. To illustrate, assume the following:

- A cotton-brokerage calls various locations in California over the PSTN
- The calls are charged at the same rate (i.e., $/minute) throughout the day, and the average monthly cost is $25,000
- During the average business day, the traffic intensity varies as shown in **Table 10.3**.
- A WATS line to California costs $1,000 per month
- For traffic carried on individual lines tried in sequence, Erlang–B gives the values shown in **Table 10.4.**

First, find the monthly cost of communication to California *per daily erlang*

Total traffic per day is 4 x 1 + 4 x 3 + 2 x 5 = 26 erlangs.

Total cost of communication to California per month = $25,000

Therefore, the monthly cost per daily erlang is 25,000/26 = $961.54.

Then, for numbers of WATS lines, calculate the offered traffic blocked each day, the cost of the number of WATS lines employed, the cost of sending overflow traf-

Table 10.4 Erlang–B Values for Traffic Carried on Each Line for Three Levels of Offered Traffic

Offered Traffic —Erlangs	Traffic Carried on Each Line—Erlangs Line Numbers									
	1	2	3	4	5	6	7	8	9	10
1	0.500	0.300	0.137	0.047	0.016*					
3	0.750	0.662	0.550	0.420	0.288	0.174	0.091	0.041	0.024*	
5	0.833	0.788	0.730	0.657	0.567	0.465	0.357	0.252	0.162	0.095

* These values have been adjusted to make the total traffic carried 100%.

fic on DDD, and the total cost. These calculations are shown in **Table 10.5**. Under a BCC policy, the least cost is $8,407 per month. It is achieved with seven WATS lines. No caller experiences an appreciable delay.

(4) Outward WATS—BCD Policy

In **Figure 10.20** we use Erlang–C to calculate the mean delay encountered by all traffic when blocked calls are delayed. Erlang–C assumes that

- Blocked calls are delayed until they can be serviced
- There is unlimited queuing capacity

Figure 10.20 shows that 5 erlangs can be completely served by six lines with a mean delay for all messages of one average holding time. If the number of lines is increased to 10, the mean delay will be reduced to 20% of one average holding

Table 10.5 Cost of Traffic to California Using BCC Policy

Number of WATS Lines	Daily Carried Traffic (erlangs)	Daily Overflow Traffic (erlangs)	Cost of WATS Lines ($/month)	Cost of Overflow Lines ($/month)	TOTAL Cost ($month)
1	4x.500+4x.750+2x.833 = 6.666	26 – 6.666=19.334	1,000	18,590	19,590
2	6.666+4x.300+4x.662+2x.788 = 12.09	26 –12.09=13.91	2,000	13,375	15,375
3	12.09+4x.137+4x.550+2x.730 = 16.298	26 – 16.298=9.702	3,000	9,329	12,329
4	16.298+4x.047+4x.420+2x.657 = 19.48	26 – 19.48=6.52	4,000	6,269	10,269
5	19.48+4x.016+4x.288+2x.567 = 21.83	26 – 21.83=4.17	5,000	4,010	9,010
6	21.83+4x.174+2x.465 = 23.456	26 – 23.456=2.544	6,000	2,446	8,446
7	23.456+4x.091+2x.357 = 24.537	26 – 24.537=1.463	7,000	1,407	8,407
8	24.537+4x.041+2x.252 = 25.205	26 – 25.205=0.795	8,000	764	8,764
9	25.205+4x.024+2x.162 = 25.625	26 – 25.625=0.375	9,000	361	9,361
10	25.625+2x.095 = 25.815	26 – 25.815=0.185	10,000	178	10,178

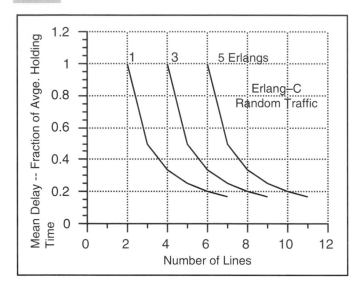

Figure 10.20 Mean Delay for 5 Erlangs Random Traffic as a Function of Lines Available When Using BCD Policy

time. Under a BCD policy, there can be no overflow to another network, so that the cost of telecommunication is limited to the cost of the WATS circuits employed.

By way of illustration, consider the cotton brokerage described above. Suppose a *blocked-calls-delayed* policy is used, and the delay encountered during the busy hour cannot exceed one average holding time. For the delay encountered by traffic that will wait as long as necessary for a circuit, Erlang–C gives the values shown in **Table 10.6**.

For 5 erlangs offered traffic (busy hour intensity), the number of WATS lines that will carry it with a mean delay of one average holding time is 6. They will cost $6,000 per month. Also, the table shows that, for seven WATS lines, the delay is half an average holding time, and for eight WATS lines, the delay is one-third an average holding time. Combining the results of the two calculations, the cost of communications can be reduced substantially.

- Using a **BCD** policy and
    - 6 WATS lines, the cost is $6,000 per month, and the average waiting time for service during busy hour will be one average holding time
    - 7 WATS lines, the cost is $7,000 per month, and the average waiting time for service during busy hour will be one-half an average holding time
    - 8 WATS lines, the cost is $8,000 per month, and the average waiting time for service during busy hour will be one-third an average holding time.
- Using a **BCC** policy and
    - 6 WATS lines, the cost is $8,446 per month, and there is no waiting time.

Table 10.6 Erlang–C Values of Waiting Time for Three Levels of Offered Traffic

Offered Traffic —Erlangs	Delay Encountered—Fraction of Average Holding Time Line Number									
	1	2	3	4	5	6	7	8	9	10
1		1.000	0.500	0.333	0.250	0.200				
3				1.000	0.500	0.333	0.250	0.200	0.167	
5						1.000	0.500	0.333	0.250	0.200

   – 7 WATS lines, the cost is $8,407 per month, and there is no waiting time

   – 8 WATS lines, the cost is $8,764 per month, and there is no waiting time.

The numbers give an indication of the cost of immediate service. If waiting for

- 1 average holding time during busy hour is deemed unacceptable, an extra $2,446 (i.e., 41%) per month will reduce the waiting time to zero
- 0.5 average holding time during busy hour is deemed unacceptable, an extra $1,407 (i.e., 20%) per month will reduce the waiting time to zero
- 0.33 average holding time during busy hour is deemed unacceptable, an extra $746 (i.e., 9%) per month will reduce the waiting time to zero.

Of course, real situations require real traffic distributions and real tariffs. These results are for demonstration only.

(5) Effect of Rate of Retries

Usually, callers to 800 and 888 numbers have no other way to complete their calls. If all lines are busy, they must hang up and retry. The rate at which they do this will depend on the enthusiasm they have for the function they wish to perform. If a viewer is anxious to obtain a bargain offered on a television shopping program, the recall rate may be very high. Less than one average holding time may elapse before retry. If a consumer is seeking weather or sports information, several holding times may elapse between trials. Finally, if a well-disposed person is responding to a radio or television appeal for funds, the caller may lose interest and not attempt to call again.

Retries cause blocking to increase significantly over the level that will prevail if there are no second attempts. Most inward WATS arrangements are provided for customer convenience, and many are an important link in selling a product. Therefore, it is important that the number of access lines be selected so as to achieve a blocking percentage commensurate with the likely rate of retries and the value of the service dispensed over them.

10.3.5 IMPLICATIONS FOR PRIVATE NETWORKS

Those who manage private networks employ several strategies to minimize the cost of using public facilities.

(1) Use of Tie Lines

Tie lines connect telephone facilities for the exclusive benefit of the lessor 24 hours a day. Spanning from point to point, they are used when the traffic pattern includes large volumes between two company locations, or between a company location and a specific local exchange area [when they are known as foreign exchange (FX) lines], or when immediate availability of a connection is important. Tie lines are not switched, and their cost depends only upon the mileage between the facilities they connect, the characteristics of the channels provided, and the types of termination arrangements. Because tie lines represent a fixed, usage-insensitive charge, they must be occupied for as long as possible so as to drive down the cost per call. If kept full, the cost per hour of calling on a tie line can be less than the same service over WATS, and both are cheaper than the same service over the DDD network.

(2) Automatic Route Advancement and Least-Cost Routing

Given a set of circumstances that results in routes served by combinations of tie lines, WATS lines and Direct-Distance-Dialing, attempting traffic should try tie lines first, WATS lines next, and then, if permitted, overflow to the public (DDD) network. The action of automatically changing from one category of service to another is known as *automatic route advancement* (ARA). Coupled with the requirement to start with the cheapest service and advance to the most expensive, the action is known as *least-cost routing* (LCR).

(3) Time-of-Day Routing for Incoming Traffic

Depending on the pattern of calls, it is possible that the busy hour in one section of a network occurs when another section is relatively quiet (because work has only just started, or is ending for the day). With a backbone of tie lines, incoming busy-hour attempts that cannot be handled at the target location can be directed to other locations. Called *time-of-day routing*, this technique shares the calling load across the entire network and reduces the need for each location to be equipped to handle its own busy hour demand exclusively. Airline reservation systems, and similar networks, employ this technique to even the load presented to reservation centers. Time-of-day routing is a time-sensitive case of alternate routing.

(4) Data Traffic Patterns

Because the traffic patterns for voice are different from those for data, carrying voice and low-speed data through a PBX may cause difficulties. During busy hours, PBXs are susceptible to blocking so that call attempts fail entirely (BCC policy) or the connection setup process is delayed (BCD policy). However, once a connection is established, no amount of additional traffic can impair its performance. Data services such as electronic mail are characterized by a large number of sources; they generate traffic that is not greatly different from voice traffic to which we can apply the Erlang–B formula. Other data services such as scientific timesharing are characterized by few sources that generate long-holding time traffic. To this situation we may apply the Erlang–Engset formula. Yet other data services such as batch transfers between mainframes may represent smooth traffic that can be queued; to these situations, we can apply the Erlang–C formula.

REVIEW QUESTIONS FOR SECTION 10.3

1 List and define three quantities related to the number and duration of calls that are used to measure telephone traffic.

2 What is an Erlang?

3 What is a ccs?

4 Give the logic for the following statement: 36 ccs are equal to 1 erlang.

5 With respect to telephone calls, define blocking.

6 Define the terms offered traffic, carried traffic, and blocked traffic. How are they related?

7 What happens to blocked traffic?

8 Explain Figure 10.12.

9 Define blocking probability. Give common values for this quantity in public and private networks.

10 List the quantities that can be measured to determine grade of service.

11 Define daily, average, and peak busy hours.

12 Describe smooth, peaked, and random traffic patterns.

13 Comment on Figures 10.15 and 10.16.

14 To calculate his formulas, what assumptions were made by Erlang?

15 In what sorts of situations is the Erlang B formula applied?

16 In what sorts of situations is the Erlang C formula applied?

17 Comment on the following statement: The optimum number of WATS lines will be the number for which the cost of calls that overflow to the public network is less than the cost of an additional WATS line.

18 What is the major tradeoff when a BCD policy is employed? Comment on the following statement: If waiting for one average holding time during busy hour is deemed unacceptable, an extra $2,446 (i.e., 41%) per month will reduce the waiting time to zero.

19 Comment on the following statement: it is important that the number of WATS lines be selected so as to achieve a blocking percentage commensurate with the likely rate of retries and the value of the service dispensed over them.

20 What strategy should be employed to minimize the costs of using tie lines?

21 Explain automatic route advancement and least-cost routing.

22 Explain time-of-day routing for incoming traffic.

10.4 INTEGRATED SERVICES DIGITAL NETWORK

After many years of study by ITU–T (and its predecessor organizations), many European telephone administrations are implementing all-digital networks.

- **Integrated Services Digital Network** (ISDN): switched, digital network that provides a range of voice, data, and image transport services through standard, multipurpose, user network interfaces based on 64 kbits/s clear channels.

In North America, ISDN deployment is moving more slowly. In part, this is due to the sophistication of the existing public networks and the availability of competing service providers. In part, it is because digital transmission and digital switching by themselves do not make a network that is attractive to sophisticated users. What does is the exploitation of the capabilities of common channel signaling in a software-driven, digital-switching, and transmission environment. Accordingly, North American carriers have concentrated on Intelligent Networks (INs).

At the risk of oversimplification, ISDN can be considered to be a provider-driven version of an all-digital network with powerful, software-based capabilities, and IN can be considered to be the market-driven version. In this section, I describe ISDNs and in Section 10.5, I describe INs.

10.4.1 REFERENCE CONFIGURATION

A formal user network interface has been defined for ISDN. It is shown in the top, left-hand corner of **Figure 10.21**. Each block represents the physical implementation of certain functions

- **Network termination 1** (NT1): the digital interface point between the network and the customer's equipment. It facilitates functions principally related to the physical and electrical termination of the network at the customer's premises.

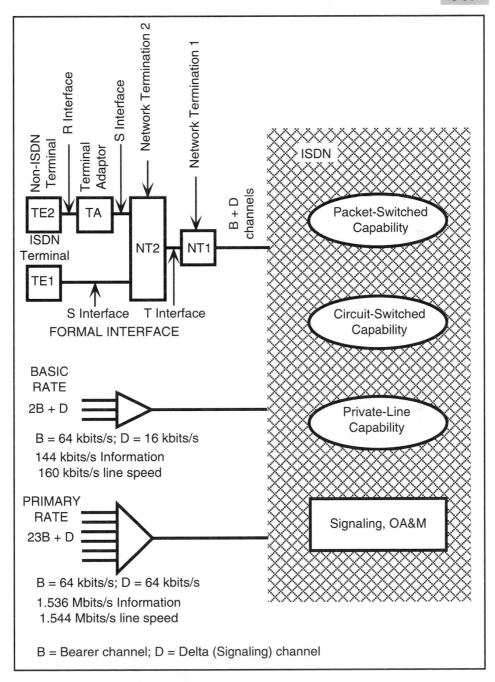

Figure 10.21 ISDN Reference Configuration

- **Network termination 2** (NT2): includes enhanced functions associated with the data link layer and the network layer. It facilitates protocol handling, switching, concentrating, and multiplexing functions associated with PBXs, cluster controllers, local networks, and multiplexers.

- **Subscriber terminal 1** (TE1): terminal that is fully compliant with ISDN requirements. Can connect directly to NT1. Includes digital telephone, data terminal, and similar equipment.

- **Subscriber terminal 2** (TE2): terminal that does not fully comply with ISDN requirements. Must connect to NT1 through a terminal adapter.

- **Terminal adapter** (TA): provides the interface and protocol conversions needed for TE2 to connect to NT1.

ISDN supports two interfaces that connect customer premises equipment to the network. Shown in the lower left-hand quadrant of Figure 10.21, they consist of bearer channels (B-channels) and delta channels (D-channels).

- **Primary rate interface** (PRI): 23 B-channels + 1 D-channel. In North America, the PRI is supplied over a single connection that operates at 1.544 Mbits/s. This is the standard T-1 rate; it contains 1.536 Mbits/s of user and signaling data, and 8 kbits/s for framing the data stream.

- **Basic rate interface** (BRI): 2 B-channels + 1 D-channel. In North America, the BRI is supplied over a single connection that operates at 160 kbits/s. This speed contains 144 kbits/s of user and signaling data, and 16 kbits/s for overhead functions.

These arrangements are service-independent, can carry voice, data, and image traffic simultaneously, and can support both switched and private line services. Moreover, data can be transferred in packet or in circuit mode.

10.4.2 ISDN Services

The services provided by ISDN cover a wide range of signals and speeds.

(1) Information Channels

The B-channel supports digital transmission at 64 kbits/s, is circuit-switched, and employs user-specified protocols. If desired, subrates of 2.4, 4.8, 8, 16, and 32 kbits/s may be multiplexed on to the channel. Other information channels, called H-channels (for higher) have been defined. They are H0, at 384 kbits/s; H11, at 1,536 kbits/s (T–1 user's information rate); and H12 at 1,920 kbits/s. The last channel is used on networks that employ ETSI standards where the first level (E–1) multiplexing rate is 2.048 Mbits/s.

(2) Signaling and Supervision Channel

The D-channel supports a packet-mode, ISDN protocol for call-control signaling, user-to-user signaling, and packet data. In addition, it is used for operations, administration, and maintenance messages. In the BRI, the speed of the D-channel is 16 kbits/s; in the PRI the speed is 64 kbits/s. When there is no signaling or network-related traffic, the entire channel can be used for customer data.

(3) Bearer Services

Information services available at the user network interface are called bearer services. They are defined by information transfer mode (circuit or packet), information transfer rate (bits/s), and information transfer capability (speech, audio, etc.).

- **Switched circuit mode**: services include
  - 64 kbits/s data,
  - 64 kbits/s PCM voice
  - 3.1 kHz audio band data.
- **Private line circuit mode**: switched circuit mode services and, in addition
  - 384 kbits/s data
  - 1,536 kbits/s data
  - 1,920 kbits/s in Europe
- **Packet-mode services**: employ X.25 protocol
  - B-channel supports 64 kbits/s synchronous data transfer
  - D-channel supports 16 kbits/s synchronous data transfer.

(4) Switched Digital Services

For applications that require point-to-point, limited time connections to one of several sites, carriers have introduced switched digital services

- **SW56**: 56 kbits/s switched digital service; this is switched DDS
- **SW64**: 64 kbits/s switched digital service; this is switched DS–0 or ISDN B channel
- **SW384**: 384 kbits/s switched digital service; this is ISDN H0
- **SW1536**: 1536 kbits/s switched digital service; this is ISDN H11.

ISDN-configured switches support SW64, SW384, and SW1536.

10.4.3 SIGNALING SYSTEM 7

ISDN uses Signaling System 7 (SS7). The network structure and protocol architecture of SS7 for public network applications are shown in **Figure 10.22.** The architecture is a quad structure consisting of signaling end-points (SEPs) and signal transfer points (STPs). The STPs are responsible for routing signaling messages from one SEP to another. To provide added reliability, each SEP is connected to the SS7 network by duplicate STPs.

(1) Signaling Network

The signaling network consists of nodes and links that connect Signaling End-Points (SEPs) by high-speed data links to a network of Signal Transfer Points (STPs). The SEPs can be Switching Points (SPs), Service Switching Points (SSPs), Service Control Points (SCPs), ISDN Service Nodes (SNs) or Operator Services Systems (OSSs). The major functions these entities perform are

- **Switching point** (SP): an end-office (or other) switch at which the customer gains access to the network. Using common channel signaling, the switch provides normal call-switching functions—establish, maintain, and disconnect.
- **Service switching point** (SSP): embedded in the SP, the SSP processes calls that require remote database translations. The SSP recognizes special call types, communicates with the Service Control Point (SCP), and handles the calls according to SCP instructions.
- **Service control point** (SCP): provides database and call-processing procedures for special calls
- **ISDN service node**: supports provision of ISDN bearer services and supplementary services (see below)
- **Operator services system**: provides operator assistance, particularly directory assistance services
- **Signal transfer point** (STP): performs as a link concentrator and message switcher to interconnect SEPs. STPs are stand-alone packet switches. For convenience, they may occupy the same site as other network equipment, including circuit-switching nodes
- **Integrated STP**: when a signaling point has an STP capability and also provides OSI transport-layer functions, it is said to have an integrated STP functionality
- **Stand-alone STP**: when a signaling point provides only STP capability, or STP and SCCP capabilities, it is commonly called a stand-alone STP.

With the exception of STPs, these entities are implemented in software at the switching nodes.

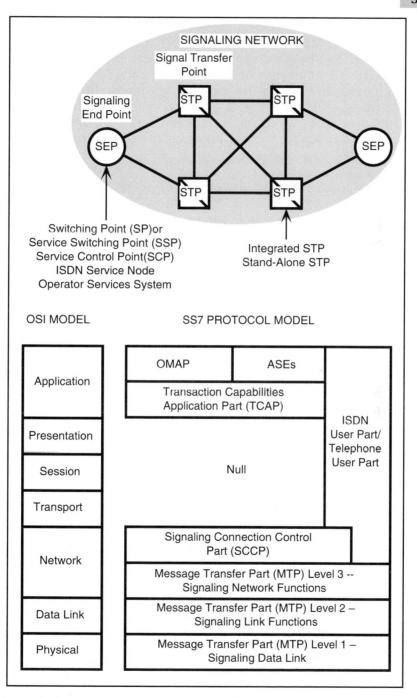

Figure 10.22 Network Structure and Protocol Architecture of Signaling System 7

(2) Protocol Layers

SS7 protocols consist of a Network Services Part (NSP) that corresponds to layers 1, 2, and 3 of the OSI model, a Transaction Capabilities Part (TCP), and an Operations, Administration, and Maintenance Part (OMAP). Together, the last two parts correspond to layer 7 of the OSI model. Because of the structure of the signaling environment (it is a homogeneous network), procedures that correspond to the presentation, session and transport layers of the OSI model are unnecessary—in the terms of the model maker, they are *null*. The remaining parts are described as follows

- **Message transfer part** (MTP): provides reliable transport of signaling information across the signaling network. This includes the capability to respond to network and system failures to ensure that reliable transfer is maintained. Briefly, the functions performed by the three levels are

  - *Signaling Data Link*. The physical layer, the signaling data link consists of two similar data channels operating in opposite directions. They provide full-duplex signaling at rates from 4.8 kbits/s to 56/64 kbits/s. In North America, all signaling links are 56 kbits/s; in many countries they are 64 kbits/s.

  - *Signaling Link*. The data link layer, the signaling link transfers signaling messages between two signaling points in variable length messages. Known as signal units (SUs), the formats of message signal units (MSUs), link status signal units (LSSUs) and fill-in signal units (FISUs) are shown in **Figure 10.23**. Error detection is accomplished through the use of a 16-bit CRC. Two forms of error correction are used. The basic method employs go-back-N techniques. A method known as preventive cyclic retransmission (PRC) uses forward error correction and is employed with satellites and other long propagation delay links. A signal unit error rate monitor (SUERM) is employed to monitor the error performance of the link and determine if it is of sufficient quality to remain in service. Flow control is accomplished through the use of LSSUs. The congested receiver notifies the sender of its condition with an LSSU that indicates busy. At the same time, it ceases to acknowledge incoming signal units. When there is no message traffic, FISUs are sent so as to sustain the error monitoring task and keep up to date with link performance.

  - *Signaling Network Functions*. The lower half of the network layer, the signaling network functions level performs signal unit handling and management tasks. They include routing and recovery from failures.

- **Signaling connection control part** (SCCP): the upper-half of the network layer, SCCP enhances the services of the MTP to provide full OSI network layer capabilities. In particular, SCCP extends MTP's addressing and routing capabilities and provides connectionless (CL-SCCP) and connection-oriented (CO-SCCP) classes of services.

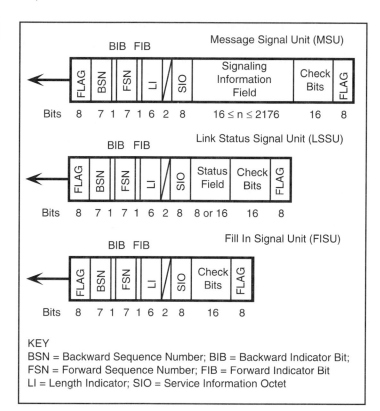

BIB FIB   Message Signal Unit (MSU)

| FLAG | BSN | FSN | LI | SIO | Signaling Information Field | Check Bits | FLAG |

Bits 8   7 1 7 1 6 2   8   16 ≤ n ≤ 2176   16   8

BIB FIB   Link Status Signal Unit (LSSU)

| FLAG | BSN | FSN | LI | SIO | Status Field | Check Bits | FLAG |

Bits 8   7 1 7 1 6 2   8   8 or 16   16   8

BIB FIB   Fill In Signal Unit (FISU)

| FLAG | BSN | FSN | LI | SIO | Check Bits | FLAG |

Bits 8   7 1 7 1 6 2   8   16   8

KEY
BSN = Backward Sequence Number; BIB = Backward Indicator Bit;
FSN = Forward Sequence Number; FIB = Forward Indicator Bit
LI = Length Indicator; SIO = Service Information Octet

Figure 10.23 Signaling System 7 Signal Unit Formats

- **Transaction capabilities part** (TCAP): provides a set of tools in a connectionless environment that can be used by an application at one node to invoke execution of a procedure at another node, and exchange the results. TCAP consists of two sublayers—the component sublayer and the transaction sublayer. The component sublayer is concerned with requests for action at the remote end, or the provision of data in response to such a request. The transaction sublayer is concerned with the exchange of messages that contain these components.

- **Operations, maintenance, and administration part** (OMAP): provides the application protocols (ASEs, Application Service Elements, see Figure 8.11) and procedures required to monitor, coordinate and control all the network resources needed to make communication based on SS7 possible.

- **ISDN user part/telephone user part** (UP/TUP): represents the location of model layers invoked when the network employs the transport capabilities of the MTP and SCCP to provide call-related services to the user. UP/TUP controls the circuit-switched connections between the originating and terminating stations. Flow control may be exercised at the UP level.

When a sender receives a busy LSSU from a congested receiver, the MTP passes the information to the UP/TUP. The UP reduces the traffic to the congested receiver in steps until a normal condition is restored. TUP represents an interim arrangement employed between carriers until ISDNUP is defined and stabilized.

REVIEW QUESTIONS FOR SECTION 10.4

1   Define ISDN.
2   Use Figure 10.21 to describe the physical functions implemented at ISDN UNI.
3   Distinguish between PRI and BRI.
4   What is a B-channel? What is a D-channel? What is an H-channel? At what speeds do they operate?
5   Describe the switched digital services offered by carriers.
6   Comment on the reasons for the network structure and protocol architecture of SS7 shown in Figure 10.22.
7   List and define the various types of signaling end-points used in SS7.
8   Justify the following statement: Because of the structure of the signaling environment, procedures that correspond to the presentation, session and transport layers of the OSI model are unnecessary.
9   What are the message transfer part, the signaling connection control part, and the transaction capabilities part of SS7 protocol?

10.5 INTELLIGENT NETWORKS

ISDN originated with European telephone administrations seeking to provide networks that could carry voice and data. For several years the concept was not well supported. What progress was made, was made in technology, and technology *push* not market *pull* was the dominant motivation. Eventually, the digital network described above was defined and is enjoying limited implementation. In the meantime, some countries permitted alternative providers to be established to assuage the demands of business users for data communication.

   In the United States, local exchange carriers, equipment vendors, enhanced services suppliers, and large users bypassed the pure ISDN development in favor of a network that uses standard interfaces to provide new services in a multivendor environment

- **Intelligent Network** (IN): based on ISDN, it
  - distributes call-processing capabilities across multiple network modules

   – employs reusable network capabilities to create a new feature or service
   – employs standard protocols across network interfaces.

10.5.1 IN ARCHITECTURE

IN is an evolving set of capabilities that permits the creation of custom services to suit specific customer needs. Mostly market-driven, IN is a major success for telecommunications consumers.

(1) Functional Components

   Unique to IN architecture, functional components (FCs) are elementary call-processing commands that direct internal resources to implement specific IN services. Combined in different ways, they create a range of network services and features. Some functional component categories are

* **Transfer control**: transfers control of service logic processing between IN modules
* **Connection control**: requests SSP (services switching point) to complete any permitted connection between users, or between users and network resources
* **Network participant interaction**: invokes specific resources in order to interact with users—e.g., speech synthesizer or DTMF digit receiver
* **Network information management**: on request, permits an SLI (service logic interpreter) to augment, modify, or remove information from other entities so as to change service limits or regime
* **Processing**: assigns resources to perform actions that satisfy IN service requests
* **Information collection**: enables collection of network usage data from network modules.

Building network services on these service-independent, functional components ensures that the network will not be dominated by the characteristics of a particular service.

(2) IN Modules

   Through the interaction of specially equipped switching systems (end-offices and/or local tandems) with systems specifically programmed to provide the logic (service logic) needed to support new services, network operators can introduce them in response to market demands.

   Figure 10.24 shows the physical and logical systems that constitute the modules of an intelligent network. The service logic—logic that defines the operation of

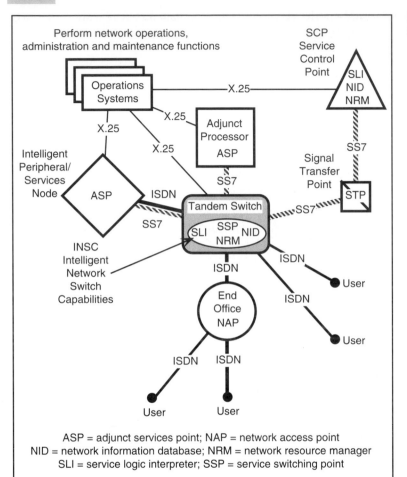

Figure 10.24 Intelligent Network Elements

Perform network operations, administration and maintenance functions

Operations Systems

SCP Service Control Point

SLI NID NRM

X.25

X.25

X.25

Adjunct Processor ASP

X.25

SS7

Intelligent Peripheral/ Services Node

ASP

ISDN

Signal Transfer Point

SS7

STP

Tandem Switch

SS7

SS7

SLI SSP NID NRM

INSC Intelligent Network Switch Capabilities

SS7

ISDN

ISDN

User

End Office NAP

ISDN

User

ISDN ISDN

User User

ASP = adjunct services point; NAP = network access point
NID = network information database; NRM = network resource manager
SLI = service logic interpreter; SSP = service switching point

specific network features or service applications—is contained in service logic programs (SLPs) that reside at service control points (SCPs), adjunct processors (APs), intelligent peripherals (IPs), and service nodes (SNs), or within the network switches themselves (SSPs). In more detail, the important elements of Figure 10.24 are

- **Service control point** (SCP): consists of processor-based service logic interpreter (SLI), network information databases (NIDs), signaling network interface, and network resource manager (NRM)
    - *service logic interpreter* (SLI): executes SLPs and handles exchanges between IN modules
    - *network information database* (NID): contains customer and network information

> – *network resource manager* (NRM): determines which IN module provides resources to continue call processing. May give routing instructions to establish connection to IN module.

Fully interconnected to IN switches through STPs, SCPs support distributed (network-wide) services. They provide database information and processing procedures for IN calls, and invoke service logic programs in response to messages from IN switching systems. For services such as 800 calling, SCPs may be service-specific—i.e., they support only 800 calling.

- **Signal transfer point** (STP): a packet switch that provides routing for signaling messages between IN-equipped switches and SCPs
- **Intelligent network switch capabilities** (INSC): identify calls associated with intelligent network services, formulate requests for call-processing instructions from service logic elements, and implement instructions. Provide IN services to subtending users and lower-level switches equipped as network access points.

These capabilities are supported by SLI, NID, and NRM systems that perform the operations described above, and by service switching points (SSPs) that contain reference call models. They permit recognition of IN call-handling requirements at trigger points, query IN modules for processing instructions, and implement them. SSP software works with modules in SCPs, SNs, APs, and IPs to provide advanced call-processing capabilities within the switch.

Intelligent network switch capabilities are deployed at access tandems and selected end-offices. (In Figure 10.24, we show deployment at a tandem switch.) As the demand for specific services grows, control migrates from the associated SCP to the logic modules in the IN equipped switch.

- **Network access point** (NAP): switches (such as the end-office in Figure 10.24) not equipped with IN switching capabilities may be equipped as network access points. NAP capabilities permit access to a limited set of IN services by users served by the switches. The NAP routes the call to an IN switching system, identifies the user, and requests IN treatment.
- **Adjunct processor** (AP): contains an adjunct service point (ASP) that responds to requests for service processing. Directly connected to an IN switching system (to assure timely communication), the adjunct processor supports services that require rapid response to users' actions. Examples are services that control provision of dial-tone, or that may require rapid reconfiguration of call connections.
- **Intelligent peripheral** (IP): contains an adjunct service point (ASP) that responds to requests for service processing. Provides and manages resources such as speech recognition, voice synthesis, announcements, and digit collection equipment as required by IN service logic.

- **Service node** (SN): provides service logic and implementing resources in a single network entity. Communicates directly with a single IN-equipped switching system using ISDN basic- or primary-rate interfaces. Contains adjunct service point (ASP) that responds to requests for service processing. Provides access to service logic programs that support a specific service, or group of services, or a full range of service functions (including attendant services). Allows network interaction with user, including collecting dialed digits, receiving spoken inputs, and providing customized announcements.

- **Operations systems**: perform network operations functions such as resource administration, surveillance, testing, traffic management, and data collection.

- **Network interfaces**: using conventional analog or ISDN channels for access, they support standard protocols:

  - *SS7 Transaction Capabilities Application Part* (TCAP). The application layer protocol used between switching systems and SCPs or adjunct processors

  - *ISDN interfaces*. Used between switching systems and intelligent peripherals or service nodes

  - *X.25-based protocols*. Used for interfaces between network components and operations systems.

IN calls are identified at the originating SSP by means of a trigger table. When a quantity associated with the call matches a quantity carried by the table, information in the table is used to generate a request for instructions from the SLI designated by the NRM (the SLI may be in the switch or in the SCP). Normal call processing is suspended until the SSP receives the reply. The SLI will respond with instructions for the SSP to execute a series of FCs that provide the service requested. IN architecture is designed to be flexible. Because the same FCs can be performed in different modules, configurations can be customized to suit traffic and performance requirements.

10.5.2 IN Services

IN services are provided by manipulating connections between nodes and addressable entities—such as other nodes or a user's station—anywhere in the network. For instance, three-way calling, a service available among extensions served by a PBX or Centrex, can be extended to remote stations using IN capabilities. During a conversation between party A in New York and party B in Chicago, suppose they wish to consult party C in Atlanta. The major steps to create the three-way call are as follows

- **Three-way calling**: a flashhook signal from A triggers the SSP serving A to request instructions from the assigned SLI for a three-way connection. As a result

- B is split from A temporarily and placed on hold
- A receives dial-tone and enters the directory number for C
- the SCP provides routing information so that A can be joined to C
- when A has established contact with C, A gives a second flashhook signal
- the serving SSP joins B to A and C to produce a three-way talking state.

When A and B have consulted with C, A gives another flashhook signal, and the connection reverts to a two-way talking state between A and B.

Other examples of IN services are

- **800/888-database service**: provides toll-free service to the caller (the 800/888 service subscriber purchases service by serving area bands and is billed for all costs incurred). The associated SCP has information on how to handle the dialed 800/888 number; it can vary by time of day, day of week, originating number, designated IXC, and other items. Both LECs (intra-LATA) and IXCs (inter-LATA) offer 800/888-database service.

- **Network ACD**: enables subscribers to control the destination to which 1–800 calls are routed to match peak loads with availability of attendants and provide relief for overflow from congested automatic call distributers (ACDs). The associated SCP maintains knowledge of the idle/busy status of a stored list of terminations and directs overflow traffic to a network address on the basis of instructions provided by the subscriber.

- **Area-wide Centrex**: based on a SCP that is connected to all switches within a LATA, provides PBX-like operations from designated stations across the LATA.

- **Custom local-area signaling services** (CLASS, also known as TouchStar): a group of switch-controlled, common channel signaling features that use existing customer lines to provide end users with call management capabilities. Based on the ability to know the calling party's number (CPN) from messages on the signaling channel, they include
  - *automatic call back* (also known as call return). Allows the customer to automatically call the last incoming caller.
  - *selective call forwarding* (also known as preferred call forwarding). Allows a customer to forward calls to another station based on the CPN.
  - *automatic recall* (also known as repeat dialing). Allows the customer to recall the last party called.
  - *distinctive ringing* (also known as call selector). Allows a customer to assign a distinctive ringing pattern to incoming calls on the basis of the CPN.
  - *selective call rejection* (also known as call block). Allows a customer to reject calls on the basis of the CPN.
  - *customer-originated trace* (also known as call tracing). Allows the terminating party to request a trace of the last call received.

- *calling number delivery* (also known as caller ID). Allows specially equipped terminating station to display CPN. The number is sent during ringing.
- *calling name delivery*. Allows specially equipped terminating station to display the directory listing for the CPN. Directory information is sent during ringing.
- *bulk calling line identification* (also known as call tracking). Allows PBX and Centrex customers to receive call-related information on incoming calls.
- *voice messaging*. Enables a subscriber to route incoming calls to a remote voice storage system.
- *call waiting*. Provides an audible alert to a party that another caller is seeking contact. By using the hookswitch, the terminating party can place the original party on hold and speak to the interrupting caller. A second hookswitch signal restores the original connection.
- *call waiting identification*. Presents the directory listing (and number) when the call waiting alert is given.

- **Alternate billing service**: through the use of a line information database, enables the calling party to bill a call to a number other than the called number (i.e., calling card or third-party number).
- **Mobile communication services**: IN can be extended to mobile cellular communications, paging, and other personal communications systems. Using SS7 for call control and database transactions, services such as selective call forwarding can be provided to these subscribers (see Figure 11.15).

Through the use of functional components and the exploitation of common channel signaling, these services will evolve as social and business environments change.

10.5.3 INTERNATIONAL STANDARDS

Regional IN activity in Australia, Japan, North America, and Europe has resulted in ITU–T efforts directed to international IN standards. The objectives are to develop an architecture that, among other things, will

- Be applicable to all telecommunications networks (i.e., PSTNs, narrowband-ISDNs, broadband-ISDNs, packet-switched data networks, and mobile networks)
- Enable service providers to define services independent of service-specific developments by equipment suppliers
- Evolve to reflect implementation experiences, new technological opportunities, and market evolution.

(1) Conceptual Model

Figure 10.25 shows the IN Conceptual Model developed by ITU–T working groups. The four planes are

- **Service plane**: describes IN services and service features (SFs) from the user's perspective. Does not describe implementation. May include management services.

- **Global functional plane**: describes IN functions from a service design perspective that treats the network as a single entity. Does not describe distribution of IN functions in the network. Contains call-processing model and service independent building blocks (SIBs). SIBs are combined with service logic (SL) to produce services in the service plane.

- **Distributed functional plane**: describes IN-distributed functions from a network design perspective. Describes the IN network in terms of functional entities (FEs, i.e., units of network functionality) and the relationships (i.e., the information that flows) between them. Does not describe how the functionality is implemented and deployed. Sequences of functional entity actions and associated information flows are used to create SIBs in the global functional plane.

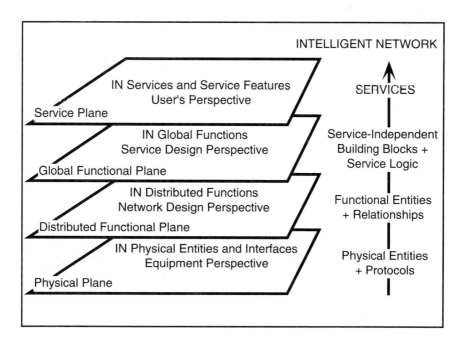

Figure 10.25 ITU–T's Conceptual Model of Intelligent Network

- **Physical plane**: describes the types of physical entities, the functions they implement, and the protocols by which they communicate. The combination of physical entities and protocols are used to create functional entities and relationships in the distributed functional plane.

(2) IN Capability Set 1

In IN Capability Set 1 (CS–1), ITU–T addresses the first stage of IN evolution. CS–1 includes services that are of high commercial value and can be implemented without significant impact on existing installations. They are Type A services; that is, they are single-user, single-ended, single point-of-control, and single-bearer services. Invoked by a single user, they carry no requirement for end-to-end messaging or control and are provided over the same medium as the communication. Some examples are call forwarding, 1–800 calling, abbreviated dialing, customized announcement, custom ringing, virtual private networks (VPNs), universal personal telecommunications (UPT), etc. VPN services provide a subscriber with private network capabilities (such as private numbering plan and dialing restrictions) over the PSTN. UPT services provide a subscriber with a unique personal number that can be used across any number of networks and with various access arrangements.

Type B services—that is services that can be invoked at any point during the call and may require end-to-end messaging and control—will follow in other Capability Sets.

(3) Distributed Functional Plane

Figure 10.26 shows the architecture of the distributed functional plane in CS–1. It consists of the following functional entities

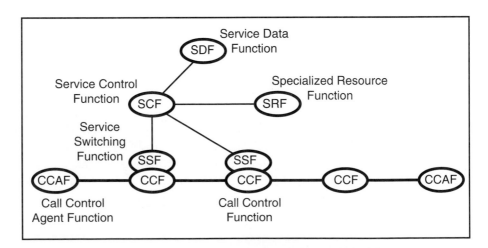

Figure 10.26 Architecture of ITU–T's CS–1 Distributed Functional Plane

- **Call control agent function** (CCAF): provides end-user access to IN call processing over analog lines, ISDN BRI and PRI, analog multifrequency signaling circuits, and SS7 interfaces.

- **Call control function** (CCF): provides call connection, maintenance, and termination.

- **Service switching function** (SSF): recognizes calls requiring IN features. Interacts with call processing and service logic to provide the services requested.

- **Service control function** (SCF): contains service logic that controls the implementation of IN services. Interfaces with service switching function, specialized resource function, and service data function.

- **Specialized resource function** (SRF): controls resources used by end-user to access network (e.g., DTMF receivers, protocol conversion, announcements, etc.).

- **Service data function** (SDF): provides access to service-related data.

For a call that requests IN service, the process can be described as follows. When the CCAF receives call activity from an end-user (DTMF digits, for instance), it passes it to the CCF/SSF for processing. If they constitute a request for IN service, SSF passes the request to an SCF. The SCF invokes the appropriate service logic and interacts with SSF, SRF, and SDF as necessary to provide the service requested by the end-user.

(4) Physical Plane

Figure 10.27 shows the architecture of the physical plane in CS–1. It consists of physical entities (PEs) that implement functional entities (FEs). Allowing for the difference in nomenclature, it is similar to Figure 10.24.

REVIEW QUESTIONS FOR SECTION 10.5

1. What is Intelligent Network?
2. Define a functional component. What functions do they perform?
3. Comment on the following statement: Building network services on service-independent functional components ensures that the network will not be dominated by the characteristics of a particular service.
4. Describe the elements of IN shown in Figure 10.24.
5. What is a trigger table? What does it do?
6. List and describe some of the services offered by INs.
7. Describe ITU–T's conceptual model of IN. Define the four planes.
8. Explain Figure 10.26.
9. Explain Figure 10.27.

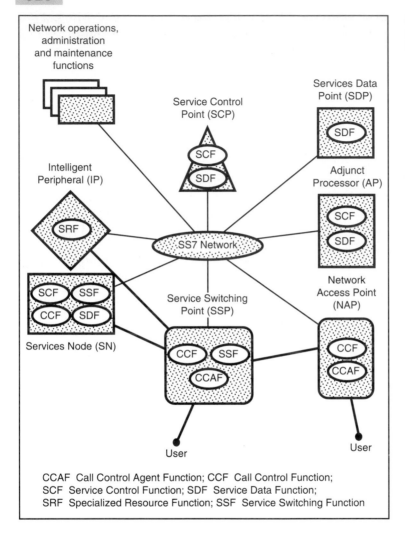

Figure 10.27 Architecture of ITU–T's CS–1 Physical Plane

CCAF Call Control Agent Function; CCF Call Control Function;
SCF Service Control Function; SDF Service Data Function;
SRF Specialized Resource Function; SSF Service Switching Function

10.6 BROADBAND ISDN

Based on a 64-kbits/s channel speed, with customer access up to the primary rate (1.544 Mbits/s), ISDN is a narrowband network. It cannot meet the requirements of a future information age for broadband services—for switched video, high-resolution graphics, television distribution, and highways between supercomputers.

Figure 10.28 shows a wide range of communication services arrayed according to channel speed, and the fraction of time that they are likely to occupy the channel (normalized holding time). Thus, television and music generate continuous signals that fill a channel for an extended period of time. On the other hand, data and voice services occupy the channel less intensely, and can share channels. ISDN covers the segment that extends to channel speeds equal to the speed of the PRI, and to channel occupancies of less than 50%.

10.6.1 BROADBAND SERVICES

ITU–T divides broadband ISDN services into two major categories

- **Interactive broadband services**: those in which there is an exchange of video and/or high-speed data information at rates between 64 kbits/s and 150 Mbits/s. They are divided into
  - *conversational services*. They provide the means for real-time, broadband dialogue between two users, or a user and a service provider. Important services in this category are video telephony and videoconference.
  - *messaging services*. They provide user-to-user, store-and-forward, broadband communication. An important service in this category may be video mail—i.e., storing a video message until the recipient requests to see it.
  - *retrieval services*. They provide the ability to retrieve broadband information from remote centers. An important service in this category may be broadband videotex; on request, the user is able to receive a mixture of sound, video and data that describes a topic of interest.

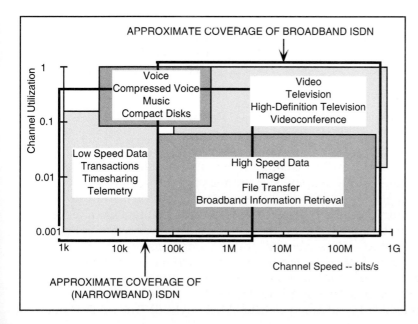

Figure 10.28 Channel Speed and Channel Utilization for a Range of Communication Services

- **Distributed broadband services**: those in which the information transfer is one way, from distributor to customer at rates up to 600 Mbits/s. They are further divided into

  - *broadcast services*. They distribute broadband information without individual presentation control. A common example is broadcast television.

  - *custom services*. They distribute broadband information with individual presentation control. A common example is the electronic delivery of movies, or similar items, on request, such as pay-per-view television service.

To satisfy the communication needs of the future, a higher-speed network is needed. Called broadband ISDN (B–ISDN), it covers the segment of Figure 10.28 that extends from channel speeds of 64 kbits/s to 600 Mbits/s and channel occupancies up to 100%. B–ISDN will distribute information to subscribers at rates up to 622 Mbits/s (STM–4), and support interactive, switched services at 155 Mbits/s (STM–1). Not intended to replace (narrowband) ISDN, B–ISDN supplements and extends the range of services that can be offered.

Figure 10.29 shows basic user interfaces and network access arrangements for ISDN and B–ISDN. On the subscriber side, the maximum digital rate for B–ISDN is at least 100 times the digital rate for ISDN, making the use of optical fiber necessary. On the network side, for transport, B–ISDN will use optical fibers or digital microwave radios that employ SONET and SDH multiplexing. **Figure 10.30** shows the formal interfaces for equipment that provides B–ISDN services. Except for the prefix B– for broadband, they are the same as ISDN.

For switching, B–ISDN will employ ATM (see Section 9.4.3). While it is relatively simple to see how this is done for voice and data traffic, sending video coded as MPEG [see Section 3.5.3(4)], is more difficult. Yet it must be done if B–ISDN is to distribute video programming to customers. Also, there must be a reliable technique for duplicating video signals [copy function, see Section 9.4.3(3)] so that the same program can be distributed to many customers.

10.6.2 VIDEO DISTRIBUTION FOR BROADBAND RESIDENTIAL SERVICES

Because the individual P, B, and I frames of an MPEG signal [see Section 3.5.3(3)] are not equally complex, MPEG coding produces a variable-bit-rate (VBR) datastream. In applications that employ synchronous transmission (e.g., over ADSL, T–1 and SONET/SDH), it is converted to a constant bit rate (CBR) stream through the use of a buffer. To prevent overflow or exhaustion of the buffer, feedback is provided to vary the quantizing level of the codec (short for coder and decoder). Thus, CBR is achieved at the expense of varying video quality.

MPEG streams are divided into 188-octet *transport* packets. In the ATM adaptation layer (AAL 5), several packets ($2 \leq n < 12$) may be grouped into a single CSPDU (Convergence Sublayer PDU, see Figure 9.20) before segmentation into 53-octet cells (5-octet ATM header and 48-octet ATM payload). The result is bursts of

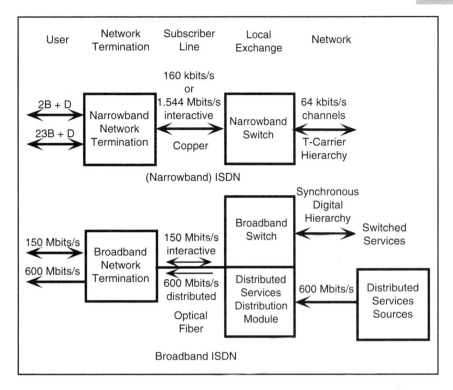

Figure 10 .29 Comparison of ISDN and B–ISDN at User, Network Termination, Subscriber Line, Local Exchange, and Network Interfaces

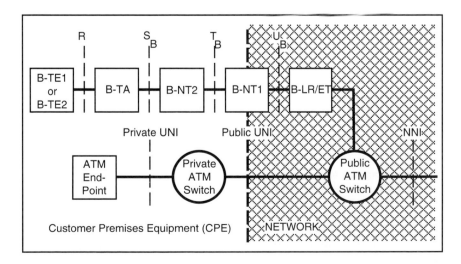

Figure 10.30 B–ISDN Reference Configuration

from 8 to as many as 50 cells per CSPDU. (An I frame may consist of several hundred cells.) At the receiver, the signals are reconstructed to give a VBR MPEG datastream.

10.6.3 B–ISDN Protocol Reference Model

B–ISDN protocols can be represented in the fashion of the model shown in **Figure 10.31**. It is divided into planes.

- **User plane** (U plane): facilitates the transfer of user application data (messages) to and from the physical layer. The U plane contains
  - physical layer
  - ATM layer
  - ATM adaptation layer
  - higher layers, if required.
- **Control plane** (C plane): facilitates switched services by implementing call setup and other connection control functions. The C plane contains
  - physical layer
  - ATM layer
  - ATM adaptation layer
  - higher-layer signaling protocols.
- **Management plane** (M plane): facilitates the exchange of data between U and C planes and provides other management functions. The M plane is divided into two sections:
  - *layer management*. Performs specialized layer management tasks
  - *plane management*. Performs management and coordination of complete system

ATM bearer services include the definition of a connection-oriented cell transfer service that maintains cells in sequence between the sender and the receiver. In addition, the sender may request a specific performance level (number of lost cells, limit delay). The physical layer is defined for rates between DS–3 and STM–4 (44.736 and 622.08 Mbits/s). It includes two sublayers

- **Transmission convergence sublayer**: implements physical layer functions that are independent of the transmission medium (e.g., containers, sequence numbers)
- **Physical medium dependent sublayer**: implements physical layer functions that depend on the medium selected (e.g., timing, coding).

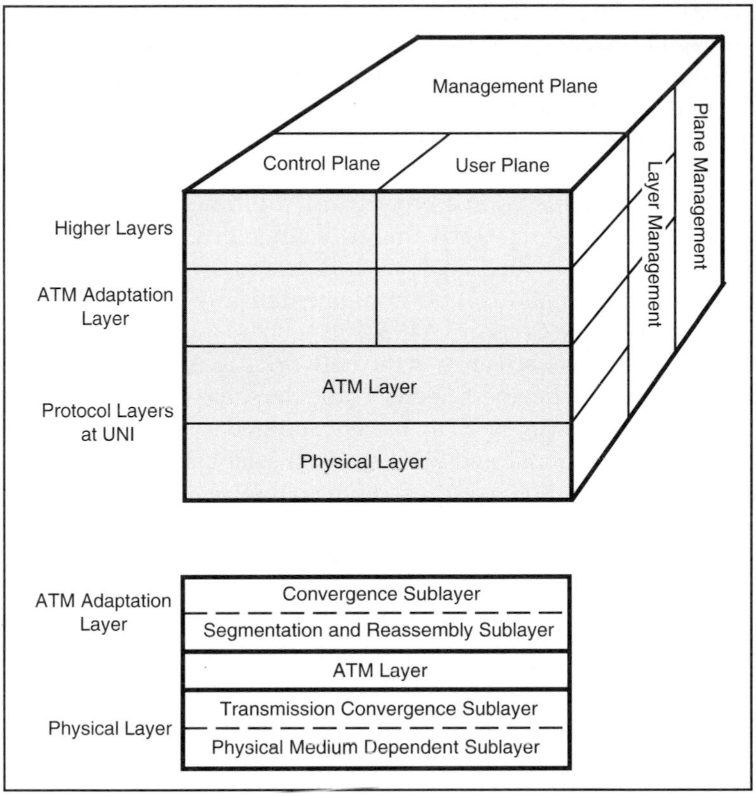

Figure 10.31 B–ISDN Protocol Reference Model

So as to guide the interoperability of different systems, ITU–T has developed a series of standards for the rapid development of global ATM connections that may employ the networks of several administrations.

10.6.4 PROMISE OF B–ISDN

In nations whose governments provide and control electronic communication, B–ISDN is a natural extension of the PTT's role. In the United States, it remains to be seen whether B–ISDN will become reality. Broadband services are already provided to households by independent television broadcasters, cable television, and satellite system operators. Will carriers and cable system operators form alliances to transport data, voice, and television services to homes? Will they build what some are calling an information highway to every residence? Will they be part of the national information infrastructure (NII, see Section 14.3)? Only time will tell—but proposals for wired cities and a wired nation have been around since the late-60s without spectacular results.[2]

In order to increase its capability, efforts are being made to integrate IN functions with B–ISDN to form an intelligent broadband network (IBN). A further development is the organization of a worldwide consortium to develop Telecommunications Information Networking Architecture (TINA).[3] TINA is an *open* architecture for services to be delivered in a broadband, multimedia, information super-highway environment. TINA takes advantage of advanced distributed software concepts, including object-oriented analysis and design, to create networks that provide multimedia services and can be implemented in a wide variety of technologies. TINA is described in Section 12.10.2.

Over the years, virtually every advance in the network capabilities of the carriers has been made in response to business needs. Later, the same capabilities have been adapted to residential uses. What B–ISDN has encouraged is the development of new, high-speed technology. SONET and SDH are well established, and ATM is a viable, high-speed switching technology. Ways in which they are used in the public and private networks that support commerce and industry are described in Sections 9.4 and 13.3.4. It is here that immediate revenues can be generated, and the ideas for an information superhighway can be put to the test. When fully worked out, they will be extended to residences. Like ISDN before it, commercial interests that are anxious to be first to market with new technical capabilities have overtaken B–ISDN. Further, in Sections 11.1.2 and 11.1.3, constellations of satellites are described that have the potential of *overbuilding* global B–ISDN.

REVIEW QUESTIONS FOR SECTION 10.6

1 Describe the wide range of communication services that B–ISDN is targeted to provide.

2 Distinguish between interactive broadband services and distributed broadband services.

3 Describe Figure 10.29.

4 Use Figure 10.30 to compare ISDN and B–ISDN.

5 Describe the components of Figure 10.31.

6 Comment on the follwoing: In countries where governments provide and control electronic communication, B–ISDN is a natural extension of the PTT's role. In the United States, it remains to be seen whether B–ISDN will become reality.

7 Describe the features of the B–ISDN protocol model shown in Figure 10.31.

[2] *On the Cable: The Television of Abundance*, Report of the Sloan Commission on Cable Communications (New York: McGraw-Hill, 1971).

[3] www.tinac.com

11

WIRELESS NETWORKS

In 1901, using the wireless telegraph, Guglielmo Marconi (an Italian physicist, 1874–1937) demonstrated radio communication across the Atlantic Ocean. Shortly thereafter, his invention was being employed for ship-to-ship and ship-to-shore communication. Using Morse code, it provided a slow-speed data channel that linked ocean liners to fixed installations on the mainland, or to other ships, and laid the foundation for today's wireless networks. Paced by improvements in portable telephones and personal computers, these networks provide pervasive connections to a wide variety of mobiles (mobile stations). In this chapter I describe some of the ways in which those who would create information anywhere and use it everywhere, without delay, are using radio.

11.1 COMMUNICATION SATELLITE NETWORKS

Because they made it possible for persons to communicate without regard to geographical barriers (such as mountains and oceans), the invention of communication satellites caused people to describe the world as a *global village*, in which inhabitants can communicate readily, and talk of *spaceship Earth*, on which inhabitants form a communicating crew. (The terms are attributed to Buckminster Fuller, a U.S. philosopher and mathematician, who lived from 1895 to 1983.) From geostationary satellites in synchronous orbit 22,300 miles (approximately 36,000 km) above Earth's surface, satellite carriers were able to provide point-to-point and point-to-multipoint voice, video, and data services to large fixed Earth stations—services that caused the satellites to be described as *repeaters* in the sky.

Since then, the growth in demand for sophisticated personal communication services has caused satellites to move closer to Earth. In lower orbits, transmission delay is reduced (see Figure 5.17) so that data throughput is increased. However, to provide continuous coverage of Earth's surface and uninterrupted communication with a wide range of fixed and mobile stations, the move has required the use of *constellations* of satellites.

To cope with the handoffs needed to route messages from satellite to satellite as the constellation moves over Earth-bound customers, a significant level of on-board processing is required. In contrast to earlier geostationary models, these satellite constellations may be described as *switching networks* in the sky. In rural areas, they could eliminate the need for terrestrial cell sites. In city cores, reception may be more difficult. Based on Figure 5.16, **Figure 11.1** shows the relative height of geostationary, medium, and low-Earth orbit (GEO, MEO, and LEO) constellations of satellites, and the number of revolutions per day required to keep them in orbit.

11.1.1 GEO SATELLITE SYSTEMS

First-generation systems consisted of large, high-power, C-band satellites parked in geostationary orbit that illuminated approximately one-third of Earth's surface. They gave rise to international consortiums that saw the advantage of alternative communication channels to span oceans and continents. Later the capability was employed to provide commercial services.

(1) International Telecommunications Satellite Consortium

INTELSAT (International Telecommunications Satellite Consortium), a consortium of national communications satellite organizations, owns and operates over 20 spacecraft in geostationary orbit. Managed by Communications Satellite Corporation (COMSAT, a company created by act of Congress in 1962), INTELSAT provides repeater-in-the-sky services to most of the countries in the world for international (and some domestic) telecommunications. Terrestrial connections are provided by approximately 300 Earth stations that are owned and operated by local administrations and operating companies. Information on INTELSAT spacecraft is given in Section 5.5.2.

(2) International Maritime Satellite Organization

Established in 1979 to provide worldwide mobile satellite communications for the maritime community, INMARSAT (International Maritime Satellite Organization) has 79 member countries and operates global mobile satellite services on land, at sea, and in the air. To cover the globe, three Inmarsat-3 GEO satellites each

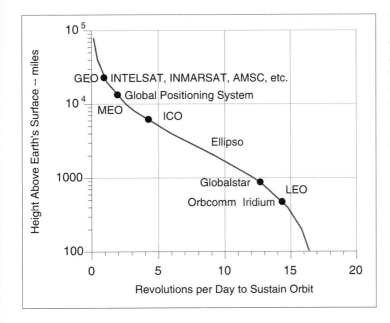

Figure 11.1 Number of Revolutions per Day Required to Maintain Orbit

GEO satellites are stationary relative to a fixed point on Earth. MEO satellites make 2 to 3 passes a day relative to a fixed point on Earth. LOE satellites make 11 to 13 passes a day relative to a fixed point on Earth.

employ a global beam and five spot beams. Stationed to cover the Pacific, Indian, and Atlantic oceans, they support a wide range of services that include

- **Maritime**: direct-dial telephone, telex, facsimile, electronic mail, and data connections
- **Land mobile**: mobile telephone, facsimile, two-way data communications, position reporting, and electronic mail
- **Air**: flight deck voice and data, position and status reporting, and direct-dial passenger telephones, facsimile, and data communications.

Calls are beamed to the satellite and returned to Earth stations that route them to the appropriate terrestrial telecommunication facilities. Using spot beams, power and bandwidth can be adjusted to follow variations in demand in different areas of the world. A wide range of terminals (down to pocket-size messaging units) is supported. Information on INMARSAT spacecraft is given in Section 5.5.3.

(3) American Mobile Satellite Corporation

A pioneer in the use of integrated satellite and terrestrial systems that provide commercial trucking fleets with two-way data communications and vehicle location services (using Global Positioning System), American Mobile Satellite Corporation (AMSC) supports seamless mobile communications coverage of

North America. Other applications include maritime and aeronautical communications, pipeline maintenance, utility and service vehicle dispatching, and emergency services.

(4) Very Small Aperture Terminal Networks

Very small aperture terminal (VSAT) networks consist of large numbers of geographically dispersed terrestrial microstations that, through geostationary satellites, are connected to hub stations, or connected among themselves. Over a VSAT network, data communications can be provided among the scattered locations that comprise a retail chain, a major manufacturing firm, or a national distribution organization. Private networks consisting of hundreds of Earth stations and a geostationary satellite have been implemented. To reduce the cost of each Earth station, small antennas (approximately 2 to 7 feet or 0.75 to 2.4 meters in diameter) and low-power transmitters (operating in C and Ku bands) are employed, giving rise to the name *microstation*. The networks can be configured in two basic ways

- **Star with terrestrial Hub**: provides multipoint communication between a large number of microstations and a central hub. The principle is shown in **Figure 11.2**. Outbound transmissions (i.e., downstream, hub-to-satellite to VSATs) employ a TDM channel (time-divided on a packet basis) broadcast to all VSATs by the satellite. In addition, with a station unique addressing scheme, the hub can send messages to individual microstations. Inbound transmissions (i.e., upstream, VSAT to satellite to hub) from individual VSATs use TDMA, Aloha, or CDMA access procedures. Because of the small antennas and low powers employed at the microstations, the signals received by the hub are likely to be seriously distorted by noise. To regenerate them, the hub employs sophisticated signal-processing techniques. Messages from microstation to microstation are sent through the hub. While the principal use of VSATs is data communication, voice communication (hub ↔ VSAT) is possible.
- **Point-to-point**: with more complex equipment at the ground stations, two-way communications (data and voice) can be supported between VSATs through the satellite (i.e., VSAT ↔ VSAT). Because the path involves one satellite hop, voice communication (with echo cancellation) can occur without major problems on the part of the users.

For those applications in which users cannot justify a private hub, sharing a hub with others is possible. System suppliers will lease space on a common hub to several users.

(5) Support for Mobile Systems

On the spacecraft, higher-power transmitters and larger antennas have made it possible for GEO satellites to communicate with hand-held ground units.

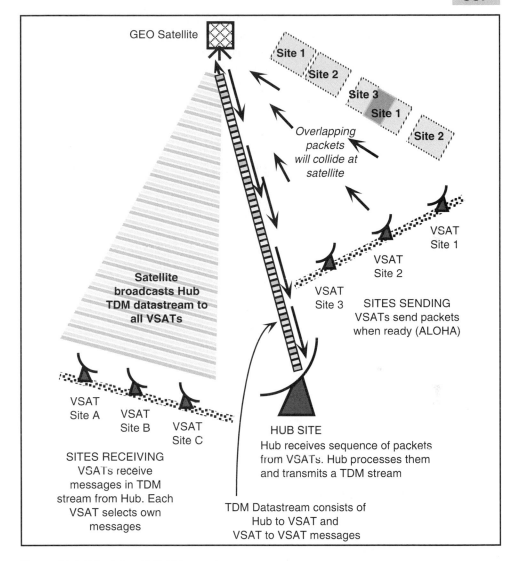

GEO Satellite

Site 1
Site 2
Site 3
Site 1
Site 2

Overlapping packets will collide at satellite

VSAT Site 1
VSAT Site 2
VSAT Site 3

SITES SENDING
VSATs send packets when ready (ALOHA)

Satellite broadcasts Hub TDM datastream to all VSATs

VSAT Site A
VSAT Site B
VSAT Site C

SITES RECEIVING
VSATs receive messages in TDM stream from Hub. Each VSAT selects own messages

HUB SITE
Hub receives sequence of packets from VSATs. Hub processes them and transmits a TDM stream

TDM Datastream consists of Hub to VSAT and VSAT to VSAT messages

Figure 11.2 Principle of VSAT Operation

Illuminating a large area of Earth's surface, synchronous orbit satellites can support regional mobile communication systems. Two recent entrants are

- **Asian Cellular System** (ACeS): illuminates the western Pacific Ocean and Southeast Asia (including the Philippines, Japan, China, India, Pakistan, Indonesia, and Thailand). With one active GEO satellite and one spare satellite (in orbit), service is expected to begin in 1999. Frequencies between satellite and mobile are 1.6265 to 1.6605 GHz uplinks and 1.525 to 1.559 GHz downlinks. Between satellite and gateways the frequencies are 6.425 to 6.725 GHz uplinks and 3.400 to 3.700 GHz downlinks. A consortium of companies from the Philippines, Indonesia, and Thailand is developing ACeS.

- **Thuraya**: illuminates the Arab States, central Asia, India, Turkey, and Eastern Europe. With one active GEO satellite and one spare satellite (in orbit), service is expected to begin in 2000. Handling 13,750 voice channels, Thuraya will route calls to stations using 256 reconfigurable spot beams. A consortium of companies led by United Arab Emirates Telecommunications Corporation is developing Thuraya.

11.1.2 MEO SATELLITE SYSTEMS

Figure 11.3 illustrates the concept of using a satellite constellation to provide mobile communications services. Groups of satellites are used to provide continuous global coverage. The U.S. Department of Defense developed the pioneering MEO system. Called Global Positioning System (GPS), it is now available to civilian users worldwide. Building on the satellite technology, several consortiums are developing global telecommunication systems to compete with terrestrial telecommunication providers. Briefly, I summarize the operation of two medium Earth orbit systems—GPS and ICO—and a third system—Ellipso—that operates as a combination low-MEO and high-LEO system.

(1) Global Positioning System

The GPS space segment consists of 24 satellites (21 in use plus 3 spares) distributed in six planes inclined at 55°. They orbit Earth every 12 hours at 12,554 miles (20,200 km) above Earth. The primary function is to assist in the navigation of mobile stations by providing position, velocity, and exact time. Signals from four satellites are required to compute position (x, y, and z) and determine time. For civilian use, navigation information is transmitted at 1,575.42 MHz. A pseudo-noise code with length 1,023 bits and period 1 millisecond is used to produce a spread spectrum signal with bandwidth 1 MHz. Each satellite employs a different *pn*-code. At the receivers, code synchronization is provided by time signals from the satellites.

(2) ICO

With 10 satellites orbiting in two 45° planes at approximately 6,500 miles (10,458 km) above Earth's surface, ICO offers mobile telephone service to existing users of terrestrial cellular systems when they travel to places where coverage is incomplete or does not exist. Calls to the mobile terminals are routed over existing terrestrial facilities to one of 12 Earth stations, directed to satellites, and redirected to the intended mobile receiver. Calls from mobiles are picked up by satellites and directed to an Earth station. There, they enter existing facilities for further routing.

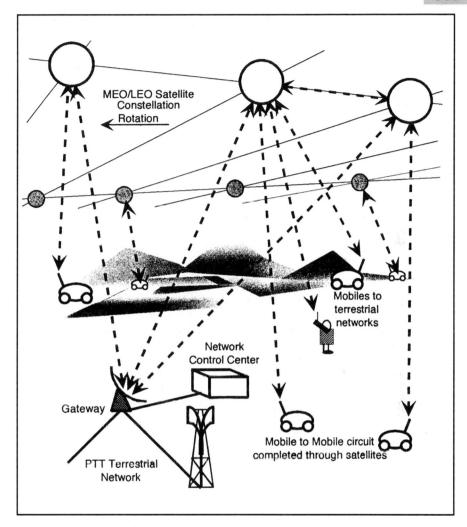

Figure 11.3 Principle of Satellite-Aided Mobile Communications

Operating in the bands 2,170 to 2,200 MHz (user to satellite) and 1,980 to 2,100 MHz (satellite to user) most of the time, each user terminal sees at least two satellites, and communicates with one of them. The best link is selected at call setup time on the basis of visibility. The number of voice circuits per satellite is 4,500. The satellites communicate with Earth stations in the bands 5 GHz (gateway to satellite) and 7 GHz (satellite to gateway). Narrowband TDMA is used in the space segment (user to satellite to Earth station) to achieve efficient use of satellite resources and to reuse as much as possible of the technology from Global System for Mobile Communications GSM, see Section 11.2.2). Mobile terminals that employ other technology will be able to use ICO. The anticipated in-service date is 2000. ICO Global Communications is London-based; investors include international telecommunications entities and Hughes Electronics Corporation.

(3) Ellipso

Global coverage is divided into two zones and is designed to match the distribution of major landmasses in the Northern and Southern hemispheres. Ellipso employs a ring of six satellites arranged in the equitorial plane that orbit Earth at a distance of 4,836 miles (8060 km). They serve a band between 25° north latitude and 55° south latitude. Ten satellites occupy two elliptical orbits that provide coverage of the northern temperate latitudes (North America, Europe, and Asia). In the Northern Hemisphere, the apogee is approximately 4,600 miles (7,600 km), and in the Southern Hemisphere, the perigee is approximately 380 miles (633 km). To serve major urban complexes, each satellite is able to generate 61 spot beams and handles approximately 3,000 simultaneous telephone calls. Voice telephony will be the primary service. Approved by the FCC in 1997, Ellipso is expected to be fully operational in 2001. The Boeing Company is system integrator and space segment prime contractor. Other participants include Lockheed Martin and Harris Corporation, and telecommunication service providers in Australia and South Africa.

11.1.3 LEO SATELLITE SYSTEMS

Other consortia are constructing low-Earth orbit satellite systems. I describe four of them, briefly—Orbcomm, Globalstar, Iridium, and Teledesic. Because they are closer to Earth, LEOs need a greater number of satellites to provide global coverage.

(1) Orbcomm

With a constellation of 26 satellites orbiting at approximately 500 miles (800 km) above Earth's surface, Orbcomm provides full-time, global, two-way digital data services for messaging, emergency alerts, position determination, and remote data collection. Twenty-four satellites are positioned approximately 470 miles (780 km) above Earth in three orbital planes inclined at 45°. Two satellites are positioned in near-polar orbits.

Uplink communication employs 148 to 149.9 MHz, and downlink communication employs 137 to 138 MHz and 400.05 to 400.15 MHz. Hand-held user units transmit at burst rates of 2,400 bits/s and receive at burst rates of 4,800 bits/s. The effective throughput is about 300 bits/s. Messages are sent by satellite to gateway stations on Earth, where they are forwarded to their destination directly, or stored and forwarded on demand. Communication between satellites and gateways can take place at up to 57.6 kbits/s. Each satellite can handle approximately 50,000 messages per hour.

The FCC granted Orbital Sciences Corporation license to construct, launch, and operate the system in October 1994. In full operation, Orbcomm is a joint venture of Orbital Sciences Corporation and Teleglobe Incorporated.

(2) Globalstar

With 48 satellites orbiting in eight planes inclined at 52° approximately 900 miles (1,450 km) above Earth's surface, Globalstar offers worldwide telephone and other telecommunication services to users equipped with hand-held or vehicle-mounted terminals (including airplanes). The constellation provides single satellite coverage between $\pm$ 70° latitude, and at least two-satellite coverage between 25° and 50° north and south latitude. Calls to mobile terminals are routed over existing terrestrial facilities to Earth stations, directed to satellites, and redirected to the intended mobile receiver. Calls from mobiles are received by satellites and directed to an Earth station, where they enter existing facilities for further routing. The number of calls per satellite is 2,000 to 3,000.

Operating in the bands 1,610 to 1,626.5 MHz (user to satellite) and 2,483.5 to 2,500 MHz (satellite to user), each mobile terminal is connected with as many as three satellites at a time. The satellites communicate to Earth stations in the bands 5,091 to 5,250 MHz (gateway to satellite) and 6,875 to 7,055 MHz (satellite to gateway). Spread-spectrum technology (CDMA) is used in the space segment. The FCC awarded a full operating license in January 1995. Owned by a group of 12 companies, Globalstar will initiate global commercial service in early 1999. Loral Aerospace Corporation and QualComm Corporation are general partners.

(3) Iridium

With 66 satellites (plus six orbiting spares) distributed equally in six planes (86.4° inclination) at an altitude of approximately 470 miles (780 km) above Earth's surface, Iridium offers global personal satellite-based communications from hand-held terminals. Using GSM architecture, it is expected to have 1 million subscribers. Calls to the mobile terminals are routed over existing terrestrial facilities to 15 to 20 Earth stations, directed to satellites, and redirected to the intended mobile receiver. Calls from mobiles are received by satellites and directed to Earth stations, where they enter existing facilities for further routing. Alternatively, calls between mobiles may be handled directly. This is made possible by intersatellite links operating at 25 Mbits/s between 22.55 and 23.55 MHz; each satellite is connected to its four nearest neighbors. The on-board processing and satellite crosslink capability that this activity requires makes the implementation of Iridium much more complex. Each satellite can handle up to 3,840 duplex circuits.

Operating in the bands 1,616 to 1,626.5 MHz, each mobile terminal may communicate with as many as three satellites at a time. The satellites communicate to Earth stations in the bands 27.5 to 30 GHz (gateway to satellite) and 18.8 to 20.2 GHz (satellite to gateway). TDMA will be used in the space segment. Iridium is authorized to operate in more than 30 countries and, to provide access to terrestrial systems, as many as a dozen gateways will be located around the world. The

FCC awarded an operational license in January 1995. Global commercial telephone service began in late 1998. Facsimile and data services will be added in 1999. Motorola Incorporated is the largest investor in a consortium of 20 companies that form Iridium LLC.

(4) Teledesic

With 288 satellites orbiting at approximately 840 miles (1,400 km) in 12 polar planes, Teledesic will offer high-capacity broadband global services to mobile terminals. Most users will have asymmetrical connections with up to 2 Mbits/s (uplink) and 64 Mbits/s (downlink). Special broadband terminals will offer 64-Mbits/s duplex service. Each satellite is a node in a fast packet switched network. It is connected to its eight nearest neighbors by STM-1 (155.52 Mbits/s) intersatellite links operating at 59 to 64 GHz. Communication between satellites uses asynchronous transfer mode (ATM) cells. Gateways are capable of bit rates from 155.52 Mbits/s to 1.24416 Gbits/s (STM–8). Each satellite is expected to support 100,000 16-kbits/s channels.

All radio operations occur in Ka band. A combination of TDMA, FDMA and space-division multiple access (SDMA) will be used in the space segment. It is widely believed that a major application will be supplying high-speed, tandem data channels to PTTs. The anticipated in-service date is 2002. Teledesic LLC is located in Kirkland, WA. The FCC issued a construction license in 1997. With major private investment, Motorola is responsible for the system implementation.

11.1.4 DISCUSSION OF SYSTEMS

With the bulk of these systems yet to be deployed, it is difficult to compare them. By far the most sophisticated of the systems described above, Teledesic has the capability of providing an independent, high-speed digital channel between points anywhere in the world. The least sophisticated proposal is ICO. Its services are not overly ambitious, it makes use of GSM architecture and technology, it does not challenge the PTTs, and it works with existing mobile systems. For these reasons it is likely to be successful. With their on-board processing, Globalstar and Iridium are more ambitious than ICO, but less sophisticated than Teledesic. Ellipso is seeking to provide a relatively unsophisticated, reliable global service.

While the most important communication mode is generally considered to be two-way voice, in global operations, nonoverlapping working days (due to time differences) limit the opportunities for conversation. In addition, religious differences must be respected. For instance, Friday is a Holy Day for Moslems, Saturday for Jews, and Sunday for Christians. For many commercial applications, store-and-forward messaging is a more valuable capability than real-time voice. Thus, Orbcomm will find a ready market (and is likely to experience significant competition).

The global systems we have described are the precursors of another generation that will operate in Ka band (17.7 to 21.2 GHz downlinks and 27.5 to 31 GHz

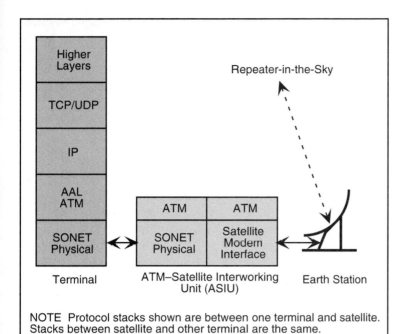

Figure 11.4 Simplified Protocol Stacks for ATM–Satellite Network

NOTE Protocol stacks shown are between one terminal and satellite. Stacks between satellite and other terminal are the same.

uplinks) and include on-board processing, switching, and intersatellite links, making them *packet exchanges* in the sky. More than 50 systems are in various stages of definition, design, or development.[1] A major application may be high-speed Internet access and the dissemination of catalogs and other mass-market information.

(5) Multimode Telephones

When all of the above systems (and probably several others) are operational, there will be a wide diversity of telephone types. Some will use CDMA, others TDMA. Some may provide handover with terrestrial systems; others will not. Handset manufacturers are likely to settle on a few basic designs with interchangeable modules to fit them to the system to be used.

11.1.5 SATELLITE ATM NETWORKS

With ATM designated as the transfer mode for B–ISDN, the question arises: Can a satellite channel provide reliable transport for ATM cells? If certain precautions are taken, the answer appears to be yes. **Figure 11.4** shows a simplified protocol stack for terminal-to-terminal communication over ATM networks and a repeater-

---

[1] www.spotbeam.com/kapress.html

in-the-sky satellite. Successful operation is achieved through the use of ATM satellite interworking units (ASIUs). The sequence of functions they perform is shown in **Figure 11.5**.

For convenience I have assumed that the transport mode is SONET. At the SONET interface ATM cells are extracted from the payload envelope. The stream of cells is then coded in various ways. Forward error correction (see Section 7.4.2) may be employed to protect against multiple random errors. Reed-Solomon coding and block interleaving [see Section 7.4.2(5)] may be employed to protect against bursts of errors. If several types of traffic are present (data, compressed video, or digital voice, for instance) they are segregated in buffers so that time-dependent signals can be afforded priority over those for which delay is not critical. At the receiver, the sequence of operations is reversed.

Finally, the cells are fed to the satellite link. One transport technique uses PLCP (physical layer convergence protocol). Defined by IEEE 802.6 and the ATM Forum, it describes how cells are to be carried in a DS–3 stream. **Figure 11.6** shows a PLCP frame that carries 12 ATM cells. Each cell is preceded by 4 overhead bytes, and followed by a trailer that contains stuffing bits to ensure the 125-µsecond rhythm of the frame stream. To combat burst errors, the frame overhead bytes are interleaved prior to transmission.

REVIEW QUESTIONS FOR SECTION 11.1

1   Why did the invention of communication satellites cause people to describe the world as a *global village* and talk of *spaceship Earth*?

2   Why has the growth in demand for sophisticated personal communication services caused satellites to move closer to Earth?

3   Describe the operations of INTELSAT, INMARSAT, and AMSC.

4   Describe a VSAT system.

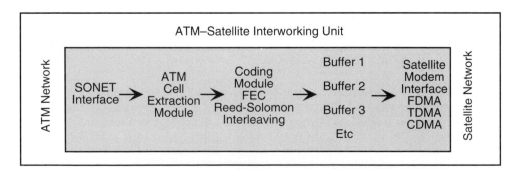

Figure 11.5 Functions Performed by ATM–Satellite Interworking Unit

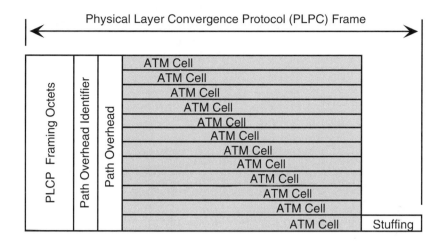

Figure 11.6 12 ATM Cells Arranged in a PLCP Frame

5   Describe two GEO systems that will support regional mobile communications.

6   Describe GPS.

7   Compare Ellipso and ICO.

8   Distinguish between MEO and LEO satellite systems.

9   Compare Orbcomm, Globalstar, and Iridium.

10   Describe Teledesic.

11   Comment on the following statement: Nonoverlapping working days and differences in the days of the working week limit global operations across several time zones and between groups with different cultures.

12   A recent report identifies more than 50 MEO and LEO, next-generation, Ka-band satellite systems in various stages of definition, design, or development. Speculate on the uses for so large a number of systems.

13   Describe the functions performed by an ASIU.

14   Describe how ATM cells are carried in a PLCP frame.

11.2 CELLULAR MOBILE RADIO NETWORKS

Cellular mobile radio telephone service (CMRTS) has shown it is possible for individuals in motion to communicate with one another, and with fixed locations. No longer is it necessary to be attached to a cable to send and receive information.

CMRTS facilities are experiencing rapidly growing demand—and the technology is changing from relatively simple FDMA systems that employ frequency modulation to spread spectrum CDMA systems. I describe four of them,[2] progressing from analog (frequency division) to digital (time division) to digital (code division), they increase in complexity and performance. They are

- **Advanced Mobile Phone Service** (AMPS): the first North American system, it operates in the 800-MHz region (824 to 849 MHz, and 869 to 894 MHz). AMPS employs a channel bandwidth of 30 kHz, uses FDMA for access, and frequency modulation. Later versions of AMPS (AMPS/IS-41) incorporate transparent roaming techniques adapted from GSM (see below).

- **Global System for Mobile Communication** (GSM): a pan-European system, GSM is a second-generation digital, cellular, land mobile telephone system. Operating in the 900 MHz region (890 to 915 MHz, and 935 to 960 MHz), GSM–900 employs 200 kHz channels separated by 45 MHz. Each channel is time-divided to create eight subchannels (FD/TDMA). GSM–1800 and GSM–1900 employ the same technology and operate in the PCS bands (Personal Communication Service, around 1.8 GHz in Europe and 1.9 GHz in North America). GSM pioneered automatic roaming techniques, mobile-assisted handover, subscriber authentication, and user privacy (across the air interface)—important features that have been adopted by North American systems.

- **Digital AMPS** (D–AMPS and NA–TDMA):
  - *D–AMPS.* Defined by Telecommunications Industries Association Interim Standard 54, D–AMPS employs AMPS frequencies. It provides three digital channels (TDMA) in a 30-kHz channel, and employs a combination of analog and digital control channels. *Dual-mode* terminals operate with AMPS or D–AMPS to facilitate the evolution of AMPS to a higher-capacity digital system.
  - *NA–TDMA.* Defined by Telecommunications Industries Association Interim Standard 136, NA–TDMA operates in two frequency bands. At AMPS frequencies, it provides six digital channels in a 30-kHz channel, and employs a combination of analog and digital control channels. Dual-mode terminals operate in analog or digital modes. At PCS frequencies (around 1.9 GHz), NA–TDMA provides six digital channels in a 30-kHz channel, and employs digital control channels exclusively. In the digital mode, automatic roaming, mobile-assisted handoff, terminal authentication, and user privacy are provided. DQPSK (differential quaternary phase-shift keying) is employed.

---

[2] For an in-depth description of AMPS, NA-TDMA, NA-CDMA, and GSM see David J. Goodman, *Wireless Personal Communications Systems* (Reading, MA: Addison-Wesley, 1997), chaps. 3, 5, 6, 7.

- **Spread Spectrum Cellular System** (NA–CDMA): defined by IS–95, NA–CDMA operates at AMPS frequencies with 1.3-MHz channels, and at PCS frequencies. Dual-mode terminals operate as AMPS terminals or CDMA terminals. Automatic roaming, mobile-assisted handoff, terminal authentication, and user privacy are provided.

11.2.1 ADVANCED MOBILE PHONE SERVICE

In the United States, frequencies in the 800-MHz region are allocated to cellular mobile radio telephone service:

- **Mobiles to cell sites**: 824 to 849 MHz.
- **Cell sites to mobiles**: 869 to 894 MHz.

(1) Number of Channels

In these bands, AMPS systems provide 30-kHz, FM channels to individual users. The 832 radio circuits that are supported consist of pairs of one-way channels separated by 45 MHz. In each *cellular geographical serving area* (CGSA), half the channels are allocated to an operating telephone company (wireline company), and half the channels to an independent company (nonwireline company) so that each provider operates 416 circuits. The majority (around 395) are used to carry analog voice or low-speed data (up to 2,400 bits/s).

Forward channels transmit information from cell site to mobile station(s); reverse channels transmit information from mobile station(s) to cell site. Some circuits are used for control purposes; thus

- **Forward control channel** (FOCC): employed in one-to-many mode to signal individual mobile units
- **Forward voice channel** (FVC): employed one-to-one to provide specific signaling information to a particular mobile. An in-band signaling channel that operates in *blank-and-burst* mode; when sent, control messages interrupt user information
- **Reverse control channel** (RECC): employed in many-to-one mode by mobile units to request cell site actions
- **Reverse voice channel** (RVC): employed one-to-one to provide specific information from particular mobile to a cell site. An in-band signaling channel that operates in *blank-and-burst* mode; when sent, control messages interrupt user information

(2) Identification Codes

Each mobile is assigned four permanent identification codes

- **Mobile Identification Number** (MIN): a 10-digit directory number assigned to mobile subscriber. It is stored as a 34-bit code that consists of
  - *Area code*. Three digits that identify the area code of the mobile's home service area
  - *Exchange code*. Three digits that identify the mobile telephone serving office (MTSO) of the home service provider within the given area code
  - *Subscriber number*. Four digits that identify the mobile subscriber.

  The MIN is the number callers use to place a call to the mobile station.
- **Electronic Serial Number** (ESN): a 32-bit code assigned to the mobile station equipment by the manufacturer
- **Station Class Mark** (SCM): a 4-bit code that describes the capabilities of the mobile station
- **System Identifier** (SID): a 15-bit code assigned by the FCC that identifies the cellular geographical service area in which the service provider is licensed to operate. Stored in the mobile station and cell sites within the CGSA, it is used to determine if the mobile station is operating within the boundaries of its home system. If the SIDs of the mobile and the serving cell site do not match, the mobile is in *roaming* status.

(3) Operations

Figure 11.7 shows the principle of AMPS. In this figure, cells are shown as overlapping circles, not neatly fitted hexagons, to make the point that a cell is defined by the mobiles that are communicating with the base station. In a practical system, the geographical arrangement of cell sites, the topology of the area served, and the decision thresholds built into the electronics dictate when a mobile communicates with one base station, and when it communicates with another.

The operation of AMPS depends on a central switch that serves the CGSA. This Mobile Telephone Switching Office (MTSO) is an electronic switching system programmed to provide call processing, system fault detection, and diagnostics. The MTSO is connected directly by wireline, or microwave radio, to a set of base stations, called *cell sites*, that communicate with the mobiles in their cells by radio.

Small programmable controllers at each cell site perform call setup, call supervision, mobile location, handoff, and call termination. A microprocessor within each mobile implements signaling, radio control, and customer alerting functions. Two types of channels are provided

- **Setup channels**: transmit data used to set up all calls
- **Voice channels**: provide message paths and carry signals needed for call supervision—i.e., to signal changes in state and to maintain the connection. The supervisory signals are
  - *Supervisory audio tone* (SAT). An out-of-band, continuous tone (5,700, 6,000 or 6,300 Hz), the SAT is present *whenever* a call is in progress. Its absence

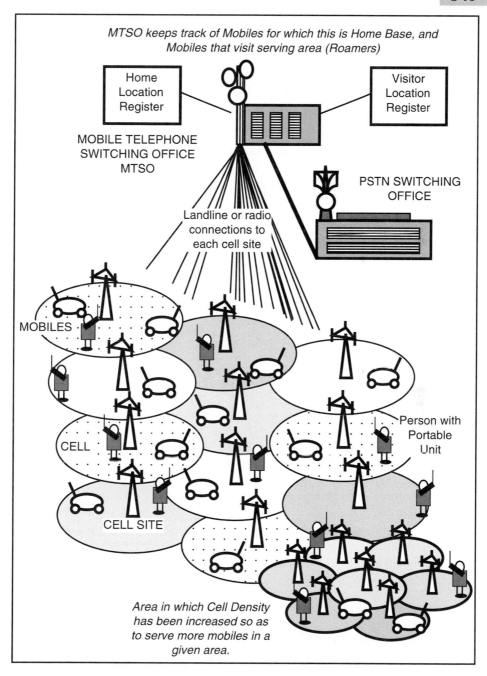

MTSO keeps track of Mobiles for which this is Home Base, and Mobiles that visit serving area (Roamers)

Home Location Register

Visitor Location Register

MOBILE TELEPHONE SWITCHING OFFICE MTSO

PSTN SWITCHING OFFICE

Landline or radio connections to each cell site

MOBILES

CELL

Person with Portable Unit

CELL SITE

Area in which Cell Density has been increased so as to serve more mobiles in a given area.

Figure 11.7 Essential Elements of a Mobile Telephone System

AMPS uses frequencies between 824 and 829 MHz for transmissions from mobiles to cell sites, and frequencies between 869 and 894 MHz for transmissions from cell sites to mobiles.

indicates that one of the parties has gone on-hook, and the call is terminated. Nearby cells operating on the same radio frequency are assigned different SATs to assist in distinguishing between signals.

– *Signaling tone* (ST). An out-of-band tone burst (10 kHz). Among other things, the signaling tone is used to initiate ringing, to signal hand off to another cell site, and to disconnect.

In addition, on command from the cell site, AMPS mobile stations can adjust their level of radiated power. Two classes of stations are in use

- **Class I Mobile**: radiated power is adjustable in eight steps from 6 mW (8 dbm) to 4 W (36 dbm)
- **Class III Mobile**: radiated power is adjustable in six steps from 6 mW (8 dbm) to 600 mW (28 dbm).

A typical Class I station is likely to be vehicle mounted and powered by the vehicle's battery. A typical Class III station is likely to be hand-held and powered by a portable, rechargeable battery. From the cell site, the radiated power may be as much as 25 W per channel.

To set up a call to a mobile, the called number is broadcast on the outgoing setup channels of the cell sites. When the called mobile detects its number in the incoming stream, it sends its identification over the return setup channel. The nearest cell site (selected by the MTSO because it has the strongest signal) responds by assigning the mobile a pair of operating frequencies. The mobile tunes to the assigned receive frequency, prepares to transmit on the assigned sending frequency, and the mobile telephone is rung. When a mobile places a call, the nearest cell site designates a pair of frequencies for the call, notifies the mobile of the frequencies assigned, and prepares to receive the called number.

Using the setup channels is not as simple as described above. The mobiles are in contention for the attention of the base station. When it is processing one mobile it places a busy signal (i.e., sets every 11th bit of the paging stream to 1) on the outgoing setup channel. This notifies other mobiles they must wait to send their request for service. To prevent several waiting mobiles from sending immediately after the base station changes from busy to idle, each mobile waits a random time before attempting seizure of the base station. If the initial attempt at gaining the attention of the base station is not successful, the mobile retries automatically. If several consecutive attempts are unsuccessful, the mobile invokes a timeout period before trying again.

As the population of a mobile system grows, it becomes increasingly important to ensure the veracity of the signaling data on the setup channels and its error-free receipt. In a large system, when a paging message is broadcast, many idle mobiles receive it. If the probability of a false received message is 1 message in 100,000 messages, and the system population is 1,000 idle mobiles, on average, 1 page in100 may result in a false response. If the probability of a false received message is 1 in 10,000, 1 page in 10 may result in a false response. Higher error rates are likely to result in false responses to almost every page. Over a fading radio channel, such error rates are quite possible. To prevent the system from choking on replies to false messages, the setup channels employ forward error correction coding, and paging calls are repeated. Before responding to a page, a mobile must receive its number several times.

(4) Security

To provide some protection against improper use, when requesting service a mobile terminal must transmit its mobile identifier (MIN) and its electronic security number (ESN). At the serving MTSO, the MIN is examined to confirm it is a valid number, and the ESN is compared with the ESN stored there on service inception. If there is a match, the request for service is judged valid and is honored. Because the ESN is transmitted with the MIN, by simply listening for service requests on the AMPS control channel, eavesdroppers are able to obtain valid pairs of numbers to use for fraudulent purposes. Known as *cloning*, this activity is of great concern to AMPS system operators.

Of concern to system users is the fact no provision is made to encrypt (or otherwise disguise) voice messages while in the air interface. They can be intercepted readily by listeners equipped with scanners tuned to the frequencies allocated to mobile service.

(5) Moving from Cell to Cell

As the mobile moves from cell to cell, the MTSO arranges for the talking path channels to switch from the cell site serving the cell the mobile is leaving to the cell site serving the cell it is entering. This is done when the cell site that is handling the call reports that the supervisory audio tone signal strength has dropped below a preset threshold. In response to this information, the MTSO instructs the mobile to transmit on the setup channel, and instructs all cell sites surrounding the reporting cell site to monitor the setup channel and report the signal strength of the transmission. From the reports received, the MTSO determines which cell site is closest to the mobile and instructs the cell site currently handling the call, the cell site that will assume responsibility for the call, and the mobile, to prepare for handoff. At a given instant, the connection between the MTSO and the first cell site is broken, a connection is made between the MTSO and the second cell site, and the mobile retunes to the new frequencies to complete the talking channels through the new cell site. The entire process takes place in milliseconds, and the mobile voice user perceives no interruption in the call. If the mobile is transmitting data, the stream will be interrupted for at least the period between ceasing to transmit on one channel and beginning to transmit on the other, and data will be lost.

(6) Data Transmission over AMPS

The unreliable radio link (due to fading and/or handoff) is a difficult connection to use for data. Nevertheless, cellular modems are available that provide

2.4 kbits/s data service. Based on V.22bis standards, they employ error correction, as in V.42 and MNP classes 2–4 [see Section 7.1.3(4)], and data compression, and the terminals are likely to employ ARQ techniques (see Section 4.2).

Some carriers support an overlay data service called *Cellular Digital Packet Data* (CDPD). It provides 19.2 kbits/s connectionless service on a 30-kHz cellular phone channel. Designed for broadcast information, electronic mail, dispatch, and similar services, packets are transmitted on idle cellular channels. When the channel is required for voice service, CDPD seizes another idle channel; if there is none, it waits for one to occur.

(7) Moving from System to System

When the SID number of a mobile station does not match the SID number of the cell site receiving requests for service, the mobile is *roaming*. Without arrangements to facilitate intersystem operations, services to the roaming subscriber are interrupted. Telecommunications Industries Association's (TIA's) Interim Standard 41 (IS–41) addresses this situation by adapting procedures first developed for GSM (see below). IS–41 establishes procedures that permit handoff to a foreign system (i.e., a system that is not the roamer's home system) and initiation and receipt of calls over the facilities of a foreign system. To make these actions transparent to the calling and called parties, the systems involved must exchange information so that the foreign system can know whether the roamer is currently authorized to obtain service, and what services are permitted, and the home system can know which system the roamer is visiting and how to route calls there. IS–41 manages these activities through communications between MTSOs (foreign and home) and associated databases. They are

- **Home location register** (HLR): database associated with each MTSO that contains subscription information on the set of terminals for which the MTSO is home system switch. For each terminal that is reported roaming by a foreign MTSO, the home MTSO records the current location and status in its HLR.

- **Visitor location register** (VLR): database associated with each MTSO that contains information provided by HLRs on terminals currently roaming in CGSA.

A prerequisite for automatic roaming is a data network that links MTSOs and routing tables that direct traffic among the MTSOs. To do this, IS–41 employs a packet network and Signaling System 7 (see Section 10.4.3).

To illustrate the use of these capabilities, consider the case of a roaming mobile that requests service from a foreign cell site. Upon receipt of an origination message, the foreign cell site passes it to its MTSO (foreign MTSO). Since the roamer is not recorded in the HLR (foreign HLR), the information concerning the roamer is stored in the VLR (foreign VLR) and a request for information is composed and sent to the home MTSO. From the home HLR, the home MTSO extracts information concerning the roamer (MIN, ESN, SCM, and SID), which is sent to the foreign system, and notes in the home HLR that the mobile is roaming in the foreign CGSA. On receipt of information on the roamer, the foreign MTSO can authenticate the roaming station and make resources available to it. With the information that the station is roaming in the foreign CGSA, the home MTSO can direct incoming calls to the area. If the roamer should pass into another foreign CGSA and request service, the home MTSO sends information concerning the roamer to foreign MTSO2, notes in the home HLR that the mobile is now roaming in foreign CGSA2, and notifies CGSA1 that the roamer has left the area.

11.2.2 GLOBAL SYSTEM FOR MOBILE COMMUNICATION

At the time AMPS was installed in North America, proprietary national networks using cellular concepts were in use in many developed countries. The result was a number of similar national systems that could not work together. In Europe, the European Community saw the need for a fully compatible, pan-European mobile telephone system that would permit mobiles to roam throughout Europe initiating and receiving calls from anywhere over the same telephone.[3] Under the aegis of the European Telecommunications Standards Institute (ETSI) specifications were developed for an open system that is compatible with ISDN. This second generation digital, cellular, land mobile system, called *Global System for Mobile Telecommunications* (GSM), uses different nomenclature from AMPS for some common facilities and actions. For example, mobile telephone switching office (MTSO) becomes mobile switching center (MSC), cell site becomes base station, mobile becomes mobile station, and handoff becomes handover.

The first ETSI development was GSM 900 (i.e., operates around 900 MHz). Phase 1 (essentially voice only) was introduced in Europe in 1991; Phase 2, which includes facsimile, video, and data communications over GSM 900, was introduced in 1995. With the establishment of PCS (Personal Communications Services) frequencies in the 1,800 (Europe) and 1,900 (North America) MHz bands, GSM 1,800 and GSM 1,900 were created. They employ the same technology as GSM 900 and were placed in service in 1997.

(1) An Open System

To emphasize the *open* system aspect of GSM, **Figure 11.8** shows a simplified reference model that includes eight interfaces (U_m, A through F). In a manner similar to ISDN, the specification for GSM defines these interfaces, not the equipment or the technology needed to implement them. Thus, GSM can be built with equipment from different manufacturers that conform to the appropriate interface requirements. To emphasize compatibility with ISDN, **Figure 11.9** shows the protocol stacks between a mobile station and its mobile switching center. Transport depends on layers borrowed from ISDN and SS7. At the physical level, with the exception of the radio link, ISDN B-channels are used.

---

[3] www.gsmworld.com/history

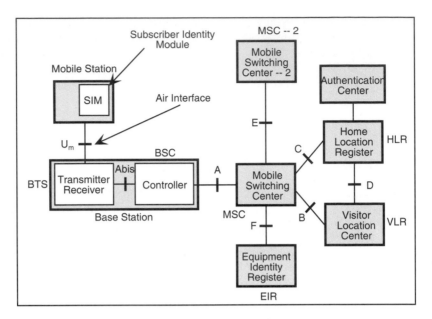

Figure 11.8 Simplified Reference Model for GSM
Global System for Mobile Communications (GSM) is an open system based on ISDN.

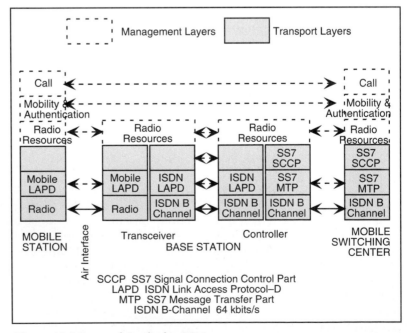

Figure 11.9 Protocol Stacks for GSM

The data link layer protocol across the air interface (U_m) is a modified version of ISDN's LAP–D. For instance, because the operation is paced by the slots and needs no additional synchronization, start and stop flags are eliminated. Further, because forward error correction (FEC) is employed at the physical level, no frame check sequence (FCS) is used to detect errors.

(2) Radio Channels

Initially, GSM 900 operated in two limited frequency bands

- **Mobile station to base station**: 905 to 915 MHz.
- **Base station to mobile station**: 950 to 960 MHz.
- As GSM displaces first-generation systems, the frequency bands are being expanded to 890 to 915 MHz and 935 to 960 MHz.

Individual circuits consist of 200-kHz channels separated by 45 MHz; they employ Gaussian minimum-shift keying modulation [see Section 4.2.1(5)]. Base stations are allocated from 1 to 15 circuits (2 x 200 kHz channels)—i.e., 8 to 120 digital mobile circuits. The GSM carrier may have a fixed frequency, or may be a set of frequencies in a hopping pattern in which it moves from one frequency to another every frame. When signal distortion is severe and sustained, frequency hopping is employed so that transmission impairments are distributed over the channels. To avoid interference with one another, all channels in a cell hop in a coordinated fashion.

(3) Time slots

Each channel is divided into time slots (1,733 slots/s, so that each slot occupies 577 μseconds) that are allocated in sequence to eight subchannels. A sequence of eight slots is known as a frame. Each time slot contains 2 x 57-bit segments of data separated by a 26-bit training sequence, together with flags and tail bits. The training sequence is used by the receiver (at base station and mobile station) to estimate the time-varying characteristics of the radio channel and train an adaptive equalizer to compensate for the effects of multipath propagation. The 1-bit flags indicate whether the 57-bit data fields are user's data or control data. In addition, there is a guard time at the end of each slot of 30.5 μseconds when the transmitter is silent. **Figure 11.10** shows the internal composition of a GSM time slot.

The slots are organized in two multiframe structures of 26 frames and 51 frames that can contain 13-kbits/s speech and 2.4-, 4.8- and 9.6-kbits/s data. **Figure 11.11** gives an overview of the 26-frame structure. Frames 0 through 11 and 13 through 24 support eight *traffic channels* (TCHs). As required, slots are robbed in each TCH for *fast-associated control channel* (FACCH) purposes—such as handover arrangements. Frame 12 contains 8 *slow-associated control channels* (SACCHs), one for each TCH, that perform call supervision and other functions. Frame 25 is reserved for future use. To separate the actions of sending and receiving, the start

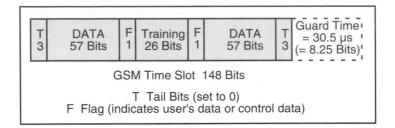

Figure 11.10 Bit Structure of GSM Time Slot across Air Interface

Because of the uncertain radio environment, equalizers in the channel are trained with each frame.

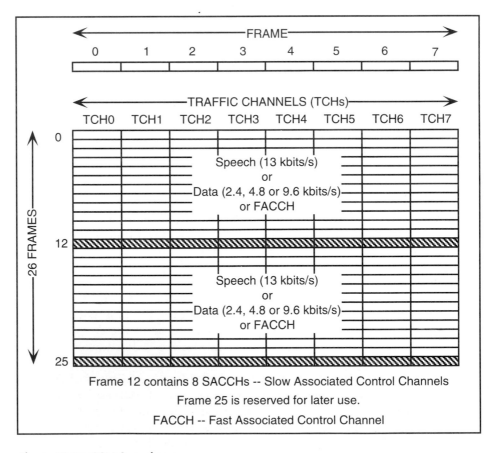

Figure 11.11 GSM Superframe

A frame contains samples from 8 traffic channels. 24 traffic frames, plus one control frame, plus one frame in reserve, make a 26-frame superframe.

of upstream frames (reverse direction frames) is delayed by three time slots from the start of downstream frames (forward direction frames).

(4) Signaling Channels

A GSM base station employs three types of control channels

- **Broadcast channels**: used to transmit the same information to all mobile stations in a cell
  - Frequency correction control channel (FCCH)
  - Synchronization control channel (SCH)
  - Broadcast control channel (BCCH)
- **Common control channels**: used to carry information to and from idle stations as required
  - Power control channel (PCCH)
  - Access grant channel (AGCH)
  - Random access channel (RACH)
- **Dedicated control channels**: channels assigned to specific terminals as required to transfer the information necessary to set up a call
  - Stand-alone dedicated control channel (SDCCH)
  - Slow-associated control channel (SACCH)
  - Fast-associated control channel (FACCH)

  and two types of traffic channels:
- **Full-rate traffic channel** (TCH/F): with a bit-rate of 22,800 bits/s, occupies 24 slots in each 26-frame multiframe (see Figure 11.11)
- **Half-rate traffic channel** (TCH/H): with a bit-rate of 11,400 bits/s, occupies 12 slots in every 26-frame multiframe.

Which traffic channel is used depends on the form of speech coding employed. GSM was designed to use *linear prediction coding with regular pulse excitation* (LPC-RPE) at 13 kbits/s. This requires a full-rate channel. Half-rate channels are used with newer speech encoding techniques that require fewer bits per second.

(5) SIM and Other Identifiers

A Subscriber Identity Module (SIM) identifies GSM subscribers. Implemented as a smart card, and containing identification and encryption information, it can be inserted into any terminal to create a personalized mobile station on which a subscriber can request service from any GSM base station. SIM identifies the subscriber so that the carrier can grant communications access if warranted, provide custom features if permitted, and bill for service (through the home system if the subscriber is roaming).

GSM employs several identifiers to facilitate operations. They are

- **International mobile subscriber identity** (IMSI): a 15-digit subscriber's directory number (stored in SIM)

- **Temporary mobile subscriber identity** (TMSI): assigned by the visitor location register to a subscriber (32 bits stored in mobile station). This identifier is used in subsequent call management actions. Effectively, by shielding the identity of the subscriber, it aids privacy and security.

- **International mobile equipment identifier** (IMEI): a 15-digit serial number assigned by the manufacturer to the mobile station (stored in the mobile station)

- **Mobile station classmark** (MSCM): describes the properties of the mobile station to the operating company (32 bits stored in the mobile station). Items include the version of the GSM specification to which the terminal conforms, the radiated power levels available, and the manner in which the station encrypts messages and control information

- **Base station identity code** (BSIC): assigned by the operating company to identify the base station (6 bits stored in the base station)

- **Location area identity** (LAI): assigned by the operating company to identify the location of the base station (40 bits stored in the base station). It includes a country code, a network code, and a local code.

(6) System Databases

GSM employs four databases to manage mobiles and control calls

- **Home location register** (HLR): maintains and updates the subscriber's location and service profile

- **Visiting location register** (VLR): obtains the visitor's service profile from the home system HLR. Allocates a temporary mobile subscriber identifier (TMSI). Informs the mobile's HLR of location

- **Equipment identity register** (EIR): lists the station's authorized equipment

- **Authentication center** (AC): secure database that supports encryption of radio channel signals. Stores the subscriber's secret key (Ki).

As mentioned above [Section 11.2.1(7)], HLR and VLR were the basis for similar registers in AMPS/IS–41. Note that, in GSM, equipment identity is an issue for the *station*, not the *subscriber*; thus, EIR is separate from HLR. Also, since all communications are encrypted, both the mobile subscriber and the authentication center must have knowledge of the encryption key.

(7) Messages

User's messages and control data are divided into 184-bit segments. **Figure 11.12** shows their composition. To alleviate the effects of errors due to the air interface, they are converted into 456-bit segments by the addition of error-correcting block coding and rate 1/2 convolutional coding. Each 456-bit segment is broken into four 114-bit segments that fill the data fields in 4 time slots.

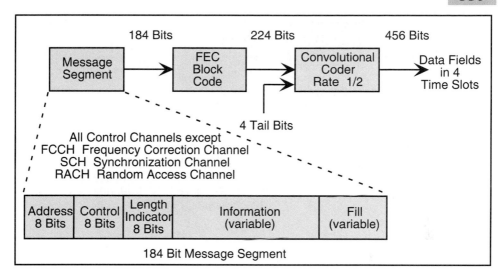

Figure 11.12 Forward Error Correction Coding is used in Message Segments across GSM Air Interface

FEC is used to combat errors due to fading and other impairments of the radio channel.

(8) Encryption and Authentication

Encryption is a cooperative activity between a mobile station and the base station/authentication center that makes it impossible for eavesdroppers to understand the user's messages. In GSM, both speech and data are encrypted over the air interface to prevent eavesdropping or fraud. GSM employs the following keys

- **Authentication key** (Ki): secret key assigned by the operating company to the subscriber (stored in SIM and the authentication center)
- **Cipher key** (Kc): computed by the network and the mobile station prior to encryption.

Figure 11.13 shows the encryption and authentication process. It begins with a 128-bit random number that is sent to the mobile station. Together with the user's authentication key (Ki, stored in SIM) and the appropriate GSM encryption algorithms, a 32-bit *signed response* word (SRES) and a 64-bit cipher key (Kc) are generated. SRES is returned to the base station. If it agrees with the word calculated by the authentication center using its version of the mobile user's authentication key (Ki), the base station knows it is communicating with the correct mobile user and the connection is *authenticated*. Meanwhile, the cipher key and frame number are employed to produce a 114-bit encryption mask that is added to 114 bits of cleartext data (one GSM time slot) to produce 114 bits of ciphertext. Because

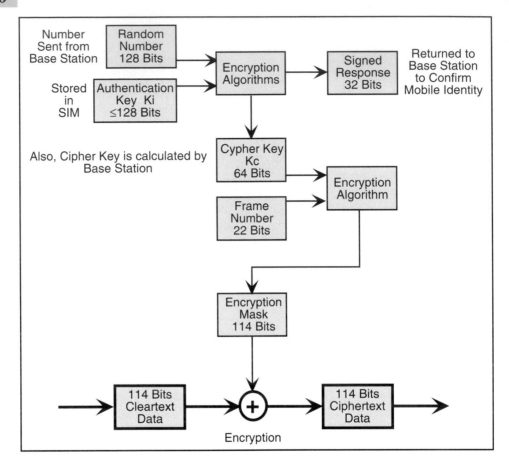

Figure 11.13 GSM Data Encryption Process
Encryption is used to prevent eavesdropping on radio channel for the purpose of hearing conversations or stealing equipment identifiers.

the encryption key is derived from the subscriber's secret authentication key, other mobile stations or eavesdroppers cannot decipher the message.

(9) Mobile-Assisted Handover

When active, mobile stations regularly measure the quality of the traffic channels they are using; in addition, they monitor conditions on active channels in different sectors of the same cell and in surrounding cells. They report their findings to the network on slow-associated control channels (SACCHs) in the active channels. (This activity contrasts with AMPS in which the MTSO assembles measurements of the strength of the mobile's signal from different cell sites.) Because quality information collection is the responsibility of the mobile, it has given rise to the description *mobile-assisted* handover (MAHO).

When the network determines that a call must be moved to a better circuit, it issues a *handover command* to the mobile station on the fast-associated control channel (FACCH). This message provides an 8-bit handover reference identifier and, on the basis of the quality measurements made by the mobile station, identifies the new circuit. It may be in the same cell, in an adjacent cell but under the control of the same MSC, or in an adjacent cell that is under the control of another MSC. Upon receipt of the command, the mobile station tunes to the new frequencies, and transmits a series of *handover access* messages. In return, the (new) base station transmits a *physical information* message that contains timing information so that the mobile can adjust its transmission to fall within the new time slot. When done, the mobile station sends a *handover complete* message on the FACCH of the new channel and begins to transmit information and quality measurements to the (new) base station.

(10) General Packet Radio Service

Inevitably, once the possibility of reliable mobile voice (and low-speed data) communications was established, attention turned to providing reliable higher-speed data (and video) communications to support the mobile office. The addition of General Packet Radio Service (GPRS) to the basic voice services allows GSM to provide data distribution, messaging, data retrieval, and conferencing services at speeds to around 170 kbits/s. GPRS supports connectionless and connection-oriented (point-to-point and point-to-multipoint) packet transport services. The length of the messages varies from less than 100 bytes to several thousand bytes.

Packet routing and transport within the GSM network is achieved through the use of a new network node. Called a *GPRS support node* (GSN), it enables autonomous operation of GPRS in the bursty environment created by data communications. User-related information needed to perform routing and other data transfer functions is stored in a section of the GSM HLR (home location register) called the GPRS register (GR).

To protect existing voice-oriented GSM services, GPRS is implemented as a separate network to which GSM radio channels are allocated as required. These packet data channels (PDCHs) are mapped to time slots and uplinks and downlinks are used independently of one another. Thus, an uplink PDCH will carry data from a mobile to base while the associated downlink is carrying data from base to another mobile. PDCHs are divided into master channels (MPDCHs) and slave channels (SPDCHs). Generally, their duties are as follows

- **MPDCHs**: common control channels that carry the signaling information required to initiate packet transfer
- **SPDCHs**: channels on which user data and dedicated signaling information is transferred.

PDCHs are organized in 51-frame multiframes that carry GSM voice-oriented information as well as GPRS information.

11.2.3 Digital AMPS and NA–TDMA

In many urban locations, AMPS rapidly ran out of capacity; in addition, subscribers wanted to send higher-speed data. Consequently, an evolutionary system was developed that allows AMPS providers to serve more users and to transport higher-speed data yet use the 30-kHz channel spacing already established. Called *digital* AMPS (D–AMPS), and using some of the concepts incorporated in GSM, it is a hybrid frequency division, time-division multiple access system in which each 30-kHz channel is divided into three digital channels. Defined by TIA's Interim Standard IS–54, digital AMPS provides three 16.2 kbits/s TDMA channels on a 30-kHz cellular phone channel (FD/TDMA). Each channel contains 9.6 kbits/s coded voice, 3.4 kbits/s error control, and 3.2 kbits/s signaling. Carriers can gradually substitute three-slot, TDMA channels for existing single-user analog channels so as to achieve graceful evolution from AMPS to D–AMPS. Under present frequency allocations, systems can operate up to 1,248 digital circuits over 416 radio circuits. IS–54 specifies *dual-mode* terminals that are capable of being used with AMPS, or with D–AMPS.

NA–TDMA provides further system expansion. Defined by IS–136, it operates at AMPS frequencies, or at PCS frequencies, and provides six digital channels in a 30-kHz channel. With only 8.1 kbits/s assigned to each user, the method of speech coding is important.

(1) Speech Coding

The recommended speech coding is VSELP (vector sum excitation linear prediction) at a rate of 7 ,950 bits/s. The coded voice is error protected in three ways

- **Cyclic redundancy check (CRC)**: in each vocoder frame of 159 bits, 77 bits (designated Class 1 bits) are especially vulnerable to transmission errors. Of these, the 12 *most perceptually significant* bits are protected by a 7-bit cyclic redundancy check.

- **Convolutional coding**: a one-half rate convolutional coder encodes the Class 1 bits plus the CRC word. They are multiplexed with the 82 unprotected bits (Class 2 bits) in a vocoder frame to form a block of 260 bits.

- **Interleaving**: to achieve some protection against fading and burst noise, the last half of one block (bits 130 to 259) is interleaved bit by bit with the first half of the next block (bits 0 to 129).

At the receiver, the arriving block (260 bits) of coded speech is deinterleaved, and decoded, and the CRC word is recalculated. If the block fails this check, it is discarded and replaced by the most recent block of decoded speech received without a CRC failure (this substitution is known as *bad frame masking*). Should six blocks in succession fail the CRC check, the receiver mutes the signal. Speech is not restored until good blocks arrive (i.e., the interference clears or the channel is changed).

In 1996, an advanced speech coder was adopted for use in full-rate channels. Called ACELP (algebraic code-excited linear prediction) and operating at 13 kbits/s, it produces better voice quality than VSELP.

(2) Time slots and Channel Speeds

In IS–136, four physical channels are defined

- **Half-rate channel**: 1 slot per frame, 8.1 kbits/s
- **Full-rate channel**: 2 slots per frame, 16.2 kbits/s
- **Double full-rate channel**: 4 slots per frame, 32.4 kbits/s
- **Triple full-rate channel**: complete frame, 48.6 kbits/s.

The frames occupy 40 milliseconds and contain six time slots with 324 bits in each slot. Thus, the duration of a time slot is 6.67 milliseconds, and the bit rate per carrier is 48.6 kbits/s. To allow the stations to transmit *or* receive, but not transmit *and* receive, a base-to-mobile slot begins approximately 1.9 milliseconds after the end of the corresponding mobile-to-base time slot.

(3) Identifiers

NA–TDMA employs the following identifiers

- **Mobile Identification Number** (MIN): a 10-digit directory number (34 bits) assigned to the mobile subscriber that consists of
  - *Area code.* 3 digits that identify the area code of the mobile's home service area
  - *Exchange code.* 3 digits that identify the mobile telephone serving office (MTSO) of the home service provider within the given area code
  - *Subscriber number.* 3 digits that identify the mobile subscriber

  The MIN is the number callers use to place a call to the mobile station
- **International Mobile Subscriber Identification** (IMSI): a directory number that conforms to international numbering conventions (50 bits)
- **Electronic Serial Number** (ESN): a 32-bit code assigned to the mobile station equipment by the manufacturer
- **Station Class Mark** (SCM): a 5-bit code that describes the capabilities of the mobile station
- **Protocol Version** (PV): a 4-bit code that indicates the capability of a station (mobile or base)
- **System Identifier** (SID): a 15-bit code assigned by the FCC that identifies the cellular geographical service area in which the service provider is licensed to operate
- **Supervisory Audio Tone** (SAT): one tone selected from three tones that is present when the mobile is active to assist the mobile to identify signals from the base station. Different tones are assigned to contiguous base stations

- **Base Station Manufacturer's Code** (BSMC): an 8-bit code that identifies the manufacturer of the base station
- **Location Area Identifier** (LOCAID): a 12-bit code that identifies the geographical area in which a base station is located
- **Digital Color Code** (DCC): a 2-bit code assigned to a base station to assist the mobile to identify signals from the base station. Different codes are assigned to contiguous base stations.
- **Digital Verification Color Code** (DVCC): a 12-bit code assigned to each base station with digital channels to assist the mobile to identify signals from the base station. Different codes are assigned to contiguous base stations.
- **A–Key**: a 64-bit secret key for encryption and authentication.

(4) Signaling Channels

NA–TDMA supports all of the AMPS channels listed in Section 11.2.1 and the digital control and traffic channels defined by IS–136. In brief, the latter comprise

- **Digital traffic channels** (DTCHs):
  - *DATA*. Contains 260 bits of user's information per time slot
  - *FACCH*. Fast-associated control channel. As required, interrupts the speech signal to seize a slot for urgent control message.
  - *SYNCH*. Contains signals that maintain synchrony between the mobile and the base station. Occupies specific bit positions in DTCH time slots. Includes information to train the adaptive equalizer
  - *DVCC*. Digital verification color code. Occupies specific bit positions in DTCH time slots. Protects against reception of the wrong signal
  - *SACCH*. Slow-associated control channel. Occupies specific bit positions in DTCH time slots. Out-of-band signaling channel that carries information while a call is in progress
  - *DL*. Digital control channel locator (forward channel only). Occupies specific bit positions in DTCH time slots. Provides the location of a carrier that contains a DCCH.
- **Digital control channels** (DCCHs):
  - *SPF*. Superframe phase information. Occupies specific bit positions in DTCH or DCCH time slots. Indicates position of the current block in a control superframe
  - *SYNCH*. Synchronization. Provides frame synchronization and locks the terminal to correct time slot
  - *F-BCCH*. Fast broadcast control channel. Occupies own slot on forward DCCH. Carries the same information to all terminals in a cell.

- *E-BCCH*. Extended broadcast control channel. Occupies own slot on forward DCCH. Carries the same information to all terminals in a cell.

- *S-BCCH*. Short message service broadcast control channel. Occupies own slot on forward DCCH. Carries the same information to all terminals in a cell.

- *SMSCH*. Short message service channel. Occupies own slot on forward DCCH

- *PCH*. Paging channel. Occupies own slot on forward DCCH

- *ARCH*. Access response channel. Occupies own slot on forward DCCH

- *SCF*. Shared channel feedback

- *RACH*. Random access channel. Terminals contend for access under the control of SCF information transmitted on forward DCCH.

(5) Authentication and Privacy

To achieve positive user identification and provide message privacy, D–AMPS and NA–TDMA employ 64-bit secret keys designated SSD–A and SSD–B. How they are calculated is shown in the top diagram in **Figure 11.14**. A 56-bit random number is generated by the user's home system and transmitted to the mobile. With the 64-bit A-key [stored in the mobile station and in a secure database (authentication database) in the user's home system], and the 32-bit Electronic Serial Number assigned to the mobile, it is presented to a cryptographic algorithm known as CAVE (Cellular Authentication and Encryption) to produce two 64-bit words—secret shared data A (SSD–A) and secret shared data B (SSD–B). SSD–A and SSD–B are computed independently by the mobile station and the base system.

In the middle diagram of Figure 11.14, the mobile and the current base station use a 32-bit random number (generated by the mobile and passed to the base station), SSD–A (existing in the mobile, and sent by the mobile's home system to the current base station), the mobile's EIN, and the first 24 bits of the mobile's MID, to produce two authentication words—AUTHBS(M) and AUTHBS(B). If they match, the mobile is the authentic mobile station and is authorized to send and receive messages.

To ensure privacy, user's data is encrypted using SSD–B, as shown at the bottom of Figure 11.14. For this activity, the mobile's home system sends SSD–B to the base station currently serving the mobile.

The strength of the authentication and privacy process depends on the fact that the A-key is never transmitted in the system, and cryptographic procedures performed at the mobile and the base station use secondary keys derived from the A-key. Further, only the random numbers used in calculating SSD–A and SSD–B appear in messages across the air interface. In addition, SSD–A and SSD–B are recalculated at suitable intervals to make the task of an eavesdropper even more difficult. Also, from time to time, using a new random number, the base station will *challenge* the mobile to prove that it is still the authentic mobile authorized to receive services. If it cannot calculate a new AUTHBS(M) that matches AUTHBS(B) calculated by the base station, transmission is halted.

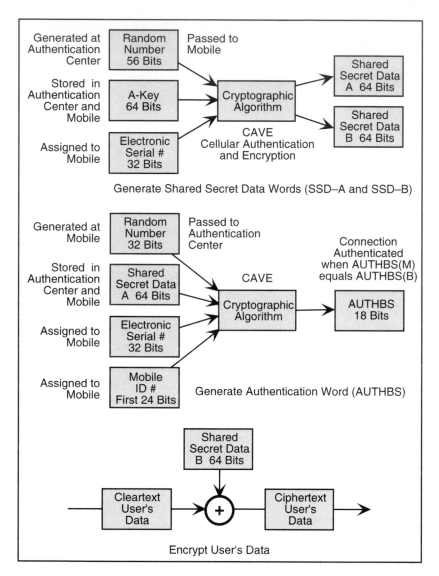

Figure 11.14 NA–TDMA Authentication and Encryption Processes

(6) Handoff

Because NA–TDMA terminals are dual-mode, they must support four types of handoffs—NA–TDMA to NA–TDMA (digital circuit to digital circuit), NA–TDMA to AMPS (digital circuit to analog circuit), AMPS to NA–TDMA (analog circuit to digital circuit), and AMPS to AMPS (analog circuit to analog circuit).

To move within a NA–TDMA environment, terminals employ mobile-assisted handoff procedures similar to those used in GSM. When a call is in progress, the terminal monitors the quality of the signal on the active channel and, during the idle time in each frame (between assigned time slots), measures the strength of signals received from other base stations. The channels that are monitored are identified in a message from the serving base station; it contains a list of idle channels available to handle new calls. The terminal reports the measured values to the base station over the slow-associated control channel. When handoff is indicated, the terminal tunes to the new frequency and cooperates with the new base station to establish a suitable power level and achieve time-slot synchronization.

Within an AMPS environment, the terminal employs the procedures described in Section 11.2.1. Within a mixed environment, the terminal employs procedures that depend on the capabilities of the base stations involved.

(7) Call Management

Services provided by intelligent network (IN) capabilities (see Section 10.5) such as *selective call forwarding, calling number exclusion, call waiting, adding third party*, etc., can be used at mobile terminals to manage calls just as the users manage calls at fixed sites. **Figure 11.15** shows some possible connections between an MTSO and the intelligent network facilities of the public switched telephone network. To initiate IN actions during a call in progress, the terminal user presses a button that sends a *flash* message to the base station.

NA–TDMA systems can implement *group* calling arrangements in which a number of mobile terminals answer calls to one number. To do this, the terminals are supplied with a user group identifier that is associated with the single telephone number. When called, the base station provides a distinctive alerting signal to each telephone; the call is transferred to the mobile that first answers.

When terminals move from system to system, the base stations and switches employ the home location, visitor location, and equipment identification registers described in Section 11.2.1(7), and the authentication database described in Section 11.2.2(6).

11.2.4 NA–CDMA

Defined by IS–95, NA–CDMA uses spread-spectrum modulation with code-division multiple access to implement *digital* mobile radio networks that operate at AMPS frequencies or PCS frequencies. Like NA–TDMA, NA–CDMA terminals are dual-mode and can operate as AMPS terminals. Unlike NA–TDMA, NA–CDMA terminals do not operate with 30-kHz radio channels. As a consequence of the employment of spread-spectrum technology, the channel bandwidth is 1.23 MHz—the bandwidth of 41 contiguous AMPS channels. Thus, the evolution of an AMPS system to a NA–CDMA system must occur in relatively large steps of approximately 10% of system capacity.

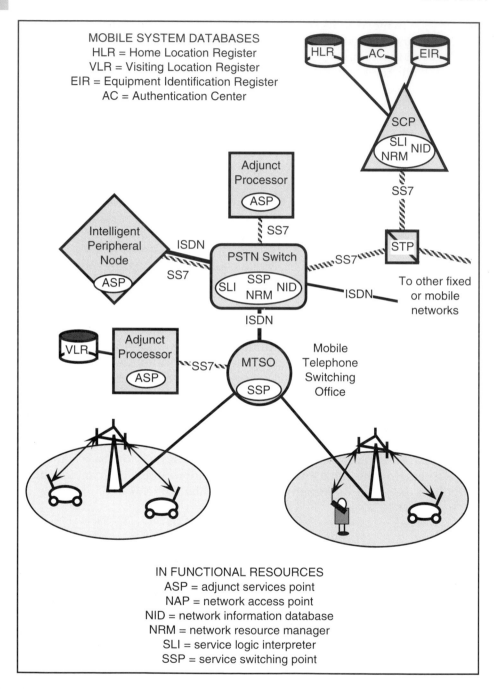

Figure 11.15 Principle of Use of Intelligent Network to Support NA–TDMA Mobile Telephone System

(1) Identifiers

NA–CDMA uses a large number of identifiers; many of them are dedicated to providing detailed information concerning the configuration of the mobile terminal or the base station. Thus

- **Mobile Identification Number** (MIN): a 10-digit directory number (34 bits) assigned to the mobile subscriber. MIN is the number callers use to place a call to the mobile station
- **Electronic Serial Number** (ESN): a 32-bit code assigned to the mobile station equipment by the manufacturer
- **Station Class Mark** (SCM): an 8-bit code that describes the capabilities of the mobile station
- **Mobile Manufacturer Code** (MOB_MFG__CODE): an 8-bit code that identifies the terminal manufacturer
- **Mobile Model Number** (MOB_MODEL): an 8-bit code assigned by the manufacturer that identifies the terminal model
- **Mobile Protocol Revision** (MOB_P_REV): an 8-bit code that indicates the version of IS–95 supported by the terminal
- **Mobile Firmware Revision** (MOB_FIRM_REV): a 16-bit code that identifies the manufacturer's version of the terminal firmware
- **System Identifier** (SID): a 15-bit code assigned by the FCC that identifies the cellular geographical service area in which the service provider is licensed to operate
- **Network Identifier** (NID): a 16-bit code assigned by the system operator that identifies a set of base stations; they form a network within the operator's system
- **Base Station Identifier** (BASE_ID): a 16-bit code assigned by the system operator to identify the base station
- *Pn-Code* **Offset** (PN_OFFSET): a 9-bit code that identifies the relative delay applied to the spreading code (*pn*-code) by the base station transmitter. The delay identifies the base station, differentiates it from the other base stations in the system, and is a means of reducing cochannel interference
- **Base Station Classification** (BASE_CLASS): a 4-bit code that identifies the type of network to which the base station is connected—i.e., public cellular, wireless business system, residential cordless telephone, etc.
- **Registration Zone** (REG_ZONE): a 12-bit code assigned by the operating company that defines the geographical area in which the base station is located
- **Base Station Latitude** (BASE_LAT): a 22-bit code that identifies the latitude of the base station

- **Base Station Longitude** (BASE_LONG): a 23-bit code that identifies the longitude of the base station

- **A–Key**: a 64-bit secret key for authentication (and encryption, if required). NA–CDMA employs the same authentication and privacy techniques as NA–TDMA.

(2) Signal Spreading and Recovery

In the reverse channel the spreading signal is a logical combination of a 42-bit channel identifier word and the output of a 42-stage shift register pseudo-random number generator; they produce a chipping rate of 1.2288 Mchips/s. The *pn*-code is $2^{42} - 1$ chips long so that it repeats every 41.4 days. The channel identifier includes the electronic serial number (ESN) of the terminal and a fixed 10-bit field. **Figure 11.16** shows the creation of a spread spectrum reverse traffic channel.

In the forward channel the digital carrier is selected from 64 orthogonal binary sequences containing 64 chips (64 x 64 Walsh-Hadamard matrix). The channel identifier contains

- **BASE_ID**: the base station identifier
- Access channel number

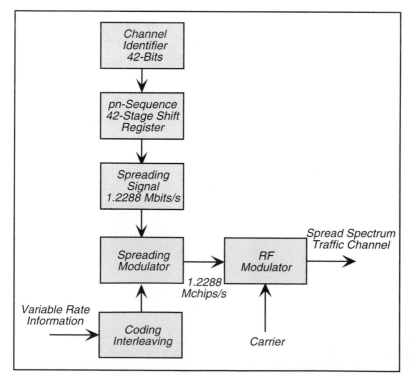

Figure 11.16 Creation of Reverse Traffic Channel in NA–CDMA

- Number of paging channels monitored by the terminal
- **PN_OFFSET**: the relative delay of the spread spectrum signals transmitted by the base station (which differentiates its signals from those from other base stations).

To recover the information embedded in a spread-spectrum signal, the receiver must generate a synchronized replica of the spreading signal [see Section 4.3.4(1)]. This is achieved with the use of global positioning system receivers at each base station. They acquire universal coordinated time accurate to within 1 microsecond and use it to synchronize the chipping signals in the entire network.

Producing a synchronized replica of the spreading signal is complicated by the likelihood of multiple paths between the base station and the mobile station (multipath propagation) and the probability that the motion of the mobile station will cause signal paths to disappear and new paths to appear. Each path will have a different length and propagation delay, and the signals are likely to have different magnitudes. To provide reliable reception under these circumstances, the receiver is equipped with several correlators whose despreading signals are synchronized with signal paths of different propagation delays. Internal circuitry ensures that the correlators drop weaker signals and pick up stronger signals as they appear. Called a *RAKE* receiver, the polarity of each received bit is determined by the sum of the outputs of its correlators.

(3) Information Signals

IS–95 specifies 9,600-bits/s speech, and data from 1,200 bits/s up to 9,600 bits/s. On the reverse channel, error correction coding, repetition, and interleaving are employed to produce 20-millisecond frames with a constant bit rate of 28.8 kbits/s. On the forward channel, similar processing produces 20-millisecond frames with a constant bit rate of 19.2 kbits/s.

(4) Radio Channels

With a channel bandwidth of 1.23 MHz, the number of carriers available in 25 MHz (the bandwidth allocated to AMPS operation) is approximately 20. IS–95 specifies 64 codes (i.e., channels) per forward channel. Reserving some for synchronization and paging tasks, the number available for user's traffic is likely to be between 55 and 60, so that the traffic channels per base station could be between 1,100 and 1,200. Because we are dealing with spread-spectrum CDMA, this is also the number of channels that can be used in every cell. Compare this with a maximum of 119 channels per cell for AMPS (with a reuse factor of 7). Calculating system capacity is much more complex; in reality, what these numbers indicate is that NA–CDMA can serve a significantly greater number of users than AMPS.

NA–CDMA forward radio channels perform the following tasks

- **Pilot channel**: transmits a continuous sequence of 0s at 1.2288 Mbits/s. The mobile terminals assigned to the base station use the signal to acquire timing information and carrier phase. Mobiles assigned to other base stations use the signal to obtain signal strength measurements for handoff purposes.

- **Synch channel**: at 1,200 bits/s, transmits system time, time delay (PN_OFFSET), system identifier (SID) and network identifier (NID) for the base station, and MIN_P_REV to specify the protocol revision level supported by the base station.
- **Paging channel**: at 4,800 or 9,600 bits/s, transmits paging messages. To conserve mobile battery power, idle terminals are programmed to listen to the channel for only a fraction of the paging cycle. When they are not listening, they power down in *sleep* mode.
- **Traffic channel**: selected from 64 possible channels and defined by a specific 64-bit code, the channel is established at the beginning of a call and will change if the mobile moves to a new cell.
  - *Variable-bit-rate user's information messages*. To reduce interference produced by each channel, the encoder examines the pattern of information activity and employs the lowest coding rate consistent with its measurements. For speech, the rates may range from 1,200 to 9,600 bits/s.
  - *Signaling messages*. While a call is in progress, network control messages may be inserted in the traffic channel.
  - *Power control messages*. Optimum operation of a spread-spectrum CDMA system requires precise control of the power radiated by each mobile so that signals appear with equal power at the base station. When a call is in progress, the base station monitors the received power level. Should adjustment be required, it instructs the mobile transmitter to increase or decrease power in 1-db steps. The power control signal is multiplexed with the user's information.

NA–CDMA reverse radio channels perform the following tasks:

- **Access channel**: used by an idle terminal to originate a call, to respond to a page, or to register its location
- **Traffic channel**: a fixed channel defined by the mobile's channel identifier (which includes its ESN)
  - *Variable-bit-rate user's information messages*. To reduce interference produced by each channel, the encoder examines the pattern of information activity and employs the lowest coding rate consistent with its measurements. For speech, the rates may range from 1,200 to 9,600 bits/s.
  - *Signaling messages*. While a call is in progress, network control messages may be inserted in the traffic channel.

(5) Soft Handoff

Within an NA–CDMA environment, terminals employ mobile-assisted soft handoff procedures to move from one cell to another under the control of the same switch. Initiated by the mobile, *soft handoff* is unique to NA–CDMA/IS–95 systems.

The terminal makes measurements on pilot channels from neighboring cells, analyzes them, and informs the base station when handoff is indicated. In turn, the controlling switch selects a forward channel from a new base station and directs it to prepare to receive signals from the mobile. As the terminal moves from cell to cell, it

- Receives synchronizing information from the new base station
- Assigns at least one correlator to the new forward channel
- Establishes communication with the new base station over its fixed reverse traffic channel (defined by the mobile's channel identifier) and the new forward channel
- Maintains communication with the old base station on the same reverse channel and the original forward channel.

When communication with the new station is judged to be established and the strength of the signal from the original base station has dropped below a given threshold, the terminal

- Sends signal strength information to the switch.

In response, the switch sends messages to both base stations to transmit handoff direction messages to the terminal. As a result

- The old base station releases the correlators tuned to the mobile's reverse channel
- The terminal assigns all correlators to the new forward channel, thus breaking connection with the old base station.

During the time that both base stations are receiving signals from the mobile, each station forwards the digital frames to the switch. There, the pairs of frames are compared, and the one with the least errors is sent on to maintain message flow.

Because NA–CDMA terminals are dual-mode, they must support NA–CDMA to AMPS, AMPS to NA–CDMA, and AMPS to AMPS handoffs. These are hard hand offs in which the terminal breaks the connection with one base station before connecting with another. In addition, hard handoffs are necessary when a CDMA call is directed from one frequency band to another, or the terminal moves to a cell that is not controlled by the same switch as the previous cell.

11.2.5 ATM OVER MOBILE RADIO

As the transfer mode of choice for B–ISDN, it is reasonable to ask two questions:

- Can the air interface transport ATM cells?
- Can a connection that includes one or two air interfaces, and may require one or more handoffs, support efficient ATM network operations?

To the first question the answer is the air interface can support cells in ATM format. If the bit rate is adjusted to match the available channel bandwidth, they can be transported without interfering with other mobiles. To the second question the answer is probably not. As discussed in Section 9.2.4, ATM relies on very low error rate transport facilities to move all checking to the end-points of the connection. At best, the error performance of the air interface can be described as fairly good, and handover is almost certain to generate additional errors. Before ATM can be extended to mobile facilities, solutions must be found for these problems.

REVIEW QUESTIONS FOR SECTION 11.2

1 Name four important, active, in-place, cellular mobile radio telephone systems.
2 Describe the operating parameters of AMPS.
3 Show AMPS has a potential of 832 radio circuits. Why are operating companies limited to 416 channels?
4 Distinguish between forward and reverse channels.
5 List, and explain the purpose of, the four permanent identification codes used in AMPS.
6 Use Figure 11.7 to describe AMPS operations.
7 Distinguish between setup channels and voice channels. What do the supervisory tones in the voice channel signify?
8 Distinguish between a Class I mobile and a Class III mobile.
9 Describe the call setup procedures used in AMPS.
10 Describe the AMPS procedures for mobile handover from one cell to a contiguous cell.
11 Discuss the passage of data over AMPS.
12 Describe the AMPS procedures when a mobile moves from system to system.
13 Why is GSM described as an open system?
14 Distinguish among GSM 900, GSM 1,800, and GSM 1900.
15 Describe the parameters of a GSM channel.
16 Why does a GSM time slot contain training bits?
17 What is the purpose of the GSM superframe?
18 List, and describe, the three types of control channels employed in GSM.
19 List, and describe, the two types of traffic channels employed in GSM.
20 Describe SIM. What are its advantages?
21 List, and describe, the six identifiers used by GSM.
22 Why does GSM require four databases (registers)?

23 Discuss Figure 11.12.

24 Explain how encryption and authentication are achieved in GSM.

25 Describe the process for mobile handover in GSM.

26 Explain how digital AMPS is compatible with AMPS and how D–AMPS can be used to expand the capacity of AMPS.

27 What is the relationship between D–AMPS and NA–AMPS?

28 Describe how NA–AMPS coded voice is protected from errors. Why is this important?

29 List the four physical channels defined for NA–AMPS.

30 List the 12 identifiers used by NA–TDMA.

31 List the six digital traffic channels used in NA–AMPS.

32 List the 10 digital control channels used in NA–AMPS.

33 Describe how NA–AMPS achieves authentication and privacy.

34 Describe how NA–AMPS achieves handoff.

35 Discuss Figure 11.15.

36 Describe the parameters of an NA–CDMA channel.

37 List the 16 identifiers employed by NA–CDMA.

38 Describe the procedures employed in NA–CDMA for spreading and recovering signals on the reverse and forward channels.

39 What is a RAKE receiver? Why is it used?

40 What tasks are performed by NA–CDMA forward radio channels?

41 What tasks are performed by NA–CDMA reverse radio channels?

42 What is soft handoff?

43 Can a mobile be an originating or terminating point for an ATM network?

11.3 WIRELESS INFORMATION SYSTEMS

Principally, the CMRTS systems described in Section 11.1 and Section 11.2 serve a rapidly moving subscriber population. With (relatively) high-power radio transmitters and cell sizes of several kilometers, they are described as *high-tier* systems. Their success has stimulated demand for *wireless information systems* that operate in the low mobility environment of the workplace, at home and downtown where peripatetic users can stay in touch. Known as *low-tier* systems, they operate with base stations separated by short distances and transmitter powers 10 to 20 db less than their high-tier counterparts. Cost and performance are optimized to provide network access comparable to *wireline* for pedestrians (inside or outside of buildings) and persons in city traffic.

11.3.1 PCS, PCN, AND UPT

Developments in low-tier systems are encapsulated in the following acronyms

- **PCS** (personal communication services): a term employed by FCC to mean a family of services provided over mobile or portable radio communications facilities for individuals and businesses on the move. Divided in three categories, the services will be provided in an integrated manner over a variety of competing networks:
  - *Narrowband PCS.* Applications include two-way digital paging and data services for interactive television
  - *Broadband PCS.* Applications include voice and data services that will compete with terrestrial mobile telephone systems
  - *Unlicensed PCS.* Applications include cordless telephones and wireless LANs.
- **PCN** (personal communication networks): first used in the United Kingdom, and then adopted in Europe, a term employed to identify forms of metropolitan-area portable radiotelephony networks.
- **UPT** (universal personal telecommunication): ubiquitous telecommunication capability that fills a perceived need for a communication system that follows the user anywhere and provides a range of sophisticated services on demand. Users are unaware of the means employed to consummate communications across town or around the world.

Because PCS and PCN overlap significantly, I will use PCS to denote both concepts, and divide the discussion of wireless information networks into current (PCS) and future (UPT) systems.

11.3.2 ELECTROMAGNETIC SPECTRUM

In 1992, ITU–R allocated 230 MHz around 2 GHz for use by wireless network services. Specifically, the allocation is in two bands—1,880 to 2,025 MHz and 2,110 to 2,200 MHz. In addition, two subbands—1,980 to 2,010 MHz and 2,170 to 2,200 MHz—were designated for mobile satellite links. National administrations and agencies are allocating frequencies in this band to wireless network applications. Within the United States, the FCC has allocated frequencies in the 1,850- to 1,970-MHz and 2,130- to 2,200-MHz bands for personal communication systems.

11.3.3 PERSONAL COMMUNICATION SERVICES

The major focus of PCS development has been to use wireless to extend the voice and data communication capabilities of the public switched telephone network to pedestrians and moderate-speed vehicles in urban and suburban environments.

(1) Cordless Telephone, Second Generation (CT2)

Cordless Telephone Second Generation (CT2) was described in British Standards issued in 1987 and 1989. In the original concept, base stations (CFP, cordless fixed part) provided one-way wireless, payphone access to the telephone network from cordless, customer-owned handsets (CPP, cordless portable part). Known as *telepoints*, the fixed stations are located in shopping malls and downtown areas and operate over a cell-like area several hundred meters in diameter. Later, the capabilities were expanded to allow CPPs situated within a CFP serving area to register and receive incoming calls. The concept expanded further to cover its use as a wireless PBX for business applications, and as a wireless serving area for the distribution of public network traffic to residential subscribers. **Figure 11.17** shows the basic air interface arrangement of CT2.

CT2 uses FDMA and operates on frequencies between 864 and 868 MHz. The band is divided into 40 100-kHz channels over which the portable and fixed parts communicate using time-division duplex. In a frame of 2 milliseconds duration, using binary frequency-shift keying, CFP transmits to CPP for 1 millisecond, and CPP transmits to CFP for the other millisecond. In each time slot, there are 64 bits of user's information and 4 bits of system control information. To the 136 bits, CT2 adds guard times equivalent to 8 bits, for a frame total of 144 bits and a speed of 72 kbits/s. Voice signals are coded as 32 kbits/s ADPCM.

When idle, CPP and CFP scan the 40 channels seeking signaling messages and measuring signal (noise) strength. To establish a link, CPP or CFP seize a channel at random from among those found to have satisfactory noise levels, and transmit a signal for up to 1.4 seconds. If a response is not received within 1.4 seconds, they select another channel and try again. The technique is vulnerable to *call deadlock*—a state that exists when CPP1 attempts to call CFP at the same time that CFP is calling CPP1. Because they send and listen on the same channel (a consequence of time-division duplex operation) CPP1 will not *hear* CFP, and vice versa.

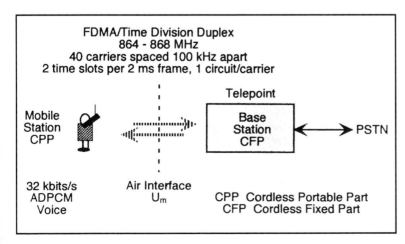

Figure 11.17 Cordless Telephone (Generation 2) Air Interface

A pedestrian can communicate through a fixed telepoint to the PSTN. The caller uses the cordless portable part of CT2; the telepoint is equipped with the cordless fixed part of CT2.

During a call, CPP and CFP transmit CPP's 19-bit identification code once per second over bits in the 4-bit system control information field (see above). If transmission errors cause one or both receivers to fail to detect this *handshaking* code, CT2 may attempt to reestablish the connection on the same channel, or on a different channel. If these actions are unsuccessful, and the handshaking code is missing for 10 seconds, communication is discontinued.

In Canada, CT2Plus was standardized in 1990. It operates on frequencies between 944 and 952 MHz and augments voice operation with the ability to transmit data and facsimile. Further, it employs four channels for common control functions that include channel allocation and terminal registration and authentication.

(2) Digital European Cordless Telecommunications

Standardized by ETSI in 1991, Digital European Cordless Telecommunications (DECT) is designed to deliver a wide range of communication services in the PCS band using FD/TDMA with time-division duplex. **Figure 11.18** shows the basic air interface arrangement. Operating between 1.88 and 1.90 GHz, DECT employs 10 carriers with a spacing of 1.728 MHz. Each carrier is time-divided in 10-millisecond frames that contain 24 time slots. Normally, slots 1 through 12 in each frame are used for portable to fixed-site communication, and slots 13 through 24 are used for fixed-site to portable communication. (However, it is possible to assign different numbers of slots to the two directions.) With a bit rate of 1.152 Mbits/s, each slot transmits 420 bits and is protected by a guard time equivalent to 60 bits. Of the 420 bits, 32 are used for synchronization, 320 contain user's information, and the remainder are used for system control functions.

Voice communication employs 32-kbits/s ADPCM in *unprotected* channels. Error protection of user's information is realized by attaching a 16-bit frame-check sequence to every 64 user's bits. This reduces the throughput of *protected* channels to 25.6 kbits/s. Besides error detection, DECT provides a range of data communication capabilities, including connection-oriented and connectionless message ser-

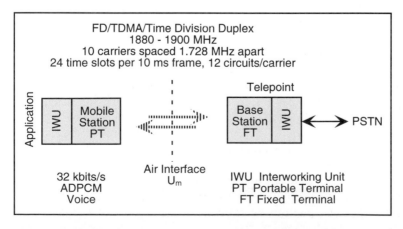

Figure 11.18 Digital European Cordless Telecommunications Air Interface

DECT employs FD/TDMA/TDD to provide digital channels to pedestrians.

vices, and supplementary IN services; consequently, it must execute actions at the network and data link levels. Initiating and coordinating these functions are the responsibility of the internetworking units (IWUs) shown in Figure 11.18. The connection-oriented message service supports point-to-point packet data communication; the connectionless service permits portable units and base stations to exchange short data messages over idle physical channels for system management.

To place a call, a portable termination (PT) seizes an idle channel at random and initiates communication with the fixed termination (FT, i.e., base station). Should transmission quality deteriorate, the combination of PT and FT transfers the call to another physical channel served by the FT. Communication is maintained on the original physical channel until a reliable connection is established on the new channel (soft handover). Should a PT be moving out of range of the serving FT and into the range of another FT, the call is switched to the new FT using local network facilities (hard handover). Coordination of these actions is performed by the IWUs.

DECT includes authentication and encryption operations. Similar to the procedures described in Section 11.2.2(8), authentication is required before an FT will provide communication services to a PT and permit access to local and global networks. Within these networks, databases and associated procedures for registering the location of PTs are maintained so that PTs can receive incoming calls. This capability, however, is not within the scope of the DECT standard.

(3) Personal Handyphone System

Developed by NTT (Nippon Telephone and Telegraph), and placed in commercial service in 1995, Personal Handyphone System (PHS) provides wireless access to public and private communication networks for pedestrians with portable terminals, and, unique among PCS systems, it provides direct wireless communication between portable terminals. Designed to work with ISDN-based networks, PHS supports 32-kbits/s ADPCM voice, 9.6-kbits/s voiceband data, and facsimile, and digital data at 32 and 64 kbits/s. The scope and arrangement of the air interface is summarized in **Figure 11.19**. The R and S interfaces are the same as those shown in the ISDN reference configuration of Figure 10.21.

With 77 carriers spaced 300-kHz apart between 1,895 and 1,918.1 MHz, PHS divides each channel in frames of eight time slots. Four slots (0 through 3) are used for transmission in one direction (personal station to cell station, or PS1 to PS2), and the remaining four slots (4 through 7) are used for transmission in the reverse direction (cell station to personal station, or PS2 to PS1). All 77 carriers are available for public use; in addition, the lower 37 carriers can be used for connections to private networks, and the lowest 10 carriers can be used for direct connection of personal stations. The frame duration is 5 milliseconds, and each frame provides 4 (physical) circuits.

PHS employs four logical control channels that are shared among all the personal stations using the same cell station. They are

- **Broadcast control channel** (BCCH): transmits the superframe structure and other system control information to all portable stations

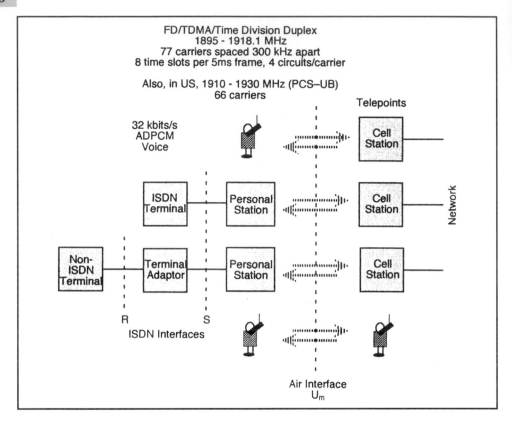

Figure 11.19 Personal Handyphone System Air Interface

Developed in Japan, PHS is unique in supporting direct, pedestrian to pedestrian channels.

- **Paging channel** (PCH): pages individual terminals for incoming calls
- **Signaling control channel** (SCCH): two-way channel employed to set up calls. Personal stations use slotted-Aloha procedure [see Section 6.5.3(5)] to gain access to cell station
- **User-specific packet control channel** (USCCH): two-way packet data channel for short messages between cell and idle portable stations.

PHS employs six service channels. They are

- **Synchronization burst** (SYC): bursts consist of a 62-bit preamble and a 32-bit unique word. They are exchanged by portable station and cell station before user's communication begins
- **Traffic channel** (TCH): with other control and identification signals carries user's information in 160-bit fields

- **Fast-associated control channel** (SACCH): similar to NA–TDMA; as required, interrupts user's information on TCH for urgent network control messages

- **Slow-associated control channel** (FACCH): similar to NA–TDMA; occupies specific bit positions in TCH time slots and carries control information while a call is in progress

- **User-specific packet channel** (USPCH): user-specific packet data channel; carries user's information in 176-bit fields

- **Voice activity channel** (VOX): when the speech activity detector detects a pause in voice activity and turns off the personal station's transmitter, the VOX channel transmits a short message in one out of every four frames that describes the background noise at the personal station.

PHS includes authentication and encryption operations. Authentication is required before a cell station will provide communication services to a personal station and permit access to the local network. Also, encryption services are available. As with DECT, the necessary procedures are executed by resources contained within the local network.

In PHS, both personal station and cell station monitor quality during calls. When appropriate, either station can initiate handoff—called *channel switching*. If the personal station initiates channel switching, and it is still within the range of the serving cell station, it sends a switching request to the cell station over FACCH. The cell station responds with information on a new physical channel. The stations exchange synchronization bursts over the new channel and resume communication. If channel switching is necessary because the personal station is going out of range of the serving cell station, the personal station selects a new cell station and requests that a channel be established. The new cell station responds with information on a new channel, and the two stations synchronize operations. What happens next depends on the way in which the serving cell station routes the call. If the serving station routes the call to the new cell station without taking itself out of the connection, communication resumes as soon as the new air connection is synchronized. If the serving station transfers the call to the new cell station and takes itself out of the connection, communication is not resumed until the new station has authenticated the personal station.

(4) Personal Access Communications System

The original objective for Personal Access Communications (PACS, circa 1990) was to provide residential subscriber radio access to the PSTN through base stations separated from one another by 1,000 to 1,500 feet. Radio terminals in each home would communicate with the base stations and obviate the need for (increasingly expensive) wire distribution plant. In the changing telecommunications environment of the 1990s, the objective was expanded and operation modified to serve pedestrians with inexpensive, lightweight terminals and vehicles moving at moderate speeds in urban and suburban areas. The air interface for this version of PACS was standardized in 1996 (ANSI J-Std-014-1996). Operating in the 1.9-GHz PCS band, it employs FD/TDMA/Frequency Division Duplex. A second standard (ANSI J-Std-014b-1996) describes the air interface for PACS–UB. Operating in the unlicensed PCS band (1,910–1,930 MHz), it employs FD/TDMA/Time Division Duplex.

The equipment arrangement and the parameters of the air interfaces in PACS and PACS–UB are summarized in **Figure 11.20**. The base station consists of two parts—the radio port (i.e., transmitter and receiver) and the radio port controller (which performs logical functions on the signal stream). Reflecting its origin as a provider of radio access to PSTN, the base station employs an ISDN interface to couple to the local telephone network switch. To manage the radio/mobility functions, the switch is augmented by a unit called an access manager. The air interfaces of the two systems differ in important ways.

- **PACS**: in the PCS frequency band, 80 MHz separates two bands of 30 carriers. Within each band, the carriers are spaced 300 kHz apart. Each carrier is time-divided into 2.5-millisecond frames that contain eight time slots. A forward channel consists of one time slot in each frame on a carrier in one band of carriers; the reverse channel consists of one time slot in each frame on a carrier in the other band of carriers. Thus, we can describe the PACS air interface as operating FD/TDMA/Frequency Division Duplex. Slots are synchronized on all carriers. Forward channel slots precede reverse channel

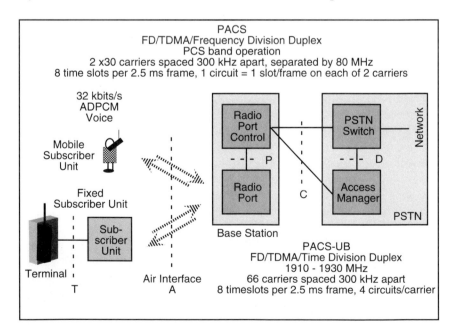

Figure 11.20 Personal Access Communications System Air Interface

PACS provides FD/TDMA/FDD channels to pedestrians and vehicles moving at moderate speeds.

slots by 375 μseconds (1 time slot = 312.5 μs) so that a terminal does not transmit and receive simultaneously.

- **PACS–UB**: in the unlicensed PCS frequency band, 66 carriers are spaced 300 kHz apart. Each carrier is time-divided into 2.5-millisecond frames that contain eight time slots. A forward channel consists of one time slot in slots 4 through 7 of each frame on a carrier; the reverse channel consists of one time slot in slots 0 through 3 of each frame on the same carrier. Under normal circumstances, a carrier has a capacity of four circuits.

In both systems, the 2.5 millisecond frames are divided into eight time slots of 312.5-μsecond duration. With 120 bits per slot, the bit rate is 384 kbits/s—the same as PHS. Further, with 80 bits in each slot assigned to user's data, the user's data rate is 32 kbits/s—for ADPCM voice—the same as PHS (and GSM and CT2).

PACS employs a system broadcast channel (SBC) that is shared among all the subscriber units using the same base station. The SBC operates at 32, 16, 8, or 4 kbits/s and occupies every slot 5, one in two slot 5s, one in four slot 5s, or one in eight slot 5s on the carrier. When not occupied by SBC data, slot 5 may be used for user's traffic. Five control channels are statistically multiplexed on the SBC

- **Synchronization** (SYC): transmits timing and framing information (14 bits/slot)
- **Channel rate** (CR): indicates bit rate available on channel (10 bits/slot)
- **System information channel** (SIC): carries general system information to all subscriber units (80 bits/slot)
- **Alerting channel** (AC): paging channel (80 bits/slot)
- **Priority request channel** (PRC): facilitates set up of emergency and other priority calls (80 bits/slot).

PACS traffic channels carry the following logical channels

- **Synchronization** (SYC): transmits timing and framing information (14 bits/slot)
- **Channel mark** (CM): transmitted by the radio port to facilitate channel assignment and automatic link transfer (handoff) functions. It indicates whether the channel is carrying user's information (i.e., busy), idle (and available), or idle (and only available for priority calls) (10 bits/slot)
- **Word error indicator** (WEI): indicates whether the radio port (or subscriber unit) has successfully decoded the information transmitted by the subscriber unit (or radio port). Each time slot is protected by a 15-bit FCS (1 bit/slot)
- **Cochannel interference control** (CCIC): the subscriber unit transmits a 4-bit code at the beginning of a call. At intervals during the call, the base station or the subscriber unit repeat the code to confirm that the signal is coming from the correct unit, and not from another unit operating nearby (4 bits/slot)

- **User information channel** (UIC): voice or data up to 32 kbits/s (80 bits/slot)
- **Slow message channel** (MC–S): out-of-band signaling channel that carries information while a call is in progress (8 bits/time slot)
- **Fast message channel** (MC–F): as required, interrupts user's data to seize slot for urgent control message (80 bits/time slot)
- **Power control channel** (PCC): commands the subscriber unit to raise or lower power in 1-db increments (1 bit/time slot).

To establish communication (in response to a page from the base unit, or at the initiative of the subscriber), a subscriber unit searches for a carrier with adequate signal strength on which a channel is marked as available at the required bit rate. Then, in this time slot, it sends an access request to the radio port. If the received signal strength is adequate, the radio port control unit accepts the request, and the units proceed to synchronize transmissions and begin communication. After a fixed interval, if the subscriber unit does not receive confirmation from the base station, it searches for another channel and tries again.

In PACS, handoff is called *automatic link transfer* (ALT). When a call is in progress, the subscriber unit monitors the received strength of the carrier signal, and the word error indicator bits associated with the channel employed. At the same time, it monitors signal strengths and channel marker codes for suitable alternate channels. When conditions warrant, the subscriber unit initiates ALT to a better channel by sending a request to use the new channel to the base station that serves it. If the received signal strength at the base station serving the new channel is adequate, the radio port control unit accepts the request, and sends an execute command over the new channel. The units proceed to synchronize transmissions and begin communication. After a fixed interval, if the subscriber unit does not receive confirmation of its request from the serving base station, it searches for another channel and tries again.

PACS provides security features intended to prevent cloning and eavesdropping. They include authentication and encryption using the CAVE algorithm described under NA–TDMA [Section 11.2.3(5)]. Authentication is required before a base station will provide communication services to a subscriber station and permit access to the local network.

11.3.4 UNIVERSAL PERSONAL TELECOMMUNICATIONS

Looking to the future, Universal Personal Telecommunications (UPT) seeks to fill the perceived need for a communication system that can follow the user anywhere and provide a range of sophisticated services on demand.

(1) Network Model

Figure 11.21 shows an abstract model of a wireless network that supports UPT. Divided into three planes, it consists of

- **User plane**: contains description of network in terms of user services
- **Access plane**: contains those facilities and procedures that contribute to user mobility and provide access to the transport levels of the model
- **Local plane**: contains facilities that complete local connections, provide IN features for personal call management, and connect to long-distance facilities to provide a global, UPT network.

We use this division to organize discussion of UPT networks and services.

(2) User's Perspective

In Section 2.2.3(2), we introduced an empirical measure called *richness* to characterize personal messages that cater to an extensive range of human interests. It is reasonable to ask what richness will UPT users expect: Will it be the same in both directions; will multimedia documents be included? The answers to these questions will have a profound effect on the implementation of UPT, and on the terminals that the subscribers will employ. According to ITU, the *future public land mobile telephone system* (FPLMTS) will provide global coverage for speech and low- to medium-bit-rate data (≤ 64 kbits/s), with high-bit-rate data (≤ 2.048 Mbits/s)

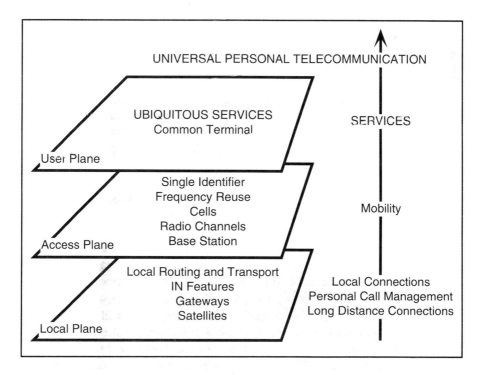

Figure 11.21 Concept of Universal Personal Telecommunications

available in limited areas. FPLMTS will be compatible with ISDN and SS7. FPLMTS services are identified as

- **Dialog services**: anticipated to include voice, facsimile, and low-resolution videotelephone
- **Message services**: anticipated to include paging, voice mail, and text mail
- **Information services**: anticipated to include voice, music, and data.

These are not the requirements of virtual reality—but they are a significant jump from AMPS and PACS. We use them to guide discussion.

(3) UPT User's Terminal

UPT requires a truly portable, user-friendly voice and data terminal. Better flat screens, higher-density (i.e., more logic functions per chip) integrated circuits, and higher-capacity, lower-weight, faster-recharging batteries are but a few of the technical improvements required to implement such a terminal. To function in a UPT environment, in addition to a transmitter and a receiver, it must contain audio transducers for speech, voice mail, and music; a (color) display for data-related functions and videotelephone; a videotelephone camera; a facsimile scanner and reproducer; a hard-copy printer; and a keyboard or equivalent device (perhaps speech recognition) for sending hard-copy messages.

(4) UPT Service Set 1

Global and regional standards organizations are working to define a set of service features for UPT. Service Set 1 defines basic capabilities needed to support personal mobility. Some of them are

- **UPT user identity authentication**: intended to protect the user and the service provider, this feature verifies the identity of the user
- **Incall registration**: to enable personal mobility, this feature allows the user to register for incoming calls at any terminal address
- **Outgoing UPT call**: to enable personal mobility, this feature allows the user to make outgoing calls from any terminal address (with UPT user identity authentication)
- **Outcall follow-on**: allows the user to continue to make outgoing calls from the terminal address (without further UPT user identity authentication)
- **Incall delivery**: calls to the user's UPT number are delivered to the terminal address registered in incall registration
- **Variable default incall registration**: permits the user to establish a default routing matrix that forwards calls on the basis of caller ID, time of day, etc.
- **UPT service profile**: allows the user to change passwords, default services, etc.

(5) Access Arrangements

The way in which users gain access to the network, and the way the network locates and gains access to the users, will determine how successful it is in providing UPT. Easy access depends on many things, including

- **Single identifiers**: users of telecommunication systems are likely to have telephones, facsimile machines, video terminals, etc., at home and in the office. Some may have mobile phones, and some may also have pagers. Most likely, each of these devices has a separate number. If they work in two or three different places on different days of the week, there are more numbers. For originating users, accessing the network is simply a matter of picking up the phone or switching on the modem. However, finding terminating users who may be in any number of locations is a different matter. As a goal, UPT requires the use of unique personal identifiers with which the network can track them down. Today, two approaches come part way to this goal

  - *1–500 numbers*. Assigned to specific persons, they are used by those who wish to call them. The callers' calls are transferred over SS7 to network control points (NCPs) which maintain listings of the numbers at which the called persons can be reached. The called parties (who register their new numbers each time they change location) have entered them. Of course, if the called persons forget to enter their new locations, they are as inaccessible as before. Safeguards, such as only authorizing the use of specific network numbers at the NCP and the inclusion of default numbers if there is no answer, can be built in to the system.

  - *Subscriber identity module (SIM)*. In GSM, the user can have a subscriber identity module that is inserted at the mobile station. SIM carries the personal number assigned to the user. GSM associates it with the registered number of the mobile station and thus can route calls to the subscriber.

- **Frequency reuse**: to accommodate a large number of users, frequency reuse is an essential ingredient of UPT, and, if users are to be able to move about freely, an area system is required. Frequency reuse is provided by the terrestrial cellular arrangements incorporated in today's systems, and by satellites (GEO, MEO, and LEO)

- **Cell size**: cell sizes must be matched to the traffic density, traffic speed, and the potential number of active users. If the traffic is mixed, with high-speed mobiles (cars) and low-speed mobiles (pedestrians) sharing the same space (as in an interstate highway traversing a downtown commercial area) the density of pedestrian users may be sufficient to require microcells within the macrocell that serves the speeding vehicles.

- **Radio channels**: at the heart of providing mobility, radio channels form a flexible link between the user and a base station. As described earlier, the channels may be based on discrete carriers that are

- assigned to mobiles on demand (FDMA or SCPC–DAMA)
- divided up among several mobiles on the basis of time (FD/TDMA)
- spread into a collection of wideband, noiselike signals (FD/CDMA).

Because of its essential simplicity, FDMA was the technique chosen for early mobile systems. Because of its inefficiency (in terms of mobiles served per Hertz) it is being replaced by digital techniques. For a future UPT system, FD/TDMA or FD/CDMA will be employed, and the radio channels will be provided by a combination of terrestrial facilities and satellite constellations. A continental or worldwide paging system could be the solution to universal callability.

(6) Connecting Arrangements

The local network provides connections between mobiles in the same CGSA, between mobiles and local fixed facilities, and transfers mobile traffic intended for other areas to long-distance facilities. Within the local network, operations conform to PSTN/ISDN procedures and employ SS7. Through the network infrastructure, the user can invoke IN features. The major elements associated with UPT are

- **Mobile switching center**: the MSC is the end-office of the UPT network. Located between the radio loop and the higher levels of the network, it serves as the interface between the facilities and procedures of wireless technology and ISDN-based, SS7-directed facilities.
- **IN features**: when equipped with services switching point (SSP) capability, the MSC can invoke IN features from anywhere in the network to assist users to manage their calls. Combinations of SCP/NCP, AP/NCP, and IP/SN provide support that implements 1–500 number calling, corporate credit card service, virtual networks, voice mail, announcements, calling number identification, call-forwarding, call-blocking, and call waiting. While some of these features may not be available to users at present, they can be implemented without great difficulty if demand warrants them. Skillful use of IN capabilities should go a long way toward providing network functionality that permits UPT users to receive calls addressed to their personal numbers and manage calls according to their personal preferences.

At the beginning of this chapter, I described several ambitious approaches to providing universal communications. Using constellations of satellites, the systems are expected to provide voice, facsimile, and data services to special handsets and regular mobile terminals.

If users can be free to roam outside, and communicate with the world, why should they not be free to roam inside buildings and communicate with their colleagues, servers, hosts, and sundry other processors of information? Such things are happening. Based on the success of digital cordless phones, wireless PBXs and wireless LANs have been developed. Employing TDMA on multiple carriers or CDMA and Aloha access techniques, they operate around 900 or 1,800 MHz in portions of the spectrum set aside for unlicensed, low-power operation. Using the same cellular principles as the larger CMRTS, the base stations are distributed throughout the facility. For multibuilding facilities, traffic can be run in cables between the buildings to connect from one wireless subarea to another.

Indoor wireless networks can be implemented with omnidirectional, infrared transmitters and receivers. Because IR radiation is absorbed or reflected by solid structures, cells are limited to the confines of rooms and other closed areas. Because of its short wavelength, IR has the advantage that it is not subject to fading. However, it has the disadvantage that sunlight and the radiation from incandescent and fluorescent lights contain IR so that there are many interfering sources. IR works best when it can be used at relatively high power over relatively short distances.

11.3.5 FUTURE OF UPT

Universal personal telecommunications has caught the imagination of many in the technical community. Given the necessary investment, technologists probably have the tools to finish the job of providing convenient, flexible, pervasive communication links between portable telephones or portable computers and fixed or mobile facilities anywhere in the world—and they will continue to push toward this goal.

We are in a period known to business majors as *technology push*. It is the first major stage in bringing a product to market. Continuation requires demonstrations that arouse latent demand and build a constituency of potential subscribers that translates into *market pull*. Completion requires agreement on standards so that manufacturing efficiencies can be brought to bear to produce final products that offer services that people want at prices they are willing to pay.

It will be several years before the answers are in—and during that period the competitive environment will change constantly as established providers seek to retain or increase market share despite entrepreneurial challenges.

REVIEW QUESTIONS FOR SECTION 11.3

1 Distinguish between high-tier and low-tier systems.
2 Distinguish among PCS, PCN, and UPT.
3 In the United States, what frequencies are allocated to PCS?
4 Describe CT2.
5 Describe DECT.

6 Describe PHS.

7 Describe PACS.

8 Discuss Figure 11.21. What are the critical issues for the establishment of UPT?

9 Comment on the services ITU has included in FPLMTS.

10 List, and describe, the capabilities included in UPT Service Set 1.

11 Comment on the things on which easy access will depend.

12

LANs, MANs, AND INTERNET

For many persons, their first introduction to data communications is a local-area network (LAN) in the workplace, or the Internet. Using them, it is possible to collect data, create information, and distribute it within offices, over floors of buildings, in entire buildings, across campuses, throughout urban cores, and around the world. In this chapter, I address some of the networks that make distributed data processing a reality.

12.1 LOCAL-AREA NETWORKS

The pervasive deployment of desktop processing has encouraged the development of

- **Local-area network** (LAN): within a limited area, a data communication network that interconnects information processing devices so as to serve groups of users who may be engaged in a common enterprise. LANs serve the needs of diverse communities of users by employing a
  - common bearer shared by all stations
  - precise channel access procedure
  - high operating speed.

The individual stations they connect may be in close proximity to each other, scattered over one floor of a multistory building, distributed throughout the entire building, or situated on a commercial or academic campus. A typical network may be used to link as many as a hundred clients and servers together.

LAN bearers employ megabit speeds. A speed of 10 megabits per second is common, and systems are rapidly moving to 100 Mbits/s. To illustrate the meaning of these speeds, suppose subscriber data units are carried in frames of 500 bytes, then the maximum capacity of a 10 Mbits/s bus is 2,500 frames/s.

- If 25 stations are connected, but only one has frames to send, it can transmit 2,500 frames/s so that station throughput is 10 Mbits/s.

- If 25 stations are connected, they all have frames to send, and they send in an orderly sequence, the maximum frame rate each can attain is 100 frames/s so that station throughput is 0.4 Mbit/s.

- If 25 stations are connected, they use a random access technique [such as Aloha, see Section 6.5.3(4)], and the offered traffic is 1,250 packets/s (i.e., 50% full load), on average, 450 frames/s will be delivered without collisions, so that the throughput per station is 72 kbits/s.

- If 25 stations are connected, they use Aloha, and the offered traffic is 125 frames/s (i.e., 5% full load), on average, 100 frames will be delivered without collisions so that the throughput per station is 16 kbits/s.

Thus, the throughput of individual stations can vary dramatically. It depends on the access procedure used and on the number of stations that have something to send. For this reason many installations are turning to 100-Mbits/s systems.

12.1.1 IMPLEMENTATION OPTIONS

A basic difference among LANs is the way in which signals are carried on the common bearer.

(1) Baseband vs. Broadband

- **Broadband** (or wideband): signals are carried on parallel channels over coaxial cable using frequency-division multiplexing.

The technique is similar to CATV. Devices on the network are assigned to channels they share with other devices. In large systems, two cables may be used to increase the number of channels and/or segregate the channels by class of application. Using quasi-analog techniques to supply digital signal streams, broadband LANs find limited use compared to baseband LANs.

- **Baseband** (or narrowband): digital signals are carried directly on the common bearer.

In a baseband network, a limiting factor is the data rate of the common channel. It affects the rate at which frames can be transported and the distances over which they are sent. Because of this, many baseband networks limit the total number of stations, the distance between stations, and the overall length of the bearer. Baseband systems are less expensive to buy and easier to implement and maintain than broadband systems. In addition, because they are digital, they can take advantage of a continuing stream of device developments that are improving speeds and decreasing costs. They are the dominant networks for today's data applications.

(2) Bus vs. Ring

Another basic difference among LANs is the configuration of the common bearer. It may assume the form of a bus or a ring.

- **Bus**: all stations are connected to the same bearer—the *bus*—that is open-ended.

The loss of a station should have no effect on the others. Failure of the bus, however, will affect all stations and may cripple the network. In a bus configuration, each station is connected to all other stations, and a protocol is required to ensure that only one station sends at a time [called medium-access control (MAC) protocol].

- **Ring**: each station is connected to two other stations so that there is a continuous, single-thread connection among all stations.

A station receives messages from the station that precedes it on the ring and sends messages to the station that follows it on the ring. In turn, the stations receive permission to transmit messages through the receipt of a *token*. The failure of a link or a station will affect all stations and will degrade the performance of the network.

(3) Contention vs. Authorization

Yet another basic difference among LANs is the way in which access is gained to the bearer so that a station can transmit data without interference. It is obtained by contention, or by authorization.

- **Contention**: when using a contention technique, each station competes for transmission resources. They may
  - send when ready (Aloha)

- send when ready *and* the bus is not occupied (carrier-sense multiple access, CSMA)
- send when ready *and* the bus is not occupied *and* listen for a collision (carrier-sense, multiple access, collision detection—CSMA/CD).

Under conditions of light traffic, the probability of a collision between two messages is low, and contention works well. Under heavy traffic, the probability of a collision between two messages is high, and so much time is spent in resending messages that almost nothing gets through.

- **Authorization**: each station sends when ready *and* authorized.

Usually, authorization is granted through possession of a token—a data message that is passed from station to station. Operating in this manner, the possibility of a collision is zero; however, the speed of operation is slowed by the need to pass the token between stations and for stations that have nothing to send to handle the token. Under conditions of light traffic, token passing is slower than contention. Under conditions of heavy traffic, token passing is faster and more reliable than contention.

(4) Practical Topologies

Rings or busses are convenient descriptors. They have an intellectual appeal that makes the differences between LANs readily understood. Also, they have an historical appeal. Until the mid-1980s, all local-area networks were laid out in some fashion that mimicked these basic figures and the common highway ran to within a short distance of every user.

With the development of better technologies, however, the LAN function has gradually been concentrated at a *hub*. Here, the LAN is constructed in miniature, and coaxial cables or unshielded twisted pairs (UTPs) connect to the users' terminals. The result is facility wiring that exhibits a starlike topology and a central point that performs all of the functions that used to be performed by a distributed LAN. As you read the remainder of this chapter, remember the concepts are bus and ring, but the reality may be a hub in a wiring closet to which all users are connected. Under this circumstance, the media access routines all take place on a backplane (or equivalent) that supports individual connections to users.

12.1.2 IEEE Standards

The Institute of Electrical and Electronics Engineers (IEEE) has pioneered in the development of standards for local-area networks. As a result of the work of Committee 802, three LAN specifications have been developed [see Section 1.3.3(2) for a listing of Committee 802's activities]. Their scope is illustrated in **Table 12.1**. In addition, IEEE Committee 002.6 has developed standards for metropolitan-area networks (MANs, described in Section 12.5). The IEEE standards are organized around the layer hierarchy shown in **Figure 12.1**. In the IEEE LAN model, layer 2 of the OSI model is divided into the Logical Link Control (LLC) sublayer and the Medium Access Control (MAC) sublayer

- **Logical link control sublayer**: defines the format and functions of the protocol data unit (PDU) that is passed between service access points (SAPs) in the source and destination stations.

- **Medium access control sublayer:** defines the format and functions of headers and trailers that are added to PDUs so that entire frames can travel between source and destination addresses over particular style LANs.

Because the decision to use a specific LAN defines the routing, station addressing, error control, and flow control techniques that will be used locally, only the LLC/MAC layer and the physical layer are involved in message transfers across it.

Table 12.1 Local Area Network Standards

OSI Layers	IEEE 802 Layers	Network Implementation				
Data Link Layer	Logical Link Control LLC	IEEE 802.2 Unacknowledged connectionless service Connection-mode service Acknowledged connectionless service				
	Medium Access Control MAC	IEEE 802.3 (Ethernet) CSMA/CD	IEEE 802.4 Token-bus	IEEE 802.5 Token-ring	ANSI FDDI Token-ring	IEEE 802.6 MAN DQDB
Physical Layer	Physical Layer	Baseband Coaxial cable or UTP 10 Mbits/s or 100 Mbits/s	Broadband Coaxial cable	Baseband UTP/STP 1, 4 or 16 Mbits/s	Baseband Optical fiber 100 Mbits/s	Baseband 1.5 Mbits/s

|←————BUS————→|←————RING————→|← BUS →|
or
RING

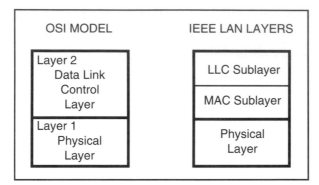

Figure 12.1 Comparison of Layers in OSI Model and IEEE 802 LAN Model

IEEE divides the Data Link Layer into two sublayers. Logical Link Control Sublayer defines the format of the protocol data unit . Medium Access Control Sublayer defines information that is added to the PDU to facilitate transfer using a specific access technique.

(1) Protocol Data Unit Format

Based on HDLC, IEEE 802.2—Logical Link Control (LLC) specifies the format in which data will be exchanged across the medium and how the stations will be addressed. The term Protocol Data Unit (PDU) is used to identify the frames that are employed. Their format is shown in **Figure 12.2**. Some points to be noted are the following:

- The addresses in the PDU are service access point addresses, not physical device addresses
- The addresses of the sending and receiving (physical) devices are included in the MAC header
- The control field in the PDU may be 1 or 2 bytes long, depending on the function the PDU performs
  - I (information) format is used for numbered information frame transfers between SAPs
  - S (supervisory) format is used for numbered supervisory functions. As in SDLC, a supervisory PDU contains no information field
  - U (unnumbered) format is used for unnumbered supervisory functions and unnumbered information transfers.

(2) Classes of Service

Three classes of services are provided to stations attached to a LAN that employs LLC.

- **Connection-oriented service**: a logical connection is set up between originating and terminating stations.

Acknowledgments, error and flow controls, and other features are employed to ensure reliable data transfer between stations using connection-oriented services.

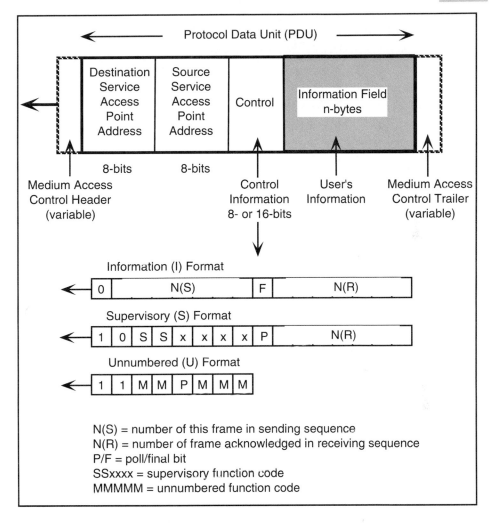

Figure 12.2 LLC Protocol Data Unit

- **Acknowledged connectionless service**: the receiver acknowledges messages, but a logical connection is not established.

This technique is used for large applications in which the overhead associated with connection-oriented service would make the operation too slow, yet it is important to know that the message was received.

- **Unacknowledged connectionless service**: the receiver does not acknowledge messages.

Error control and flow control are not employed. This is a simple style of communication that is used in applications where the occasional loss or corruption of a PDU is corrected by other procedures.

REVIEW QUESTIONS FOR SECTION 12.1

1 Define a LAN. To serve the needs of its users for specialized data processing, what does a LAN employ?

2 Explain why the throughput of stations on a LAN is likely to be much less than the speed of the common medium.

3 Distinguish between baseband and broadband.

4 Distinguish between a ring configuration and a bus configuration.

5 Distinguish between access by contention and access by authorization. Name one, or more, techniques for each class.

6 Describe the practical way to implement a LAN.

7 Explain the following statement: In the IEEE LAN model, layer 2 of the OSI model is divided into the Logical Link Control (LLC) sublayer and the Medium Access Control (MAC) sublayer. Describe them.

8 Why are only the LLC/MAC layer and the physical layer involved in message transfers across a LAN?

9 Use Figure 12.2 to describe the components of the Protocol Data Unit.

10 What classes of service does LLC support?

12.2 BUS-CONNECTED NETWORKS

Connecting all stations to a common point may be simpler than connecting all stations to a common signal path. In one case the descriptor *bus* applies; in the other it is a *star*. Either way, physically, each station is connected to all other stations.

12.2.1 IEEE 802.3 CSMA/CD Bus

Based on *Ethernet*, a LAN pioneered by Xerox Corporation and developed further by a team consisting of Xerox, Digital Equipment Corporation, and Intel Corporation, IEEE 802.3 describes bus-based LANs that employ carrier sense, multiple access, collision detection techniques. While there are minor differences between the original Ethernet and IEEE 802.3, they are compatible, and most persons describe either of them as Ethernet.

(1) CSMA/CD Operation

The principle of operation is shown in **Figure 12.3**. Key to successful data transfer is the continual monitoring of activity on the bus by each DTE. This is necessary so that

- Each station will receive messages addressed to it
- Periods of no bus activity can be detected
- During transmission by the station, the appearance of another signal can be detected and a collision condition declared.

When a period of no activity on the bus is detected, the station with something to send may send it. Once this station has begun transmission, the other stations should detect the activity and withhold their own messages.

If two, or more, stations begin to transmit at the same time, they will interfere with each other, a collision will be declared, and they must try again at a later time. To avoid a further collision, IEEE 802.3 defines a backoff algorithm for use by colliding stations to calculate some random time interval during which they will refrain from attempting to retransmit their messages. In a production environment, satisfactory performance is obtained when the offered traffic is < 20% of the bus capacity.

An alternative configuration is shown at the bottom of Figure 12.3. It performs the same functions as the bus but is easier to install and maintain. The principal feature is the concentration of the connections in a hub to produce a star topology. As a further development, communities of stations can be connected to hubs, and the hubs connected together. In addition, switches can replace the hubs [see Section 12.2.1(3)].

(2) CSMA/CD Frame Format

Details of the frame employed in CSMA/CD operation are given in **Figure 12.4**. The medium access control (MAC) header consists of the following:

- 7-octet preamble used to establish bit synchronization
- 1-octet flag that locates the beginning of the data fields
- 2- or 6-octet network address that identifies the terminating station
- 2- or 6-octet network address that identifies the originating station
- 2-octet-length field that contains the number of octets in the logical link protocol data field that follows.

The two network addresses must be the same length (2 octets or 6 octets); they identify the physical units that originate and terminate the communication.

The MAC trailer consists of a 4-octet frame check sequence (FCS) and, if necessary, a field of octets that fills the frame out to a minimum total length of 64

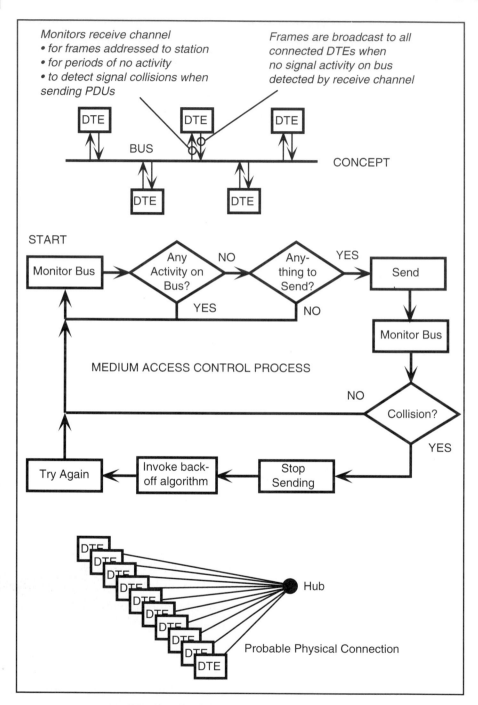

Figure 12.3 CSMA/CD Medium Access Control Technique

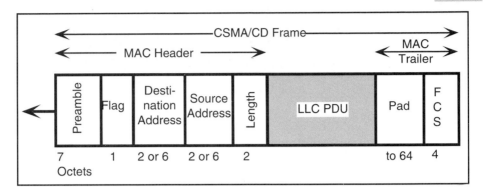

Figure 12.4 CSMA/CD Frame Format

octets (to stabilize the operation of the LAN). The 4-octet FCS is created through the use of a 33-bit generating function [see Section 7.3.2(2)].

(3) Ethernet Physical Layer

Originally, Ethernet employed two cable sizes:

- 0.40-inch diameter coaxial cable bearer limited to 500 meters when operating at 10 Mbits/s with Manchester signaling (see Figures 3.8 and 3.9)
- 0.25-inch diameter coaxial cable bearer limited to 185 meters when operating at 10 Mbits/s with Manchester signaling.

Longer runs require repeaters between cable segments (of 500 or 185 meters). The nomenclature has been standardized. Thus, the two systems described above are known respectively as 10BASE5 and 10BASE2. The first number indicates the LAN speed (in Mbits/s); BASE indicates it is a baseband LAN; and the trailing number is an indicator of the segment length for reliable operation (in 100s of meters). Other configurations are

- **10BASE-T**: employs unshielded twisted pair (UTP) connected in a star. Each UTP supports a single station that is no more than 100 meters from the hub.
- **10BASE-F**: employs optical fiber to connect hubs separated by long distances (to connect between floors, for instance). Fibers are run between the hubs. However, it is likely that each hub is connected to its community of users by UTP.
- **10BROAD36**: employs coaxial cable in segments up to 3,600 meters. Operates with broadband channels of 6 or 12 MHz to deliver digital signal streams.

In addition, a class of *fast* Ethernet LANs has been introduced.

- **100BASE-T**: operates at 100 Mbits/s, employs standard Ethernet features, and uses multiple UTPs, shielded twisted pairs (STPs) or optical fibers to interconnect hubs. Stations are limited to less than 100 meters from a hub, and a total span length of no more than 250 meters.

Another development is

- **Switched Ethernet**: a switch replaces the hub so that the two stations involved in a message transfer can be connected directly over a high-speed channel. Thus, collisions are eliminated, stations do not have to wait for the bus to be quiet, and stations can operate at full bit rate.

To operate at the speed of the stations (usually 10 Mbits/s), the switch fabric is likely to be one of the candidates described in Section 9.4.3(2). Because switched Ethernet no longer contains a shared bus, it might be called fast packet switching! It marries the best of the Ethernet protocol to data switching fabrics and eliminates the worst aspect of the protocol—collisions and procedures for collision avoidance—thereby achieving operation without the burden of the medium access control procedures. In addition, it allows the stations to operate at the bit rates of which they are capable.

12.2.2 IEEE 802.4 TOKEN-PASSING BUS

On a token-passing bus operating under IEEE 802.4, each station with something to send transmits only when permission to transmit has been received.

(1) Token-Passing

Permission is conveyed by means of a token that is passed from station to station, in sequence. On a bus, all of the stations are assigned a position in the token-passing sequence, and the final station returns the token to the beginning station to provide the ability to repeat the process. Each station must maintain a record of its neighbors in the sequence, and the addition of stations to the bus will require modifications to the sequence. The basic medium access technique is shown in **Figure 12.5**.

(2) What Constitutes a Token on a Bus?

The token is an addressed supervisory frame that grants access to the bearer for a specific period during which the addressed station may send frames. At the end of the period, the station with the token generates a token frame addressed to the next station in the prearranged sequence. The time for which the station is in possession of the token can be adjusted to give stations that generate more traffic more time to transmit material than stations that generate less traffic. As the total traffic increases, it becomes important to *tune* the system to optimize the flow of data.

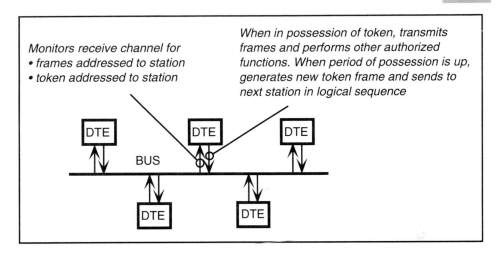

Figure 12.5 Token-Passing Bus Medium Access Control Technique

(3) Need for a Master Station

What if a station fails? It will not be able to regenerate the token frame when its turn comes, and operations will cease. What if a station erroneously addresses the token to itself? If it is a one-time fault, probably no harm is done. However, if it is a permanent fault, something must be done to break the transmission monopoly the station enjoys. What if the token-passing station addresses the token frame not to the next in sequence, but to some other station? One, or more, of the stations will be cut out of the sequence and will not receive permission to transmit. And so on. A bus is prone to many conditions that can defeat the smooth operation of the token-passing strategy. For this reason, one station monitors the operation of the entire bus and is endowed with the ability to detect and correct whatever irregularities arise.

(4) Token-Passing Bus Frame Format

Details of the frame employed in token-passing bus operations are given in **Figure 12.6**. The medium access control header consists of

- **Preamble**: 1 octet, or more, used to establish bit synchronization
- **Flag**: 1 octet that locates the beginning of the data fields
- **Frame control field**: 1 octet that defines the frame as a
  - *MAC* frame, in which case the data field contains information concerning token management or recovery from abnormal conditions
  - *User's data* frame, in which case the data field contains the information assembled by the logical link control sublayer.
- **Network address:**

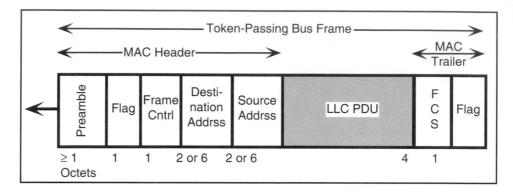

Figure 12.6 Token-Passing Bus Frame Format

- 2 or 6 octets that identify the destination (terminating) station
- 2 or 6 octets that identify the source (originating) station.

The MAC trailer consists of a 4-octet frame check sequence (FCS) and a flag (1 octet).

(5) Token-Passing Bus Physical Layer

IEEE 802.4 defines a broadband LAN that operates over a coaxial cable bearer and employs a modulated carrier for the transmission of information. Using frequency modulation, one technique provides a data rate of 1 Mbit/s over a single channel. Using phase modulation, another technique achieves data rates of 5 Mbits/s or 10 Mbits/s over a single channel. Using frequency division multiplexing, a third technique supports multiple channels that can operate at 1, 5, and 10 Mbits/s.

12.2.3 IEEE 802.6 METROPOLITAN AREA NETWORK

Metropolitan Area Networks (MANs) are facilities that span urban cores, metropolitan areas, or major industrial campuses. Connecting the premises of many organizations, they provide the opportunity to employ public, switched, high-speed facilities to transport data, voice, and video. MANs cover a significantly larger area than the local-area networks discussed so far. Described by IEEE 802.6, they employ a *distributed-queue dual-bus* (DQDB) access protocol.

Figure 12.7 shows the dual-bus network topology in both open-bus and looped-bus versions. Each node serves as the collection point for local traffic and is connected to its partners by point-to-point connections that are operated to provide a transmission channel in each direction. Labeled *A* and *B*, these channels transmit data in fixed-size units. Called *slots*, they consist of 53 octets (52-octet

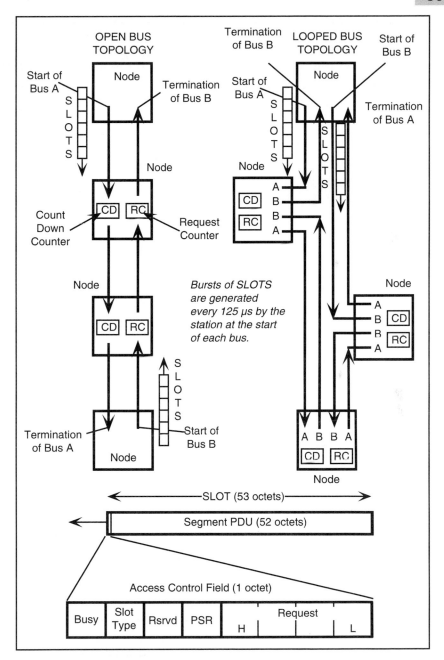

Figure 12.7 MAN Topologies

In either configuration, two busses (A and B, one for each direction) handle information transfer between stations. Stations gain access to one or the other through the use of the distributed-queue, dual-bus (DQDB) protocol.

payload + 1-octet header). Speeds on the busses range from 44.736 Mbits/s (DS–3) over coaxial cable or fiber, to 155.52 Mbits/s (STS–3, STM–1) over single mode fiber. Every 125 microseconds, the slot generator at the node from which a bus starts transmits bursts of slots. The number of slots depends on the data rate (13 slots at T–3, 45 slots at STS–3). When permitted, nodes write data to the slots and read and copy data from them.

(1) DQDB Access Protocol

The distributed-queue dual-bus access protocol provides connectionless, connection-oriented, and isochronous transfer services. Each node gains access to a bus by joining a queue of nodes waiting to transmit on the bus (there are two queues, one for each bus). For one direction (sending on bus A), the procedure is shown in **Figure 12.8**. When a node has a message to transmit on bus A, it sets a request bit on bus B. (Each node keeps a tally of downstream requests by moni-

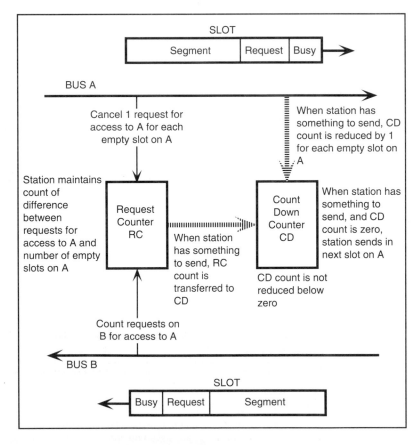

Figure 12.8 DQDB Access Procedure

toring the request field of slots that pass by on bus B.) It joins the queue for bus A by transferring the request count to the countdown counter and resetting the request counter to zero. The countdown count is reduced by one each time an empty slot on bus A passes the node. When the countdown counter is zero, all current requests to transmit from downstream stations have been satisfied, and the node transmits in the next empty slot.

The countdown counter is not reduced below zero. The number of stations in the distributed queue on bus A is obtained by counting requests passing on bus B and counting free slots passing on bus A. The difference between the two counts is the number of outstanding requests for slots on bus A. Because MANs carry voice and video traffic (i.e., isochronous traffic), arrangements must be made to ensure that a station with voice (or video) to send obtains regular access. To do this, a number of *prearbitrated* slots are included in the slot stream generated at the beginning of each bus. Within these slots, octets are assigned to the stations requesting isochronous service.

(2) From LLC PDU to Slot PDU

Figure 12.9 shows the components of the DQDB medium-access control sublayer. Divided into two blocks, its operation is directed by a layer management entity. The convergence block adds a 24-octet header (including destination address, source address, and class of service) and a 4-octet trailer to the LLC PDU

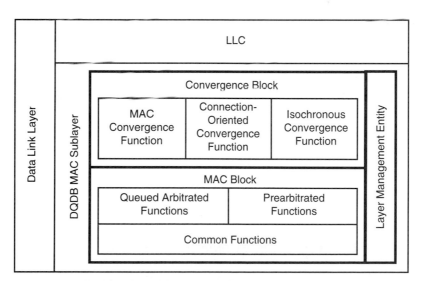

Figure 12.9 DQDB Medium Access Control Sublayer

The medium access control sublayer for DQDB is divided in two. The convergence layer adds information to the PDU, divides it into 44-octet fragments, adds information to aid reassembly, and creates 48 octet PDUs. The MAC block adds information concerning routing and control to create 53-octet slot PDUs.

to form the *initial* MAC PDU. If it is greater than 44 octets, the convergence block divides the PDU into 44-octet fragments, adds 2-octet headers (including segment type and message identifier), and 2-octet trailers (including error checking) to form 48 octet, *derived* MAC PDUs. In turn, by adding 4-octet headers (including virtual channel identifier), they are formed into 52-octet *segment* PDUs and then into 53-octet *slot* PDUs.

Routing information within the MAN is provided by the virtual channel identifier in the segment header; physical (source and) destination information is contained within the initial MAC PDU; and (source and) destination service access point information is provided within the LLC PDU. The essential steps in this transformation are shown in **Figure 12.10**. For illustration, an initial MAC PDU of *xx* octets is assumed, where 88 < *xx* < 132 octets. More details of the information contained in the headers and trailers are given below each diagram. When receiving, the sublayer functions reverse the procedure and assemble the LLC PDU from the incoming segments.

(3) Deployment

MANs may be deployed as individual open bus structures, as single rings, or as interconnected rings. In **Figure 12.11**, we show the way in which an open bus can provide high-speed connections between the individual locations of an enterprise centered on an urban core. Traffic from many nodes passes over each bus, raising questions of information security. With operations restricted to a single company, this may not be an important issue. For public networks, such is not the case, and separate access lines are provided between each customer and the MAN node that is located in a central office or other protected location.

Figure 12.12 shows the use of a ring-connected MAN to service several independent (perhaps competing) enterprises. Individual customer locations are connected through central office equipment to the high-speed node. In high-density urban areas, public MANs are likely to consist of orthogonal rings with common nodes. One of these structures is shown in **Figure 12.13**. Known as the Manhattan Street Network (MSN), it consists of orthogonal rings that pass information in alternating directions. A disadvantage of the topology is that routing decisions must be made at each node. The advantages are that connections between two nodes are likely to be shorter than in a conventional network, and alternate routes exist so that individual failures need not be catastrophic.

12.2.4 COMMENTS ON BUS-CONNECTED NETWORKS

In all kinds of establishments, Ethernet LANs operate successfully. They are the largest class of LANs in existence today. Yet there are problems achieving higher speeds, and the most popular development exchanges the shared bus for a high-speed switch. By borrowing from ATM technologies, 1 Gbit/s switched Ethernet systems appear probable.

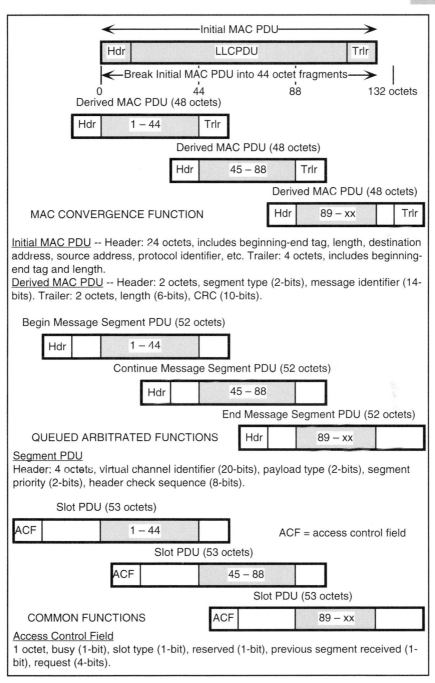

Figure 12.10 Formation of Slot PDUs
Detailed look at the process of dividing the initial MAC PDU into 53-octet slot PDUs.

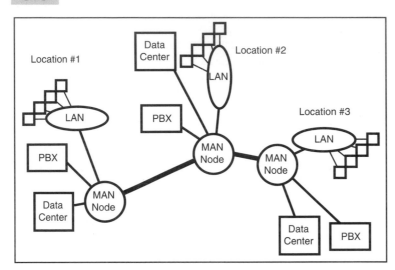

Figure 12.11 Single-Enterprise MAN

A DQDB open bus MAN that provides high-speed connections between the separated locations of an enterprise.

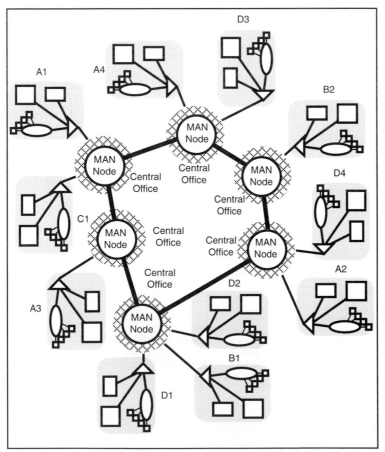

Figure 12.12 Public MAN

In a public MAN, provisions are made to safeguard each customer's traffic. The MAN nodes are located in central offices, and individual customers are connected through the distributing frame. The internode busses do not go through the customer's premises.

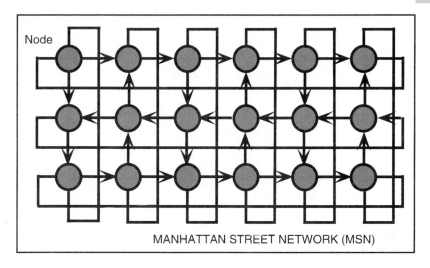

Figure 12.13 Urban Core MAN
The network consists of orthogonal rings that pass information in alternating directions.

Token bus LANs were developed by a consortium of manufacturers anxious to have a robust device for factory automation. Without other users, the demand for token bus LANs has evaporated.

Metropolitan area networks support LAN-to-LAN data communications across significant distances. In addition, MANs are compatible with SMDS (see Section 9.5), and the combination MAN and SMDS can carry LAN traffic over continental distances. They both employ cells and benefit from ATM technologies.

REVIEW QUESTIONS FOR SECTION 12.2

1   Use Figure 12.3 to describe CSMA/CD operation. Discuss what happens when two stations transmit at the same time.
2   Describe the frame used in CSMA/CD operation. Distinguish between LLC and MAC portions.
3   Identify the meanings of 10Base-T, 10Base-F, 10Broad-T, and 100Base-T.
4   Describe switched Ethernet. Is it really a LAN? Justify your answer.
5   What constitutes a token on a bus?
6   Why must there be a master station on a token bus LAN?
7   Describe the frame used by a token-passing bus LAN.
8   Define a Metropolitan Area Network.
9   Describe the open-loop and closed-loop configurations employed by MANs.

612

10   Describe the DQDB access protocol employed by MANs.
11   Explain Figure 12.9.
12   Explain Figure 12.10.
13   How is privacy protected on a public MAN?

12.3 RING-CONNECTED NETWORKS

Connecting networks in a ring requires their order to be defined so that they handle messages in sequence.

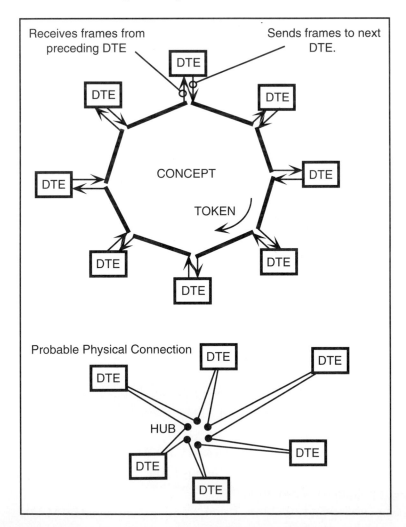

Figure 12.14 Principle of Token-Ring LAN

12.3.1 IEEE 802.5 TOKEN-RING

In a token-ring LAN operating under IEEE 802.5, each station is connected to two others so as to form a single-thread loop that connects all of the stations. Data are transferred around the ring from station to station; each station

- Receives the datastream from the station preceding it on the ring
- May add to it
- Regenerates it
- Passes the datastream along to the station next on the physical ring.

(1) Token-Ring

The principle of token-ring operation is shown in **Figure 12.14**. It is continued in **Figure 12.15,** where some of the more important decisions are sketched.

(2) What Constitutes a Token in a Token-Ring?

In a token-ring network, the token available frame (empty token) consists of 3 octets, a start flag, an access control octet, and an end flag. An empty token is shown in **Figure 12.16**. The access octet contains

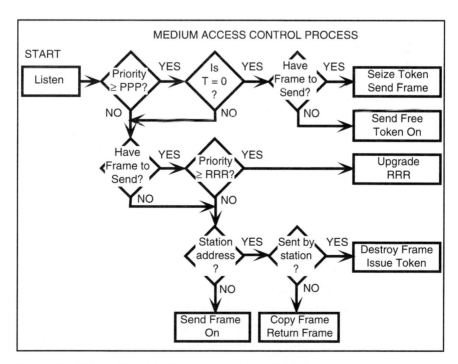

Figure 12.15 Token-Ring Medium Access Control Technique

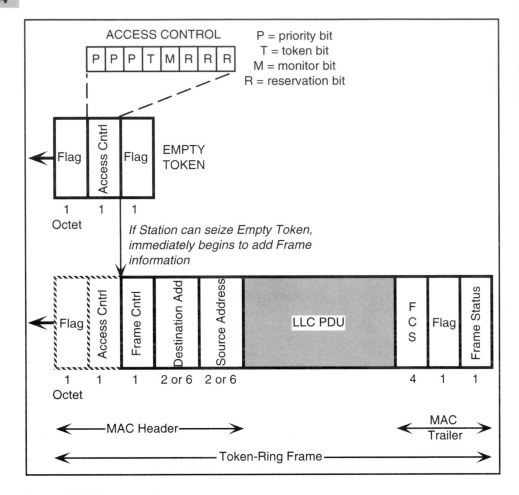

Figure 12.16 Token-Ring Frame Format

- 3 **priority bits** (PPP): they identify the level of priority a station must have to seize the token
- 1 **token bit** (T): it gives the token status. If it is 0, a station that has sufficient priority may seize the token. If it is 1, the token has been seized by another station and the frame is in use
- 1 **monitor bit** (M): it is used to detect unclaimed frames
- 3 **reservation bits** (RRR): they provide a mechanism for lower-priority devices to request the opportunity to transmit.

(3) Basic Operations

Some of the basic operations involved in the functioning of a token-ring LAN are

- **Seizing the token**: when a station with data to transmit has priority equal to or higher than the priority bits in an empty token frame, it
  - seizes the token
  - changes the token bit to 1
  - adds data to it
  - transmits the datastream to the next station on the ring.

  The formation of a transmission frame from an empty token frame is shown in Figure 12.16.

- **Receiving and acknowledging a frame**: when a station receives a filled frame that is addressed to it, it
  - copies the information
  - reverses the source and destination addresses so that the frame will return to the originating station
  - adds an acknowledgment to the trailer
  - sends the frame on to the next station.

- **Regenerating an empty token**: when the returned frame (above) is received by the originating station, it
  - recognizes it is a frame that it sent some time ago
  - destroys the frame
  - regenerates an empty token frame in which T = 0
  - replaces the priority bits in the access control octet with the priority expressed in the reservation bit space
  - sends the empty token to the next station.

- **Using reservation bits**: when a station with something to send receives an empty token, it examines the priority level. If the token priority is greater than the priority the station enjoys, the station next examines the priority number contained in the reservation field. If it is less than the station priority, the station upgrades the R-bits to reflect its level of priority. In this way, the station protects the next empty token from being seized by a station with lower priority. However, it cannot prevent a station with a higher level of priority from upgrading the R-bits further. Nor can it prevent a station with the same priority, but located earlier in the sequence, from seizing the next empty token.

- **Using the monitor bit**: when a station fails, messages addressed to it are likely to continue to circulate around the ring. To prevent this, one station is designated to monitor traffic. When filled frames pass the monitoring station, it sets the M-bit to 1; then, if a frame passes the monitor station with M = 1, the station knows it has made a complete circuit without being received and processed. The monitoring station destroys the frame and generates an empty token.

- **Using priority levels**: with three P-bits, eight levels of priority can be used. By assigning different levels to each station (or class of station) on the network, individual opportunities to send can be made to match the amount of traffic they have and the delay they can tolerate. Proper assignment of priority levels is key to the successful operation of a token-ring LAN. Adjusting each station's priority level so that all have an opportunity to transmit their information is a formidable challenge.

(4) Station/Ring Interface

Because all stations in the ring handle each message, it is important that the messages pass through each access point as quickly as possible. At the same time, the receiving station must interrogate each bit, recognize particular bits (T- and M-bit), and bit patterns (PPP, RRR, and destination address, for instance), and take action as necessary. When an empty token frame is first received, the flag is used to synchronize the station receiver and identify the beginning of the control octet. The next octet, the control octet, contains priority level, token status, and reservation status. If the station has something to send, it compares the priority bits with its own priority level. If the station has a higher priority than the three P-bits received, the next bit is crucial.

- If it is 0, the station is receiving an empty token (T = 0) and can seize it. It sets the T-bit to 1 and, as soon as the reservation bits have cleared, proceeds to transmit the frame control octet and the remainder of the message.
- If it is 1, the station is receiving a filled token (T = 1). There is no opportunity to send.

With a filled token, the next matter of interest is whether the PDU is addressed to the station. If it is, *and*

- It is not the PDU the station just sent, it copies the data it contains, performs a cyclic redundancy check, reverses the source and destination addresses, adds a frame status octet, and transmits the PDU to the next station
- It is the PDU the station just sent, it generates an empty token frame, transfers the R-bits to the P-bit field, sets the R-bits to zero, and transmits the empty token to the next station. If the frame status octet indicates retransmission is required, the station waits until the next time it can seize an empty token to do this.

(5) Physical Medium

IEEE 802.5 is strongly supported by IBM and is the basis for the IBM token-ring product line. IBM's cabling system uses twisted pairs with Manchester signaling at speeds of 1, 4, and 16 Mbits/s. A *multistation access unit* (MAU) provides the ability to connect stations by UTP wiring to a central device in which the token ring is implemented. Further, MAUs can be connected in a ring to connect communities of stations, and the ring can be made *self-healing* by arranging for it to reverse itself on the failure of one link. Some of these alternative configurations are shown at the bottom of Figure 12.14.

12.3.2 FIBER-DISTRIBUTED DATA INTERFACE NETWORK

Intended for use within a campus, Fiber-Distributed Data Interface (FDDI) Network provides the backbone for a distributed computing environment.

(1) FDDI Configuration

Standardized by ANSI, FDDI is a dual-ring, optical-fiber network that connects up to 500 stations in a ring of ≤ 100 kilometers. With multimode fibers, point-to-point links are limited to ≤ 2 kilometers; using single-mode fibers, individual links can be extended up to 50 kilometers. Data are passed at a speed of 100 Mbits/s using a token-passing protocol. One hundred Mbits/s is defined by ANSI as the base data rate. It can be increased in increments of 6.144 Mbits/s. FDDI employs 4B/5B coding (i.e., 4-bit binary substituted by 5-bit binary—4 signal bits are represented by a 5-bit symbol). Thus, the signaling rate for 100 Mbits/s is 125 Mbits/s. FDDI frames (up to 4,500 octets) can be transported over SONET connections as STS–3c modules [STS–3c = concatenated payloads of 2,322 octets, see Section 6.4.1(4)].

Figure 12.17 shows the concept of an FDDI network. Stations are connected to the fiber rings by channel interface units (CIUs). Normally, CIUs receive frames from their upstream neighbors and send frames to their downstream neighbors over the primary fibers. Should a link in the ring be interrupted, using the standby fibers in the manner shown in the lower diagram bridges the gap. Although conceived as a fiber-based network, versions of FDDI are available that employ shielded or unshielded twisted pairs. They can be used in offices and buildings that do not require long distances between stations (meters not kilometers), and are designed to employ existing wiring.

(2) FDDI Operation

Permission to send is derived from possession of a token. This message is circulated in the ring and can be seized by a station that has a frame to send. Possession of the token permits the station to send for a specific period. When the period is up, or the station has no more frames to send, the station generates a new token and sends it on to its downstream neighbor. The format of the token and

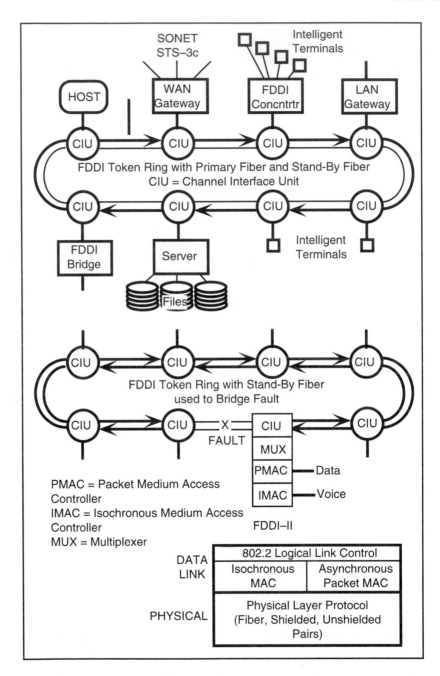

Figure 12.17 FDDI Network

In the lower diagram, two separate points are illustrated. (1) When a fault occurs, operation is maintained through the use of the stand-by fiber. (2) In FDDI–II, the channel interface unit is augmented with a multiplexer and controllers to carry both voice and data.

information frames is shown in **Figure 12.18**. For all frames, the CIU performs an error check by computing the frame check sequence. If an error is detected, it sets the error symbol in the frame status field. For frames that are not addressed to one of the logical entities supported by the CIU, it sends the frame forward to its downstream neighbor. If the frame is addressed to one of the logical entities sup-

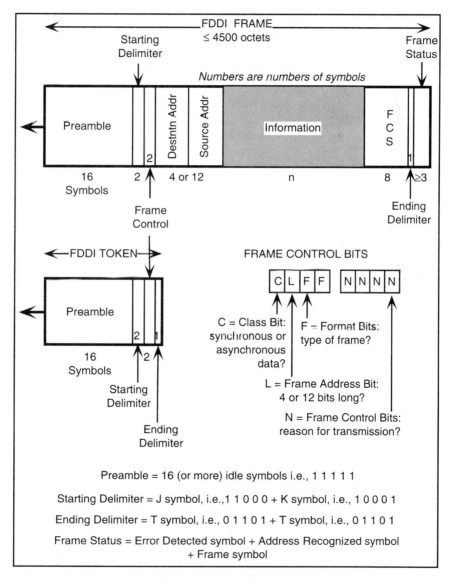

Figure 12.18 FDDI Frame Format

ported by the CIU, it copies the relevant information, forwards it to the logical entity addressed, sets the address recognized bit, and sends the frame on to its downstream neighbor to complete a circuit of the ring. On arriving back at its source, the originating station notes that the frame has (or has not) been received correctly and removes the frame from the ring. If retransmission is required, it occurs next time the station has the token.

(3) FDDI–II and FFOL

FDDI is a data-only network. With the appearance of techniques such as SONET and ATM that handle asynchronous packet data and isochronous voice and video, FDDI has been enhanced to provide voice and data transport. Called FDDI–II, it includes a synchronous, two-way transmission capability. To achieve this state, each node is equipped with two medium access controllers. One, known as PMAC (packet medium access controller) supports data applications. The other, known as IMAC (isochronous médium access controller) supports voice applications. Asynchronous data or synchronous voice signals are carried in standard FDDI frames. FDDI–II operates in two modes

- **Basic mode**: the same as FDDI—i.e., data only
- **Hybrid mode**: asynchronous data and synchronous voice are carried in FDDI frames.

When receiving, the mixed frames are demultiplexed and directed to the appropriate MAC. When sending, the frames are directed to the multiplexer by their respective controllers for transportation over the ring.

In the lower diagram of Figure 12.17, I show the physical arrangement of MUX, PMAC and IMAC, and the data link and physical layers for FDDI–II. The combination of multiplexer and IMAC is known as the hybrid ring control (HRC). Another extension of FDDI is known as FFOL (FDDI Follow-On LAN). As the name suggests, work is already under way to address a gigabit version of FDDI that will be compatible with higher-speed SONET levels (STS–48, 2.488 Gbits/s, for instance) and B-ISDN. FFOL will be compatible with FDDI and FDDI–II, and will transport the full range of signals required by B-ISDN.

REVIEW QUESTIONS FOR SECTION 12.3

1 Describe the connections and functions of a station on an IEEE 802.5 token-ring LAN.

2 Describe the principle of token-ring operation. Explain Figures 12.14 and 12.10.

3 What constitutes a token in an IEEE 802.5 LAN?

4 Describe the basic operations involved in the functioning of a token-ring LAN.

5 Discuss the ways in which the receiving station determines the purpose of an arriving frame.

6 Describe the configuration of an FDDI network.

7 Describe the modes of operation of an FDDI network.

8 Distinguish among FDDI, FDDI-II, and FFOL.

12.4 LAN INTERCONNECTIONS

Most often, LANs are employed to connect stations in a building or a building complex. Within the same local area there may be other LANs to which interconnection is necessary to share the same databases, or for other purposes. What is more, there may be other LANs in other buildings remote from the complex that require interconnection, and some of them may be a considerable distance away. As the effectiveness and efficiency of LANs improve, there are more and more reasons to want to interconnect them with wide-area networks so that the pockets of information activity that make up the modern corporation can communicate on any level. To achieve transparent information exchanges between users anywhere, differences between the networks must be reconciled.

12.4.1 USE OF REPEATERS, BRIDGES, ROUTERS, AND GATEWAYS

From the discussion of Ethernet, Token-bus, and Token-ring LANs, it is obvious that two networks can differ greatly in access methods, frame sizes, frame formats, sequence numbering techniques, addressing conventions, and other items. In connecting two networks the management information that originates in one network must be processable in the network in which it terminates, and in any intermediate networks that it traverses. Also, the user's data must be understandable in the network in which it terminates.

As explained in Section 8.5, the device used to effect interconnection depends on the highest layer in which differences exist. Examples of the use of repeaters, bridges, routers, and gateways in connecting one LAN to another are given in **Figure 12.19**. Their application must be worked out in exact detail for each specific combination of networks. Not only must normal operations be supported, but also abnormal operations, including recovery from failures, must be assured. An example of the interconnection of two dissimilar, geographically separated LANs by a packet network is given in **Figure 12.20**. Each host is assumed capable of direct connection to their respective LANs, and TCP and IP are used as the transport and network protocols. The hosts generate frames that are matched to the requirements of an Ethernet LAN and a Token-Ring LAN. At the gateways, the frames are converted to frames that are compatible with the packet network.

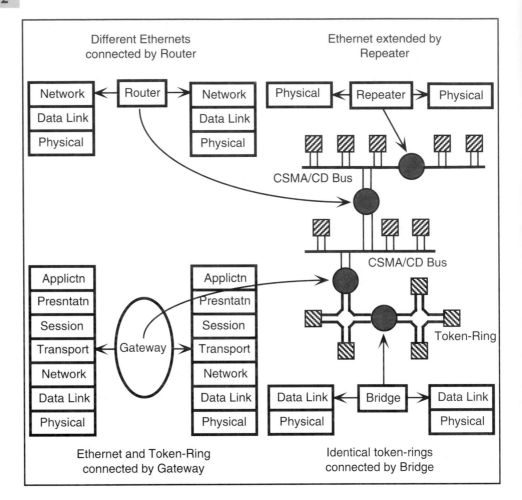

Figure 12.19 LAN Interconnection Techniques

Repeaters are used to connect the same LANs at the Physical Layer. Bridges are used to connect similar LANs at the Data Link Layer. Routers are used to connect similar LANs at the Network Layer. Gateways are used to connect dissimilar LANs at the Transport Layer.

12.4.2 NOVELL'S NETWARE PROTOCOL STACKS

Implementing these devices in a LAN interconnection environment can be done in many ways. A popular approach uses software developed by Novell, a leading supplier of local-area networks. Called NetWare, Novell's network operating system includes proprietary network and upper layers that facilitate routing between networks. **Figure 12.21** shows the protocol stack. It includes

- **Internet packet exchange** (IPX): the network layer

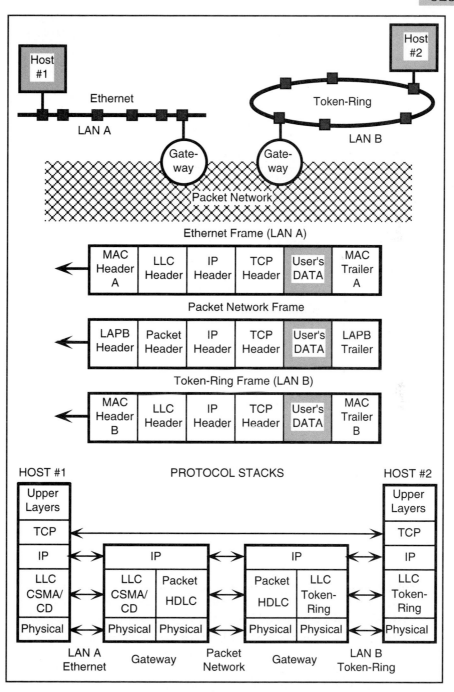

Figure 12.20 Connecting an Ethernet LAN to a Token-Ring LAN by Means of a Packet Network

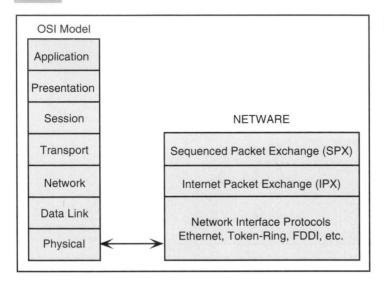

Figure 12.21 Novell's Netware Protocol Stack

- makes a best-effort attempt to deliver packets to their destination
- uses a 12-byte address that includes the network address of an individual work group, the node address of the individual workstation in the workgroup, and a socket address that identifies a particular application on the station
- **Sequenced packet exchange** (SPX): the transport layer
  - provides virtual circuit service
  - uses checksum for error detection
  - sends positive acknowledgements
  - on timeout, retransmits packet
  - ensures sequencing
  - provides flow control.

An IPX/SPX router/gateway uses the destination address in the IPX header to select the best path for the packet and sends it on its way through the network. SPX provides connection-oriented services, establishes virtual circuits, provides error control, and assures correct sequencing. In addition, NetWare includes an interface to IBM's NetBIOS (network basic input/output protocol) programming interface for IBM PC networks, and two protocols that support the routing process

- **Service advertising protocol**: means for a router to acquaint neighboring routers with its capabilities
- **Routing information protocol**: means to select an optimum route for a particular destination.

The uses of these classes of protocols are discussed later in Section 12.7.4.

12.4.3 Use of ATM to Produce a Seamless Network

With the ability to handle multimedia signals, and to provide connection-oriented or connectionless service, a single ATM node can function as the central hub of a star-connected, local-area network. Individual equipment with ATM adapter cards (see Figure 12.22) can be directly connected to it, and connections can be made between the hub and other public or private facilities.

Figure **12.22** shows the concept of an ATM wide-area network that connects high-speed workstations, servers, LANs, and hosts over a fiber backbone. Using the multiplexing capabilities of the segmentation and reassembly sublayer of the ATM adaptation layer (AAL), traffic from LANs (which provide connectionless services) is carried from ATM gateways to servers or other LANs as multiplexed connection-oriented streams. Across the UNIs, AAL type 1 or type 2 may be used for video and voice, and AAL type 5 for data services [see Section 9.4.1(2)]. A regional private network can be established with small ATM nodes interconnected by SONET fibers. In turn one or more of the private switches can be connected to a public ATM network.

Without distance limitations, ATM is a routing technology that can be applied universally—i.e., to local, regional, and wide-area networks. Its use frees multimedia communications from a proliferation of special devices and creates a seamless network that can extend from the very local to the worldwide.

REVIEW QUESTIONS FOR SECTION 12.4

1 Explain Figure 12.19.
2 Describe the protocol stack for Novell's NetWare.
3 Explain Figure 12.22.

12.5 EVOLUTION OF INTERNET

Growing at an extraordinary pace, Internet provides public access to libraries, data archives, bulletin boards, directories, databases, games, and much more; it supports electronic mail services, facilitates the transfer of data files, and presents material in colorful, attractive ways to the accompaniment of sound and moving pictures. Recent additions include CommerceNet (provides commercial services that include billing and secure payments), and the World Wide Web. Created by researchers in Switzerland, the *Web* is a collection of on-line documents resident on Internet servers around the world. In response to requests, software agents (known as Web *browsers*) seek out information for clients and direct the servers on which it resides to send the documents to (servers serving) the requesters. Today, users of the World Wide Web generate the largest individual component of traffic on Internet (around 25%). The success of Internet has stimulated plans for national and global information infrastructures (NII and GII, see Section 12.10).

12.5.1 RECENT HISTORY

In 1990, ARPAnet (see Section 1.2.2) was formally abolished.[1] Changes in organization and mission made ARPAnet resources available for commercial and public uses and transferred responsibility for operations to a government-university-industry consortium. With NSFnet[2] as backbone, ARPAnet became the Internet as we know it today. In the National Science Foundation 1991, (NSF) upgraded NSFnet to DS–3, several national data network utilities were formed or expanded to provide data transport at DS-3 and SONET speeds, and commercial use of the network was permitted.

In 1995, NSF relinquished its interest in Internet; the network was *privatized*. It became a network of interconnected resources owned and operated by commercial communications companies (MCI, Sprint, GTE, ATT, etc.). To provide the nation's research and engineering (R&E) community with a high-performance network, NSF (partnered by MCI) turned to the creation of vBNS (very high performance Backbone Network Service). Between a limited number of R&E institutions it operates at STS–12 (622.08 Mbits/s).

In 1997, a group of agencies of the U.S. government began to sponsor the development of Next Generation Internet (NGI). Meanwhile, concerned by the growing congestion of Internet, a group of over 100 universities and corporate partners is studying ways to address the challenges facing the next generation of university networks. Known as Internet2 (I2), the group's objective is to focus energy and resources on ways to meet emerging academic requirements in research and teaching.

12.5.2 DIRECTION

Three organizations provide loosely integrated direction to Internet

- **The Internet Society**: an international organization that promotes cooperation and coordination. It includes

---

[1] For an interesting account of the major events from 1961, see Richard H. Zakon, *Hobbes' Internet Timeline v3.3*, http://info.isoc.org/guest/zakon/Internet/History/HIT.html

[2] NSFnet was created in 1986 to provide data communications between five supercomputing centers.

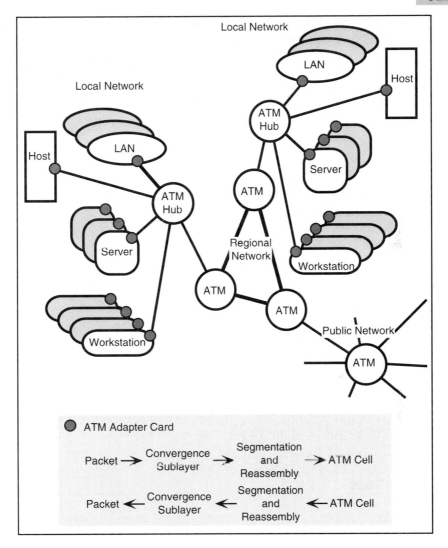

Figure 12.22 Application of ATM to LANs and Regional Networks

ATM hubs connect high-speed terminals, servers, and hosts in local networks that are joined into a regional network and connected to the public network. Use of ATM frees multimedia communications from a proliferation of special devices designed to make the transition from one network to another.

- Internet Activities Board
- Internet Engineering Task Force
- Internet Assigned Numbers Authority

They are concerned with network architecture, the evolution of TCP/IP protocols, and assigning IP addresses.

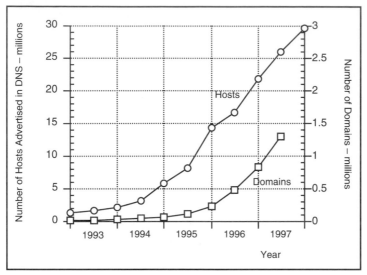

Figure 12.23 Estimated Number of Hosts and Domains on Internet

Data produced by Network Wizards and published at http://www.nw.com.

- **Internet Registry**: administers *Domain Name System*. Associates domain names with IP addresses. Works through InterNIC (Internet Network Information Center) in the United States, RIPE Network Coordination Center in Europe, and APNIC (Asia-Pacific Network Information Center) in the Asian-Pacific region.
- **World Wide Web Consortium** (W3C): an industry consortium managed by MIT's Laboratory for Computer Science; develops common standards for the evolution of the World Wide Web.

12.5.3 SIZE

Estimates of Internet's size are hard to make. **Figure 12.23** shows the exponential growth of the number of domains and number of hosts through the end of 1997.[3] With certainty it can be stated that today's network contains more than 1.5 million domains populated by more than 30 million hosts. The adoption and gradual introduction of IPv6 [see Section 12.6.2(2)] ensures that Internet will continue to grow to serve a burgeoning computer-literate community.

---

[3] Network Wizards (http://www.nw.com), which update its estimates every 6 months, provides the figures, and are kind enough to permit them to be reproduced on condition of acknowledging the source.

12.5.4 ARCHITECTURE

Internet consists of local networks that are connected into regional networks (autonomous networks, see Section 12.7). In turn they connect to a high-speed backbone to form a global network. **Figure 12.24** shows a representation of Internet. It contains local networks operated by Internet service providers (ISPs) that are interconnected to form a regional network. Through a hub they connect to network access points (NAPs) on a high-speed backbone. In addition, Internet contains on-line service providers (OLSPs, such as America Online and Compuserve) that operate nationwide networks. They are likely to connect directly to a network access point.

Internet service providers and on-line service providers attract subscribers by leasing access to Internet through their facilities. Information providers (IPs) operate servers that dispense information files on request. The entire network is permeated by routers that direct packets between clients and servers by reading the destination address in the IP header, consulting the routing table they contain for instructions, and sending the packets on to their destination.

Internet users share a layered set of rules by which computer processes are able to communicate effectively across this network of networks. Called TCP/IP, the name is shorthand for a suite of protocols whose major members are TCP (transmission control protocol), a transmission layer procedure, and IP (Internet protocol), a network layer procedure. The number of TCP/IP installations is growing at an exponential rate. They provide decentralized networks with dynamic routing and easy connectivity—the sort of environments that can support client-server applications. However, for transaction-oriented applications that require a high degree of certainty of execution (such as airline reservation systems and banking systems) the informality of TCP/IP is a deterrent to its use. [For these applications, IBM's Systems Network Architecture (SNA) (see Sections 13.4 and 13.5) is the implementation of choice.]

REVIEW QUESTIONS FOR SECTION 12.5

1   What is CommerceNet?

2   What is the World Wide Web?

3   Describe the relationship between Internet, Next Generation Internet, and Internet2.

4   What is The Internet Society? Name three subsidiary organizations and describe what they do.

5   What is the Internet Registry?

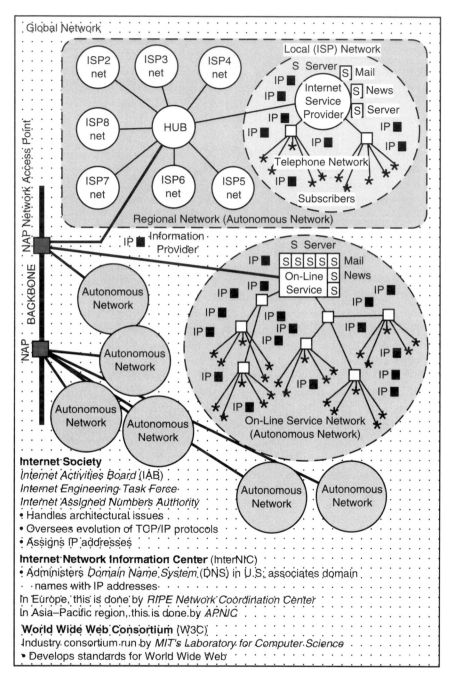

Figure 12.24 Representation of Internet

The Internet is a global network of networks. Because there is no central managing authority, it has been described as the purest form of electronic democracy.

6 What is the World Wide Web Consortium?

7 What is the size of Internet?

8 Explain Figure 12.24.

9 For what applications is IBM's SNA better than TCP/IP?

12.6 TCP/IP

The transport layer protocol used by the Internet community can operate with a variety of network layer protocols. Known as TCP, and most often used with IP datagram service, it was the first, successful, general-purpose, transport layer protocol. TCP[4] is a connection-oriented protocol responsible for the reliable transfer of information between source and destination hosts. In spite of possible problems in the network layer protocol that may cause packet duplication, packet loss, or the delivery of packets out of sequence, TCP guarantees delivery of the exact datastream passed to it by the source. Sequence numbers, acknowledgments that track datagrams, and timers that look for acknowledgments within specified time periods accomplish this.

In 1988, the Department of Defense specified TCP/IP as an interim protocol set to be used in the migration to an open systems (OSI) environment for government-purchased products. Because of the difficulties encountered in standardizing an OSI network protocol (see Section 8.4.1), in 1994 the National Institute for Standards and Technology (NIST) recommended that the U.S. Government OSI Profile (GOSIP) incorporate TCP/IP.

12.6.1 TRANSMISSION CONTROL AND USER DATAGRAM PROTOCOLS

By sending a window value to the source host, TCP performs flow control at the destination host. It authorizes the source to send just so many octets. When that number is reached, the source ceases to transmit until another authorization is received. **Figure 12.25A** shows the makeup of a TCP data unit (TCPDU). While most of the entries are readily understood, the checksum needs an explanation. It is computed by placing the 32-bit source address (contained in the IP Header), the 32-bit destination address (contained in the IP header), and other information ahead of the TCPDU and performing the calculation over these items and the entire TCPDU. The inclusion of the addresses in the checksum is a means of ensuring that the complete PDU reaches the correct host attached to the correct network. TCP employs virtual circuits (defined by the combination of source and

---

[4] Pete Loshin, *TCP/IP Clearly Explained*, 2nd ed. (San Diego, CA: Academic Press, 1997).

A: TCP Data Unit Format

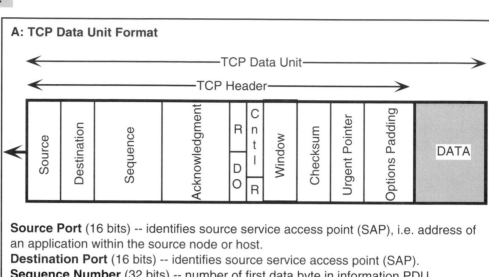

Source Port (16 bits) -- identifies source service access point (SAP), i.e. address of an application within the source node or host.
Destination Port (16 bits) -- identifies source service access point (SAP).
Sequence Number (32 bits) -- number of first data byte in information PDU.
Acknowledgment Number (32 bits) -- sequence number of next byte expected by TCP entity.
Cntrl = Control Bits (6 bits) -- bits are set to indicate urgency, request immediate delivery, request reset, synchronize sequence numbers, and end of datastream.
R = Reserved (6 bits) -- reserved for future uses.
Window (16 bits) -- for flow control, number of bytes that receive port will accept before acknowledgment.
Checksum (16 bits) -- see text for explanation.
Urgent Pointer (16 bits) -- indicates sequence number of byte following urgent data.
Options (variable) -- requests specific receive buffer size, and other services.
Padding (variable) -- ensures that Header is a multiple of 4 octets long.

B: UDP Data Unit Format

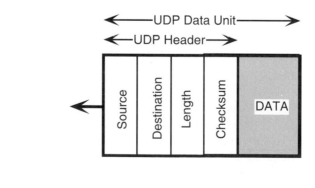

Figure 12.25 TCP and UDP Data Unit Formats

destination identifiers) and ensures that all, in process, data operations are completed before breaking a connection.

For many applications, the capabilities of TCP are more than are required. This is particularly true for relatively short messages. For them, a connectionless procedure called the user datagram protocol (UDP) can be used. Without flow control and error recovery procedures, UDP employs the PDU shown in Figure 12.25B. It contains a destination port number that identifies the point of termination in the upper layer protocol, and may contain a source port number that identifies the port in the source host to which any reply is sent. The checksum, which is calculated in the same manner as the checksum for TCP, is used to check incoming datagrams. If the checksum is invalid, the datagram is destroyed.

12.6.2 INTERNET PROTOCOL

IP version 4 (IPv4) provides a robust, connectionless service between hosts in which data from the sending host is joined to an IP header to form an IP datagram.

(1) IP Operation

Shown in **Figure 12.26**, the datagram is passed to a subnetwork in which it is encapsulated in the appropriate data link layer frame. For the particular arrangement shown, the datagram is passed across the subnetwork to an IP gateway that links to a second subnetwork. In turn, this network is connected to the destination host. At the gateway, the subnetwork information is stripped off the datagram, and it is encapsulated in a data link frame appropriate to the next leg of the journey. The gateway uses the destination address in the datagram header to route it to the intended host. Eventually, the datagram arrives at its destination and is delivered to the host. Practically, as long as the gateways know how to access them, the route can traverse any number of subnetworks and the subnetworks can employ different data link control protocols (DLCPs). To prevent datagrams from circulating indefinitely, the header contains a *time to live* number that is reduced by one every time the datagram traverses a gateway. If the number becomes zero, the datagram is destroyed. For lost, missing, or destroyed datagrams, it is up to the transport layer to provide corrective action.

(2) IP Addressing

The leading bits in the 32-bit Internet address block identify the network on which a host resides (called the network number or network prefix). The remaining bits identify a particular host on a given network (called the host number). To accommodate different size networks, three address formats are employed. Called *classful* addressing, they are

- **Class A or /8 Network**: address consisting of 8-bit network prefix beginning with 0, and 24-bit host number. Because they have an 8-bit network prefix,

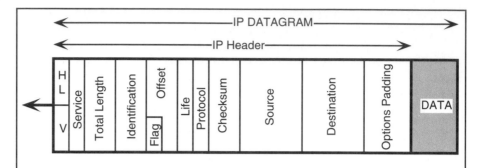

V = Version (4 bits) -- identifies version of IP in use.
HL = Header Length (4 bits) -- specifies length of IP Header in multiples of 4 octets.
Service = Type of Service (8 bits) -- specifies parameters such as desired reliability and throughput.
Total Length (16 bits) -- identifies length of datagram (or current fragment). Maximum length 65 535 bits.
Identification (16 bits) -- sequence number.
Flag (3 bits) -- permit, or prohibit, fragmentation of datagram.
Offset (13 bits) -- i.e. fragment offset. Indicates where in PDU this fragment belongs.
Life = Time to Live (8 bits) -- measured in gateway hops. Ensures fragment does not loop indefinitely.
Protocol (8 bits) -- identifies next-level protocol to receive data at destination.
Header Checksum (16 bits) -- performs error check on Header.
Source Address (32 bits)
Destination address (32 bits)
Options (variable) -- requests specific routing, handling, and other services.
Padding (variable) -- ensures that Header is a multiple of 4 octets long.

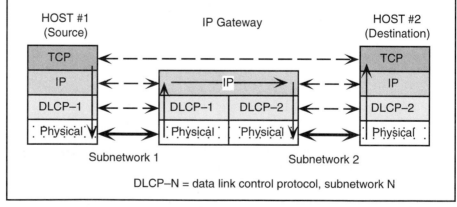

Figure 12.26 IP Datagram Format
The lower diagram shows the protocol stacks employed by two hosts using IP to communicate through an IP-configured gateway (or router).

Class A Networks are called *slash eights* or *eights* (written /8s). The address block contains 2^{31} individual addresses; they identify $2^7 - 2$ (i.e., 126 /8) networks each of which can support 16,777,214 (i.e., $2^{24} - 2$) hosts. (Two is subtracted because all-0s and all-1s in bits 1 through 7 are employed for other purposes.) slash eight addresses account for one-half of all Internet addresses.

- **Class B or /16 Network**: address consisting of 16-bit network prefix beginning with 10, and 16-bit host number. Because they have a 16-bit network prefix, Class B Networks are called *slash sixteens* or *sixteens* (written /16s). The address block contains 2^{30} individual addresses; they identify 16,384 /16 networks each of which can support 65,534 hosts. Slash sixteen addresses account for one-quarter of all Internet addresses.

- **Class C or /24 Network**: address consisting of 24-bit network prefix beginning with 110, and 8-bit host number. Because they have a 24-bit network prefix, Class C Networks are called *slash twenty-fours* or *twenty-fours* (written /24s). The address block contains 2^{29} individual addresses; they identify 2,097,152 /24 networks each of which can support 254 hosts. Slash twenty-four addresses account for one-eighth of all Internet addresses.

Two other classes are defined

- **Class D**: has a 32-bit address block with leading bits 1110. It is used for multicasting.
- **Class E**: has a 32-bit address block with leading bits 1111. It is reserved for experimental use.

To make them easier to read and write, IP addresses are expressed as four-decimal numbers. Called *dotted-decimal* notation, it consists of the decimal equivalents of the 4 bytes in the 32-bit address block separated by dots. Finally, to facilitate the division of networks into smaller pieces, the host number field can be separated into a subnet number and a host number for that subnet. In this way, specialized subnetworks can be established within the overall network. **Figure 12.27** shows examples of /8, /16, /24, subnet, and dotted-decimal addresses.

(3) IPv6

Internet is characterized by phenomenal growth—the number of networks doubles every 9 to 12 months—so that

- The present Internet Protocol (IPv4) is rapidly running out of available addresses
- The size of Internet routing tables is growing exponentially, making their timely computation and efficient management impossible.

Figure 12.27 Examples of Internet Addresses

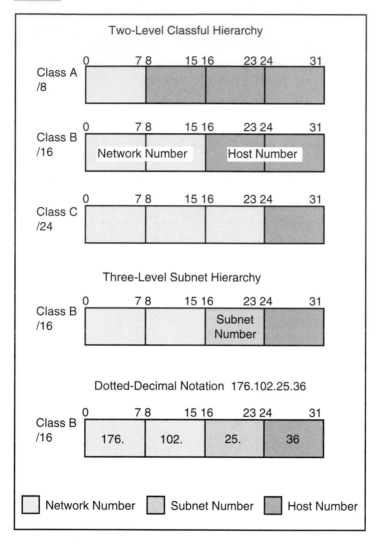

Accordingly, a next-generation protocol (now standardized as IPv6[5]) has been developed and is being deployed with a 128-bit address space. Intended to be interoperable with IPv4, IPv6 is designed to serve high-performance networks, yet be efficient for less demanding applications. Early adopters of IPv6 are creating an overlay backbone network called *6bone* (i.e., IPv6 backbone). It is likely that Internet will be a mixed IPv4/IPv6 network for the foreseeable future.

---

[5] Christian Huitema, *IPv6: The New Internet Protocol* (Upper Saddle River, NJ: Prentice-Hall, 1996).

12.6.3 TCP/IP Suite

Thanks to TCP/IP, ARPAnet has evolved into an international *network of networks* that serves data users around the world. That so large a set of networks can communicate is a tribute to the wisdom of the early designers of these protocols and the addressing scheme. Servers and clients convert their data into TCP/IP packets for transmission, then convert them back to their native languages for further processing. De facto, TCP/IP has become the universal data communication procedure that permits members of the public, with machines made by different manufacturers, in different lands, to exchange data on an ad hoc basis, among themselves, and with organizations of all kinds.

Probably the first segmentation and reassembly protocol, TCP converts messages into streams of packets at the sender and reassembles them into messages at the receiver. IP (Internet Protocol), a connectionless, best efforts, packet switching protocol, is used as the network layer. It is responsible for maintaining the terminating address as the packet is routed over several networks to its destination.

Shown in **Figure 12.28**, the major members of the TCP/IP protocol suite are

- **Application protocols**: interface user processes with lower-level protocols. Some specific capabilities are
  - *TELNET*. Virtual terminal service that provides access to remote hosts for terminals attached to a local host
  - *File transfer protocol* (FTP). Transfers files between remote servers
  - *Simple mail transfer protocol* (SMTP). Transfers mail between servers, and servers and users
  - *Domain name service* (DNS). Provides numerical IP address and other information concerning remote server identified by domain name
  - *Hypertext transfer protocol* (HTTP). Facilitates file transfers, and communications between browsers and Web servers, for HTML (HyperText Markup Language) documents
  - *NetBIOS*. Supplements IBM's NetBIOS with internetworking capabilities of TCP/IP
  - *Network news transport protocol* (NNTP). Facilitates the operation of Usenet, a message service for newsgroups. (Newsgroups are e-mail discussion groups that focus on particular topics.)
- **Sockets**: a software entity called a *socket* serves as the interface between the client and TCP/IP protocols. Called Winsock (for PCs) and MacTCP (for Macintoshs), it sits on top of the transport protocol and performs a type of *session* layer function. A secure sockets layer (SSL) has been developed.
  - It establishes a private connection
  - It authenticates the identities of client and server
  - It guarantees message integrity.

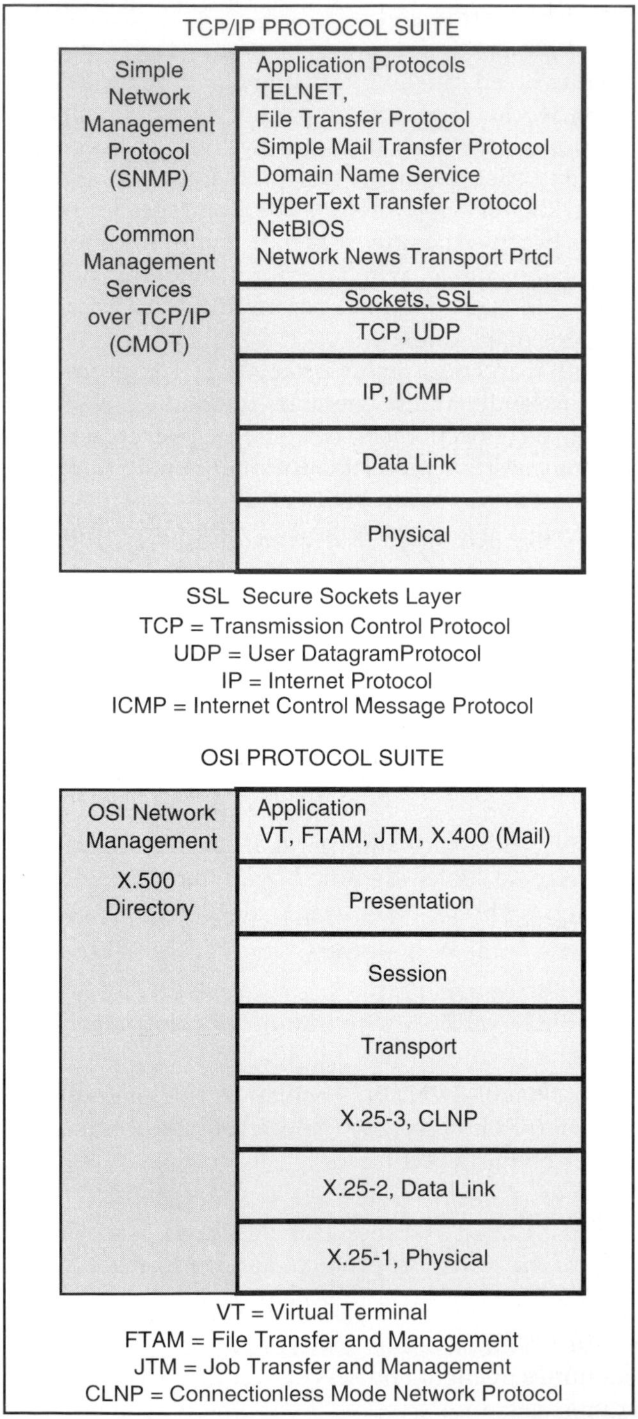

TCP/IP PROTOCOL SUITE

Simple Network Management Protocol (SNMP) Common Management Services over TCP/IP (CMOT)	Application Protocols TELNET, File Transfer Protocol Simple Mail Transfer Protocol Domain Name Service HyperText Transfer Protocol NetBIOS Network News Transport Prtcl
	Sockets, SSL
	TCP, UDP
	IP, ICMP
	Data Link
	Physical

SSL Secure Sockets Layer
TCP = Transmission Control Protocol
UDP = User DatagramProtocol
IP = Internet Protocol
ICMP = Internet Control Message Protocol

OSI PROTOCOL SUITE

OSI Network Management X.500 Directory	Application VT, FTAM, JTM, X.400 (Mail)
	Presentation
	Session
	Transport
	X.25-3, CLNP
	X.25-2, Data Link
	X.25-1, Physical

VT = Virtual Terminal
FTAM = File Transfer and Management
JTM = Job Transfer and Management
CLNP = Connectionless Mode Network Protocol

Figure 12.28 TCP/IP Protocol Suite Compared with OSI Protocol Suite

- **Transport protocols**: establish, control, and terminate network connections between data structure end-points on source and destination servers. Specific capabilities are as follows:
  - *Transmission control protocol* (TCP). Provides reliable, connection-oriented network communications between peer processes. Maintains and controls the state of the connection and ensures that all data have been received by the destination server.
  - *User datagram protocol* (UDP). Provides connectionless network communications between peer processes.
- **Network protocols**: implement addressing that supports communications between network devices and provides routing for transmitting data between networks. Specific capabilities are as follows:
  - *Internet protocol* (IP). Provides the Internet address to identify network devices and physical networks. Fragments or reconstructs data segments as needed to suit diverse network requirements.
  - *Internetwork control message protocol* (ICMP). Returns error messages to source when appropriate.
  - *Address resolution protocol*. Translates 32-bit IP addresses and 48-bit Ethernet addresses.
- **Data link protocol**: employs HDLC, LAP-B, LAP-D, and IEEE 802.2 (LLC) protocols.
- **Physical layer**: incorporates a wide range of physical interfaces such as EIA232/449 [see Section 7.1.2(1)], ISDN (see Section 10.4), IEEE 802 (see Section 12.1.2) and ITU-T V.35.

12.6.4 COMPARISON WITH OSI MODEL

In Figure 12.28, the TCP/IP suite is compared with protocols contained in the OSI model of a server working into a packet network. In the physical layer, both suites accommodate a range of connections that includes EIA232/449, ISDN, IEEE 802 standards, SONET, and SDH. In the data link layer, both suites can employ versions of HDLC (such as LAP-B in X.25 networks and LAP-D in ISDNs) and IEEE 802.2 (logical link control).

It is in the OSI network layer that the two begin to separate. X.25 is a connection-oriented network interface procedure, while IP is a connectionless procedure. IP shares some features with CLNP (connectionless network protocol), but the latter is significantly more sophisticated and the protocol data unit formats are different.

The Internet control message protocol (ICMP) is an essential adjunct to IP. It provides information to the servers about communications problems that exist at the gateways between networks. When sending messages, it constructs a data unit and passes it to IP for encapsulation in an IP PDU.

In the transport layer, the transport protocols share the same objective of assuring end-to-end transport but differ in the way it is achieved and the functions that are performed.

TCP/IP does not have segregated procedures that map into the session and presentation layers of OSI. Because of this, TCP/IP applications are limited in scope and flexibility; they cannot invoke many of the services that the OSI model envisions. A similar comment can be made to describe the network management standards that support the two systems (see Sections 8.6 and 12.10).

REVIEW QUESTIONS FOR SECTION 12.6

1   Describe TCP. What does it use to guarantee delivery of the datastream?
2   How does TCP exercise flow control?
3   Explain how the checksum is formed in the TCPDU. Why is it done this way?
4   Describe UDP. When is it used?
5   Describe IPv4.
6   Explain the protocol stacks in Figure 12.26.
7   What is classful addressing? Describe the address formats for Class A, B, and C networks.
8   Describe the address formats for Class D and E networks. For what are these networks used?
9   What is the purpose of a subnet number?
10   Why is IPv6 needed?
11   List the major members of the TCP/IP suite, and explain their applications.
12   Use Figure 12.28 to compare the protocols in the TCP/IP suite with the protocols in the OSI model.
13   What is the purpose of the Internet control message protocol?

12.7 SYSTEMS, GATEWAYS, AND ROUTERS

In Section 8.5, four types of internetworking devices were introduced—repeater, bridge, router, and gateway. Gateways and routers are key to managing Internet traffic. For this reason, it is important that they contain up-to-date routing tables and that they have current information on domain names and IP addresses. For backbone (core) routers, this is a formidable task, since they must remain current with new names and routes throughout the network. In this section, I describe how routers and gateways inform each other about the resources in their attached local networks, and route traffic between servers and clients, or terminals and hosts.

12.7.1 SYSTEMS

Internet is a loosely organized collection of networks to which hosts, servers, clients, and terminals are attached. Individual networks are connected into autonomous systems. In turn, these autonomous systems are connected to create a global system—Internet.

- **Autonomous system**: administered by a single authority, a group of networks interconnected by routers/gateways (e.g., networks that support a college campus or an industrial entity, a regional network with several ISPs, or a network operated by an on-line service provider, OLSP). A unique number identifies the system. Within each autonomous system, the administering authority decides the way in which routing information is assembled, updated, controlled, and used.
- **Global system**: administered by a global authority, a group of autonomous networks interconnected by gateways. Within the global system, authorities responsible for individual autonomous systems agree on a common way to provide routing update messages to each other.

A routing update message consists of a listing of the networks that can be reached through the gateway that originates the message and other information such as the number of hops that are required to reach each of them. Routing table information is *advertised* by routers/gateways.

- **Advertise**: action of providing information to others on the connections and resources associated with a router/gateway. An automatic, software-driven function, advertising is used to establish and update the routing tables in neighboring router/gateways.

Using it, the gateways determine how to reach one another (reachability) and thus the hosts within each network.

- **Reachability**: information concerning how to reach individual hosts in an autonomous system.

12.7.2 GATEWAYS

Within Internet, gateways are classified as follows.

- **Core gateways**: gateways administered by the Internet Network Operation Center (INOC). Principally, they are the gateways that route traffic into, along, and out of the backbone networks.

- **Noncore gateways**: gateways not administered by INOC.

In addition, gateways are divided into

- **Exterior gateways**: they exchange resource information with gateways in different autonomous systems

- **Interior gateways**: they exchange resource information with gateways in the same autonomous system.

To illustrate these distinctions, **Figure 12.29** shows three autonomous systems that are part of a global system. Within each autonomous system, reachability information is

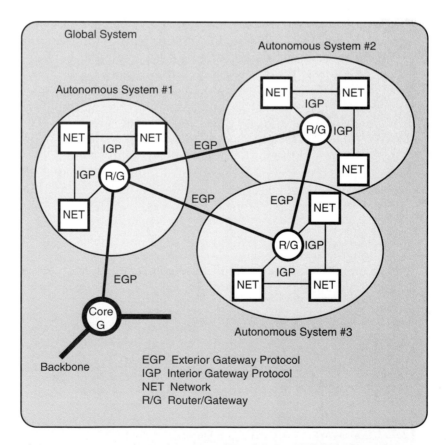

Figure 12.29 Protocols Employed in Advertising Reachability Information within a Global System

passed between router/gateways using internal gateway protocols (IGPs). Between autonomous systems, reachability information is passed between router/gateways using exterior gateway protocols (EGPs). Between a router/gateway in an autonomous system, and a core gateway in the backbone of the global system, reachability information is passed using exterior gateway protocols. Finally, between core gateways, reachability information is passed using gateway-to-gateway protocols.

In the diagram, each router/gateway (R/G) serves both as an exterior gateway (to other autonomous systems) and an interior gateway (within an autonomous system). The core gateway supports EGP between an autonomous system and the backbone. Along the backbone, gateways employ procedures based on EGP.

12.7.3 ROUTING TABLES

In a small system, each router/gateway can have complete knowledge of the locations of all resources in its associated networks, and how to get to them. However, as the system grows, maintaining comprehensive routing tables at every router/gateway for all system resources becomes a daunting task. As new resources are added, the growing volume of updates requires ever-increasing memory. Larger tables mean that the time taken to add to them and organize them becomes greater, and the time required to look up routes expands. In a system as large as, and growing as rapidly as, Internet, not all gateways need to know all there is to know about the system, and a practical compromise uses partial tables at most gateways and full tables only where they are required (principally core gateways). Most importantly, the tables must be updated automatically to keep up with the immense amount of data generated by the continually changing configuration.

To do this, each gateway *advertises* its routing table to its neighbors. For each destination that can be reached through the gateway, the routing table contains (among other things)

- Destination address
- Number of hops to reach the destination address
- Next gateway address on the way to the destination
- Timer values associated with each route.

Table look-up is the most challenging of the tasks performed by routers. The complexity involved and the speed required increases as the traffic ascends from access routers to backbone routers. As Internet gets even larger, and operates even faster, new approaches are needed.[6]

---

[6] Vijay P. Kumar, T. V. Lakshman, and Dimitrios Stiliadis, "Beyond Best Efforts: Router Architectures for the Differentiated Services of Tomorrow's Internet," *IEEE Communications Magazine*, 36 (1998), 152–64.

12.7.4 INTERIOR GATEWAY PROTOCOLS

Routing among routers/gateways in an autonomous system is known as *interior* routing. The system administrator decides the procedures employed. Two common protocols are Routing Information Protocol (RIP) and Open Shortest Path First (OSPF).

(1) Routing Information Protocol

- **Routing information protocol** (RIP): the gateway selects a route that includes a minimum number of intermediate hops.

Each gateway is responsible for knowing the status of its neighbors. Every 15 seconds, it sends an *echo* request message to each neighbor, and they return an *echo* reply. Echo request/reply messages share a simple format. They consist of the source address and a 1-byte *type* field. It is set to binary-8 for echo request, and binary-0 for echo reply. The procedure is known as *pinging*. A neighbor is classified as

- **Operational**: if at least two of the last four echo messages were answered
- **Nonoperational**: if three of the last four messages were unanswered.

Gateways number the update messages they send to their neighbors and keep track of the replies. If the reply received does not match the last update sent, the gateway resends the updates that have not been acknowledged until the reply received matches the last update. Acknowledgment messages consist of the source address, a 1-byte *type* field set to binary-2 when the acknowledgment is positive, and binary-10 when the acknowledgment is negative, and the sequence number of the message being acknowledged.

Every 30 seconds, RIP causes each gateway to advertise to its neighbors the network addresses and distances (in hops) of destinations that can be reached through it. The receiving systems compare these broadcasts to their own routing tables and update them if they contain new routes, better (i.e., shorter) routes, or unreachable gateways.

To limit the size of routing tables, any network for which the number of intermediate hops is >16 is classified as unreachable and is dropped from the listing. A *time-out* timer is set each time a route is initialized or updated. If, in 3 minutes, no additional update is received, the route is classified *obsolete*. After a further 2 minutes without an update message, the route is dropped from the table.

(2) Open Shortest Path First Protocol

- **Open shortest path first protocol** (OSPF): all participating routers are equipped with a database that includes information about system topology and the status of neighboring nodes. Routing is decided on the basis of the shortest route that is compatible with delay, throughput, security, and other requirements. Periodically, an updated database is broadcast to all routers in the system.

OSPF routers are classified according to the functions they perform

- **Internal router**: all networks connected directly to this router belong to the same area
- **Border router**: a router that is not an internal router
- **Backbone router**: any router that is connected to the backbone
- **Boundary router**: a router that is connected to another autonomous system.

OSPF permits the system to be partitioned into areas to reduce the amount of information internal routers must maintain about the entire system. OSPF routers send four types of advertisements.

- **Router links advertisements**: distributed to internal routers, they contain information on the links in the area
- **Networks links advertisements**: they list routers connected to a network
- **Summary links advertisments**: sent by border routers, they contain information on routes outside an area
- **Autonomous systems extended links advertisements**: contain information on routes in other autonomous systems.

OSFP advertisments are sent as high-priority packets over IP.

As networks have increased in size and complexity, it is becoming obvious that the minimum number of hops criterion used by RIP is far from efficient. Letting each router/gateway build its own view of its corner of the network/system makes OSPF more likely to settle on an *optimum* route (i.e., not necessarily the minimum hop route, but one that considers quality of service issues, also). Because each router is likely to have more neighbors, and must send the same routing table to each of them, as the system grows, the overhead messages in RIP increase according to some power of the number of routers. In OSFP a single routing update is sent to each gateway/router. As a result, OSFP is the preferred interior routing protocol.

12.7.5 EXTERIOR GATEWAY PROTOCOLS

Routing among autonomous systems is known as *exterior* routing. Also, because it routes traffic between domains, it can be called *interdomain routing*. Originally, it was accomplished through core gateways—a set of centrally managed routers—that employed GGP (gateway-to-gateway protocol). As Internet grew, keeping track of all possible routes, and updating them when conditions changed (as GGP required), became impossible. Furthermore, not all autonomous networks could connect directly to a core gateway, and the backbone became a network in its own right. Two approaches to exterior routing are currently in use

- **Border Gateway Protocol** (BGP): participating routers provide full routing information to each destination. The information is compared to a global network map to ensure short, optimum routing.
- **Classless Inter-Domain Routing** (CIDR): aggregates consecutively numbered Class C networks into larger networks. Because they have a common subaddress, this permits interdomain routing to all networks in the group using a single table entry. The strategy is sometimes called *supernetting*.

Exterior routing is a dynamic field that continues to engage the attention of the Internet Engineering Task Force.

REVIEW QUESTIONS FOR SECTION 12.7

1   In order to manage Internet traffic, what information must the routers and gateways contain?
2   Distinguish between an autonomous system and a global system.
3   On Internet, what do routers and gateways advertise? Why is it done?
4   What is reachability?
5   What is a core gateway?
6   What is a noncore gateway?
7   What is an exterior gateway?
8   What is an interior gateway?
9   For what is an EGP used?
10   For what is an IGP used?
11   Discuss the problems with routing tables on Internet.
12   For each destination that can be reached through a gateway, what information does it contain?
13   What is interior routing?
14   Describe RIP.
15   What is pinging? Why is it used?
16   Describe the processes used by RIP to keep routing tables current.
17   What is OSPF?
18   Give four classes of OSPF routers, and describe their application.

19 List four types of advertisments sent by OSPF routers.

20 Comment on the following statement: In OSFP a single routing update is sent to each gateway/router. As a result, OSFP is the preferred interior routing protocol.

21 What is exterior routing?

22 Describe the application of BGP and CIDR to exterior routing.

12.8 INTRANETS AND EXTRANETS

Internet, TCP/IP, and the World Wide Web have shown how to achieve the widespread dissemination of information over a system that is not limited to a specific platform or a specific manufacturer's equipment (software or hardware). Their success has inspired companies and organizations to distribute an increasing amount of internal information electronically. In a format made easy to read by incorporating the graphical interfaces and hypertext techniques of the Web, they are able to provide proprietary information to employees. Some have done this using the public Internet; others have constructed private internets with company facilities. The latter are called *intranets*.

- **Intranet**: a private network based on Internet protocols, but not connected to Internet, that distributes (proprietary) information to employees. Using a private network ensures the owner has a significant level of control over who can

  - read it (i.e., read only)
  - work with it (i.e., download)
  - change it.

  In addition, the intranet can be used to restrict electronic mail (and other traffic) to within the company.

By using their own networks the owners can employ higher access speeds than are common in the public network and can present longer, and possibly more complex, documents.

A natural extension of intranet is to place selected outsiders on the rosters of those who may read and work with the information. (Reasons for this are described in Sections 1.2.5, 1.2.6, and 1.2.7.) This arrangement is called *extranet*.[7]

- **Extranet**: an intranet that permits access by selected outsiders (e.g., vendors and customers).

---

[7] Julie Bort and Bradley Felix, *Building an Extranet—Connect Your Intranet with Vendors and Customers* (New York: John Wiley & Sons, 1997).

Without special arrangements, the additional access can compromise the level of privacy the owners sought to attain with their intranet. In particular, if the outsiders use Internet to gain access to it, the connection (i.e., Internet to intranet) will make it possible for unauthorized persons to obtain proprietary information without great difficulty.

12.8.1 PREVENTING UNAUTHORIZED ACCESS

To prevent unauthorized access to intranet, three techniques are employed.

- **Firewall**: software/hardware device that denies access to unauthorized callers. Situated between intranet and Internet, a firewall consists of a screening router and logic that implements a set of rules to determine which connections are allowed and which are not. Called *proxies*, the rules restrict the number of services available to outside connections and prevent the manipulation of services to provide unauthorized levels of access. In addition, a firewall can be used to limit the flow of specific information to callers from within the intranet.

A firewall guards and isolates a corporate network so as to maintain privacy and ensure the integrity of data communications that pass through the network boundaries.

Because there may be uncertainty as to the identity of the outsider, messages between the outsider and the firewall can be encrypted to provide both privacy and authentication [see Section 13.7.2(4)]. This action is said to produce a *tunnel*.

- **Tunnel**: an arrangement in which the outsider and the firewall use public key encryption to encipher the packets they exchange. Thus, information is secure no matter what network the outsider chooses to use to connect to the firewall, and the firewall can authenticate the outsider. The process is known as *tunneling*.

Should further assurance be required, the connection between firewall and outsider can be made exclusive.

- **Private line**: the outsider is connected to the firewall by a private line (or permanent virtual circuit).

12.8.2 DEFENSE MECHANISMS

The relationship between Internet, intranet and extranet is shown in **Figure 12.30**. Several *defense* mechanisms are illustrated.

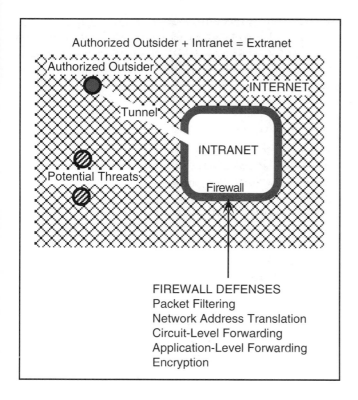

Authorized Outsider + Intranet = Extranet

Authorized Outsider

INTERNET

Tunnel

INTRANET

Potential Threats

Firewall

FIREWALL DEFENSES
Packet Filtering
Network Address Translation
Circuit-Level Forwarding
Application-Level Forwarding
Encryption

Figure 12.30 The Relationships among Internet, Intranet, and Extranet

- **Packet filtering**: after checking the header and contents in accordance with a set of rules (proxies), a firewall router permits or denies the passage of a data packet

- **Network address translation** (NAT): the firewall router is equipped with a table that lists the local IP addresses and the globally unique addresses of intranet clients and servers. Within intranet, the local IP addresses are used; outside intranet, the corresponding globally unique address is used. In this way the firewall router prevents outsiders from discovering the internal addresses and topology of the intranet.

- **Circuit-level forwarding**: by observing the grouping of packets the firewall router detects a connection between client and server. Using proxies it determines whether the source and destination are compatible (within the meaning of the criteria it has) and whether the transfer of the information is permitted. (Thus an *http* client may only access an *http* server to obtain an acceptable class of information.) Commonly, circuit-level forwarding is implemented by Unix Sockets (or Socks)—software that permits applications to access a number of communications protocols.

- **Application-level forwarding**: by observing the data contained in the packets that constitute a communication, the firewall router can determine the acceptability of the communication and permit or deny the passage of the data. Application-level proxies can be very complex. Accordingly, application-level forwarding is likely to consume a substantial time and significantly reduce throughput.

With the proliferation of new devices using different technologies and protocols, the number of potential threats to the privacy of an intranet is increasing rapidly. Maintaining a tight security fence around proprietary data in this environment is a formidable task. It is made even more difficult if persons inside the fence are also intent on obtaining information they should not have. Issues of physical security are discussed in Section 13.6.3.

REVIEW QUESTIONS FOR SECTION 12.8

1 Define intranet. For what purpose is it used?

2 Explain the following statement: By using their own networks the owners can employ higher access speeds than are common in the public network and can present longer, and possibly more complex, documents.

3 What is an extranet? Why are special arrangements required to protect proprietary information?

4 What is a firewall?

5 What is tunneling?

6 Explain Figure 12.30. Describe the defense mechanisms that can be invoked.

12.9 NEW NETWORK INITIATIVES

Several initiatives are under way to develop the best network service possible to serve the national research and engineering community (principally research universities, national laboratories, and supercomputer centers).

12.9.1 VERY-HIGH PERFORMANCE BACKBONE NETWORK

With MCI as its partner, the NSF is supporting the development of a very high speed network to serve the R&E community.[8] Called vBN (very-high performance

---

[8] http://www.vbns.net

Backbone Network), it is a stand-alone network intended to support substantial high-speed computing applications. The experience gained in operating the network is shared with others seeking to upgrade Internet (Internet2 and Next Generation Internet, see Section 12.9.2). Through its High Performance Connections (HPC) program, NSF provides grants to research and engineering institutions to upgrade campus infrastructures and connect to vBNS. Some 100 institutions are connected.

(1) Architecture

The backbone network consists of 12 nodes collocated with MCI's 622.08 Mbits/s national ATM network. Each node contains the following:

- ATM switch (operating at 622.08 Mbits/s)
- one or more IP (and some ATM) routers
- ports that operate at DS–3, OC–3, and OC–12 for user-network access
- OC–12 node-network connection.

The nodes are fully connected by permanent virtual paths on SONET fibers operating at OC–12. A representation of vBN is shown in **Figure 12.31**.

Users of vBNS are divided into four classes

- **vBNS approved institutions** (vAIs): supercomputer centers and universities with grants from HPC program
- **vBNS partner institutions** (vPIs): institutions served by similar networks— e.g., Energy Sciences network, NASA R&E network, Taiwan network, etc.
- **vBNS collaborative institutions** (vCIs): other nonprofit institutions—e.g., Argonne National Laboratory, Fermi National Accelerator Laboratory, etc.
- **vBNS industrial partners** (vIPs): commercial research institutions working with VAIs.

vAIs are known as primary peers. They can access all routes. vPIs, vCIs, and vIPs are known as secondary peers. Connected to a vAI, they request access to vBN from the vAI. Secondary-peer-to-secondary peer connections are not made over vBNS.

vBNS provides high-speed best-effort delivery of IPv4 datagrams between vBN-attached sites with only two backbone router hops. End-to-end delivery requires four or more router hops.

(2) Science, Technology and Research Transit Access Point

To facilitate international connections to vBN and other high-performance networks, the NSF funded the development of STAR TAP (Science, Technology and Research Transit Access Point). Located in Chicago, it connects to vBN through Ameritech's NAP and enables high-speed communication between the

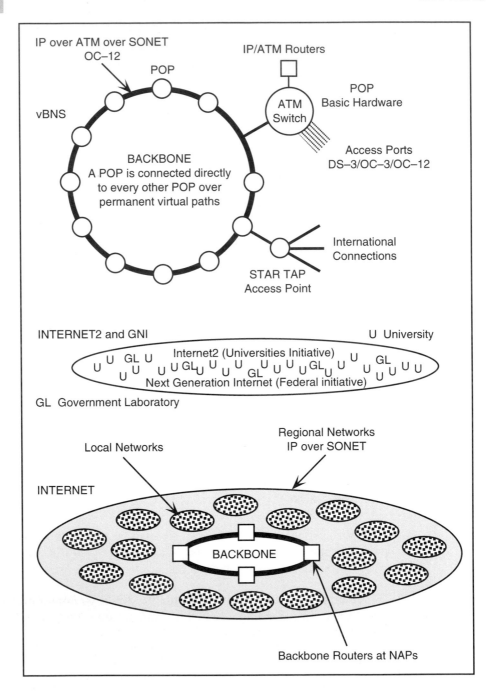

Figure 12.31 Abstract Representation of Internet and Initiatives for Higher-Performance Networks

approximately 100 U.S. universities and supercomputer centers that have access to vBNS, and international collaborators. At various speeds, connections have been established with universities in Taiwan, Singapore, Canada, Western Europe, and Russia.

(3) IP over ATM over SONET

Employing IP, ATM, and SONET at layers 3, 2, and 1 of the protocol stack,requires each packet to be divided into cells and inserted into an envelope (synchronous payload envelope, see Section 6.4.1). At STS–12 (622.08 Mbits/s) the payload transports 9,288 octets and can accommodate 175 ATM cells. With a payload of 48 octets per cell, an IP frame of 148 octets (20 octets overhead plus 128 octets payload) requires 3.08 cells. Thus, an STS–12 SPE can carry some 62 IP PDUs. If the data link layer employs an LLC format (see Figure 12.2) with a 3-octet header and a 2-octet trailer, the PDU is 133 octets long and an STS–12 SPE can carry almost 70 PDUs, an 11% improvement. Thus, there is a penalty to pay for the use of ATM at layer 2. However, if high-speed switching is required, ATM is the only technique available.

12.9.2 INTERNET2 AND NEXT GENERATION INTERNET

Internet2 (I2) and Next Generation Internet (NGI) are initiatives aimed at improving communications between academic and scientific institutions.

- **Internet2**: a private effort by the nation's universities and corporate partners (University Corporation for Advanced Internet Development)[9] to facilitate the development and deployment of advanced network-based applications and network services to support the research and education objectives of the higher education community. Internet2 will employ existing networks to connect members together.

- **Next Generation Internet**: announced in 1997, a multiagency, U.S. government initiative to build a network 100 times as fast as the present Internet so as to upgrade connections between research universities and national laboratories. vBNS is an important part of this effort.

Figure 12.31 attempts to represent the concepts of vBNS, I2, NGI, and Internet. vBNS, I2, and NGI are driven by a need to provide high-performance connections to universities and research laboratories. Internet allows anyone who has a personal computer, modem, and Internet access to communicate with any source.

---

[9] http://www.internet2.edu

12.9.3 NETWORKS FOR THE 21ST CENTURY

The successful evolution of Internet, and the certainty that the world has entered *information society*, have combined to convince operating companies and entrepreneurs of the need for the rapid development of sophisticated communication services. Driven by the belief that the 21st century will be shaped by the availability of advanced communication facilities that permit, indeed encourage, the generation of information anywhere and its use everywhere, national and global bodies have begun to study and create standards for national (NII) and global (GII) information infrastructures.

(1) NII and GII

In 1993, in the United States, the President's Information Infrastructure Task Force (IITF)[10] was formed under the aegis of the White House Office of Science and Technology and the National Economic Council; it undertook to develop policy and plans for a National Information Infrastructure. In 1994, ministers of the Group of Seven countries endorsed the importance of a Global Information Infrastructure and proceeded to plan demonstration projects. In 1996, with members from a broad range of organizations and commercial ventures, the Advisory Council on the National Information Infrastructure delivered its final report concerning the architecture for the NII. **Figure 12.32** shows the NII Functional Services Framework and the NII Architecture Reference Model they recommend.

(2) Functional Services Framework

The Functional Services Framework describes the NII by component layer and key aspects. The three component layers are

- **Physical infrastructure**: contains the computer and communications resources of the infrastructure
- **Enabling services**: generic and domain-specific services that create an environment in which new services and applications can be introduced and integrated with existing services and applications. Examples of classes of enabling systems are
  - *Distributed computing services*. Link multiple separate nodes into a distributed system
  - *Information management services*. Organize, store, and retrieve information
  - *Application cooperation services*. Enable applications to work together to create multiple end-user activities
  - *User interface services*. Provide the links between NII and its users
  - *Financial support services*. Support commercial and personal financial transactions

---

[10] www.iitf.nist.gov

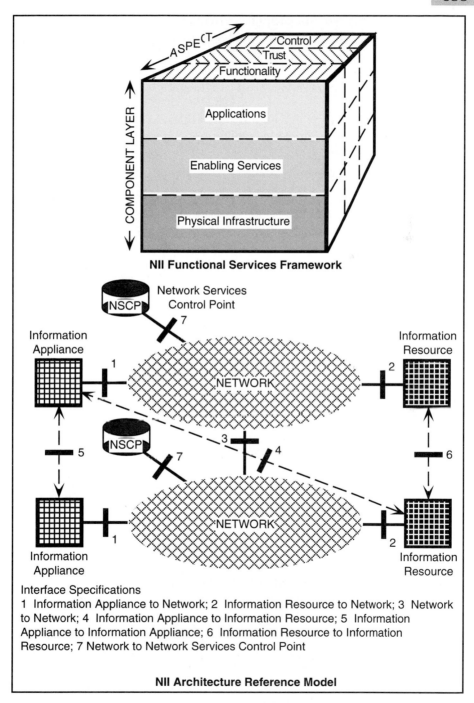

Figure 12.32 Two Views of National Information Infrastructure

  – *Utility services.* Facilitate system uses.

- **Applications**: support users' objectives; may require the capabilities of several enabling services.

Each of the component layers must be considered in three aspects

- **Functionality**: the combination of interfaces, usability, and localization that ensures each component will work with other components to achieve the users' objectives
- **Trust**: consists of three major elements
  - *Security.* Ensures adequate protection, accessibility, and accuracy of information
  - *Integrity.* Ensures information protection at time of failure, assures recovery after failure, and fault tolerance
  - *Performance.* Ensures acceptable performance at reasonable cost.
- **Control**: consists of four major elements
  - *Manageability.* Ability to control components under normal conditions
  - *Serviceability.* Ability to replace and repair components
  - *Measurement.* Ability to record performance statistics and equipment states and to collect accounting data
  - *Adaptability.* Ability to evolve with changing technologies.

(3) Architecture Reference Model

The Architecture Reference Model describes the infrastructure in terms of four major elements and the interfaces between them

- **Information appliance**: equipment employed by end-user to access the infrastructure. Will likely contain data entry and display devices as well as processing and communication capabilities.
- **Communication network**: individual network of communication facilities that interconnects with other communication networks. Among others, may be broadcast radio and television, cable television, public telephone, value-added and private networks.
- **Information resource**: a database, or other form of organized data, available to the end-users. Among others, may be electronic mail system, airline reservation system, virtual library, and directory and listing information. Information resources are developed and maintained by organizations wishing to provide services on the infrastructure.
- **Network services control point**: manages the provision of services over the infrastructure to the end-users.

The bottom half of Figure 12.32 shows seven interfaces for which hardware and software parameters must be defined and universal standards must be developed. Information crossing the interface must be formatted to reflect the procedures (protocols) employed to effect communication, and the physical parameters of the connection must be defined. Focusing on the connections between the functional elements permits them to evolve as technologies are perfected and ensures that equipment from one manufacturer will function properly with equipment from any other manufacturer. ANSI has established the Information Infrastructure Standards Panel (IISP)[11] within the National Voluntary Standards System to facilitate development of standards critical to NII and GII.

(4) Virtual Enterprises

In what is regarded as an important step toward NII, a team of companies led by IBM has entered into a cooperative agreement with the U.S. Government to develop open-industry software protocols to enable the formation of *virtual* enterprises. Called the National Industrial Information Infrastructure Protocols (NIIIP) Consortium,[12] its objective is to allow systems from different manufacturers to be linked together and work as a single, integrated virtual enterprise to address a business opportunity. This will make it possible for geographically separated organizations to collaborate efficiently on common projects. The NIIIP protocols will be building blocks for the National Information Infrastructure. It will lend support to the developments sketched in Sections 1.2.6 through 1.2.8.

REVIEW QUESTIONS FOR SECTION 12.9

1 Describe vBN.

2 Describe four classes of vBNS users. To what connections are they entitled?

3 What is STAR TAP?

4 Discuss why is it necessary to use IP over ATM over SONET in vBN? What penalty does the use of ATM bring? Why is ATM used?

5 Describe the goals of Internet2 and NGI.

6 Discuss the Functional Services Framework shown in Figure 12.32.

7 What is meant by functionality, trust, and control with respect to the Functional Services Framework?

8 Discuss the Architecture Reference Model shown in Figure 12.32.

9 What is an information appliance?

---

[11] www.ansi.org/iisp/iisphome/html

[12] www.niiip.org

10 What is the National Industrial Information Infrastructure Protocols Consortium? Why was it formed?

12.10 NETWORK MANAGEMENT

For networks operating under TCP/IP, network management support is provided by SNMP.[13]

- **Simple Network Management Protocol** (SNMP): support software that monitors and manages network elements (i.e., router/gateways, servers, etc.) in an internetwork environment. (The term network element corresponds to OSI's managed object.) SNMP collects network statistics, changes network routing tables, and modifies the states of links and devices (elements).
  - SNMPv1 runs over a connectionless transport layer (UDP), describes a framework for formatting and storing management information, and defines management information elements (variables)
  - SNMPv2 adds a wider range of commands
  - SNMPv3 adds improved privacy and authentication features.

In addition, for connection-oriented or connectionless operation, network management support can be provided by CMOT.

- **Common Management Information Services and Protocol over TCP/IP** (CMOT): support software based on OSI network management standards, it incorporates ACSE and ROSE. CMOT runs with TCP or UDP.

Figure 12.33 compares the components of SNMP, CMOT, and OSI management protocols.

12.10.1 SIMPLE NETWORK MANAGEMENT PROTOCOL

The simplicity of SNMP makes it the management software of choice for TCP/IP networks. The *network manager* is a client process that communicates through SNMP with server processes (called *agents*) in the network elements. Each managed element contains vendor-designed software that responds to enquiries from the network management stations. The information collected is stored in a Man-

---

[13] Marshall T. Rose, *The Simple Book: An Introduction to Networking Management*, 2nd ed., (Upper Saddle River, NJ: Prentice-Hall, 1996).

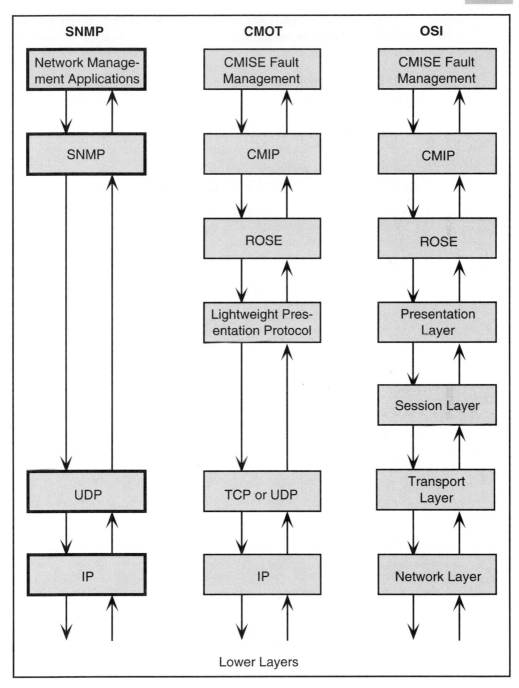

Figure 12.33 Components of Simple Network Management Protocol Compared with CMOT and OSI Protocols

agement Information Base (MIB) and can be used by several clients. SNMP contains hierarchically organized data in a standard representation (irrespective of vendor or network boundaries).

(1) SNMP Components

SNMP employs

- **Management Station**: serves as the interface between the network manager and the network. It provides status information in human terms and includes
  - analysis and fault recovery applications
  - database of information on all managed elements in the network.
- **Management Agent**: responds to requests for information or action from one management station. In addition, it provides the management station with important unsolicited information. Key network devices—e.g., hosts, gateways, routers, bridges, etc.—are equipped with management agent software.
- **Management Information Base**: maintained by the management agent software, it is the collection of operating information describing each managed element in the network.
- **Network management protocol**: SNMP is an application-level protocol implemented on top of UDP, IP, and network-dependent protocols such as X.25, Ethernet, and FDDI.

(2) Messages

Network managers interact with MIB and the network elements through a set of messages.

- **From manager:**
  - *Get request.* Fetch value of one, or more, variables from agent or MIB list
  - *Get-next-request.* Fetch next value of one, or more, variables in MIB tree. May be used to loop through variables or tables.
  - *Set-request.* Perform an action on a network element such as set value of variable to specific value
  - *Get-bulk-request.* Fetch values of variables for group of elements
  - *Inform-request.* Request for information exchanged between managers.
- **From agent:**
  - *Response.* Reply to get-request, get-next-request, or get bulk request
  - *Trap.* Reports the occurrence of an event at a network element such as a fault condition or return to service.

The *trap* message reports the occurrence of an unusual event—something the manager should know about—such as

- **Cold start**: agent is initializing self; states or conditions may change
- **Warm start**: agent is reinitializing self; states or conditions may change
- **Link down:** agent reports failure of a communication link
- **Link up**: agent reports restoration of a communication link
- **Authentication failure**: agent reports packet received from unknown manager
- **EGP neighbor loss**: agent reports router/gateway in peer autonomous network is down; affects agent's ability to communicate with peer network
- **Enterprise specific**: agents report unusual event peculiar to enterprise network.

The top half of **Figure 12.34** shows the protocol architecture employed between the management station and an agent that manages TCP/IP network resources.

(3) Management of Non-TCP/IP Equipment

To accommodate equipment that does not support UDP, a *proxy* agent is used. The arrangement is shown in the bottom half of Figure 12.34. The agent software includes the protocols required to communicate with the non-TCP/IP device. Element information is passed from the proxy to the management station in SNMP format using UDP.

(4) Trap-Directed Polling

As a network grows, it becomes impractical for the management station to poll all agents continuously and collect all their element data. In addition, at times of equipment failure, the sudden rush of trap reports may overwhelm the management station. A strategy to reduce routine management information traffic is called *trap-directed* polling. The network manager establishes a performance baseline by polling every known agent for key information on all managed elements. Once the baseline is established, agents use trap messages to notify the manager of unusual events or major deviations from the base. When sufficient changes have been reported, or at regular intervals, the manager repolls the agents and establishes a new baseline. In this way, reporting traffic is limited mostly to conditions that fall outside predetermined limits, or that tell of the failure or initialization of elements.

(5) Delegated Agent Software

Another technique that reduces the amount of network traffic consumed by management messages extends the capabilities of management software embedded in managed elements. Through the use of *delegated agent* software, network devices can be programmed to analyze the data they collect and to perform control functions as needed. Furthermore, as the network and operating environment change, the capabilities of the delegated agent software can be modified to ensure continuation of correct and efficient operation.

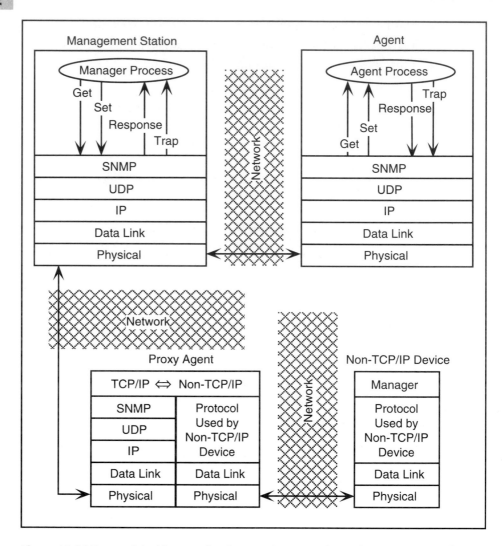

Figure 12.34 Protocol Architecture for Communication with TCP/IP Equipment and Non-TCP/IP Equipment

12.10.2 TELECOMMUNICATIONS INFORMATION NETWORKING ARCHITECTURE

TINA[14] (Telecommunications Information Networking Architecture) is an ongoing effort. In 1997, the TINA Consortium (TINA–C) delivered a set of baseline specifications that describe the software architecture for advanced multimedia networking.

---

[14] www.tinac.com

(1) TINA Specifications

TINA seeks to create a telecommunications software component marketplace that will address the requirements for existing established services and future new services. The objective is to provide customers with the ability to select and control telecommunications services for themselves—the ultimate goal of intelligent networks (see Section 10.5.2). TINA permits graceful evolution from current services and network elements and will accommodate future technologies.

The set of specifications is divided into four major components.

- **Business Model**: defines the roles of, and interfaces between
    - *Consumer*. Consumes services
    - *Retailer*. Provides services
    - *Broker*. Provides information about how to find services
    - *Third-Party Service Provider*. Provides services to other stakeholders rather than consumers.
    - *Connectivity Provider*. Provides transport facilities
- **Service Architecture**: defines a set of concepts, principles, rules, and guidelines for the construction, deployment, operation, and withdrawal of services
- **Network Architecture**: provides a generic, technology-independent model for the control and management of network resources
- **Computing Architecture**: specifies a distributed computing environment that can support telecommunication applications.

The TINA Consortium is continuing to refine these specifications. It predicts large-scale deployment of TINA-conforming products and TINA-based services by 2000.

(2) Mobile Software Agents

Historically, communication networks have been dependent on central hubs. Their switches, signal transfer points, and other equipment are designed to serve a large number of users. A fair question to ask is: Is there a limit to the size of centrally managed networks? As the number of users increases, will congestion cause choking and failure, or can the capabilities of the hubs be continually upgraded? Common sense replies: There are limits to the capabilities of all physical things. A related question is: Will the limits be reached in the foreseeable future? A partial answer is: The *scalability* of many communication systems is limited by the impossibility of expanding the capacity of their hubs indefinitely.

Internet has shown the way to construct an enormous network without central controls; the *Web browser* has shown how to collect information from a globally distributed database; and *Java* has shown how to invoke routines and perform software-based tasks from afar. Furthermore, the development of *intelligent networks* has established the utility of various levels of service-related capabilities distributed among network nodes and devices.

The convergence of these concerns and capabilities has produced[15]

- **Mobile agents** or **mobile objects** or **mobile executing units**: self-contained programs designed to perform specific tasks for users, other agents, and other entities. A mobile agent may be a continuously executing program whose execution is interrupted momentarily when transported from one location to another; or it may be a program that runs to completion before being moved. In addition, an external device may initiate movement, or it may be triggered from within the program. Mobile agents travel between mobile agent systems.

- **Mobile agent system**: a combination client-server system situated on a host that runs, sends, and receives the programs that are mobile agents. It provides access to the environment in which the mobile agent runs. (Agents execute tasks in an *agent execution environment* that provides the basic rules and underlying intelligence.) Several techniques are used to achieve mobility. Three of them are

  - *Remote execution or remote invocation.* In principle, similar to submitting programs by RJE (Remote Job Entry) or the transmission of macros embedded in documents

  - *Cloning.* On command, action of duplicating the program in another location

  - *Agent Transfer protocol.* Uses an application-level protocol that communicates via a socket over TCP/IP to send an agent to another location.

Mobile agent systems must be deployed at all locations to which a mobile agent may travel. Most mobile agents are written in an interpreted machine-independent language so as to run in diverse environments. Many are implemented in Java and include custom extensions to achieve enhanced performance. Currently mobile agents are being applied to tasks in

- **Intelligent networks,** where they download customized service scripts to provide intelligence on demand. Agents may be

  - *smart network* agents. Located at network nodes, they execute tasks autonomously and exchange *control scripts* with other agents.

  - *smart message* agents. They travel between nodes to perform tasks.

Using them, IN services can be provided that are timely, customized, and distributed.

---

[15] Vu Anh Pham and Ahmed Karmouch, "Mobile Software Agents: An Overview," *IEEE Communications Magazine*, 36, 7 (1998), 26–37.

- **Telecommunications information networking architecture** (TINA), where some 10 characteristics have been identified for agents to aid in the distribution of computational objects in a distributed processing environment.
- **Client/server-based networks,** where they are authorized to perform network management tasks to free the management station for other tasks.

In concept, mobile agents are powerful tools that can solve many of the problems associated with network scalability. Because they are mobile and execute independently, they can be deployed to relieve congestion (or other problems) at any facilities in the network. However, precisely because they are powerful and self-executing, mobile agents pose substantial risks to the stability of the network. For instance, moving a mobile agent to the wrong location may cause it to modify or erase memory and destroy or change services. A mobile agent in the wrong execution environment may not be able to complete its task and is prevented from moving. An executing mobile agent may overload its host's buffers, creating deadlock. An error in an agent may produce unanticipated actions such as denial of services. Each of these situations may affect the operation of other mobile agents supported by the same mobile agent system, or other mobile agents on other systems. For sure, mobile agents must be employed with extreme care.

REVIEW QUESTIONS FOR SECTION 12.10

1   Describe SNMP.

2   What is CMOT?

3   List and explain the set of messages used in SNMP to communicate with the management information base and the network elements.

4   What events do trap messages report? List some examples.

5   Explain how non-TCP/IP equipment is managed.

6   What is TINA–C?

7   Describe the major components of TINA.

8   Is there a limit to the size of centrally managed networks?

9   Comment on the following statement: The scalability of many communication systems is limited by the impossibility of expanding the capacity of their hubs indefinitely.

10   What is a mobile agent?

11   What is a mobile agent system?

12   For what applications are mobile agents being considered?

13   What are the risks inherent in the deployment of mobile agents?

13

ENTERPRISE NETWORKS

Global competition has changed the way corporations operate [see Section 1.2.8(4)]. For many of them, a strategic response has been to take advantage of as many technical developments as possible. In this environment, because the timely flow of data and information is essential to increasing productivity and boosting profitability, the networks over which they send data, voice, and video signals become increasingly important to them. In the United States, annual expenditures on enterprise networks account for as much as one-third of the total spent on communications facilities.

13.1 FORCES THAT CREATE, GROW, AND SHAPE ENTERPRISE NETWORKS

The government of the United States maintains the largest private network. Called FTS, the Federal Telephone System serves the needs of federal government offices across the country and provides citizens with toll-free calling to Social Security, Internal Revenue, and other government departments. Individual corporations maintain smaller networks.

13.1.1 COMPANY NETWORKS

If better communications can make the difference between success and stalemate, many corporate managers feel the need to control the facilities they use.

(1) Do-It-Yourself Telecommunications

Managers justify taking responsibility for telecommunications within a company because

- *Telecommunications is a* **critical resource**: if telecommunications is recognized as a vital part of present and future operations, common sense dictates that management protect and strengthen this capability, and managers may say that communications is just too important a part of the business to be left to others—that management must control this resource just as they do engineering, manufacturing, marketing, and distribution.

- *Telecommunications is an* **expensive resource**: if telecommunications expenses are perceived to be high, management must attempt to contain and to reduce them.

- *Telecommunications is a* **competitive resource**: if telecommunications is seen as vital to maintaining and expanding market share, or to maintaining customer loyalty, management is likely to conclude that they must be the first to have the latest technology. Further, they will argue that creating a competitive edge through the use of telecommunications should be easier to achieve with their own facilities than by sharing the same public facilities as their competitors [see Section 1.2.6(5)].

One of their first proposals is likely to be to establish a private (enterprise) network.

(2) Demands on Existing Networks

For most companies, the question of whether or not to invest in telecommunication facilities is not an issue. Already, they have PBXs, LANs, and corporate tie lines that connect some of their business functions together. The question for the future is how *much more to invest in our network; not, whether we should have a network.* Much more activity is devoted to improving, enhancing, and evolving existing networks than creating new ones.

Today's enterprise networks must satisfy several communities. Some of their needs are as follows

- **Corporate** needs:
  - Serve more locations effectively (as corporations are acquired, merge, or expand)
  - Accommodate equipment from different vendors that are of different technology vintages (particularly in the wake of acquisitions and mergers)

- Increase traffic (as volume grows and business becomes more information intensive).

- **User** needs:
  - Satisfy more diverse transmission and connection requirements (as product lines and services become more specialized)
  - Improve responsiveness (faster turnarounds beget demands for even faster turnarounds)
  - Improve network services (more sophisticated, more user-friendly applications)
  - Expand applications services delivered by current capabilities (every improvement cannot be implemented with new hardware and software)
  - Improve support services (as the network becomes more intelligent, users understand it less).

- **Operating staff** needs:
  - Improve cost-effectiveness (managers must show a continual increase in performance and a continuing decrease in cost per employee served)
  - Increase network flexibility (keeping up with organizational changes requires the ability to reconfigure facilities quickly, and surely)
  - Add intelligence (diminishing the number of routine tasks performed by persons)
  - Expand network information and network control capabilities (knowing more about the utilization of current resources, and optimizing their use)
  - Improve feature and service transparency (adding new features and services should not change the way existing functions are performed).

Fortunately, rapidly evolving computer-based technologies provide ways to meet many of these requirements; however, they increase the sophistication and the complexity of the operating environments in which users, support staffs, and network managers must perform.

(3) Needs Assessment

The orderly evolution of enterprise telecommunication facilities should be based on a continuing, comprehensive assessment that addresses

- **Network requirements**: a reliable assessment must be based on a thorough review of the users' future requirements and the capabilities of existing facilities. With this knowledge, likely gaps in coverage can be discovered.

Defining future requirements is not easy, particularly if wide spectrums of users and uses are involved. When asked, each category of users may express future requirements in different terms, and individuals will vary greatly in their understanding of the scope of the inquiry and in their ability to describe and to estimate their future activities. Further, it is in the nature of work that individuals are likely to overstate the importance of their activities to the corporation, and to overestimate the growth of their activities by large amounts. When all of the information is gathered, there remains a significant level of editing that must be done by the assessors.

- **Value of information**: all information is not equally valuable, and all communications are not equally important.

An analysis of the relative values of messages with respect to system failures, or system compromises, can give insight into the importance of some capabilities vis-à-vis others. Will the firm really be unable to function if so-and-so does not have a working terminal for an hour or two? If response time is increased by x% does that really mean that everyone will be less productive by at least x%, or does it mean only that they will have less waiting time between jobs? Etc.

- **System architecture**: architecture is required to provide the framework for future networking decisions. Without architecture to guide its growth, even the best network is only a set of connections created to solve yesterday's problems.

In large measure, the geographical distribution of users, and the communities of interest that form around certain applications, establish the architecture, and bound the set of alternatives to be considered in the evolution of a network. Properly conceived, the architecture will separate the network into independent parts that can be changed to meet the demands of new markets or to incorporate new technologies. It will permit change and growth in one network service segment without unduly affecting other segments.

(4) Owning the Switches and Leasing the Lines

For most corporations, having an enterprise network means *owning* the switches (or gateways/routers) and *leasing* the lines from several carriers. Some popular arrangements for specific classes of enterprise network facilities are

- **Terrestrial transmission equipment**: most corporate entities lease from carriers
- **Satellite transmission equipment**: many corporate entities own Earth stations and lease satellite transponders from satellite carriers
- **Network switching** or **Gateway/Router equipment**: corporate entities own, or lease from carriers
- **Customer premises equipment** (CPE): corporate entities own or lease from
  - Manufacturers
  - Third parties that provide maintenance
  - Carriers.

As carriers see opportunities to provide new public network options, the amount of equipment owned directly by users is likely to decrease.

13.1.2 NETWORK APPLICATIONS

Telecommunication networks support a wide range of applications that are related to the *sorts of connections* they provide, and the *types of information exchanges* they facilitate. The sorts of connections networks provide can be described as

- **Public—anyone-to-anyone**: pervasive, dial-up, interactive connections used for voice or low-speed data communication and provided by public switched telephone networks
- **Public—anyone-to-one**: pervasive, dial-up, interactive connections, provided by PSTN (e.g., 1-800 call, or low-speed data call to public database)
- **Interorganization—any vendor's equipment**: a network serving the special needs of several organizations that connects standard equipment made by any qualified vendor (e.g., FTS)
- **Interorganization—specific vendor's equipment**: a network serving the special needs of several organizations that connects equipment made by a specific vendor (e.g., a banking system clearing network)
- **Intraorganization—all departments**: a network that serves the communication needs of an entire company, or similar entity (e.g., corporate enterprise network)
- **Intraorganization—one department**: a network that serves the communication needs of a single department (e.g., a local-area network).

This range of connections covers most person-centered and machine-centered needs.

The types of information exchanges supported by these networks can be described as

- **Batch**: transmission of information to generate, or add to, files (e.g., update records of hours-worked by employees)
- **Simple transaction(s)**: question-and-answer exchanges probably using structured formats (e.g,. request for person's telephone extension, address, or driver's license number); or the transmission of information that is processed and returned (e.g., conversion of hours worked and hourly wage rate into paycheck)
- **Related transactions**: a series of simple transactions that are related to one another (e.g., inquiry that relates payroll status, vacation entitlements, medical benefits, and retirement benefits for a particular employee), including time-sharing sessions on remote hosts

- **Interactive sessions**: implementation of any style of transactions, at will (e.g., telephone conversation, computer conference, video seminar, etc.).

Examples of applications supported by networks exhibiting these styles of connections, and types of information exchanges, are shown in **Table 13.1**. Person-centered and machine-centered applications are distributed over the connection versus information exchange plane. It is divided into areas that are served mostly by public networks and mostly by enterprise networks. By no means are the categories exclusive, nor are the listings exhaustive. The three left-hand columns contain applications that employ data, and the right-hand column contains applications that employ data or voice or video, or combinations thereof.

13.1.3 COMMUNICATION REQUIREMENTS

While Table 13.1 is not a complete description of the activities supported by a private network, we can use it to define some of the categories of traffic that must be accommodated to support an enterprise.

(1) Traffic Categories

Defining the network users as hosts (servers), LANs and terminals (DTEs, clients, telephones, or video/television terminals), we can divide the network traffic into

- **Asynchronous** and **variable bit rate**:
  - *High-speed data*. Point-to-point, possibly half-duplex, between hosts. The number of connections is small, and terminations are likely to be within the enterprise. In addition, point-to-point, possibly half-duplex, between LANs. The number of connections is fairly small, and terminations are quite likely to be within the enterprise.
  - *Low-speed data*. Point-to-point, mostly duplex, between terminals and LANs, or between terminals and host. The number of potential connections can be large, although only a small fraction may be in use at any time, and most terminations will be within the enterprise.
- **Synchronous** and **constant bit rate**:
  - *Video*. Point-to-point, duplex, between principal locations. The number of connections is small, and terminations are likely to be within the enterprise. In addition, point-to-point, simplex, between major locations. The number of connections is fairly small, and terminations are quite likely to be within the enterprise.
  - *Voice*. Point-to-point, duplex, between all locations, and interconnecting with PSTN. The number of connections is large, and terminations can be inside or outside the enterprise. In addition, many-to-one, duplex,

Table 13.1 Some Activities Supported by Public and Private Networks

SORTS OF CONNECTIONS EMPLOYED	TYPES OF INFORMATION EXCHANGES SUPPORTED			
	Batch	Simple Transaction	Related Transactions	Interactive Sessions
PUBLIC anyone-to-anyone anyone-to-one	PUBLIC NETWORK DOMAIN			
	Facsimile Computer Notice-boards	Public Database		Public Switched Telephone Network 1-800 calling
INTER-COMPANY any equipment special equipment	PUBLIC or PRIVATE NETWORK			
	File Transfer or Update Electronic Mail	Industry-wide Database Automated Teller	Industry-wide Transactions Airline, Hotel, Automobile Reservations	Packet Network Supplier, Producer, Consumer Sessions
INTRA-COMPANY all units one unit	PRIVATE NETWORK DOMAIN			
	File Transfer Electronic Mail	Company Database Department Database	Time-Sharing Services	Peer group Sessions

←————Mostly Data————→

Data, Voice, Video – alone or in combination

between members of the public and an answering point. The number of connections can be large, and exchanges are likely to be from stations in the PSTN to a central facility within the enterprise.

From a different perspective we can divide the network connections into

- **Pervasive connections**: they carry voice and low-speed data and are likely to be provided over switched networks
- **Discrete connections**: they carry high-speed data or video and are likely to be provided over dedicated or routed connections (i.e., connected through a router or a digital cross-connect).

(2) Off-Net and On-Net Calling

The calling requirements of an organization are not limited to calls that can be completed within their enterprise network. For instance:

- **Vendors** and **customers** must be called—they are likely to be contacted through public networks
- **Incoming calls** originate from many facilities that are served by public networks
- At times of **high usage**, an enterprise network may become congested. Employing the public network for some of the busy-hour traffic will provide relief.

Thus, an enterprise network must have access to public networks. Those calls that originate and terminate within the firm and are completed wholly over the enterprise network are known as *on-net* calls; those calls that originate or terminate at points off the enterprise network are called *off-net* calls.

(3) Topology, Hierarchy, and Architecture

Topology, hierarchy and architecture characterize networks

- **Topology** is concerned with the ways in which nodes and links are interconnected.

Topological complexity can be expected to increase with the number of elements a network contains. In addition, within the network, multilevel structures are created to control the ways in which traffic flows.

- **Hierarchy** is concerned with the ways in which nodes (switching centers, multiplexers, etc.) are arranged to achieve efficient division and routing of different classes of communications traffic.

In addition, within the network, the total set of communications activities must be considered.

- **Architecture** is concerned with the ways in which messages interact with topology and hierarchy so that communication can take place.

To understand the functioning of a network requires an appreciation of the unique properties present in all three features.

Many factors must be weighed in establishing the topology, determining the hierarchy, and selecting architecture. Usually, cost is given greatest importance; but, if the goal is a growing, thriving, competitive business, the immediate cost cannot be the only thing to be considered—future options and opportunities must be safeguarded. The possible consequences of each alternative must be evaluated to determine both short-term savings, and potential, long-term, performance liabilities.

REVIEW QUESTIONS FOR SECTION 13.1

1 What organization maintains the largest private telecommunications network?

2 List some of the reasons given by corporate managers for having a private network.

3 List the corporate needs an enterprise network must satisfy.

4 List the user needs an enterprise network must satisfy.

5 List the operating staff needs an enterprise network must satisfy.

6 What factors should a needs assessment address?

7 Comment on the following statement: Without architecture to guide its growth, even the best network is only a set of connections created to solve yesterday's problems.

8 Comment on the following statement: For most corporations, having an enterprise network means owning the switches (or gateways/routers) and leasing the lines from several carriers.

9 Discuss Table 2.1.

10 Define some of the categories of traffic that must be accommodated to support an enterprise.

11 Distinguish between off-net and on-net calling.

12 Define topology, hierarchy, and architecture with respect to a private network.

13.2 SWITCHED NETWORKS

Recognizing that many corporations have been in the business of providing their own communication facilities for some time, I will describe a pervasive network of the sort that is maintained to transport voice and low-speed data in support of the business. It is based on specialized switches known as PBXs. Before describing the network, I discuss the characteristics of PBXs and some other custom facilities.

13.2.1 PBXs and Related Equipment

A private branch exchange (PBX) is a circuit switch that serves a community of stations. Usually, the community is located in a limited area that includes the PBX.

(1) PBX Functions

PBXs perform the following functions:

- By **interconnecting stations** (extensions), they provide communication paths among the community of users
- By **connecting users** to
  - nodes in the public network, they assist in providing communications paths to remote stations, and vice versa. These calls are *off-net* calls.
  - other nodes in the enterprise networks to which they belong, they assist in providing communications paths to remote stations, and vice versa. These calls are *on-net* calls.
- **Set up**, **maintain**, **terminate,** and **record details** of on-net calls, and assist in these activities for off-net calls
- Provide **special services** according to the capabilities with which they are equipped. Many PBXs, particularly the larger ones, support networking—in conjunction with the public network or as nodes in an enterprise network.

(2) PBX Organization

Figure 13.1 shows the major components of a generic, digital PBX. Voice, data, and trunk interface cards are attached to a nonblocking, digital switch. The digital processor

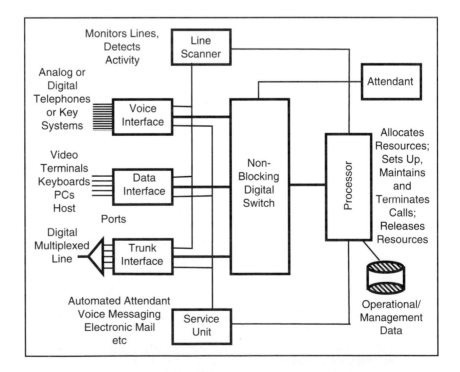

Figure 13.1 Major Components of Generic PBX

- **Responds** to activity indications from the line scanning unit
- **Allocates** switching resources and/or service circuits as needed
- **Maintains** the connection for the duration of each call
- **Returns** the resources to idle state to await other calls.

The PBX is likely to offer voice-messaging services, to contain an electronic mail center, and to have an automated attendant capability. It permits the caller to select a department or service at the time of call reception, or to be referred to a human attendant for directory assistance, call-routing instructions, and other information. At the same time, it records operational and management data for use later in billing or expense allocation activities and in system testing. In addition, it supports administration and maintenance terminals that are used to make changes to the system database, update station information, request statistics and status reports, initiate diagnostic routines, analyze test results, and so on.

(3) PBX Features

Vendors seeking to differentiate their products from competing systems employ the full power of computer control to provide customized feature sets. Some common features are

- Extension-to-extension calling, direct outward dialing (DOD), direct inward dialing (DID)
- Line selection on handset pickup, call transfer, call forwarding, camp on, call waiting, call restriction
- Last number dialed, saved number dialed, abbreviated dialing
- Conferencing, paging, answering and message services
- Automatic route selection
- Automated moves and changes, message detail reporting
- Automatic call distribution
- Modem pooling, (i.e., modems are treated as common equipment and assigned to calls as needed), etc.

These features are but a small sampling from a generic capability that is likely to include several hundred voice and data options.

(4) Station Equipment

Advanced electronic telephones offer easy access to proprietary features. They include programmable keys, multiple line appearances, calling party identifiers, time of day, call duration, etc. Usually, because they are feature-rich, and most features are not standardized, it is impossible to interchange station equipment between PBX systems produced by different manufacturers.

(5) Distributed PBX Arrangements

In common with many electronic devices, PBXs need not be implemented as centralized systems. For instance, the geographical distribution of users may make it cost-effective to place the line interfaces in the user communities and employ multiplexed facilities to connect to a central switching element. In this case, all switching activities occur in the central switch matrix. If the environment contains a few, large concentrations of users, another alternative places line interfaces and limited local switching capabilities with the communities, and employs the central switch to complete connections among the communities, and to the world. These options are shown in **Figure 13.2**.

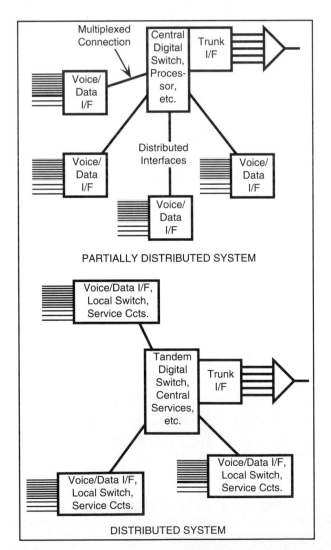

Figure 13.2 Some Distributed PBX Arrangements

(6) Connecting the Public to an Attendant

In consumer-oriented businesses, communications applications such as

- 800 and 888-number services
- Product service centers
- Reservation centers.

are supported by equipment that receives calls and directs them to one of a group of attendants waiting to provide service. Called *automatic call distribution* (ACD) systems, they may be stand-alone, integrated with a PBX, or central-office-based.

A basic configuration for a moderate-size ACD system that is integrated with a PBX is shown in **Figure 13.3**. Suppose customer calls are routed through the

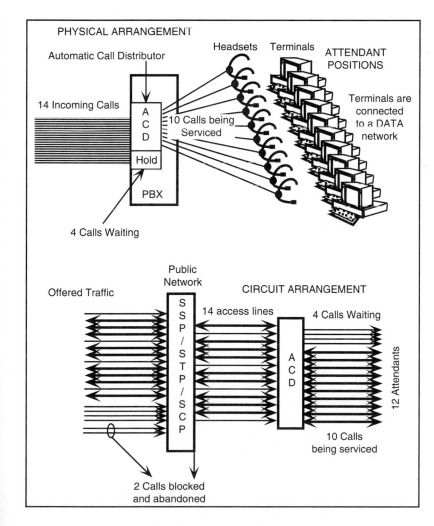

Figure 13.3 Moderate Size Automatic Call Distributor Integrated with PBX

public network to an ACD system equipped with 14 access lines. When the 14 lines are occupied, further calls are blocked and the attempting traffic receives a busy signal; the caller may retry later or abandon the endeavor. On the 14 access lines, the traffic is distributed to available attendants and to a holding device. With a pool of 12 attendants, it is possible that an average of 10 calls is being serviced and 4 calls are waiting for service.

(7) Serving Blocked Callers

If the ACD is one of several units that are employed for a common purpose, the blocked calls at one ACD may be redirected to other ACDs where service is available. Such a situation is likely to exist if units are located across the country in different time zones. Through the use of alternate routing instructions at the service control points (SCPs) in the public networks, several attempts can be made to find an open line. All possibilities will be exhausted before returning a busy signal to the caller. Successful use of *time-of-day routing* moves the concern of the system operator from the number of blocked calls to the number of callers on hold. If the wait is too long, these potential customers will hang up and may dial a competitor.

(8) Centrex

To have the use of the sophisticated services provided by a modern PBX, it is unnecessary for an organization to install and maintain PBX equipment on its premises. Called *centrex* (from central exchange) services, the local exchange carrier can provide them from equipment located at a central office. Each centrex system is assigned a set of standard, seven-digit numbers that are used to complete incoming and outgoing calls in the normal way.

In essence, the customer's extensions are multiplexed to the serving office; there, switching and other functions that mimic the actions of a PBX are performed in the LEC's equipment. Centrex services include voice mail, electronic mail, message center services, modem pooling, protocol conversion, automatic route selection, and packet switching, as well as connecting to standard trunking facilities such as WATS, FX, tie lines, and IXC terminations. Centrex facilities can function as network end-points (in the manner of the PBXs in Figure 13.5), or as tandem switching points (in the manner of the main switches in Figure 13.5).

Centrex users receive station message detail records and other cost-management information and can have access to the appropriate facilities to effect station rearrangement and to see to other matters. Even though the physical facilities are owned and operated by the LEC, centrex services provide firms with the same feelings of being in control of their local communications as are engendered by owning and operating their own PBXs. The provision of centrex services is illustrated at the bottom of **Figure 13.4**.

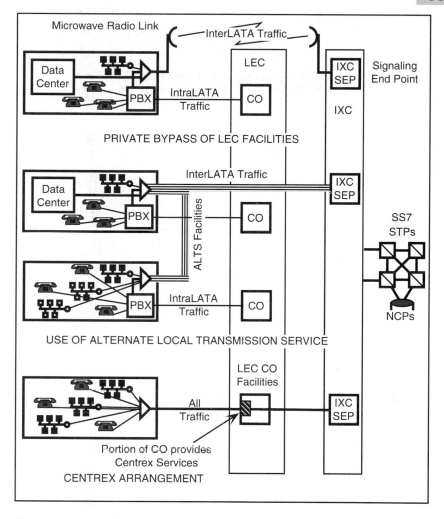

Figure 13.4 Special PBX-Related Arrangements

13.2.2 CUSTOM CONNECTIONS

Businesses with large volumes of inter-LATA calls may wish to deliver them directly to IXCs over private connections that bypass the facilities of the local exchange carriers. In this manner they circumvent access fees and may be able to take advantage of superior transmission performance or other technical capabilities.

(1) LEC Bypass

Microwave radio systems are convenient bypass facilities. One end of the link can be located on the premises of the business seeking to bypass the local exchange carrier, and the other end of the link can be located at an IXC facility. A representation of the provision of microwave bypass is given at the top of Figure 13.4. Some other options are

- **Digital termination service** (DTS): operating in the band 10.55 to 10.68 GHz, DTSs provide digital channels at speeds up to 1.5 Mbits/s and distances up to 6 miles. At the DTS node, data are transferred to high-speed terrestrial links or to other radio links for transport over DEMSNETs (digital electronic message service networks) to other cities, or other carriers. By dividing the area into sectors and reusing the assigned frequencies, several DTSs can operate in a city.
- **CATV service**: cable television system operators provide wideband, point-to-point connections for videoconferences and data services. Franchise limitations permitting, they connect directly to IXC equipment.

(2) Intracity Bypass

A second problem facing corporations may have to do with providing high-speed connections between locations in the same city. In many of the major cities of the United States, alternative local transmission service (ALTS) providers have established networks.

- **Alternative local transmission companies**: employing optical fibers, they provide point-to-point digital circuits to organizations that wish to interconnect their locations within the metropolitan area, and/or to pass traffic directly to IXCs.

Because they are independent entities, ALTS are not restricted by LATA boundaries—for instance, they carry traffic across entire urban areas, even though the areas may be divided into several LATAs (e.g., New York City). Because they are (specialized) common carriers, state commissions regulate ALTS.

A representation of the provision of alternate local transmission service to interconnect separate locations of a company, and to provide bypass of the LEC, is shown in the middle of Figure 13.4. Another option is

- **Point-to-point digital radio**: operating in the 18- and 23-GHz portions of the spectrum, digital radio links can provide wideband, high-speed connections over several miles. Provided line-of-sight is available, they may be the simplest solution to linking different locations in the same city.

(3) Intercity Bypass

Finally, corporations can implement communications independent of terrestrial carriers through the use of

- **Satellite services**: satellite carriers provide point-to-point, point-to-multi-point, and broadcast, data, voice, and video services on dedicated or on-demand connections.

Through the use of a geostationary satellite, points across town or points separated by several thousand miles can communicate at the same cost. It remains to be seen how pricing will be set for medium Earth orbit (MEO) and low Earth orbit (LEO) systems. Satellite systems are described in Section 11.1.

13.2.3 MODEL OF SWITCHED NETWORK

Figure 13.5 shows a wide-area enterprise network that supports business applications requiring pervasive voice and low-speed data connections.

(1) Wide-Area Enterprise Network

Network elements are clustered together in geographical areas served by a local network. A long-distance network connects the clusters to facilitate calling

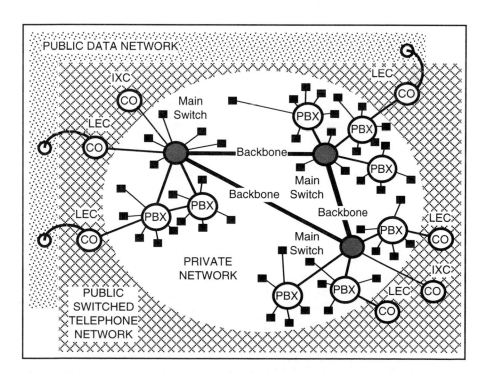

Figure 13.5 Wide-Area Enterprise Network Supporting Voice and Low-Speed Data Connections

from one community to another. The network can be described in the following terms.

- **Topology**: the resources provided to each area consist of star-connected links that terminate on local nodes (PBXs). The nodes are interconnected in star, tree, or mesh fashion to form the local network. In turn, one node in each area (main switch) is the focal point for long-distance traffic and is mesh-connected to nodes that serve the long-distance needs of the other areas. These nodes form the long-distance network (the backbone). At various points, access is provided to the public switched telephone network, and, through PSTN end-offices, to a public data network (PDN).

- **Hierarchy**: within the enterprise network there is a two-level switching hierarchy composed of
  - PBXs that handle local connections and connections to other local switches
  - PBXs (main switches) that handle local connections and connections to other local switches, and make long-distance connections.

  In addition, there is a two-level network hierarchy that is composed of
  - local networks
  - long-distance backbone network.

- **Architecture**: calling from one telephone to another requires that plans have been made, and hardware and software are in place, to handle
  - reception and interpretation of the called number
  - routing of the call to its destination
  - signaling and supervision associated with the communication session
  - provision of facilities to condition the signal (analog to digital, for instance) and to provide alternate routes if needed.

(2) Model of Enterprise Network that Includes Public Network

The model presented in Figure 13.5 is a distortion of reality. In practice, the enterprise network, the PSTN, and the PDN extend over a common geographical area, and public facilities are intermingled with the facilities that serve the entity that created the enterprise network. A way of drawing these public and private network relationships is shown in **Figure 13.6**. To the right-hand side are the signaling and message networks of the interexchange carriers, the public data carriers, and the message networks of the local exchange carriers. To the left-hand side are the PBXs and main switches that form the nodes of the enterprise network.

Each PBX is connected to one of the three main switches by tie lines (leased from the public network operators). Each main switch is connected to the other two by tie trunks (leased from public network operators) that form the network backbone. Access to the public switched network is obtained through five connections to the local exchange carriers' networks. To make these arrangements

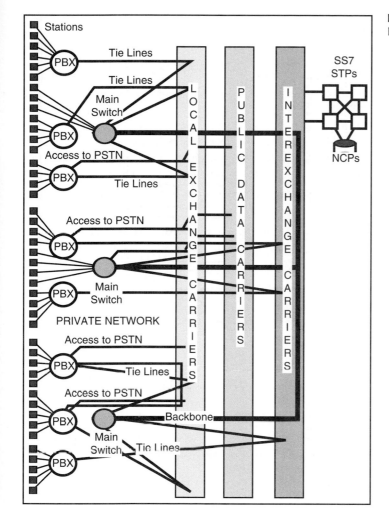

Figure 13.6 Implementation of Network Drawn in Figure 13.5

clearer, **Figure 13.7** presents four partial views of Figure 13.6; they show the backbone network, PBX and main switch connections, connections to the PSTN, and station connections. (To simplify the diagrams, the PDN is not shown. The reader is invited to extend arrangements in Figure 13.7.)

(3) Application of Signaling System 7

SS7 improves the operation of public networks by reducing the amount of call-carrying equipment that is occupied with call setup, and by providing the opportunity to access special services through information stored in a database (SCP). In the same way, it can be used to improve equipment utilization and provide enhanced services. **Figure 13.8** shows the addition of a signal transfer point

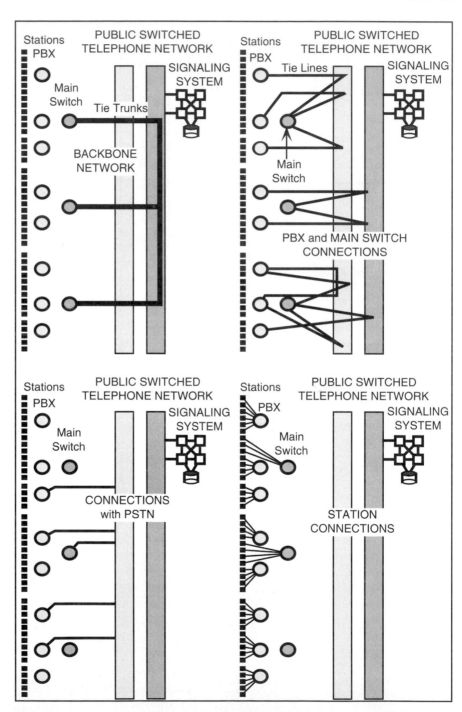

Figure 13.7 Partial Views of Components of Figure 13.6

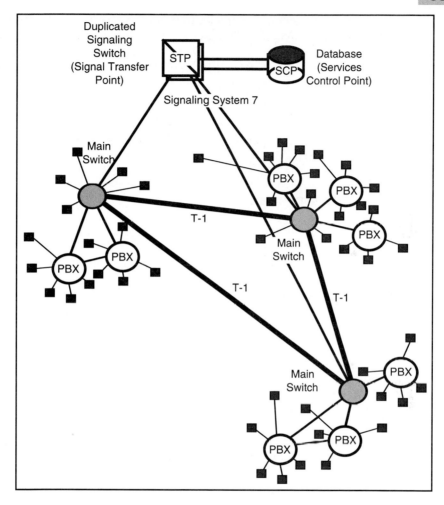

Figure 13.8 Addition of Signaling System 7 to Network in Figure 13.5

and a services control point—elements that implement SS7—to the network drawn in Figure 13.5. They enable the use of special features across the network and support database-related services such as

- **Corporate credit card service**: completes off-net calls over the corporate network for corporate credit card holders. Upon receiving a request for connection to an 800 number, the LEC routes the caller to the corporate point of presence (POP). The corporate node supporting the POP prompts the caller for additional information (called number, credit card number, authorization code, etc.), and routes the call over the enterprise network in accordance with established procedures.

- **Corporate 800 service**: completes traditional 800 service calls over the corporate network. Upon receiving an assigned 800 number, the LEC routes the caller to the corporate point of presence. Depending on time of day, and any other criteria established by the corporation, the node supporting the POP translates the 800 number into an enterprise network number, and completes the call.

- **Network ACD service**: routes calls intended for overloaded service centers to other centers that have idle agents.

- **Network attendant service**: routes calls requiring attendant assistance to a central location.

13.2.4 VIRTUAL NETWORKS

The network shown in Figure 13.8 can duplicate the enhanced signaling capabilities present in intelligent networks (see Section 10.5). In fact, an option for implementing this arrangement is to lease STP/SCP capability from a carrier. Another alternative is to create a complete network out of carrier facilities—one that is operated as a private network with the special capabilities and performance required by the corporation, but without the burden of investment, maintenance, and provision for disaster recovery.

(1) Virtual Private Network

Called *virtual* networks, they are configured to provide users with reports on calls, equipment usage, configuration status, and other information that is required to manage corporate telecommunications activities. **Figure 13.9** redraws the dedicated enterprise network of Figure 13.5 as a virtual network that is distributed throughout the carrier's network, yet under the control of the user.

The principal components of a virtual network are

- **Access arrangements**: how does the customer connect to the carrier's network?
- **Switching**: what services will the customer receive from the serving switch?
- **Record creation and keeping**: how will call-processing records (CPRs) be created and used, and what customer-specific data (CSD) will be kept?
- **Customer premises equipment**: what equipment will the customer expect the network to support?

(2) Network Access

Entry to a virtual network (VN) can be made over direct, switched, or remote connections. Egress is made over direct or switched connections. Direct access arrangements can include

- Voice-grade private lines
- DS1 lines (i.e., ISDN lines)

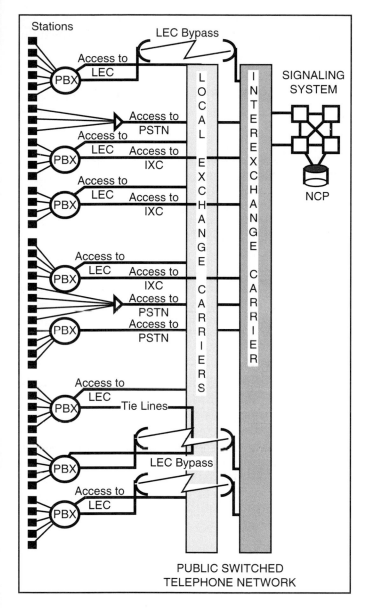

Figure 13.9 Virtual Private Network Based on PBXs and PSTN Facilities

- T-l links equipped with bit-compression multiplexers (44 or 88 channels)
- 56-kbits/s data channels
- Other speed data channels.

Switched access is obtained through the LEC serving Class 5 office. It recognizes VN calls based on automatic number identification (ANI) capability and forwards them to the IXC's, VN-equipped office. Remote access is obtained through the use of a VN access number from any remote terminal. The VN switch uses the number to set up a VN call.

(3) VN Switch

The VN-equipped, IXC switch receives VN traffic over direct, switched, or remote connections. For each event, over SS7, the switch communicates with the network element storing the customer-specific data and call-processing record. This unit verifies the call, may request additional information, and instructs the VN switch on how to process it. When it receives instructions to do so, the VN switch sets up a circuit to the terminating LATA where the call is routed to its destination.

(4) Call Processing Record/Customer-Specific Data

The call processing record (CPR) contains the tests to be applied in processing a call from the specific access location (based on ANI). If an authorization code is required, the network element sends a message to prompt for and collect the extra digits. They are verified against customer-specific data (CSD). Authorization codes are used to identify callers and define their privileges. If an authorization code is not required, privileges are defined by the calling number. After examination of the situation by the call-processing record, instructions are sent to the switch to complete the call or take other action.

(5) Customer Premises Equipment

Virtual networks operate with a variety of common types of customer premises equipment. This includes analog and digital PBXs, multiplexers, and routers.

(6) Features

Because they are a small part of a very much larger network, VNs can offer sets of features that match, or surpass, those present in private networks. For instance, small sites can be included in the corporate virtual network over switched access, and, using remote access arrangements, calls can be placed over the VN from anywhere. In addition, securing alternate facilities and coping with disasters become the carrier's responsibility not the company's. In addition, through protected access to the CPR/CSD database, the enterprise network manager can make changes to stations, authorization codes, alternate routes, and other matters, as desired. Against these advantages must be weighed the vulnerabilities produced by the larger network—perhaps less privacy and susceptibility to the effects of unrelated disasters.

REVIEW QUESTIONS FOR SECTION 13.2

1   What is a PBX? What does it do?
2   Discuss the functions performed by the major PBX components shown in Figure 13.1.

3 Describe some of the features a digital PBX is likely to have.

4 Describe an automatic call distribution system. Why is it used?

5 What strategies are used to serve callers who overflow the capacity of an ACD?

6 What is Centrex? What services can it provide?

7 Describe several arrangements that can be used to bypass carrier facilities.

8 Discuss Figure 13.5. Describe the topology, hierarchy, and architecture.

9 Discuss Figure 13.6.

10 How is signaling system 7 used with an enterprise network?

11 What is a virtual network? List the principal components.

13.3 LIMITED CONNECTION NETWORKS

With the arrival of speedier PCs and workstations, their distribution throughout the workplace, and the proliferation of LANs, demands arose for higher-speed data connections than the network described in Section 13.2 can provide. Among other things, their satisfaction produced frame relay and SMDS (see Sections 9.3 and 9.5). We begin our discussion of limited connection networks with networks based on T–carrier.

13.3.1 VOICE COMPRESSION AND STATISTICAL MULTIPLEXING

For organizations that generate significant voice traffic among several locations, tie lines are likely to be used to reduce the cost of communications. Married to voice compression multiplexers that create 16- or 32-kbits/s voice (ADPCM), or statistical multiplexers, the capacity of standard connections can be increased several-fold. Once installed, the lines are permanently assigned between the sites and carry only voice traffic. For reasons described in Section 10.1.3(3), a special fax relay must be provided for facsimile transmission.

13.3.2 T–CARRIER

Early data networks were composed of individual, point-to-point, *voice-grade lines* (VGLs) carrying multibit symbols that occupied no more than the voice passband. Some of the larger networks contained several thousand tie lines. With the tariffing of T–1 multiplexing for private line service, the number of individual VGLs providing data services has declined rapidly. In their place are networks of T-carrier facilities that carry multiplexed streams organized by digital cross-connect systems (DCSs, see Section 9.6.3) and intelligent multiplexers [see Section 13.3.2(3)].

(1) Digital Data Service

In 1973 the Bell System introduced *digital data service* (DDS), a digital transport service for private line applications. It employed existing technology to multiplex subrate streams (2.4 kbits/s and up) into 56-kbits/s channels as described in Section 6.2.2(2) (see Figure 6.18). To achieve the published goals of 99.96% availability and 99.95% error-free seconds, the DDS network is implemented as a set of interconnected hubs that perform as routing, maintenance and administration centers. In turn, the hubs are connected to digital lines that collect customer channels from serving offices. One hub serves many SOs. Customers are connected to the SOs over four-wire loops in existing distribution cables. They pay for the entire circuit. For end-points that are relatively close together but far from the hub, the line to and from the hub is the major element of cost.

Over the years, DDS service has been improved and reduced in cost. With newer equipment and better facilities, it is possible to arrange passage over the shortest practical route. For intra-LATA use, *hubless* DDS has been introduced that eliminates the back-haul to a hub. Nevertheless, the limitations of a 56-kbits/s channel remain, and DDS is being overtaken by more capable services.

(2) Fractional T–Carrier

Small users can lease a fraction of a T-1 facility from IXCs. Known as *fractional* T-1 (FT-1) service, it permits them to obtain n DS–0 channels, where n = 2, 4, 6, 8, or 12. (The remaining channels are leased to other customers.) In addition, because the individual DS–0 channels are synchronized together, multiple channels can be used to achieve speeds of n 64 kbits/s. Thus, speeds of 128 kbits/s, 256 kbits/s, 384 kbits/s, etc., can be obtained for compressed video or other purposes. Such channels are known as FT128, FT256, and FT384, respectively.

For users that need several T-1s, *fractional* T-3 lines are available. Both FT-1 and FT-3 are priced to encourage leasing a complete system when occupancy is as low as one-third. For eight DS–0 channels the lease cost is approximately equal to a complete T–1, and for nine DS–1 channels the lease cost is approximately equal to a complete T–3. The exact crossover point depends on distance and variations in tariffs from different carriers.

In the local exchange area, connections between the customers' premises and the LEC's serving offices are made by dedicated T–1 facilities, or they may be provided as channels derived from digital loop carrier systems (see Section 6.3.2). Depending on the LEC, the connections may be tariffed as full T–1s or as FT–1s.

Figure 13.10 shows the way in which three locations of an enterprise that share the same LEC serving office might be connected to the corporation's enterprise network. The backbone is assumed to consist of T–1 lines between IXC central offices where digital cross-connect capability is provided. Undoubtedly, the T–1s are part of in-place T–3 or SONET circuits maintained by the IXC. Between

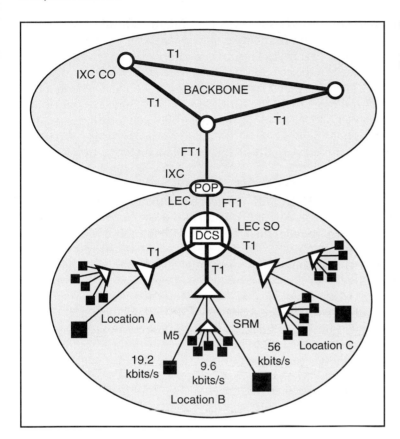

Figure 13.10 How Three Locations of an Enterprise Might Be Connected to the Corporation's Enterprise Network

the IXC office and the POP, transport service is provided by 7 or 8 DS–0 channels (= FT–1) derived from other in-place IXC circuits. At the POP they are connected to LEC circuits that provide transport to the serving office. At the SO, a DCS splits the channels and directs them to locations A, B, or C. We assume that each location is served by a dedicated T–1 facility, and that advantage is taken of subrate multiplexing to reduce the number of individual channels to the serving office. As long-term traffic distributions vary, through the use of the DCSs at the LEC SO, POP, and IXC CO, individual point-to-point connections can be changed.

(3) Intelligent Multiplexers

Today's T-1 multiplexer is a far cry from the equipment first given that name (see Section 6.2.1). Spurred on by the availability of low-cost DS-1 access lines, and augmented with cross-connect and other capabilities, it has become a digital system capable of supporting sophisticated enterprise network applications. Contemporary T-1 multiplexers, called *intelligent* (T-1) multiplexers (IMs), support large, complex, mesh networks by providing the ability to add, drop, and redistribute circuits among multiple trunks, on command. In addition, some IMs include the ability to interface with frame relay and cell relay equipment.

By way of illustration, **Figure 13.11** shows a drop and insert application among three locations. The number of circuits (12) between each pair of locations (A⇔B, A⇔C, and B⇔C) is determined by the digital cross-connect system in the intelligent multiplexer. In **Figure 13.12** the configuration is changed to permit the use of 12 DS–0 channels (FT768) for a videoconference between locations A and C, yet maintain 6 DS–0 channels between each location for other purposes. When the conference is completed, the configuration can be changed back by electronic command.

Intelligent multiplexers can be used to construct corporate data networks. An example is shown in **Figure 13.13**. With IM nodes located in San Francisco, Los Angeles, Denver, Dallas, Chicago, Atlanta, Miami, and New York City, point-to-point digital connections can be reorganized at will on commands from the network management center (NMC). The multiplexers connect to a variety of customer premise equipment including PBXs, computer hosts, terminal controllers, and routers. Figure 13.13 shows a network with significant traffic between New York City and Chicago, Chicago and Dallas, and New York City and Dallas. It employs a T-3 connection (NY–CHI), and a fractional T-3 connection (CHI—DAL). Other locations employ T–1 and FT–1 links.

In a practical arrangement, IMs could be owned by the enterprise, located at points within corporate facilities, and interconnected by transport facilities leased from the carriers. Corporate staff would operate the management center.

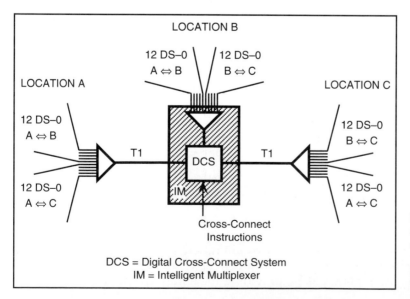

Figure 13.11 The Use of an Intelligent Multiplexer to Provide 12 DS–0 Channels between Three Company Locations

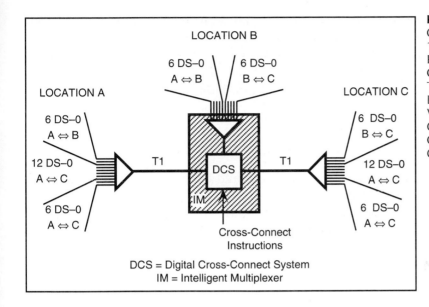

Figure 13.12 The Configuration of Figure 13.11 Is Changed to Provide 6 DS–0 Channels between Three Company Locations and a Videoconference Connection of 12 DS–0 Channels between Two Company Locations

(4) T–Carrier Networks

Private T-carrier networks exhibit a range of topologies. Different objectives, different locations, and different styles lead to many arrangements. Some points of general agreement are

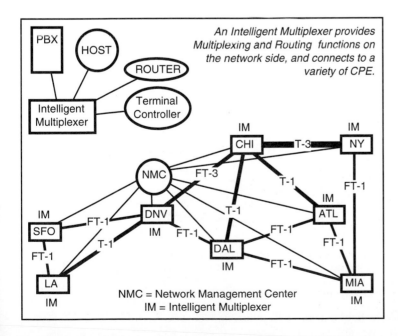

Figure 13.13 Private Network Configured from Intelligent Multiplexers and Leased T–3, FT–3, T–1, and FT–1 Facilities

- **Intra-LATA connections**: minimizing intra-LATA costs favors a star topology in which a hub connects to individual locations within the LATA
- **Inter-LATA connections**: minimizing inter-LATA costs favors one connection between each pair of LATAs. This can be a hub-to-hub connection. These arrangements are shown in **Figure 13.14**
- **Location of work sites**: the distribution of user sites will influence the topology. In smaller corporations it is likely that only one arrangement will fit the distribution, but, for many corporations, there will be several possibilities
- **Protection**: to protect against seriously degraded performance due to the loss of a line, traffic between nodes should have the opportunity to use as many alternate routes as possible
- **Daisy-chain connections**: several locations can be connected together in sequence by a single transmission system that includes T-1 multiplexers with drop and insert capability
- **Multidrop connections**: multidrop connections between one primary and several secondary stations are possible up to speeds of 64 kbits/s

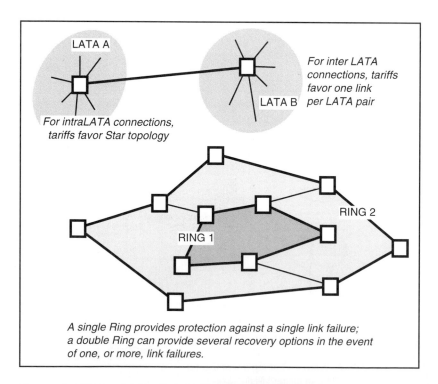

Figure 13.14 Potential Configurations of Private Networks

- **Ring connections**: several locations can be connected by one continuous transmission system so that, should the ring fail at any point, all of the stations remain connected in daisy-chain fashion. By providing concentric rings, a network is formed that can have several options for recovery from the loss of one, or more, links. This arrangement is shown in Figure 13.14

- **Mesh connections**: a fully connected mesh provides several options for recovery from the loss of one, or more, links. The expense of redundant connections must be balanced against the value of maintaining communication under adverse circumstances.

13.3.3 FRAME RELAY AND CELL RELAY

For many years it was an axiom that voice traffic (measured in bits/s) would always be substantially greater than data traffic. Consequently, many corporations were satisfied to create the sort of network shown in **Figure 13.5**—a network that serves the needs of the voice community and does not do too bad a job for most data users. However, data traffic grew at a much greater rate than voice traffic. Over some of today's *enterprise* networks the number of bits of data traffic may be as much as twice the number of bits of voice traffic. Under this circumstance, the technology of enterprise networks is changing from time-division, synchronous techniques to statistical, asynchronous techniques that can also handle some synchronous traffic.

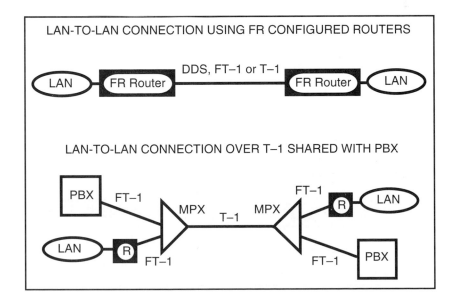

Figure 13.15 Simple Examples of Frame Relay LAN-to-LAN Connections

(1) From T–Carrier to Cell Relay

In T-carrier systems, time slots are allocated to sources on a semipermanent basis, whether they need them or not, and line speed is calculated by summing the maximum speeds of the circuits, whether they transmit continuously, or only once in a while [see Section 6.1.4(3)]. If the sources are bursty (as data sources generally are), a better approach is to employ statistical multiplexing. Frame relay and cell relay are statistical multiplexing techniques that provide bandwidth on demand up to the limit of the facilities.

(2) Frame Relay Networks

Private frame relay (FR) networks make use of FR configured routers and FR nodal processors to connect LANs and other data equipment. Two simple examples are given in Figure 13.15. The top diagram shows a direct connection between two LANs that is implemented over DDS, FT–1, or T–1 circuits. In the bottom diagram the same connection is made over a T–1 line in which some of the channels are employed by a voice PBX, and some of the channels are used to connect the two LANs using frame relay.

In **Figure 13.16**, several LANs and some other equipment are connected together in a hubbed network through a single nodal processor. (This situation was discussed in Section 9.3.3). At the hub, the nodal processor is connected to

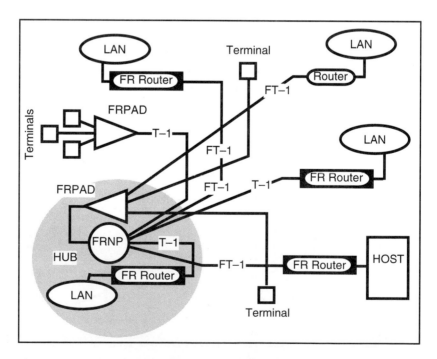

Figure 13.16 More Complex Frame Relay Network

several FR routers that serve individual LANs, and local or remote FR PADs. The latter operate very much like PADs (packet assembler and disassembler devices) in X.25 packet networks. They create FR frames for equipment that is not able to do so and convert FR frames to formats that are compatible with the end-users. The local FR PAD (at the hub) connects to multiplexed terminals and a non-FR configured router that serves a LAN. The other FR PAD creates FR frames for three terminals before data are transmitted from a remote site to the hub. Connections across the network are implemented over FT–1 or T–1 circuits.

Figure 13.17 shows the elements of a larger network. Three FR nodal processors constitute the backbone network. LANs and other terminals are connected to these processors through appropriate devices. Within the backbone network, data are transported by cell relay [see Section 9.4.2(3)].

13.3.4 ATM—the Ultimate Backbone

Figure 13.18 shows the elements of an enterprise network that serves the needs of the data community and carries high-volume voice traffic, and some video traffic. The backbone consists of mesh-connected ATM nodes that operate over SONET at 155 or 622 Mbits/s. They are fed from various data sources, from PBXs, and from

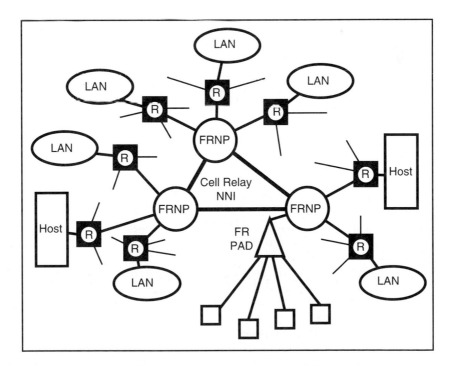

Figure 13.17 Complex Frame Relay Network with Cell Relay Core Network

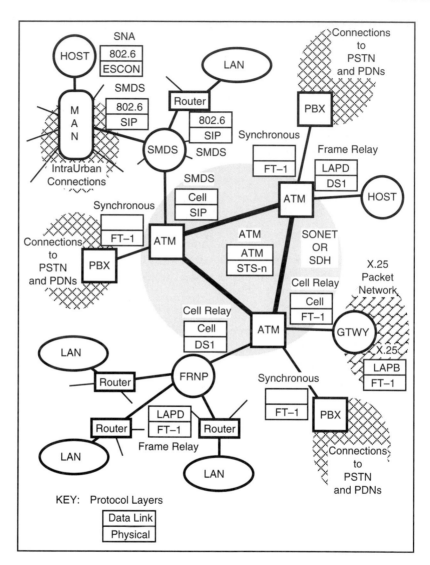

Figure 13.18 Complex Private Network that Includes Many Types of Data Transport and Routing Entities

videoconference rooms. The network transports PBX-to-PBX traffic within the enterprise but does not duplicate services that can be obtained from a virtual network created by a carrier. One cannot help but be concerned about the number of protocol changes that can occur in transferring messages from one end of the network to the other—yet all of the equipment shown is in use today, and is likely to stay in use for some time. As ATM becomes the dominant technology, the environment will be simplified (in the manner of Figure 12.22).

REVIEW QUESTIONS FOR SECTION 13.3

1 How is the composition of enterprise networks changing?
2 Explain the bottom diagram in Figure 13.15.
3 What is the function of an FR PAD?
4 Discuss Figure 13.16.
5 Figure 13.18 shows the elements of a data-oriented enterprise network. What is the major advantage of using an ATM core network?

13.4 SYSTEMS NETWORK ARCHITECTURE

Announced in 1974, SNA (Systems Network Architecture) is a blueprint for tightly integrating communication functions and computing resources. It is the architecture for some 50,000, data-only enterprise networks. Linking IBM (and compatible) data processing equipment, SNA provides a full range of networking protocols (not just the UNI, as with X.25, or UNI and NNI, as with frame relay). Over the years, SNA has been carefully upgraded and documented. Originally designed for a single-host, one-vendor environment connected by leased lines, today's SNA accommodates a multiple-host architecture that can operate over dial-up connections and connect with local-area, packet-switched, frame relay, and ATM networks.

In 1996, the introduction of APPN (Advanced Peer-to-Peer Networking) extended the scope of SNA to decentralized environments that support client/server computing. Today, SNA includes two distinct styles of networking—subarea networking and APPN.

Like the OSI model, the SNA model consists of seven layers. Each layer performs a set of functions designed to achieve reliable data transfer between logical units. The three lower layers make up the transport, or path control network. It provides basic message handling and delivery. The upper four layers provide end-to-end procedures that support communication between users.

13.4.1 SNA Model

Figure 13.19 shows the seven layers arranged in model form. Their functions are described below.

(1) End-User or Transaction Services Layer

The end-user or transaction services layer provides

- **Application services** to the end-user, including end-user access to the network, and requests for services from the network addressable unit or presentation services layer

- **Network management services**, including
  - *Configuration control* associated with the physical configuration during an SSCP–PU session (see Section 13.5.2)
  - *Session services* associated with establishing LU—LU sessions including verification of user passwords and access authority, confirmation of session parameters, and translation of network names to network addresses
  - *Monitoring*, *testing*, *tracing*, and *recording* statistics for resources used in SSCP–LU and SSCP–PU sessions.

(2) Network Accessible Unit or Presentation Services Layer

The network accessible unit (NAU, see Section 13.5.1) or presentation services layer defines and maintains communication procedures for transaction services programs. In addition, the NAU layer controls the interaction between these programs

- By implementing functions for loading and invoking other transaction services programs
- By enforcing correct syntax use and sequencing restrictions
- By processing transaction services program commands.

(3) Data Flow Control Layer

Orderly data flow in a logical unit to logical unit (LU–LU) session is the responsibility of the data flow control layer. Units of data being transmitted are

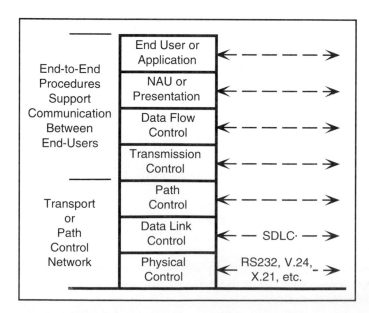

Figure 13.19 Seven-Layer SNA Protocol Stack

given sequence numbers, and an end-user request is correlated with its response. Related request units are grouped into chains, and related chains are grouped into brackets of data. This layer enforces communication procedures between LUs and coordinates which LU can send and which LU can receive at any given time.

(4) Transmission Control Layer

The transmission control layer is responsible for the synchronization and pacing of session-level data traffic. It does this by checking the session sequence numbers that were assigned to the request units from the data flow control layer. Another function it performs is the enciphering and deciphering of data.

(5) Path Control Layer

The path control layer is responsible for routing data units through a network to the desired destination. The function is performed in subarea nodes and consists of several steps:

- **Selecting** the transmission group to be used between subarea nodes for a specific session
- **Determining** which of the forward and reverse links available in the transmission group are to be used to make the explicit connection between the subarea nodes
- **Controlling** the flow of data over the virtual route formed from all of the explicit links assigned to the session.

(6) Data Link Control Layer

The data link control layer is responsible for scheduling and error recovery associated with the transfer of data across a link connecting two nodes. The data link control layer supports the link level flow of data for both SDLC and System 370/390 data channel protocols, including ESCON [see Section 7.1.2 (3)].

(7) Physical Control Layer

The physical layer defines the electrical and transmission signal characteristics necessary to establish, maintain, and terminate all physical connections of a link. SNA makes no statement concerning preferred procedures. EIA232, V.24, and X.21 or other standards may be used.

(8) Path Control Network

By routing and transmitting message units between end-users across network nodes, Layers 1, 2, and 3 of the SNA model implement the path control network. Normally, connections between nodes are implemented on parallel transmission links known as transmission groups (TGs). To the rest of the network, the group appears to be a single, logical link. To the path control layer, however, each transmission group consists of a set of individual links, some in the forward direction, and some in the reverse direction.

An individual session is allocated an explicit forward link, and an explicit reverse link on a nonexclusive basis. Together, they are known as a *virtual route*. High, medium, or low priority is assigned to each session message unit, and higher priority messages are handled ahead of lower priority messages in the nodes serving the links.

All messages carry a sequence number that identifies their position in the particular data exchange to which they belong. Using this, the node can handle them in sequence, and in order of priority. Because the connections may be used by a diversity of traffic, flow control must be employed to prevent message units arriving faster than they can be processed—an event that will lead to their destruction and subsequent requests for retransmission—and a reduction in the throughput of the network.

13.4.2 FORMATION OF MESSAGE UNIT

Between end-users, messages are called *commands* or *responses*.

- **Command**: a message containing management instructions, or end-user data
- **Response**: a message replying to a command.

As a message descends to the physical control layer, headers (and a trailer) are added to the originating end-user's information unit. They are removed as the information unit ascends through the layers of the model to the terminating end-user. The formation of a message unit is shown in **Figure 13.20**.

(1) Request/Response Unit

The request/response unit (RU) is variable in length and contains the information that is to be sent from the originating end-user to the terminating end-user. It may be end-user information or a command to control the operation of the network. For most logical units, the size of RU is ≤256 bytes. The actual number of bytes is carried in the transmission header.

(2) Request/Response Header

The request/response header (RH) is a 3-byte header that describes the information in RU. If the first bit of RH is 0, the header is a request header; if the first bit of RH is a 1, the header is a response header.

- **Request header**: provides information on the data format and the protocol that governs the session. Tells PU whether a response is required, if data brackets are being used, where in the chain of data the Basis Information Unit (BIU = RH + RU) is positioned, and if a pacing request applies to this transmission.

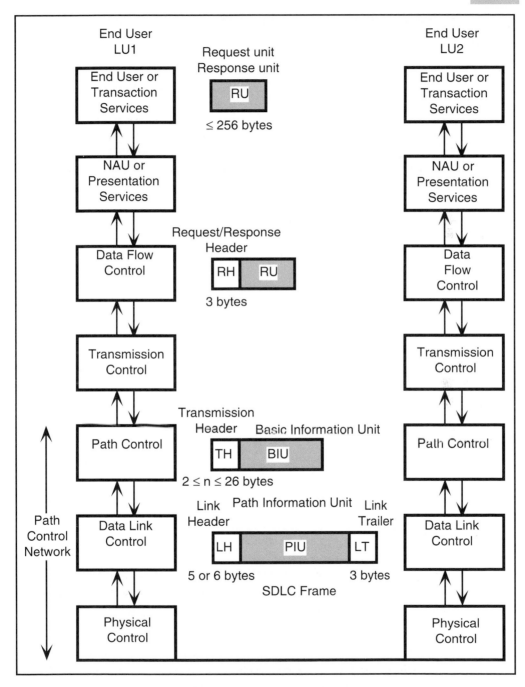

Figure 13.20 Formation of Message Unit

- **Response header**: returns a positive or negative response to request from other LU.

(3) Transmission Header

The transmission header (TH) is concerned with path control. Its format and length depend on the types of traffic and the types of nodes employed. The first byte in TH, called the format identification field (FID), is used to identify them:

- **FID Type 0** (TH = 10 bytes): identifies non-SNA traffic between adjacent subarea nodes not supporting explicit and virtual route protocols
- **FID Type 1** (TH = 10 bytes): identifies SNA traffic between adjacent subareas not supporting explicit and virtual route protocols
- **FID Type 2** (TH = 6 bytes): identifies SNA traffic between a subarea node and an adjacent peripheral node
- **FID Type 3** (TH = 2 bytes): identifies SNA traffic between an NCP subarea node and a type 1 peripheral node
- **FID Type 4** (TH = 26 bytes): identifies SNA traffic between adjacent subarea nodes supporting explicit and virtual route protocols
- **FID Type F** (TH = 26 bytes): identifies specific SNA commands between adjacent subarea nodes supporting explicit and virtual route protocols.

The remainder of TH includes

- Originating and terminating address fields
- Sequence number of this path information unit (PIU = TH + BIU) in relation to other PIUs for this request or response
- Length of the RU
- Whether it is the only, first, middle, or last, in the transmission
- Explicit and virtual route information
- Addresses of the originating subarea and element and the terminating subarea and element.

(4) Link Header and Link Trailer

The link header and trailer (LH and LT) are standard SDLC frame items. The header contains a flag (1 byte), the station address (3 or 4 bytes), and a control byte. The trailer contains a 2-byte frame check sequence and the terminating flag (1 byte).

13.4.3 COMPONENTS OF SNA NETWORKS

An SNA network consists of nodes interconnected by data links. Software resident in the nodes implements a set of SNA protocols. Nodes in subarea networks support hierarchical networking, while nodes in APPN support decentralized networking.

(1) Network Accessible Units

- **Network accessible unit** (NAU): software implementation of end-to-end protocols that implement the transmission control layer, the data flow control layer, the presentation services layer, and the transaction services layer. (NAU is not a physical unit; several NAUs may run in a single piece of hardware.)

Also known as network *addressable* units, network accessible units are software entities that support end-to-end communication. Resident in network nodes, and interconnected by the path control network, they provide services that permit end-users to send and receive data, and help perform network control and management functions. As shown in **Figure 13.21**, NAUs terminate the SNA network.

Four types of NAUs are used in SNA networks. They differ in the types of communication they support and the types of networks in which they are used. Some are used for end-to-end user communication; others are used for network management. Some are used in subarea networks, some in APPN, and some in both.

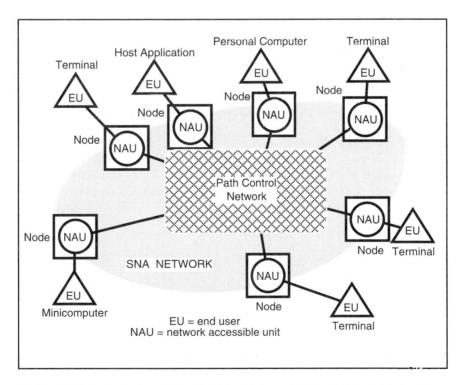

Figure 13.21 Network Accessible Units Are Software Entities in Nodes at the Edges of the Network

- **Logical unit** (LU): LUs are the units through which end-users communicate. In subarea networks, they are used only for end-user communication; in APPN, LUs are used for end-user communication and to support network management functions. LUs are of several types:
  - *Type 1.* Software that supports communication with character-oriented printers
  - *Type 2.* Software that supports communication with 3270 video terminals
  - *Type 3.* Software that supports communication with 3270 printers
  - *Type 4.* Software that supports communication with 5250 printers
  - *Type 7.* Software that supports communication with 5250 terminals
  - *Type 6.2.* Also called APPC (advanced peer-to-peer communications). Software that substitutes for all types of LUs in APPN networks.
- **System services control point** (SSCP): in subarea networks, used by management software to communicate with software in managed entities.
- **Physical unit** (PU): in subarea networks, software that enables software in managed entities to communicate with network management software.
- **Control point** (CP): in APPN networks, supports communication between network management software entities to update routing tables, provide directory services, etc.

Each NAU is assigned a unique name (address) within the SNA network. It consists of an eight-character network name and an eight-character NAU name. The combination of network name and NAU name forms a *network-qualified* name. To assist users to assign unique names, IBM maintains a name registration service.

(2) Sessions

For two NAUs to communicate, a logical connection must be established between them. Called a *session*, it exists for as long as the participating NAUs communicate. Sessions may be initiated for network management reasons, or for end-user reasons. Five specific sessions are required to implement the operation of SNA networks:

- **LU–LU session** (logical unit to logical unit): subarea and APPN end-users communicate with one another through LU—LU sessions. The LUs establish the communication procedures (roles, rules, and protocols). Data flow between them as SDLC frames (or other formats) called message units (MUs).
- **SSCP–LU session** (system services control point to logical unit): supports requests for session services in subarea networks. For instance, to connect two end-users in the same domain, the request for communication from the originating LU (logon TSO, for instance) must be converted to the network address of the terminating LU (TSO). During the logon process, an existing

SSCP–LU session is employed so that the SSCP can supply this information and direct the request to the terminating LU.

- **SSCP–SSCP session** (system services control point to system services control point): supports network management communications in subarea networks. For instance, to connect two end-users in different domains, the SSCP in the domain containing the originating LU must establish a session with the SSCP serving the domain that contains the terminating LU. The session may be established during the system generation process.

- **SSCP–PU session** (system services control point to physical unit): supports communication between network management software and software in managed entities. These sessions are used for internal network purposes such as configuration and reconfiguration, gathering performance information, etc.

- **CP–CP session** (control point to control point): supports network management functions in APPN networks.

LU types 1 through 5, and LU type 7, are called *dependent* LUs. They cannot activate a session themselves but depend on an SSCP to do it. LUs type 6.2 may be dependent or independent. They can activate a session, or not, depending on circumstances.

13.4.4 SNA Subarea Networks

Subarea networks support hierarchical communication and centralized network control. **Figure 13.22** shows a subarea network consisting of a domain (Domain 1) that is divided in two subareas, and a second domain (Domain 2). The nodes are connected by communication links under the control of the subarea nodes. They are described below.

(1) Types of Nodes

In a subarea network, the nodes are divided into peripheral nodes and subarea nodes. Peripheral nodes are of two types

- **Type 2.0**: supports end-user communication with host-based applications
- **Type 2.1**: supports end-user communication with host-based applications and communication with other type 2.1 nodes.

Peripheral nodes contain a PU dedicated to the exchange of management data with an SSCP, and up to 254 LUs to support end-user communication. They are assigned to a specific subarea

- **Subarea**: a network address associated with a node (subarea node) and its attached peripheral nodes.

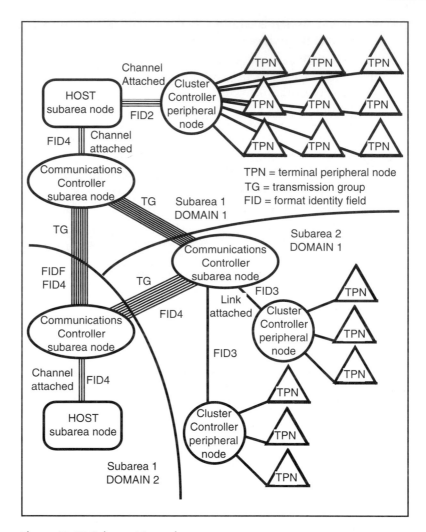

Figure 13.22 Subarea Networks

Domain 1 consists of two subareas. Domain 2 has only one subarea.

Subarea nodes contain routing tables that are generated when the controlling software (VTAM or NCP, see below) is configured. Subarea nodes are of two types

- **Communications controllers** (also called type 4.0 nodes): intelligent communications hardware that manages communication lines, cluster controllers, workstations, and the routing of data through a network. Communications controllers
  - are dedicated to communication processing and do not support local end-users

- always contain a PU to communicate with host-based network management software over a SSCP–PU session
- employ NCP to manage all of the data links assigned to the controller
- are divided into the functional categories of front-end processors, concentrators, and intelligent switches
- may support a host subarea node in the subarea assigned to a host, or operate in a subarea of their own.

- **Hosts** (also called type 5 nodes): combinations of hardware and software that comprise the main data processor in a subarea. Hosts
  - always contain an SSCP
  - support one, or more, LUs.

The LUs are used to provide access to host-based applications. The SSCP is the NAU used to support network management functions. Along with basic network management functions, it is contained in VTAM [see Section 13.4.4(2) below]. Additional network management functions are contained in IBM's NetView product.

(2) Telecommunications Software

In subarea networks, the telecommunication software resides in the subarea host and communications controller(s). The major components are the teleprocessing access method (VTAM) and the network control program (NCP). VTAM provides a software interface between the host main memory and its peripherals that moves information between the network and the host over the I/O channel—a software-driven processor that executes commands and programs.

- **VTAM** (Virtual Telecommunications Access Method): an access method that directly controls transmission of data to and from channel-attached devices. In an MVS (multiple virtual systems) environment, manages a communications network for the exclusive use of the applications it supports. To send data to, and receive data from, remote devices over SNA links, VTAM requires the use of NCP.
- **NCP** (Network Control Program):
  - provides network information to VTAM
  - controls cluster controllers and attached terminal nodes
  - routes data to communications controllers in other subareas
  - performs as a primary station for communication terminations
  - polls secondary stations, dials, and answers switched lines, detects and corrects errors, and maintains operating records
  - allocates buffers for data coming from and going to the host, and to other stations.

NCP should not be confused with the generic description, network control program which identifies any program that performs network control functions.

(3) Establishing an LU–LU Session

To illustrate some of the steps required to manage subarea networks, consider the problem of establishing an LU–LU session across domains. They are shown in **Figure 13.23**.

- The originating end-user enters a logon message that requests connection to a specific service such as TSO (step 1).

- Over an already existing LU–SSCP session, LU1 sends the message to SSCP1 (in VTAM1) for logon name-to-network address translation (step 2).

- Having determined that LU1 is permitted access to TSO, and that TSO is not within its domain, SSCP1 consults a listing of cross-domain resources and contacts the owning SSCP (SSCP2 in VTAM2) over an already existing SSCP–SSCP session (step 3).

- SSCP1 informs SSCP2 of the communication capabilities associated with LU1.

- Over an existing SSCP–LU session, SSCP2 delivers this information, together with the address of LU1, to LU(TSO) (step 4).

- LU(TSO) sends a session initiate request to the TSO application and receives a session initiate response (step 5).

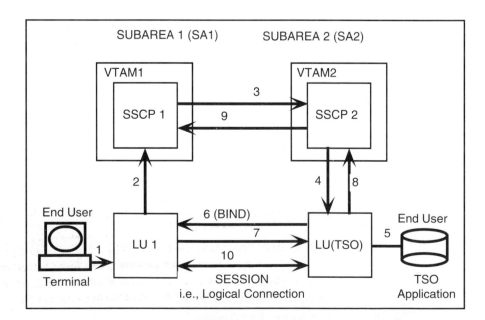

Figure 13.23 Steps Required to Establish a LU–LU Session across Domains

- At the same time, SSCP2 causes the PU in NCP2 to establish explicit paths between LU(TSO) and the boundary unit in the subarea (SA2). Route selection is done by consulting tables or using other means.

- SSCP1 causes NCP1 to establish explicit paths between LU1 and the boundary unit in its subarea (SA1).

- When a virtual route has been established between LU(TSO) and LU1, LU(TSO) sends a command (called a BIND) to describe the parameters proposed for the session. The BIND customizes the session by specifying the roles, rules, and procedures that will be used (step 6).

- If LU1 is satisfied with the BIND, it replies affirmatively (step 7).

- LU(TSO) sends a session established data unit to SSCP2 (step 8).

- SSCP2 sends a session established data unit to SSCP1 (step 9).

- As the primary LU, LU(TSO) issues a start data request to LU1, and the session proceeds over the virtual route established directly between the units (step 10).

Note that the initial communication steps are accomplished over preexisting sessions that are established when the system is generated. They form part of the communication structure required for operation. Unless an LU is connected to its owning SSCP in the original startup procedures, it cannot communicate. To connect an LU after system generation, the system operator must instruct the owning SSCP to activate a session between them.

(4) Teleprocessing Monitors

In order to implement real-time, on-line transaction processing, some applications are separated from the access method by an additional software module. In a manner that mimics the operating system, these modules supervise multiple transaction programs that run in a single partition. Called TP monitors, to the programs they appear as the operating system. For programmers, TP monitors make it possible to develop applications in a high-level language without detailed knowledge of the terminal characteristics. Two important TP monitors are

- **Customer Information Control System** (CICS): supports multiple, on-line applications in a common network. Provides interfaces to files and database systems. Handles errors, security, and priority transactions.

- **Information Management System/Data Communications** (IMS/DC): originally a database management system, IMS has been augmented by communications capabilities to provide a comprehensive, database management and data communications capability.

In **Figure 13.24,** I conceptualize the major partitions of the host and front-end processor software. In **Figure 13.25,** I show the arrangement of LUs, PUs, and

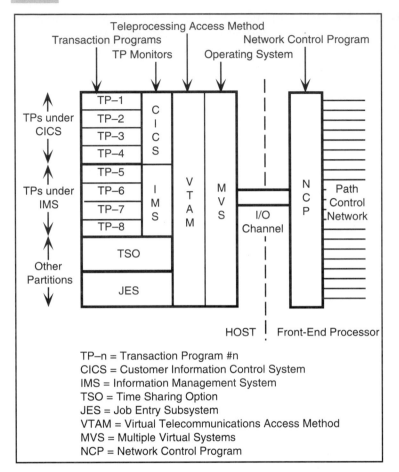

Figure 13.24 Major Partitions of Host and Front-End Processor Software

TP–n = Transaction Program #n
CICS = Customer Information Control System
IMS = Information Management System
TSO = Time Sharing Option
JES = Job Entry Subsystem
VTAM = Virtual Telecommunications Access Method
MVS = Multiple Virtual Systems
NCP = Network Control Program

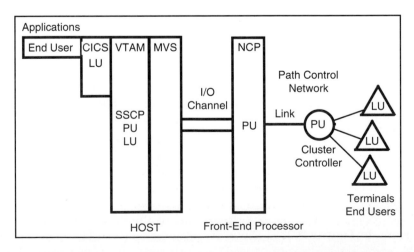

Figure 13.25 Arrangement of LUs, PUs, and SSCP in a Single-Host Subarea Environment

SSCP in a single-host environment with transaction programs running under CICS and VTAM. End-users at the terminals can access one of the applications (the other end-user) running under CICS. The SSCP in VTAM employs PUs in NCP and the cluster controller to establish, maintain, and terminate communication between the host and the terminal cluster over a link in the path control network. Also, the SSCP employs the PU resident in VTAM to activate, control, and deactivate CICS and the end-user application.

REVIEW QUESTIONS FOR SECTION 13.4

1 Give a concise description of the purpose of SNA.

2 In what environments can SNA operate?

3 Distinguish between subarea networking and APPN.

4 State the names of the seven layers of the SNA protocol model, and describe their functions.

5 What is the path control network? What does it do?

6 Use Figure 13.20 to describe the formation of a message unit.

7 What is a network accessible unit? Name four types of NAUs, and identify whether they are used in subarea networks, APPNs, or both.

8 What is a session? Name and describe five kinds of sessions.

9 Use Figure 13.22 to describe an SNA subarea network.

10 Distinguish between Type 2.0 and Type 2.1 peripheral nodes.

11 Describe two types of subarea nodes.

12 Describe VTAM and NCP.

13 Explain Figure 13.23.

14 To connect an LU after system generation, why must the system operator instruct the owning SSCP to activate a session between them?

15 What are teleprocessing monitors? Describe two TPs.

16 Explain Figure 13.25.

13.5 SNA APPN NETWORKS

APPN (Advanced Peer-to-Peer Networking) is designed to support the decentralized network operation inherent in client/server environments. Instead of predetermined routing and hierarchy (as in subarea networks), APPN networks dynamically reconfigure themselves as links and nodes are added or turned off. **Figure 13.26** shows a simple APPN network. Its operation is described below.

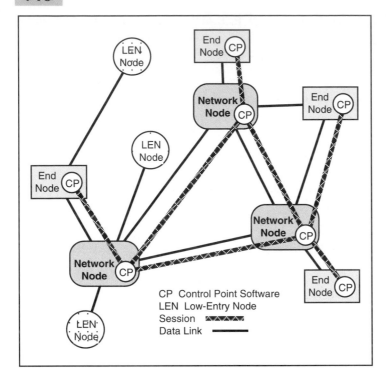

Figure 13.26 Simple APPN Network

13.5.1 APPN NODES

An APPN network contains three categories of nodes. All of them are type 2.1, that is to say, besides end-user communication with host-based applications, they support communication with other type 2.1 nodes.

- **Network node**: backbone node connected to others by transmission groups. Each network node
    - contains CP software that supports communication between network management software entities to update routing tables, provide directory services, etc. CPs in adjacent network nodes establish sessions with one another for these purposes.
    - contains a topology database that describes all of the network nodes in the network and the transmission groups that connect them. It is dynamically updated as conditions change.
    - maintains a directory with entries for each local LU, and can contain entries for all directly attached end nodes and low-entry networking nodes
    - supports local-user LUs

- acts as a communication server for directly attached end nodes and low-entry networking nodes

- provides directory and route selection services to all dependent nodes.

- **End node**: node that exists at the edge of an APPN network and provides network access to end-users. End nodes

  - contain directories of local LUs and may have entries for LUs that reside on adjacent end nodes

  - contain descriptions of the local transmission groups that connect them to the network

  - contain CPs that establish sessions with CPs in network nodes. An end node can establish a CP–CP session to only one adjacent network node (at a time).

- **Low-entry networking** (LEN) **node**:

  - supports simple point-to-point connections between adjacent LEN nodes

  - when connected to a network node, can communicate with other LEN nodes in the APPN network, and request directory searches and route calculations.

  - contains LUs that support local end-users.

13.5.2 APPN Network Operations

The routes and links that are entered into the SSCP at the time of generation of a *subarea* network cannot be changed without operator intervention. They are set for the life of the network. In contrast, each node in an APPN network activates its own resources, as the situation dictates. Each network node interacts with its peers (or superiors) to activate CP–CP sessions over which dynamic updates are shared, and connections are made, as conditions and requirements change. There are no predefined routes between LUs; they are created when a network node receives a request for a session.

(1) Learning the Capabilities of Neighboring Nodes

The activation of a node causes it to open links to its neighbors. Over them, the nodes exchange basic information so that each node learns the capabilities of its neighbors—whether they are network nodes, end nodes, or LEN nodes—and their capabilities.

(2) Activation of CP–CP Sessions

To generate the network, CP–CP sessions must be started between adjacent nodes. For network nodes, the sessions will be with peer nodes and with dependent end nodes. For end nodes, the sessions will be with one network node (although they may have data links to several). At startup, the data links over which sessions are activated are determined from information stored locally. Once the initial network has been configured, sessions will be activated or deactivated as traffic and equipment conditions demand.

(3) Updating Topology Databases

Each network node maintains a topology database that describes the network. When changes occur, they must be disseminated to all databases to ensure the accuracy of routing and directory services provided by the nodes. To do this, *topology database updates* (TDUs) are exchanged between network nodes. When a CP–CP session is activated between network nodes, they exchange TDUs. When a network change is reported to the node, it forwards the information to its neighboring peers.

To preserve some level of order and control of the number of TDUs in the network, TDUs are assigned sequence numbers (FRSNs, flow reduction sequence numbers). When a transmission group between two network nodes is activated, the CPs exchange the numbers of the last updates they received. They can then determine which updates one of them needs to transmit to the other.

(4) Directory Service

- **Directory service**: provided by a node in response to a request from an originating LU to find the location of a terminating LU prior to starting an LU–LU session.

If the originating LU (LU1) resides on an end node, the search for the location of the terminating LU (LU2) begins with the local directory that lists local LUs, and perhaps some LUs in adjacent nodes. If an entry is not found, the search moves to the directory on the end node's network node server.

Over their CP–CP session (already in place), a request is sent that identifies LU1, LU2, and describes the transmission groups between the originating end node and the network node (known as the *tail* circuit). Because the network node has information only on transmission groups in the backbone, the latter information is needed for the network node to describe the route between LU1 and LU2 completely.

First, the CP in the receiving network node searches its local directory. This contains entries for local LUs, and may contain entries for LUs on dependent nodes. If the entry for LU2 is found, the CP initiates a *directed* search to confirm that LU2 is indeed in the location listed, and obtains details of its tail circuit. If the entry for LU2 is not found in the local directory, a *broadcast* search is begun. In it, locate requests are sent to each of the neighboring network nodes. If they do not have the entry for LU2, they will send the locate request on to their neighbor, and so on. When LU2 is found, the location information, and tail circuit routing are sent back to the original network node.

To reduce the traffic generated by broadcast searches, APPN can use one, or more, *central directory servers* (CDSs). Nodes so configured are identified in the topology databases. After searching the local directory, network nodes send the locate request directly to CDS. If an entry is found, CDS initiates a directed search, collects tail circuit information, and returns a found message to the original network node. If no entry is found for LU2, CDS initiates a broadcast search.

(5) Route Selection

- **Route selection service**: provided by the CP in the network node serving the requesting LU. The CP calculates an optimal route on the basis of the locations of LU1 and LU2 and information obtained from directory services.

Following LU1's request for the location of LU2, and the return of the location of LU2 and tail circuit information, the CP in the original network node is responsible for selecting an optimal route. By comparing a set of criteria provided by LU1 to the characteristics of the available routes, the CP determines the most suitable route for the LU1–LU2 session. In the calculation, elements such as delay, security, congestion, and cost are considered. Upon its completion, a description of the route (called a *route selection control vector*, RSCV) is sent to LU1. It is an ordered list of each node and transmission group along the route to LU2. The use that is made of RSCV depends on the routing protocol employed.

(6) Routing Procedures

- **Intermediate session routing** (ISR): the path control layer in each node assigns a unique identifier for the session. Called the *local-form session identifier* (LFSID), it is different for each transmission group in the connection. A 17-bit word, it is carried in the address field of the transmission header of the BIND command, and all subsequent messages for the session. At each intermediate node, a lookup table is required to identify the next leg of the route. Because of this node-by-node checking, ISR is relatively slow but reliable.

- **High-performance routing** (HPR): the entire path is carried in the packet header. This eliminates table lookup at each intermediate node, and makes delivery much faster than ISR over low error rate links.

13.5.3 TCP/IP AND SNA/APPN

With over 25 years experience in providing time-sensitive transaction networks, SNA has an installed base of some 50,000 networks.[1] Internet, based on TCP/IP, represents a very different type of network architecture. Without particular regard to the timeliness of transactions, it connects millions of networks. In many instances, the company intranet (TCP/IP) and the company batch-processing and report-generating network (SNA) exist side by side and run to the same locations. Inevitably, businesses that seek to optimize their investments in SNA/APPN and

---

[1] Daniel Lynch, James P. Gray, and Edward Rabinovitch, eds., *SNA & TCP/IP Enterprise Networking* (Greenwich, CT: Manning Publications, Inc., 1998), 456.

TCP/IP equipment are looking for ways in which both operations can be supported on one transmission facility. Furthermore, there ought to be common ground between databases (SNA) and servers (TCP/IP) to minimize the amount of data that must be maintained.

To satisfy these demands, *multiprotocol* routers have been developed that process SNA, TCP/IP, Novell's IPX, and other frames over common transmission facilities. In effect, the router consists of separate parts that contain routing tables and logic peculiar to each protocol. In addition, software is available that adapts SNA frames to frame relay or cell relay applications. For each protocol change, the frame is encapsulated and decapsulated. An important consideration is to achieve delivery and acknowledgment before the sender times-out and resends the frame.

REVIEW QUESTIONS FOR SECTION 13.5

1 What are the unique networking properties of APPN?

2 Describe the three types of nodes employed in an APPN network.

3 Explain how an APPN network node learns the capabilities of its neighboring nodes.

4 How are CP–CP sessions activated?

5 How is the topology database updated?

6 What is directory service?

7 Describe how a search proceeds to find a terminating LU so that a session can be activated.

8 What is a broadcast search?

9 Describe route selection service.

10 Describe intermediate session routing.

11 Describe high-performance routing.

12 Explain how businesses that seek to optimize their investments in SNA/APPN and TCP/IP equipment look for ways in which both operations can be supported on one transmission facility.

13 What is a multiprotocol router?

13.6 DATA SECURITY

- **Data Security**: the action of protecting a message, a file, or a transaction from unauthorized access, usage, or alteration by an agent. (The agent is a software entity under the control of a user.)

Telecommunications technology has made it possible to generate information anywhere, and to use it everywhere, without delay. I might add that this applies to unauthorized as well as authorized persons. Unless precautions are taken, connecting an information system to a communication network makes it possible for anyone on the network to access the system. Person's who access the information system fall in two categories

- **Authorized users**: those who are connected with the full knowledge of the network managers
- **Unauthorized users**: those who connect themselves to the network for the express purpose of obtaining information to which they are not entitled.

13.6.1 BREACHES OF SECURITY

In a world characterized by terminals and hosts, or clients and servers, connected by pervasive communication networks, the problems extend far beyond traditional considerations associated with the protection of property from physical harm

- *It may not be clear who might do the harm.* The network connects sensitive data resources to the world, and any agent, for any reason—or no reason at all— may attempt to gain access and sabotage them.
- When dealing with electronic data, *generating, duplicating, and altering can be done easily* with simple software commands. In a physical sense, electronic data in an illicit message are exactly the same as data in a genuine message. Their meaning may be different, but examining the signals will not reveal anything untoward.
- *Efforts to breach security can be automated* so that thousands of attempts can be made per second.

All of these events are complicated by the *virtual world* of data communication in which those seeking to do illicit things may be in the same office as the data they wish to exploit, or separated by half a world from the host or server.

To protect data under these circumstances requires attention to

- **Authentication**: establishing the identity of agents with certainty
- **Authorization**: establishing that an agent is endowed with authority to perform requested functions
- **Data Integrity**: proving that the data are complete, uncorrupted, and genuine
- **Nonrepudiation**: ensuring that agents cannot reject an action they authorized, or declare it to have no binding force

- **Privacy**: maintaining the confidentiality of data from agents not authorized to receive them.

The last item, privacy, is discussed in Section 13.7.

(1) Objectives of Security Procedures

As they flow through the system, data messages may be altered or destroyed by accidental events and deliberate acts. At the very least, procedures must be built in to the overall communication operations to detect these happenings. The ultimate objective of these procedures should be to guarantee that the system delivers authentic messages only to the persons authorized by the owners to receive them—no more, and no less—so that messages are not

- **Stolen** by
  - competitors, to gain an advantage
  - blackmailers, to extort money or other valuables
  - terrorists to disrupt the business
- **Distorted** by changing
  - values or quantities to misrepresent facts
  - addresses to steal money or goods from the organization
- **Damaged** by deleting information fields in order to disrupt an activity or organization.

(2) Authentic, Corrupt, and Illicit PDUs

Because messages may be made up of several frames, I use the precise term protocol data unit (PDU)—the entire sequence of bits transmitted at one time—to discuss the results of these happenings. In addition, I define three states for the PDU

- **Authentic protocol data unit** (APDU): the true PDU that the source wishes to send to one, or more, receivers for a purpose approved by the network owner
- **Corrupted protocol data unit** (CPDU): an authentic PDU that has been altered by accidental means
- **Illicit protocol data unit** (IPDU): a PDU introduced into the system for an illicit purpose. It may be an APDU that has been altered deliberately, an APDU that is *received* by an illicit source, or a PDU *generated* by an illicit source.

(3) Authentic and Illicit Sources and Receivers

In addition, I distinguish between

- **Authentic sources** (ASs): units that are connected to the network with the knowledge and approval of the network owner

- normally, they send authentic PDUs (APDUs)
- sometimes, they send corrupted PDUs (CPDUs)
- sometimes, they send illicit PDUs (IPDUs)

- **Authentic receivers** (ARs): units that are connected to the network with the knowledge and approval of the network owner
  - normally, they receive authentic PDUs
  - sometimes, they receive corrupted PDUs
  - sometimes, they receive illicit PDUs
- **Illicit sources** (ISs): units that are connected to the network without the knowledge and approval of the network owner
  - always, they send illicit PDUs.
- **Illicit receivers** (IRs): units that are connected to the network without the knowledge and approval of the network owner
  - they receive authentic PDUs, corrupted PDUs, and illicit PDUs.

These definitions and concepts are diagrammed in **Figure 13.27**.

(4) Events that Affect the Proper Functioning of Communication Systems

In a communication system, PDUs may be subjected to one, or more, of the following events

- **Loss**: the APDU never reaches the AR due to
  - *Misrouting*. An authentic PDU is routed improperly. This may be accidental (due to a one-time fault at a switch or router, for instance) or deliberate (due to unauthorized alteration of the routing table).
  - *Wrong address*. The APDU is corrupted by an address that is not the address of the intended AR. This may be accidental (due to a mistake at the source, or to an error picked up in transmission that modifies the authentic address) or deliberate (due to the unauthorized alteration of the authorized address at the source).
- **Denial of service**: PDU delivery may be prevented by the failure of equipment as a component comes to the end of its life (a natural occurrence in electronic devices), or by an accident, natural catastrophe, or sabotage, that affects the entire system.
- **Alteration**: the APDU is corrupted so as to change the message from that which the sender intended. Alteration may be
  - *Accidental*. When it is due to a mistake made entering the message into the system, or it is due to errors picked up in transmission that modify the text of the PDU

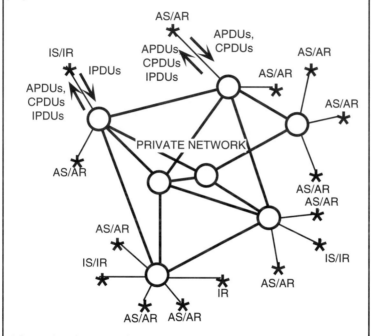

APDU = authentic PDU, i.e., the true PDU that the source wishes to send to one, or more, receivers for a purpose approved by the network owner.
CPDU = corrupted PDU, i.e., an authentic PDU that has been altered accidentally.
IPDU = illicit PDU, i.e., a PDU introduced into the system for an illicit purpose.

AS = authentic source, i.e., source connected to network with approval of network owner.
AR = authentic receiver, i.e., receiver connected to network with approval of network owner.
IS = illicit source, i.e., source connected to network without approval of network owner.
IR = illicit receiver, i.e., receiver connected to network without approval of network owner.

Figure 13.27
Representation of Types of Sources and Receivers in a Private Network

- *Deliberate.* When the authentic message is changed without authority prior to sending.

- **Insertion**: an illicit PDU is sent from an authentic source, or introduced by an illicit source, to provide information (true or false) to a receiver

- **Duplication**: copies of an authentic PDU may be sent to receivers other than the receiver(s) intended by the sender. This may be accidental (due to improper addressing, perhaps) or deliberate (desire to leak information, perhaps)

- **Disclosure**: an authentic PDU is delivered to an unintended authentic receiver, or an illicit receiver.

In each case, the events may be readily apparent or may go unnoticed. Prudent management requires that measures be introduced to detect and confound them.

13.6.2 ROUTINE MEASURES THAT DETECT CORRUPT OR ILLICIT PDUs

At all levels of the communication process, measures can be introduced to detect corrupt and illicit messages

- **Sequence number checking**: message/PDU sequence numbers can be inserted in the text portion of the PDU. Their presence makes it possible for the human receiver to discover that PDUs are missing, or that PDUs have been inserted.
- **Address check**: the use of an error-detecting device (block check character or frmae check sequence, for instance) on the PDU address will provide a measure of confidence that, if the address is corrupted in transmission, any receiver will destroy it. This may prevent accidental disclosure. For deliberate changes in address, the agent will recalculate the checking device.
- **Positive acknowledgment**: the receiver can acknowledge each PDU by message/PDU sequence number, confirming that the message reached the intended receiver. Such information is distinct from any acknowledgment scheme employed by the protocol, and is included in the text portion of the PDU.
- **Transaction journal**: by maintaining a copy of the header, message sequence number, and trailer of each PDU, and noting every communication event (such as transmit, arrive/depart intermediate nodes, and receipt) encountered by the PDU, an audit trail can be created. This will facilitate the detection and investigation of unauthorized events.
- **Time and date stamp**: the audit trail can be enhanced by the addition of time and date information showing when the PDU was generated, when it passed through each node in the network, and when it was received. If such information is added to a PDU during the communication process, it can be passed back to the originating point when the receiver acknowledges receipt of the PDU. As an alternative to collecting all of the information at the originating point, each node can maintain a log of its own activities (including message sequence number) that can be consulted by network control as necessary.
- **Periodic message reconciliation**: periodic, automatic testing of the audit trail should reveal loss, misrouting, disclosure, insertion, or duplication of a PDU and provide the basis for an investigation of unusual events.
- **Redundant information**: names, numbers, dollar amounts, etc., can be repeated within the text of the PDU, or within the text of a following PDU;

examination at the receiver should reveal whether these key items have been transmitted correctly.

13.6.3 PHYSICAL CONTROLS

To a large degree, the security of data in a communications and processing environment depends on the ways in which access to the environment is authorized and controlled.

(1) Access Controls

Three forms of access can be distinguished

- **Physical access**: access to the installation, system components, input-output terminals, communication lines, switching points, and other physical entities that comprise the environment
- **Operational access**: access to processing, storage, and input-output resources through the use of terminals, personal computers, or other devices that are connected over communications facilities to the processing entities in the environment
- **Unauthorized monitoring**: access to the data transmission path for the purpose of intercepting the messages thereon.

The security of a communications and data processing environment depends upon preventing these types of accesses.

(2) Controlling Physical Access to Major Facilities

For facilities that contain major communications and processing equipment, the most straightforward technique is to control all entrances to the facility. This can be done through the use of

- **Guards** who permit access based on a badge or other identification
- **Manual** or **electronic keys** that are carried by those authorized to gain access
- **Magnetic** or other **electronic card systems**, with or without a personal identifier
- **Biometric parameters** (voiceprints, fingerprints, etc.) that identify the person seeking entry.

These techniques may be employed singly or in combination. A basic assumption is that possession of an appropriate access method identifies an authorized person who will use the facilities for the purpose access is authorized. None of them can prevent a person who is authorized access from stealing or misusing information. The integrity of controls that limit physical access is based on the understanding that those who have access share common objectives. Because circumstances change over time, it is important to review frequently the list of those granted access, and to investigate changes in patterns of access.

(3) Controlling Physical Access to Remote Facilities

For remote facilities, such as cable vaults and switch rooms not normally occupied, or storage facilities to which persons need access occasionally

- **Alarm systems** and **intrusion detection devices** can be installed
- **Video surveillance** can be maintained
- **Access** can be restricted to authorized workers identified by keys, cards, or biometric parameters.

However, if some of the transmission and switching facilities in the communications and data processing environment are part of someone else's network, these techniques cannot be implemented easily. While the messages are outside of the jurisdiction of the owner, they are susceptible to intrusion. Under these circumstances, private messages can be protected by encryption.

(4) Controlling Operational Access

Operational access does not require physical access. The moment a communication connection is made to a data processing center, it is vulnerable to electronic intrusion from around the world. Using a personal computer and a dial-up connection, computer *hackers* have demonstrated that unauthorized operational access can be gained to any system. Preventing operational intrusion relies on the use of passwords and procedures designed to make it practically impossible to penetrate the system

- **Passwords** are used to authorize communication, access to resources, copying or altering file entries, etc.
- **Procedures** are used to ensure that certain types of data are restricted to certain locations, terminals may be designated read-only for certain classes of data, terminals may be restricted to working with one file at a time, etc.

A common technique to make dial-up penetration more difficult employs a callback routine. On receiving a dial-up request for connection to the data center, the communications processor hangs up and then dials the number listed for the person whose password(s) accompanied the dial-in request. For the connection request to be honored, the requester must be at the authorized telephone location. If this is in a protected facility, the likelihood of unauthorized use is diminished.

(5) Protection from Monitoring

Transmission facilities that employ radio can be easy targets for eavesdroppers. Protection against monitoring relies upon making the cost uneconomical

compared to other ways of obtaining the same information (compromising authorized personnel, for instance). Intrusion can be made more difficult (and more
expensive) by

- Employing **multiplexed facilities** in which each call uses a different channel
  (so that the eavesdropper must search a number of channels for the information)
- Using the **highest-speed data circuits** available (so that messages are short,
  and therefore more difficult to intercept)
- Employing **alternative transmission paths** (so that eavesdroppers must
  monitor several routes)
- Using **packet technology** (so that the message is segmented into packets that
  are mixed with other traffic)
- **Encrypting** the data.

13.6.4 THE SECURITY CHALLENGE

If the rewards are great enough, those who practice illicit activities will circumvent any procedures set up to thwart them. What one determined person constructs, a more determined person can destroy. Maintaining the integrity of a
telecommunications-based information system that transports sensitive information is a task that requires eternal vigilance and inquisitiveness.

REVIEW QUESTIONS FOR SECTION 13.6

1 What is meant by data security?
2 Distinguish between authorized users and unauthorized users.
3 Explain the following statement: In a world characterized by terminals and
   hosts, or clients and servers, connected by pervasive communication networks, the problems extend far beyond traditional considerations associated
   with the protection of property from physical harm.
4 Define authentication, authorization, data integrity, nonrepudiation, and
   privacy. Why are they important?
5 What is the objective of security procedures?
6 What actions must a security system prevent?
7 Define authentic PDU, corrupted PDU, and illicit PDU.
8 Define authentic sources, authentic receivers, illicit sources, and illicit
   receivers.

9 List events that affect the proper receipt of a PDU.

10 List the measures that detect corrupt or illicit PDUs.

11 What physical controls can be used to increase the security of facilities?

12 What physical methods can be used to prevent monitoring of signals?

13 Comment on the following statement: What one determined person constructs, a more determined person can destroy.

13.7 DATA PRIVACY

Records pertaining to any one of us can be collected, processed, and manipulated in telecommunications-based information systems to satisfy inquiries that may have to do with our performance as an employee, as a consumer, as a debtor, or as a vital statistic. They are all roles that most of us will play in our lifetimes.

13.7.1 INDIVIDUAL AND COMPANY PRIVACY

Because they involve considerable sorting, review, and validation, without telecommunications-based information systems, it is unlikely that such collections would exist. Doing anything out of the ordinary is likely to lead to bureaucratic inquiries to some of these databases.

   Because the records may be concentrated in a few data processors that are connected to the public telephone network, records on an individual can be collected together relatively easily by those authorized to do so, and with somewhat more difficulty by those without authorization to do so. From such a collection, trained persons can construct an accurate and detailed profile of an individual, or a family. Developing and using such knowledge is an invasion of an individual's privacy.

- **Invasion of individual privacy**: theft, interception, misuse, or aggregation of information concerning an individual's thoughts, sentiments, emotions, lifestyle, or status.

Depending on the individuals involved, some, or all, of this information can be obtained from computer files through the use of Social Security numbers, credit card numbers, telephone numbers, and zip codes. Many countries have laws to protect the individual against the disclosure or misuse of information stored in such databanks.

   Company records are stored in interconnected systems that are accessed from many units every working day. Thus, it may be easier to collect information on company operations and conditions than on private individuals. Developing and using such knowledge is an invasion of company privacy.

- **Invasion of company privacy**: theft, interception, misuse, or aggregation of information concerning the operation of the business, particularly that information which may give a competitive advantage to the receiver.

In many companies, this information is declared to have the status of company proprietary information.

13.7.2 Protecting Private Information

The principal protective measures are

- **Encryption** to make the meaning obscure to unauthorized persons
- **Access controls** to limit the number of persons who can use, modify, or examine data
- **Psychological deterrents** produced by laws that impose fines or imprisonment for improper use of personal information, and by company policies that proscribe dismissal for unauthorized use of proprietary information.

Of these measures, the use of encryption is discussed here, access controls were discussed in Section 13.6.3(1), and a discussion of psychological deterrents is beyond the scope of this book.

(1) Encryption Defined

- **Encryption**: the action of disguising information so it can be recovered relatively easily by persons who have the *key*, but is highly resistant to discovery by persons who do not have the key.

A message in cleartext (also called plaintext) is disguised (encrypted) with an encryption key to create a cryptogram (i.e., create a facsimile of the plaintext message in ciphertext). **Figure 13.28** shows some of these basic relationships. The

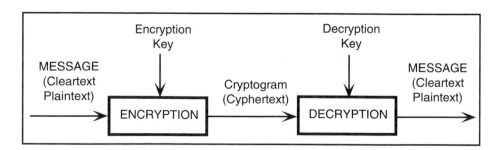

Figure 13.28 Representation of Encryption and Decryption

encryption key may be changed from time to time to make an intruder's task more difficult. Restoration of a cryptogram to cleartext is achieved by the action of decryption using a decryption key. In a symmetric or single-key cryptosystem, the encryption and decryption keys are the same. In an asymmetric or two-key cryptosystem, the encryption and decryption keys are different, and one cannot be derived readily from the other. Usually, in a two-key cryptosystem, one key is made public and the other key is kept secret.

- **Secret-key Cryptography** (SKC): uses a single secret key known only to the sender and receiver
- **Public-Key Cryptography** (PKC): uses a secret key known only to the sender and a public key that is available to anyone who wants it.

The SKC standard is the data encryption standard (DES); standards for PKC are Rivest, Shamir, Adelman algorithm (RSA) and digital signature standard (DSS). They are discussed below.

(2) Encryption Methods

Encryption is accomplished by scrambling the bits, characters, words, or phrases in the original message. Scrambling involves two activities

- **Transposition,** in which the order of the bit patterns, characters, words, or phrases is rearranged
- **Substitution,** in which new bit patterns, characters, words, or phrases are substituted for the originals without changing their order.

They are said to employ

- **Ciphers**, when the bit patterns or characters in the original text are transposed or substituted
- **Codes**, when the words or phrases in the original text are transposed or substituted.

Because the use of a code requires a large element of human judgment, computer-based encryption and decryption algorithms employ ciphers.

(3) Secret Key Cryptography

Probably the most widely used encryption algorithm is the Data Encryption Standard (DES). Pioneered by IBM, it is employed in symmetric cryptosystems. Originally developed to protect sensitive, unclassified U.S. government computer data, DES is used to provide authentication of electronic funds transfer (EFT) messages and is available in both hardware and software implementations for the protection of other private data. Based on a 56-bit cipher key, a single, randomly selected specific key (selected from the 2^{56}, 56-bit patterns afforded by the cipher key) is employed in both encryption and decryption. DES uses a complicated algorithm that thoroughly scrambles the plaintext.

In the standard version of DES, 64-bit blocks of data are scrambled using a sequence of bit substitutions and bit transpositions. Encryption begins by transposing the 64-bit block of plaintext using the specific cipher key selected. The resulting block is split into two halves. One half is transformed by a nonlinear function defined by the specific key in use. It is then added to the other half of the transposed text. Next, the halves are interchanged, and the process is repeated using an intermediate key derived from the original key. After 16 operations of this kind, a final transposition is applied to create a 64-bit block of ciphertext. Decryption is the inverse of encryption; the intermediate keys are used in reverse order. The principle of encryption and decryption using DES is shown in **Figure 13.29**. The success of DES depends on random key selection and frequent key changes. There is no known way to break DES except to test possible keys one by one. If all keys (2^{56}) are equally likely to have been used an average of $2^{56}/2$ (i.e., 2^{55}, tests will be required. While this is a daunting number, there is evidence that a parallel machine has broken a DES key in 56 hours.[2] If greater security is desired, a 112-bit cipher key is employed. The longer the cipher key, the *stronger* the encryption is said to be.

(4) Public Key Cryptography

A public key cryptosystem is an asymmetric cryptosystem. First described to the technical world in 1976, public key algorithms offer a radically different approach to encryption.[3] Based on the difficulty of factoring very large integers, the idea depends on the use of a pair of keys that differ in a complementary way. Of the several algorithms originally proposed, the RSA algorithm[4] (named for its inventors Ron Rivest, Adi Shamir, and Leonard Adelman) is considered highly secure. Patented in 1978, the input is recovered by successive transformations based on the two keys. By holding one key secret, a substantial level of protection is achieved; by making the other key public, a significant operating flexibility is achieved. Although the public key (PK) is known, and messages encrypted through its use may be available to those who would steal the information, without the use of the secret key (SK), it is impossible to decipher the encrypted message. **Figure 13.30** shows the use of PKS to achieve

---

[2] The Center for Democracy and Technology, "DES Is Dead," http:// www.cdt.org/crypto/posted July 17, 1998.

[3] W. Diffie and M. E. Helman, "New Directions in Cryptography," *IEEE Transactions on Information Theory,* 22 (1976), 995–1008.

[4] In 1982, RSA Data Security Inc. (http://www.rsa.co) was formed to commercialize RSA through product development and licensing.

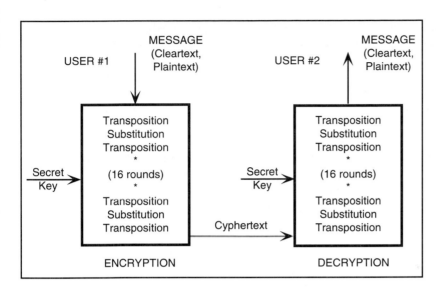

Figure 13.29
Encryption and
Decryption Using
DES

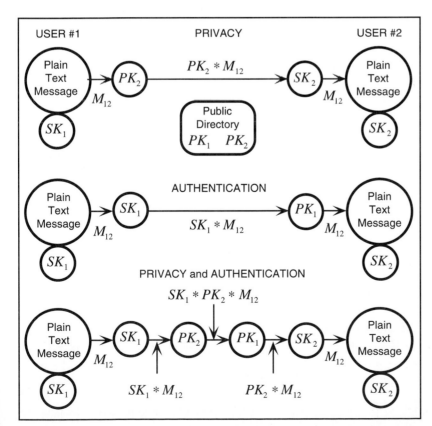

Figure 13.30 Using
PKS to Achieve
Privacy,
Authentication, and
Privacy and
Authentication

- **Privacy**: when user 1 wishes to send a private message (M_{12}) to user 2, it is encrypted as PK_2*M_{12} (i.e., M_{12} is encrypted by user 1 using the public key belonging to user 2). When user 2 wishes to read the message, it is decrypted as $SK_2*PK_2*M_{12}$ to yield M_{12} (i.e., PK_2*M_{12} is decrypted by user 2 using the secret key known only to user 2). Without SK_2, others cannot read M_{12}.

- **Authentication**: if user 1 wishes to send message M_{12} to user 2, and wishes user 2 to know without a doubt it comes from user 1, the message is encrypted as SK_1*M_{12}. When user 2 receives the message, it is decrypted as $PK_1*SK_1*M_{12}$ to yield M_{12}. In this case, all users have access to PK_1 and can read M_{12}. However, authentication is the issue for user 1 and user 2, not privacy.

- **Privacy and authentication**: if user 1 wishes to obtain both privacy and authentication, the message is encrypted as $PK_2*SK_1*M_{12}$. On receiving the message, user 2 forms $SK_2*PK_2*SK_1*M_{12}$ to yield SK_1*M_{12}, and then forms $PK_1*SK_1*M_{12}$ to yield M_{12}. Only user 2 has access to both PK_1 and SK_2.

In RSA, the keys are functions of a pair of very large prime numbers. To determine one from the other is equivalent to factoring the product of the two primes. The strength of RSA increases with the size of the prime numbers used. In 1997, IBM announced a public key system that relies on finding a *unique shortest vector* in an *n*-dimensional lattice. Increasing the number of dimensions in the lattice enhances its strength.

(5) Digital Signature Standard

In 1991, the National Institute of Standards and Technology introduced the digital signature standard (DSS). Using the principle of public-key cryptography, the DSS algorithm encrypts and verifies a short code that is attached to a data file. This digital *signature* is created from a reduced representation (called a *hash*) of the data file. The code provides electronic verification of the data and their source and leaves the entire file in cleartext. If greater security is desired, electronic documents can be signed with a digital signature and sealed with encryption.

(6) Pretty Good Privacy

PGP (pretty good privacy)[5] is a combination of SKC and PKC for protecting electronic mail. The sender encrypts the message using a randomly generated key, then encrypts the key using the recipient's public key. The recipient uses his secret key to decipher the random key, then uses the random key to decipher the message.

(7) Digital Certificates

How does the sender of a message over a PKC system know that the published public key is really that of the intended recipient? Through the use of a trusted third party *certification authority* (CA), the authenticity of both sender and receiv-

---

[5] Simon Garfinkel, *PGP—Pretty Good Privacy* (Sebastopol, CA: O'Reilly and Associates, 1995).

er can be confirmed.[6] The key lies with the CA that issues *digital certificates* that contain a secret key that is paired with a public key. In addition to creating the pair of keys, the CA certifies that the device to which the certificate is issued is the device it represents to be. Thus, authentication of the device rests on the integrity of the CA.

If the sender (device A) wishes to authenticate a receiver (device B), it sends a short message (known as the message digest), which it encrypts using the public key that the CA asserts belongs to device B. When the receiver (B) receives the encrypted digest, it decrypts it using its secret key (created by the CA). The receiver then encrypts the message digest using the public key that the CA asserts belongs to device A. When the sender (A) receives the encrypted digest, it decrypts it using its secret key (created by the CA). If the message digest is the same as the original, the device A knows it is in contact with the authentic device B. For device B to authenticate device A, the message sequence starts with device B. Once A and B are satisfied that each is authentic, they may elect to communicate using a symmetrical cryptosystem.

13.7.3 KEY MANAGEMENT

Restoration of a cryptogram to cleartext is achieved through the use of a decryption key. In a symmetric or single-key cryptosystem, the encryption and decryption keys are the same. To ensure the integrity of the single key, its distribution must be carefully protected. In an asymmetric or two-key cryptosystem, the encryption and decryption keys are different, and the public keys are available to all. To protect message integrity requires that the public keys be guaranteed as belonging to the communicating entities.

(1) SKC Keys

Key management is the greatest single challenge in administering a symmetrical cryptosystem. The keys must be protected to maintain security, yet they must be shared by at least two sites to permit mutual encryption and decryption. The basic methods of key management are employed in symmetrical systems

- **Physical transportation**: if there is limited data traffic, and only a few communicating sites, keys may be carried to all sites by trusted persons who install them at prearranged times. However, if heavy data traffic or many communicating sites are involved, and the keys are changed frequently enough to assure privacy, this option quickly becomes labor and travel intensive. A modification uses bonded delivery services, or other third-party means to distribute new keys. Obviously, the chance for compromise increases with the number of persons that knowingly handle them.

---

[6] Gail L. Grant, *Understanding Digital Signatures—Establishing Trust over the Internet and Other Networks* (New York: McGraw-Hill, 1998), 13–26.

- **Automatic updating**: the encrypting device stores two keys. One is the key used for data encryption; the other is a key-encrypting key. The key-encrypting key is entered manually at some central site (A) before the encryption device is shipped to a remote operational location (B). When the encryption key must be changed, a new encryption key is generated at A. It is encrypted with the key-encrypting key and sent to B, where it is decrypted and installed automatically. Because the key-encrypting key is used infrequently, it may be used over a longer time before being changed.

- **Using PKC**: after obtaining a digital certificate for site B, the new key is encrypted with B's public key. When the message is received, B decrypts it with its secret key.

(2) PKC Keys

The use of public key cryptosystems and trusted certification authorities eliminates most of the concerns related to key management. However, each site must protect its own secret key, and site authentication depends on the integrity of the certification authority.

(3) Escrowed Encryption Standard

The emergence of practically unbreakable, commercial encryption capabilities in integrated circuit chips that can be incorporated in digital equipment ensures that persons can use communication facilities for private purposes with confidence. Neither an ill-disposed competitor, nor the government, can understand the messages exchanged. However, in 1994, so that encryption shall not protect both law-abiders and law-breakers, the U.S. government proposed the voluntary use of a special encryption chip.

Described in a standard known as *Escrowed Encryption Standard* (EES), a secret key (Device Unique Key, KU) is created and embedded in a tamper-resistant chip at the time of manufacture. Called the *Clipper chip* (officially designated MYK–78) it is given a device-unique identifier (UID) and an 80-bit common Family Key (KF). KU is split in two key components that are encrypted and given to two Key Escrow Agents for safekeeping. With court approval, for a specific UID, government officials can obtain these keys. The combination of KU, KF, and a 128-bit Law Enforcement Access Field (LEAF) permits them to decipher intercepted messages. EES specifies the use of *SKIPJACK*, a classified encryption algorithm. It transforms a 64-bit input to a 64-bit output using an 80-bit secret key. The same key is used for decryption.

The *Key Escrow System* (KES) has created significant concern that it can be exploited or abused by government agencies. It remains to be seen if the Clipper chip (or a successor) will be incorporated in enough communications equipment to make the concept effective.

REVIEW QUESTIONS FOR SECTION 13.7

1   What is meant by invasion of individual or company privacy?

2   List the three principal methods of ensuring message privacy.

3   Define encryption.

4   Define SKC and PKC.

5   What techniques are available for encrypting data messages?

6   Why do computer-based encryption and decryption algorithms employ ciphers?

7   Describe how encryption is performed using the DES algorithm.

8   Describe how privacy, authentication, and privacy and authentication are obtained using a public-key cryptosystem.

9   What is DSS?

10   What is PGP?

11   What is a digital certificate? Why is it used?

12   Describe acceptable ways of distributing keys in symmetrical cryptosystems.

13   What is EES? Why does the U.S. government support EES?

ACRONYMS AND ABBREVIATIONS

2B1Q: two binary, one quaternary
2W: two wire
4B5B: code in which 4 bits are represented by a 5-bit symbol
4W: four wire
7B8B: code in which 7 bits are represented by an 8-bit symbol
8B10B: code in which 8 bits are represented by a 10-bit symbol
ΦM: phase modulation

AAL: ATM adaptation layer
ABR: answer-to-bid ratio; also, available bit rate
AC: authentication center; also, alerting channel
ACD: automatic call distribution; also automatic call distributor
ACELP: algebraic code-excited linear prediction
ACeS: Asian cellular system
ACK: acknowledgment
ACSE: association control service element
ADDC: advanced data communications control procedure
ADM: add/drop multiplexer
ADPCM: adaptive differential pulse code modulation
ADSL: asymmetric digital subscriber line
ADTS: automated digital terminal equipment system
AGCH: access grant control channel

AH: application header
ALT: automatic link transfer
ALTS: alternative local transmission system
AM: amplitude modulation
AMI: alternate mark inversion
AMPS: advanced mobile phone service
AMSC: American Mobile Satellite Corporation
ANI: automatic number identification
ANSI: American National Standards Institute
ANSI–T1: Committee T1–Telecommunications of ANSI
ANSI–X3: Information Technology Committee of ANSI
ANSI-X12: Electronic Data Interchange Committee of ANSI
AOR: Atlantic Ocean region
AP: adjunct processor
APCM: adaptive PCM
APCN: Asia Pacific cable network
APDU: application protocol data unit; also, authentic protocol data unit
APNIC: Asia-Pacific Network Information Center
APPC: advanced peer-to-peer communications
APPN: advanced peer-to-peer networking
APR: Asia-Pacific region
APS: automatic protection switching
AR: authentic receiver
ARA: automatic route advancement
ARPA: Advanced Research Projects Agency
ARQ: automatic-repeat-request
AS: authentic source

ASCII: American Standard Code for Information Interchange
ASE: application service element
ASIU: ATM-satellite interworking unit
ASK: amplitude-shift keying
ASP: adjunct service point
AT: access tandem
ATC: adaptive transfer coding
ATD: asynchronous time division
ATIS: Alliance for Telecommunications Industry Solutions
ATM: asynchronous transfer mode
ATSC: Advanced Television System Committee
AU: administrative unit
AVR: available bit rate

B8ZS: bipolar with 8 zeros substitution
BASE_CLASS: base station classification
BASE_ID: base station identifier
BASE_LAT: base station latitude
BASE_LONG: base station longitude
BAsize: buffer allocation size
BASK: binary amplitude-shift keying
BCC: block check character; also, blocked-calls-cleared
BCCH: broadcast control channel
BCD: blocked-calls-delayed; also, binary-coded decimal
BECN: backward explicit congestion notification
Bellcore: Bell Communications Research
BFSK: binary frequency-shift keying
BGP: border gateway protocol
B–ISDN: broadband ISDN
BISYNC: binary synchronous data link control
bit: binary digit
BIU: basic information unit
BML: business management layer
BORSCHT: battery feed, overvoltage protection, ringing, supervision, coding and filtering, hybrid, and testing
BPSK: binary phase-shift keying
BRI: basic rate interface
BS: bursty second
BSIC: base station identity code
BSMC: base station manufacturers code
Btag: beginning tag

C/R: command response
CA: certification authority
CATV: cable television
CAVE: cellular authentication and encryption
CBMS: computer-based message system
CBR: constant bit rate
CCAF: call control agent function
CCC: clear channel capability
CCF: call control function
CCIC: cochannel interference control
CCITT: Comité Consultatif International Télégraphique et Téléphonique
CCMRS: common-carrier point-to-point microwave radio service
ccs: 100 call-seconds
CCSN: common channel signaling network
CDF: cumulative distributive function
CDMA: code-division multiple access
CDPD: cellular digital packet data
CDS: central directory server
CENTREX: central exchange service
CEPT: European Conference for Post and Telecommunication
CFP: cordless fixed part
CGSA: cellular geographical serving area
CI: congestion indication
CICS: customer information control system
CICS: customer information control subsystem
CIDR: classless interdomain routing
CIM: computer-integrated manufacturing
CIR: committed information rate
CIU: channel interface unit
CLASS: custom local-area signaling services
CLLM: consolidated link layer management
CLNP: connectionless network layer protocol
CLP: cell loss priority
CLTP: connectionless transport layer protocol
CL–SCCP: connectionless SCCP
CM: channel mark
CMISE: common management information service element
CMOT: common management information services and protocol over TCP/IP
CMRTS: cellular mobile radio telephone service; also, cellular mobile radio telephone system
CO: central office
codec: coder and decoder

COMSAT: Communications Satellite Corporation
CO–SCCP: connection-oriented SCCP
COT: central office terminal
CP: control point
CPDU: corrupted protocol data unit
CPE: customer premises equipment; also customer-provided equipment
CPFSK: continuous-phase FSK
CPN: calling party's number
CPP: cordless portable part
CPR: call-processing record
CR: channel rate
CRC: cyclic redundancy check
CRS: cell relay service; also cellular radio service
CS: convergence sublayer
CSA: carrier service area
CSD: customer-specific data
CSMA/CD: carrier-sense multiple access collision detection
CSPDU: convergence sublayer PDU
CSU: channel service unit
CS–1: capability set level 1
CT2: cordless telephone, second generation

D+R: drop and repeat node
D-AMPS: digital AMPS
DA: destination address
DAMA: demand-assigned, multiple access
DAML: digital-added main line
DARPA: Defense Advanced Projects Research Agency
DCC: digital color code
DCCH: digital control channel
DCE: data circuit terminating equipment
DCF: data communication function
DCME: digital circuit multiplication equipment
DCS: digital cross-connect system
DCT: discrete-cosine transform
DDCMP: digital data communications message protocol
DDD: direct distance dialing
DDS: digital data service
DE: discard eligibility

DECT: digital European cordless telecommunications
DEMS: digital electronic message service
DEMSNET: digital electronic message service network
DES: digital encryption standard
DID: direct inward dialing
DLC: data link control; also, digital loop carrier
DLCI: data link connection identifier
DLCP: data link control protocol
DLE: data link escape
DNHR: dynamic nonhierarchical routing
DNS: domain name service
DOD: direct outward dialing
DPCM: differential pulse code modulation
DPDU: data link protocol data unit
dpi: dots per inch
DPSK: differential phase-shift keying
DQDB: distributed queue dual bus
DQPSK: differential quaternary phase-shift keying
DS: direct sequence
DSBSC: double-sideband, suppressed carrier
DSL: digital subscriber line
DSS: digital signature standard
DSU: digital service unit
DT: data trailer
DTCH: digital traffic channel
DTE: data terminal equipment
DTMF: dual-tone multifrequency
DTS: digital termination service
DVCC: digital verification color code
DVD: digital videodisc
DX: duplex

EA: extended address
E-BCCH: extended broadcast control channel
EBCDIC: Extended Binary Coded Decimal Interchange Code
EC: European Community
EDFA: erbium-doped fiber amplifier
EDI: electronic data interchange
EES: escrowed encryption standard
EFT: electronic funds transfer
EGP: exterior gateway protocol
EIA: Electronic Industries Association

EIR: equipment identity register; also, excess information rate

e-mail: electronic mail

EML: element management layer

ENQ: enquiry

EOT: end of transmission

ER: explicit cell rate

ES: errored second

ESCON: enterprise systems connection

ESF: extended superframe format

ESN: electronic serial number

Etag: end tag

ETB: end of text data block

ETSI: European Telecommunications Standards Institute

ETX: end of text

E–1: first level of voice multiplexing in Europe

FACCH: fast-associated control channel

F-BCCH: fast broadcast control channel

FC: functional component; also, fiber channel

FCC: Federal Communications Commission

FCCH: frequency correction control channel

FCS: frame check sequence

FDDI: fiber-distributed data interface

FDL: facility data link

FDM: frequency-division multiplexing

FDMA: frequency-division multiple access

FDX: full-duplex

FE: functional entity

FEC: forward error correction

FECN: forward explicit congestion notification

FEP: front-end processor

FEXT: far-end cross talk

FFOL: FDDI follow-on LAN

FH: frequency hopping

FID: format identification field

FIFO: first in, first out

FIPS: Federal Information Processing Standards

FISU: fill-in signal unit

FITL: fiber-in-the-loop

FK: family key

FLAG: fiberoptic link around the globe

FM: frequency modulation

FOCC: forward control channel

FPLMTS: future public land mobile telephone system

FPS: fast-packet switching

FR: frame relay

FRI: frame relay interface

FRNP: frame relay nodal processor

FRPDU: frame relay PDU

FRSN: flow reduction sequence number

FR–PAD: frame relay frame assembler and disassembler

FS: failed seconds

FSK: frequency-shift keying

FSS: fixed satellite service

FTAM: file transfer and access management

FTP: file transfer protocol

FTS: federal telephone system

FTTC: fiber to the curb

FTTH: fiber to the home

FT–1: fractional T–1

FT–3: fractional T–3

FVC: forward voice channel

FX: foreign exchange

GEO: geostationary Earth orbit

GFC: generic flow control

GGP: gate-to-gateway protocol

GII: global information infrastructure

GOP: group of pictures

GOS: grade of service

GOSIP: Government OSI profile

GPRS: general packet radio service

GPS: global positioning system

GR: GPRS register

GSM: global system for mobile communications

GSN: GPRS support node

HDLC: high-level data link control

HDSL: high-data-rate digital subscriber line

HDX: half-duplex

HE: header extension

HEC: header error control

HEL: header extension length

HI: highway interchanger

HIPPI: high-performance serial interface

HLPI: higher-level protocol identifier

HLR: home location register

HPC: high-performance connection

HPR: high-performance routing
HRC: hybrid ring control
HSSI: high-speed serial interface
HTML: hypertext markup language
HTR: hard to reach
HTTP: hypertext transfer protocol

I2: Internet2
IBN: intelligent broadband network
ICC: Interstate Commerce Commission
ICI: intercarrier interface
ICMP: internetwork control message protocol
ICO: ICO Global Communications
ID: identification
IDU: interface data unit
IEC: International Electrotechnical Commission
IEEE: Institute of Electrical and Electronic Engineers
IGP: internal gateway protocol
IISP: Information Infrastructure Standards Panel
IITF: Information Infrastructure Task Force
IM: intelligent multiplexer
IMAC: isochronous medium-access controller
IMEI: international mobile equipment identifier
IMS/DC: information management system/data communications
IMSI: international mobile subscriber identity
IN: intelligent network
INMARSAT: International Maritime Satellite Organization
INOC: Internet Network Operation Center
INSC: intelligent network switch capabilities
INTELSAT: International Telecommunications Satellite Consortium
IOR: Indian Ocean region
IP: Internet protocol; also, intelligent peripheral; also information provider
IPDU: illicit protocol data unit
IPOP: ISP point of presence
IPv4: Internet protocol version 4
IPv6: Internet protocol version 6
IPX: Internet packet exchange
IR: illicit receiver
IS: illicit source
ISDN: integrated services digital network

ISI: interswitching system interface; also intersymbol interference
ISO: International Standards Organization
ISP: Internet service provider
ISR: intermediate session routing
ITU: International Telecommunication Union
ITU–D: Telecommunications Development Sector of ITU
ITU–R: Radiocommunications Sector of ITU
ITU–T: Telecommunications Standardization Sector of ITU
IWU: internetworking unit
IXC: interexchange carrier

JPEG: Joint Photographic Experts Group
JTC1: ISO/IEC Joint Technical Committee 1
JTM: job transfer and manipulation

Kc: cipher key
KES: key escrow system
KF: family key
Ki: authentication key
KU: unique key

LAI: location area identity
LAN: local-area network
LAP–B: link access procedure–balanced
LAP–D: link access procedure–D channel
LAP–E: link access procedure–extended
LAP–F: link access procedure –frame relay
LAP–M: link access procedure–modem
LATA: local access and transport area
LCAM: large-carrier amplitude modulation
LCC: lost-calls-cleared
LCD: lost-calls-delayed
LCN: logical channel number
LCR: least-cost routing
LEAF: law enforcement access field
LEC: local exchange carrier
LEN: low-entry networking
LENO: line equipment number–originating
LENT: line equipment number–terminating
LEO: low Earth orbit
LFSID: local-form session identifier
LLC: logical link control

LOCAID: location area identifier
LOH: line overhead
LPC: linear prediction coefficient
LPC–RPE: linear prediction coding with regular pulse excitation
LRC: longitudinal redundancy checking
LSB: lower sideband
LSSU: link status signaling unit
LT: link trailer
LU: logical unit

MAC: medium access control
MAHO: mobile-assisted handover
MAN: metropolitan area network
MAU: multistation access unit
MC–F: fast message channel
MC–S: slow message channel
MCU: multipoint control unit
MD: mediation device
MDS: multipoint distribution service
MEO: medium Earth orbit
MF: mediation function block
MFJ: modified final judgment
MIB: management information base
MID: multiplexing identifier
MIN: mobile identification number
MN: matched node
MNP: microcom networking protocol
MOB_FIRM_REV: mobile firmware revision
MOB_MFG_CODE: mobile manufacturer code
MOB_MODEL: mobile model number
MOB_P_REV: mobile protocol revision
MOS: mean opinion score
MPDCH: master PDCH
MPEG: Moving Picture Experts Group
MSC: mobile switching center
MSCM: mobile station classmark
MSK: minimum-shift keying
MSN: Manhattan street network
MSRN: mobile station roaming number
MSU: message signal unit
MTP: message transfer part
MTSO: mobile telephone switching office; also mobile telephone serving office
MU: message unit
MVS: multiple virtual systems

NAK: negative acknowledgment
NANP: North American Numbering Plan
NAP: network access point
NAT: network address translation
NAU: network addressable unit; also network accessible unit
NCP: network control point; also, network control program
NE: network element
NEF: network element function block
NEXT: near-end cross talk
NGI: next-generation Internet
NH: network header
NID: network information database; also, network interface device; also, network identifier
NII: national information infrastructure
NIIIP: National Industrial Information Infrastructure Protocols
NIST: National Institute of Standards and Technology
NMC: network management center
NML: network management layer
NNI: node-network interface
NNTP: network news transport protocol
NPDU: network protocol data unit
NRM: network resource manager
NRZ: nonreturn to zero
NSP: network services part
NT1: network termination 1
NT2: network termination 2
NTIA: National Telecommunication and Information Agency
NTSC: National Television Standards Committee
NTT: Nippon Telephone and Telegraph

OAM&P: operations, administration, maintenance, and provisioning
OCC: other common carrier
OC–1: optical carrier level 1
OC–N: optical carrier level N
OEM: original equipment manufacturer
OLSP: on-line service provider
OMAP: operations, maintenance, and administration part
OOF: out of frame

OOK: on-off keying
OQPSK: offset QPSK
OS: operationing system, also operations system
OSA: open systems architecture
OSF: operations system function block
OSI: open systems interconnection
OSIRM: open systems interconnection reference model
OSP: outside plant
OSPF: open shortest path first
OSS: operator services system
OTA: Office of Technology Assessment
OUI: unique identifier

PACS: personal access communications system
PACS–UB: personal access communications system operating in unlicensed band
PAD: packet assembler and disassembler
PAM: pulse amplitude modulation
PBX: private branch exchange
PCC: power control channel
PCCH: power control channel
PCH: paging channel
PCI: protocol control information
PCM: pulse code modulation
PCN: personal communication network
PCR: peak cell rate
PCS: personal communication services
PDCH: packet data channel
PDF: probability density function
PDN: public data network
PDU: protocol data unit
PE: physical entity
PER: packet error rate
PGP: pretty good privacy
PH: presentation header
PHS: personal handyphone system
PID: protocol identifier
PIR: packet insertion rate
PIU: path information unit
PK: public key
PKC: public-key cryptography
PKS: public-key cryptosystem
PLCP: physical layer convergence protocol
PLR: packet loss rate
PMAC: packet medium-access controller

pn: pseudo noise
PN_OFFSET: *pn*-code offset
POH: path overhead
POP: point of presence
POR: Pacific Ocean region
POTS: plain old telephone service
PPDU: presentation protocol data unit
PRC: primary reference clock; also, preventive cyclic retransmission; also, priority request channel
PRI: primary rate interface
PRK: phase-reversal keying
PSD: power spectral density
psi: pounds per square inch
PSK: phase-shift keying
PSTN: public switched telephone network
PT: portable termination
PTI: payload type identifier
PU: physical unit
PV: protocol version
PVC: permanent virtual circuit

QA: Q adapter
QAF: Q adapter function block
QAM: quadrature-carrier amplitude modulation
QPSK: quadrature phase-shift keying; also, quaternary PSK and quadriphase PSK

R&E: research and engineering
RACH: random access channel
RARC: regional administrative radio conference
RBOC: Regional Bell Operating Company
READ: relative element address designate code
RECC: reverse control channel
REG_ZONE: registration zone
REJ: reject
RER: residual error rate
R/G: router/gateway
RH: request/response header
RIP: routing information protocol
RJE: remote job entry
RM: resource management
RNR: receive not ready

ROSE: remote operations service element
RS: recommended standard; also, Reed-Solomon
RSA: Rivest, Shamir, Adelman
RSCV: route selection control vector
RT: remote terminal
RTDTR: real-time dynamic traffic routing
RU: request/response unit
RVC: reverse voice channel
RZ: return to zero

SA: source address
SACCH: slow-associated control channel
SAI: serving-area interface
SAP: service access point
SAR: segmentation and reassembly
SAT: supervisory audio tone
SBC: subband coding; also, system broadcast channel
SCC: specialized common carrier
SCCH: signaling channel
SCCP: signaling connection control part
SCF: service control function
SCH: synchronization control channel
SCM: station class mark
SCO: serving central office
SCP: service control point
SCPC: single channel per carrier
SDCCH: stand-alone dedicated control channel
SDCU: satellite delay compensation unit
SDF: service data function
SDH: synchronous digital hierarchy
SDLC: synchronous data link control
SDLS: single-line digital subscriber line
SDMA: space-multiple division access
SDO: selective dynamic overload
SDU: service data unit
SEAL: simple and efficient layer
SEP: signaling end-point
SES: severely errored second
SF: superframe format; also service feature
SH: session header
SH: session header
SIB: service independent building block
SIC: signal information channel; also, system information channel
SID: system identifier

SIM: subscriber identity module
SIP: SMDS interface protocol
SIR: signal to interference
SK: secret key
SKC: secret-key cryptography
SL: service logic
SLC: subscriber loop carrier
SLI: service logic interpreter
SLIC: subscriber line interface circuit
SLP: service logic program
SMASE: system management application service element
SMDS: switched multimegabit data service
SMF: system management function
SMFA: specific management function area
SML: service management layer
SMPTE: Society of Motion Picture and Television Engineers
SMPX: statistical multiplexer
SMTP: simple mail transfer protocol
SN: service node
SNA: systems network architecture
SNAF: subnetwork access functions
SNAP: subnetwork access protocol
SNDCF: subnetwork dependent convergence functions
SNI: SMDS network interface; also service node interface
SNICF: subnetwork independent convergence functions
SNMP: simple network management protocol
SNR: signal-to-noise ratio
SO: serving office
SOH: section overhead; also, start of header
SONET: synchronous optical network
SP: switching point
SPC: stored program control
SPDCH: slave PDCH
SPDU: session protocol data unit
SPE: synchronous payload envelope
SPF: superframe phase information
SPX: sequenced packet exchange
SRES: signed response
SRF: specialized resource function
SRM: subrate multiplexer
SS: satellite switched
SS7: Signaling System 7
SSB: single sideband

SSBAM: single sideband amplitude modulation
SSCP: system services control point
SSD: secret shared data
SSF: service switching function
SSL: secure sockets layer
SSM: spread spectrum modulation
SSP: service switching point
ST: signaling tone
STAR TAP: Science, Technology and Research Transit Access Point
STE: signaling terminal exchange
STM–1: synchronous transport module level 1
STM–N: synchronous transport module level N
STP: signal transfer point; also, shielded twisted pair
STR: selective trunk reservation
STS: space-time-space
STS–1: synchronous transport signal level 1
STS–N: synchronous transport signal level N
STX: start of text
su: signal unit
SUERM: signal unit error rate monitor
SVC: switched virtual circuit
SW56: 56 kbits/s switched digital service
SW64: 64 kbits/s switched digital service
SW384: 384 kbits/s switched digital service
SW1536: 1,536 kbits/s switched digital service
SX: simplex
SYC: synchronization
SYC: synchronization channel
SYN: synchronization character
SYNTRAN: synchronous M13

TA: terminal adapter
TAT: transatlantic telephone (cable)
TCAM: telecommunications access method
TCAP: transaction capabilities application part
TCH: traffic channel
TCH/F: full-rate traffic channel
TCH/H: half-rate traffic channel
TCM: trellis code modulation
TCP: transmission control protocol
TCPDU: TCP data unit
TCP/IP: transmission control protocol/Internet protocol
TDM: time-division multiplexing

TDMA: time-division multiple access
TDU: topology database update
TE1: subscriber terminal 1
TE2: subscriber terminal 2
TG: transmission group
TH: transmission header; also, transport header
TIA: Telecommunications Industries Association
TINA: telecommunications information networking architecture
TM: terminal multiplexer
TMN: telecommunications management network
TMS: timing monitor system
TMSI: temporary mobile subscriber identity
TOH: transport overhead
TP: teleprocessing access method; also, teleprocessing (monitor); also, transport layer protocol; also transaction program
TP0, **TP1**, **TP2**, **TP3**, **TP4**: transport-layer protocols
TPC: trans-Pacific cable
TPDU: transport protocol data unit
TSI: time-slot interchanger
TST: time-space-time
TTC: Telecommunication Technology Committee
TTY: teletypwriter
TU: tributary unit
TUG: tributary unit group
T–1: T–carrier level 1
T–1C: T–carrier level 1C
T–2: T–carrier level 2
T–3: T–carrier level 3
T–4: T–carrier level 4
T–4A: T–carrier level 4A

UBR: unspecified bit rate
UDC: universal digital channel
UDP: user datagram protocol
UIC: user information channel
UID: unique identifier
UNI: user network interface
UPT: universal personal telecommunications
UP/TUP: user part/telephone user part
USART: universal synchronous/asynchronous receiver transmitter

USB: upper sideband
USCCH: user-specific packet control channel
USTIA: United States Telecommunications Industry Association
UTP: unshielded twisted pair

vAI: vBNS approved institution
VBI: vertical blanking interval
vBNS; very high performance Backbone Network Service
VBR: variable bit rate
VC: virtual container
VCI: virtual circuit identifier; also virtual channel identifier
vCI: vBNS collaborative institution
VCR: videocassette recorder
VDSL: very high data rate digital subscriber line
VDT: video display terminal
VGL: voice-grade line
vIP: vBNS industrial partner
VLR: visitor location register
VN: virtual network
VOX: voice activity channel
VP: virtual path
VPI: virtual path identifier
vPI: vBNS partner institution

VPN: virtual private network
VRC: vertical redundancy checking
VSAT: very small aperture terminal
VSB: vestigial sideband modulation
VSELP: vector sum excitation linear prediction
VT: virtual tributary
VTAM: virtual telecommunications access method

W3C: World Wide Web Consortium
WAN: wide-area network
WARC: world administrative radio conference
WATS: wide-area telephone service
WDM: wavelength division multiplexing
WEI: word error indicator
WS: workstation
WSF: workstation function block

X.25: ITU–T Recommendation X.25
xDSL: shorthand for family of digital subscriber lines—i.e., ISDN DSL, ADSL 1/2/3, HDSL, SDSL, and VDSL

ZBTSI: zero-byte time-slot interchange

GLOSSARY

1% bandwidth: for a frequency modulated wave, frequency interval that just contains modulation components whose amplitudes are ≥ 1% of the amplitude of the unmodulated carrier.

2B1Q digital subscriber line: digital subscriber equipment that employs 2B1Q coding to provide 160 kbits/s, duplex connections over two-wire copper loops up to 18,000 feet.

2B1Q signal format: four signal levels (± 3 and ± 1) each represent a pair of bits. Of each pair, the first bit determines whether the level is positive or negative (1 = +ve, 0 = –ve), and the second bit determines the magnitude of the level (1 = |1|, 0 = |3|).

10BASE-T: operating at 10 Mbits/s, Ethernet that employs unshielded twisted pair (UTP) connected in a star. Each UTP supports a single station that is no more than 100 meters from the hub.

10BASE-F: operating at 10 Mbits/s, Ethernet that employs optical fiber to connect hubs separated by long distances (to connect between floors, for instance). Fibers are run between the hubs. However, it is likely that each hub is connected to its community of users by UTP.

10BROAD36: operating at 10 Mbits/s, Ethernet that employs coaxial cable in segments up to 3,600 meters. Operates with broadband channels of 6 or 12 MHz to deliver digital signal streams.

100BASE-T: operating at 100 Mbits/s, Ethernet uses multiple UTPs, STPs, or optical fibers to interconnect hubs. Stations are limited to less than 100 meters from a hub, and a total span length of no more than 250 meters.

acknowledged connectionless service: receiver confirms receipt of message, but a logical connection is not established.

active: in network management, operational state in which the managed object is available for use and can accept new users.

adaptive delta modulation: in digital voice, coding technique that employs step-size adaptation in regions that contain the faster signal changes.

adaptive differential pulse code modulation: DPCM technique in which input signal samples are encoded in 14-bit linear PCM. The current sample is subtracted from a predicted value derived from the weighted average of several previous samples and the last two predicted values, and the difference is applied to a 15-level adaptive quantizer that assigns a 4-bit code.

adaptive equalizer: in digital transmission, device that continuously corrects channel-induced distortion.

adaptive transform coding: frequency domain coding technique in which the input signal is divided into short time blocks (frames) that are transformed into the frequency domain using techniques based on Fourier series. The result is a set of transform coefficients that describe each frame. These coefficients are quantized into numbers of levels based on the expected spectral levels of the input signal. The number of bits for any quantizer and the sum of all the bits per block are limited to a predetermined maximum.

add/drop multiplexer: aggregates or splits SONET traffic at various speeds so as to provide access without demultiplexing the SONET signal stream. May be inserted in a ring to provide access to/from the ring for several terminal multiplexers. Generally, has two equal-speed network connections.

adjunct processor: in intelligent network, device that contains an adjunct service point that responds to requests for service processing. Directly connected to an IN switching system, the adjunct processor supports services that require rapid response to users' actions.

adjunct service point: in intelligent network, a location that responds to requests for IN services.

administrative unit: in synchronous digital hierarchy, combination of a tributary group and pointer information.

ADSL Forum: association of competing companies committed to the promotion of applications of Asymmetric Digital Subscriber Lines.

Advanced Mobile Phone Service: cellular mobile radio telephone service operating in the United States on assigned frequencies of 824 to 849 MHz (mobiles to cell sites) and 869 to 894 MHz (cell sites to mobiles). AMPS employs a channel band-width of 30 kHz, uses FDMA as the access method, and employs frequency modulation. AMPS/IS-41 incorporates transparent roaming techniques adapted from GSM.

Advanced Peer-to-Peer Networking: in SNA, supports the decentralized network operation inherent in client/server environments. Instead of predetermined routing and hierarchy (as in subarea networks), APPN networks dynamically reconfigure themselves as links and nodes are added or turned off.

advertise: action of providing information to others on the connections and resources associated with a router/gateway. An automatic, software-driven function, advertising is used to establish and update the routing tables in neighboring routers/gateways.

agent process: in network management, that part of the application process that is concerned with performing management functions on managed objects at the request of the managing process.

aliasing: condition that exists when signal energy that is present over one range of frequencies in the original signal appears in another range of frequencies after sampling and reconstruction.

Aloha: in satellite communication, access technique in which stations send data packets as they are ready. So long as the total offered traffic from all stations is not a very large fraction of the capacity of the transponder, packets have a reasonable chance of finding the transponder unoccupied upon arrival at the satellite and can be relayed without interference.

alternate mark inversion signal: 1s are represented by return-to-zero pulses that alternate between positive and negative levels, and 0s are represented by the absence of pulses.

alternative local transmission system: a transmission system that bypasses facilities of the local exchange carrier. Many ALTSs provide point-to-point optical fiber transport services that are used to connect buildings in the same city, or connect customers directly to IXC equipment.

American National Standards Institute: organization that facilitates the development of standards by establishing consensus among qualified parties. ANSI is the U.S. representative to ISO and IEC.

American Standard Code for Information Interchange: alphanumeric code that employs a sequence of 7 bits.

amplitude-shift keying: in digital modulation, the transmitted signal is a sinusoid whose amplitude is keyed in discrete steps. When two states are employed, the ASK-wave is known as a *binary*-ASK wave. To produce more complex symbols that represent groups of bits, several amplitude levels are employed.

analog function: single-valued, continuous-state function that assumes positive, zero, or negative values. Changes in value occur smoothly, and rates of change are finite.

analog line circuit: circuit that performs analog-to-digital conversion, 2- to 4-wire conversion, signaling detection, and battery feed.

analog modulation: signal processing technique that brings together a higher-frequency sinusoidal carrier and a lower-frequency analog message signal to create a modulated wave in which the parameters change smoothly, and rates of change are finite. The energy in the modulated wave is distributed at frequencies around the carrier signal frequency.

angle modulation: result of varying the argument of a sinusoidal function in sympathy with the message signal. Gives rise to frequency modulation and phase modulation.

annual charge: annual cost associated with purchasing, owning, maintaining, and using up plant.

answer-to-bid ratio: ratio of the number of calls accepted by, and total number of calls presented to, a specific switch, in a given time period.

anti-aliasing filter: in digital voice, low-pass filter that passes frequencies up to W Hz and precedes a sampling device operating at $r \geq 2W$ samples/s.

application agent: in OSIRM, embedded in the local operating system, the application agent exists between the user and the operating system of the processing resource to provide access to specific processor capabilities.

application entity: in OSIRM, exists within the application layer. As required, provides services to the application agent by exchanging protocol data units with peer entities (application entities in other machines).

application header: in OSIRM, header added to user's data unit by application layer to create application PDU. It contains the unique identifier for the exchange agreed by the communicating hosts.

application layer: layer 7 of OSI reference model. When sending, the application layer organizes the user's data and negotiates a unique identifier with the other host. When receiving, the application layer strips of the application header and passes the information contained in the APDU to the user.

application-level forwarding: by observing data contained in the packets that constitute a communication, the firewall router can determine the acceptability of the communication and permit or deny the passage of the data.

application protocol data unit: combination of application header and user's data.

architecture: concerned with the ways in which messages interact with topology and hierarchy so that communication can take place. The blueprint for effecting communications.

area code: North America is divided into sub-areas that are assigned three-digit area codes.

area-wide Centrex: based on an SCP that is connected to all switches within a LATA, an intelligent network service that provides PBX-like operations from designated stations across the LATA.

Argand diagram: consists of a phasor of amplitude A that rotates anticlockwise about the origin at a constant angular velocity and completes f rotations a second. The horizontal axis is known as the *real* axis, and the vertical axis is known as the *imaginary* axis. Angles are measured in relation to the horizontal, right-pointing, real axis (marked 0°). As the phasor rotates, its projection on the real axis is $A\cos\phi$, and on the imaginary axis is $A\sin\phi$.

associated signaling: signaling messages are sent over the same route as the information that is exchanged between the parties.

association control service element: in network management, general software support system that works between application processes that are independent of application-specific needs. ACSE supports the establishment, maintenance, and termination of a cooperative relationship between application entities.

asymmetric digital subscriber line: transports a 1.5, 6.3, or 13 Mbits/s simplex channel downstream, a duplex 16 to 640 kbits/s channel, and a 4 kHz duplex voice channel. Originally intended to carry one, two, or four television signals encoded according to MPEG2 digital video compression standards, ADSL can be used in many applications that require a broadband, high-speed channel from network to user.

asynchronous multiplexing: digital multiplexing technique in which time slots are assigned to signals as required. Arrival times at the multiplexer, not what input line they are on, determine the order in which octets are sequenced. As they are placed on the multiplexed line, the multiplexer adds an identifier so that the demultiplexer will know to which output line the octets are to go.

asynchronous operation: data are sent or actions occur when the sender is ready. Activities occur without reference to an external timing source.

Asynchronous Transfer Mode: high-speed (155 and 622 Mbits/s), fast-packet switching technique employing short, fixed-length cells that are statistically multiplexed over virtual connections. Each cell consists of a 5-octet header and a 48-octet payload of user information (data, voice, or video).

ATM adaptation layer: level 3 protocol employed in ATM bearer service. Receives different classes of information from higher layers and adapts it to occupy fixed-length ATM cells. Receives data from ATM layer and rearranges it to fit user's application requirements. AAL is divided into two sublayers—SAR and CS.

ATM bearer service: employs a three-level protocol to provide fast-packet cell relay service.

ATM Forum: international nonprofit organization of producers and users committed to promoting the application of ATM technology through the development of interoperability specifications.

ATM LAN: local ATM node that connects high-speed terminals, servers, and other LANs using the multiplexing capabilities of the segmentation and reassembly sublayer of AAL to provide connectionless services in connection-oriented streams.

ATM layer: level 2 protocol employed in ATM bearer service. Multiplexes and demultiplexes cells belonging to different connections identified by unique VCIs. Translates VCI at switches and cross-connection equipment. Passes cell payload to and from the ATM adaptation layer without error control or similar processing. Performs flow control at UNI. Implements connection-oriented and connectionless services.

audio conference: interactive real-time service that may be augmented by text or graphics.

authentication: establishing the identity of agents with certainty.

authentication center: in mobile telephone network, database that verifies user's credentials and supports encryption of radio channel signals.

authentication key: secret key assigned by the operating company to the subscriber (stored in SIM and the authentication center).

authentic protocol data unit: in communication security, true PDU that the source wishes to send to one, or more, receivers for a purpose approved by the network owner.

authentic receiver: in communication security, receiver that is connected to the network with the knowledge and approval of the network owner.

authentic source: in communication security, sender that is connected to the network with the knowledge and approval of the network owner.

authorization: establishing an agent endowed with authority to perform requested functions.

authorization access: in local area networks, operating mode in which each station sends when ready *and* authorized. Usually, authorization is granted through possession of a token—a data message that is passed from station to station.

automated digital terminal equipment system: digital loop carrier system that multiplexes individual voice and data channels over DS–1, DS–2, or DS–3 facilities. Provides monitoring and maintenance functions, routing capability, clear–64 and ESF capability, fiber connections to the users' premises, and supports SONET connections.

automated factory: component of an automated enterprise that uses information technologies to improve the effectiveness of factory processes.

automated office: component of an automated enterprise that uses information technologies to improve the effectiveness of persons performing office-based tasks.

automatic call distribution system: equipment that receives calls and directs them to attendants waiting to provide service. ACD systems may be stand-alone, integrated with a PBX, or central office-based.

automatic-repeat-request: error correction procedure that requires the transmitter to resend the portions of the exchange in which errors have been detected.

automatic route advancement: action of automatically changing from one category of transport service to another (i.e., tie line to WATS to DDD).

autonomous system: in Internet, administered by a single authority, a group of networks interconnected by routers/gateways (e.g., networks that support a college campus or an industrial entity, a regional network with several ISPs, or a network operated by an OLSP). A unique number identifies the system. Within each autonomous system, the administering authority decides the way in which routing information is assembled, updated, controlled, and used.

autonomous systems extended links advertisements: contain information on routes in other autonomous systems.

average busy hour: 60-minute period in which the traffic intensity is the average of the traffic intensities observed in a series of daily busy hours.

backbone router: any router that is connected to the backbone.

band-limited signal: signal that contains energy in the range $0 < f \le W$ Hz. Elsewhere the signal value is zero.

band-limited white noise: noise that has a constant power spectral density over a finite range of frequencies.

band-pass characteristic: performance such that energy is transmitted between mf Hz and nf Hz $(m < n)$, but not at frequencies below mf Hz nor above nf Hz.

bandwidth: range of frequencies that just encompasses all of the energy present in a given signal.

Banyan: family of self-routing networks composed of 2 x 2 switching elements. The network is nonblocking provided: incoming cells are sorted in the order of their destination ports and presented on adjacent network inputs, and multiple inputs are not destined for the same output at the same time.

baseband: frequency range (bandwidth) associated with a message signal when first generated.

baseband LAN: digital signals are carried directly on the common bearer.

base station: in wireless networks, fixed equipment at the center of a cell. The number of base stations and the frequency reuse factor determine the number of mobiles that can be served simultaneously in a given frequency band.

basic rate interface: ISDN channel consisting of two B-channels + one D-channel. In the United States the BRI is supplied over a connection that operates at 160 kbits/s. It contains 144 kbits/s of user data, and signaling data.

Batcher–Banyan: self-routing Batcher switching fabric that sorts incoming cells on the basis of their output port sequence, connected by a *perfect* shuffle interconnection network to a self-routing Banyan network of 2 x 2 switching elements.

baud: unit of signaling rate, one symbol per second.

B-channel: ISDN information channel that supports digital transmission at 64 kbits/s. The channel is circuit-switched and employs user-specified protocols. Subrates of 2.4, 4.8, 8, 16, and 32 kbits/s may be multiplexed on to it.

bearer: physical entity that supports an electronic medium.

bearer services: information services available at the user-network interface.

Bell Communications Research: central technical organization funded by the seven RBOCs.

binary-ASK: amplitude-shift keying that employs two amplitudes.

binary-FSK: frequency-shift keying that employs two frequencies. The presence of a signal at one frequency corresponds to a 1-state, and the presence of a signal at another frequency corresponds to a 0-state.

binary keying: modulating action that creates two values of amplitude, phase, or frequency of a carrier signal in sympathy with the value of the bits in a binary signal stream. These actions are known as ASK, PSK, and FSK.

binary function: a digital function that exists in two states only.

binary-PSK: phase-shift keying that employs two phase angles. The presence of the carrier signal in one phase condition corresponds to a 1-state, and the presence of the carrier signal 180° out of phase with it corresponds to a 0-state. Also known as phase-reversal keying.

binary synchronous data link control: character-oriented data link control protocol.

biphase signal format: 1 is a positive current pulse that changes to a negative current pulse of equal magnitude, and 0 is a negative current pulse that changes to a positive current pulse. The changeover occurs exactly at the middle of the time slot.

bipolar signal format: for 1s, positive current flows, and for 0s, negative current flows.

bipolar with 8 zeros substitution: in T–1 system carrying data, technique used to provide 64 kbits/s clear channels to the user yet maintain 1s density requirements.

bit: binary digit.

bit-level multiplexing: in turn, bits from each input line are interleaved on the common bearer in a regular fashion. Timing is more critical than at the octet level, but throughput delay is reduced.

bit-oriented data link control protocol: data communication procedure that makes use of a flag character to mark the beginning and ending of the frame. Between these markers, the header and the trailer fields are of predetermined lengths. Whatever data lie between them is the text field.

bit stuffing: introduction of additional bits to adjust the speed of a datastream to match the exact speed required by switching facilities, and to adjust delays to compensate for different transmission paths.

bit synchronization: state of transmission facility when receiver can identify bits within each time slot and reconstruct the same bit sequence as the sender transmitted.

block check character: character formed from bits created by the longitudinal redundancy check process.

blocked-calls-cleared: on receiving a busy signal that indicates all facilities are occupied and the call cannot be handled, the caller hangs up so that the call is lost to the system. The caller may retry over the network at a later time. Alternatively, the caller may place the call over another network.

blocked-calls-delayed: call that cannot be processed immediately is held in the system until it can be processed.

blocked traffic: number of calls blocked times the average holding time of the carried traffic. It represents the amount of traffic that cannot be carried to its destination on the first attempt because the equipment is fully occupied with other calls.

blocking: traffic condition when a call attempt cannot be completed immediately because network resources are fully occupied serving other calls.

blocking probability: probability of the first call attempt being rejected because of the unavailability of facilities.

block interleaving: to correct for error bursts, information bits are arranged in sequential rows to form columns of bits to which forward error correction parity bits are added. The sequence of information bits is transmitted; it is followed by the sequence of parity bits. During transmission, a limited error burst will affect a sequence of bits in one or more rows. At the receiver, if the errors in a column are within the capabilities of the coding, the errors are corrected and the information bits can be read out in the original sequence.

border gateway protocol: participating routers provide full routing information to each destination. The information is compared against a global network map to ensure short, optimum routing.

border router: router that is not an internal router.

boundary router: router that is connected to another autonomous system.

bridge: operating with MAC sublayer addresses, device that links networks together when their differences reach no higher than the data link layers.

bridged tap: in the outside plant, some loops may have been used to connect to several terminating points at different times. In the process of rearrangement, the pair may have been left connected to some of them so that there are unused, open-circuited, cable pairs connected to the active pair. They represent shunting impedance at the point of connection and cause reflections.

broadband ISDN: digital network that distributes information at rates up to 600 Mbits/s and supports interactive, switched services up to approximately 150 Mbits/s.

broadband LAN: signals are carried on parallel channels over a coaxial cable using frequency-division multiplexing. Devices on the network are assigned to channels they share with other devices. A specific set of frequencies is used for sending, and another set of frequencies is used for receiving.

broadcast network: nonswitched network in which a transmitting node broadcasts information to a large number of receivers in a specific geographical area.

bursty second: in ESF error checking, any second in which from 2 to 319 CRC-6 error events are detected.

bus LAN: all stations are connected to the same bearer—the *bus*—which is open-ended.

bus network: all nodes or user stations are connected to a common bearer.

busy: in network management, operational state in which the managed object is in service but cannot accept new users.

busy hour: one-hour period in any day in which most use is made of the network.

byte-count-oriented data link control protocol: data communication procedure in which the number of bytes in the text field is carried in a byte-count field in the header.

cable television carrier: cable television company that uses some portion of the cable bandwidth to provide data or voice services.

call control agent function: provides end-user access to intelligent network call processing over analog lines, ISDN digital interface lines, analog multifrequency signaling circuits, and SS7 interfaces.

call deadlock: state that exists when a low-tier mobile and a cell site attempt to call each other on the same radio frequency.

call processing software: in switching machine, software that initiates, maintains, and terminates a message path between calling and called parties.

call request packet: precursor of a packet call. Establishes virtual circuit. Forwarding depends on its destination and the best link available to facilitate migration toward this destination at the time the packet is received.

call set-up time: time to dial tone, plus the time to enter called number, plus time to make connection.

carried traffic: the number of calls in progress in the network times the average holding time. It represents the traffic that is carried to its destination.

carrier-sense multiple access, collision detection: access technique employed with IEEE 802.3 (Ethernet) LANs. When a period of no activity on the bus is detected, the station with something to send may send it. Once this station has begun transmission, the other stations should detect the activity and withhold their own messages. If two, or more, stations begin to transmit at the same time, they will interfere with each other, a collision will be declared, and both must try again at a later time.

carrier serving area: area served from remote terminal of digital loop carrier equipment.

carrier signal: see modulation.

Carson's rule: for small values of modulation index ($\beta < 1$), the spectrum of a frequency modulated wave is limited to the carrier and one pair of side frequencies, so that the bandwidth is $2f_m$ Hz. For large values of modulation index, the bandwidth is approximately $2\Delta f$ Hz, where Δf is the frequency deviation of the modulated wave.

C-bit parity DS–3: employs the format of traditional DS–3 frame. However, the seven DS–2 signals created in the multiplexing process are generated at exactly 6.306272 Mbits/s and a stuffing bit is inserted in every subframe. The result is a DS–3 signal at exactly 44.736 Mbits/s. Because they are not needed to report the presence or absence of stuffing bits, the 21 C bits are used for other purposes such as error detection, alarms, and status information.

ccs: 1 ccs is the traffic due to a call that lasts 100 seconds, or any combination of calls for which the sum of the holding times (in seconds) is 100.

cell: in cellular wireless networks, area occupied by mobiles that communicate with a single base station; also, a 53-byte frame employed in cell relay.

cell relay: high-speed routing technique that employs short, fixed-length cells that are statistically multiplexed over virtual connections.

cell relay service: bearer service that can transport voice, video, and data messages simultaneously.

cell site: in cellular wireless networks, location of the base station.

cellular digital packet data: cellular wireless information system that provides 19.2 kbits/s connectionless service on idle 30 kHz cellular phone channels.

cellular geographic serving area: in wireless networks, area consisting of multiple cells served by a mobile telephone switching office.

cellular mobile radio telephone system: provides voice or low-speed data (up to 2,400 bits/s) communication between mobile units and fixed sites.

central exchange service: services provided by a carrier that mimic the services the customer could obtain from a PBX. The customer's extensions are multiplexed to the serving office; there, switching and other functions are performed in the carrier's equipment.

centralized routing: in a packet network, a primary (and perhaps alternate) path is (are) dedicated to a pair of stations at the time of need. For the sequence of packets that constitutes a block of user's information, each packet uses the same primary (or alternate) en-route resources.

central office: end office.

channel: one-way path between communicating entities.

channel bank: equipment that combines analog voice signals into an FDM or TDM signal. In the case of TDM: when transmitting, the channel bank performs quantization, coding, μ-law compression, and multiplexing; when receiving, the channel bank performs demultiplexing, decoding, μ-law expansion, and reconstruction of the voice signals.

channel capacity: for a signal-to-noise power ratio S/N, Shannon showed that a channel of bandwidth B Hz can transport (a maximum of) C error-free bits/s given by $C = B\log_2(1 + S/N)$.

channel interface unit: in FDDI, units that receive frames from their upstream neighbors and send frames to their downstream neighbors over the primary fibers.

channel service unit: provides impedance matching, loopback (for testing), and limited diagnostic capabilities, so that the circuit provider can monitor and adjust the performance of a digital line. Also, generates AMI signals and provides span powering for repeaters, if needed. In addition, supports special framing requirements and may generate keep-alive signals when the connection is broken on the customer side.

character-framed data: each character is framed by a start bit (0) and one or two stop bits (1 or 11).

characteristic impedance of ideal transmission line: value of the terminating resistance that turns a finite line into a circuit that mimics an infinite line

character-oriented compression: by substituting special characters, or small groups of characters, for larger strings of characters that appear in a message, the length of the datastream can be made shorter than the original.

character-oriented DLC protocol: procedure that employs special characters (reserved in the code set) to facilitate communication.

checksum: error detecting procedure. The result of adding together the characters in a message to create a total that can be checked after transmission.

chip: in spread spectrum modulation, output of a code generator during one clock period. The name given to each element of a *pn*-sequence.

cipher: in encryption, action of transposing or substituting the bit patterns or characters in the original text.

cipher key: key computed by the network and the mobile station prior to encryption.

circuit: combination of two channels, one in each direction, between communicating entities.

circuit-level forwarding: by observing the grouping of packets the firewall router detects a connection between client and server. Using proxies it determines whether the source and destination are compatible (within the meaning of the criteria it has) and whether the transfer of the information is permitted.

circuit speed: number of bits transmitted over a circuit in 1 second.

circuit switch: selectively establishes and releases connections between transmission facilities to provide dedicated paths for the exchange of messages between users. The paths are established before the information exchanges begin, and, until the session is terminated, they are maintained for the exclusive use of the calling and called parties.

circuit-switched network: network in which circuits are formed by interconnecting links at network nodes so that voice or data paths are established between pairs of users. Each connection is set up using signaling information (dialed telephone number, for instance), is maintained for the duration of the call (holding time), and is taken down upon completion of the call. The connections may be physical (i.e., actual segments of wire interconnected by contact devices to form metallic paths) derived (i.e., produced by time- or frequency-division of a transmission facility) or part physical and part derived.

Class I Mobile: radiated power is adjustable in eight steps from 6 mW (8 dbm) to 4 W (36 dbm).

Class III Mobile: radiated power is adjustable in six steps from 6 mW (8 dbm) to 600 mW (28 dbm).

Class 4 Switch: toll center switch.

Class 4/5 Switch: switch that combines toll and end office features.

Class 5 Switch: switch that serves a community of business and residential users. Located in the network end office.

Class A or /8 Network: Internet address consisting of 8-bit network prefix beginning with 0, and 24-bit host number.

Class B or /16 Network: Internet address consisting of 16-bit network prefix beginning with 10, and 16-bit host number.

Class C or /24 Network: Internet address consisting of 24-bit network prefix beginning with 110, and 8-bit host number.

Class D: Internet address consisting of a 32-bit address block with leading bits 1110. It is used for multicasting.

Class E: Internet address consisting of a 32-bit address block with leading bits 1111. It is reserved for experimental use.

classless interdomain routing: aggregates consecutively numbered Class C networks into larger networks. Because they have a common subaddress, this permits interdomain routing to all networks in the group using a single table entry. Also called supernetting.

clear 64: customer has full use of a 64-kbits/s channel.

clear channel capability: customer has full use of a 64 kbits/s channel.

client: terminal with significant computing capability.

Clipper chip: asymmetric encryption chip proposed for use by the U.S. government so that the pervasive encryption of communication channels shall not protect both the law abiding and law breakers. Designated MYK–78, a secret key is created at the time of manufacture and placed in escrow. With court approval, government agencies can obtain this key and employ it to decipher encrypted messages obtained through wiretaps.

code: in encryption, action of transposing or substituting the words or phrases in the original text.

code-division multiple access: spread spectrum technique in which all stations in the network transmit at the same carrier and the same chip rate. Each station employs a code that is orthogonal to the codes used by the other stations. Each receiver sees the sum of the individual spread spectrum signals as uncorrelated noise. It can demodulate an individual signal if it has knowledge of the spreading code and the carrier frequency.

code-division multiplexing: to each channel, the entire bandwidth of the bearer is assigned so that the signals exist simultaneously on the bearer. Modulating them with orthogonal codes differentiates user's messages. The receiver uses these codes to demodulate the individual message signals.

command: in SDLC, a frame that passes from the primary station to a secondary station.

committed burst size: maximum number of bits the network commits to transfer.

committed information rate: average rate, in bits/s, at which the network agrees to transfer data.

Committee T1–Telecommunications: sponsored by ECSA, accredited by ANSI, and approved by the FCC as an appropriate forum, an organization that addresses telecommunications networks standards issues in North America.

common-channel signaling: signaling messages relating to many circuits are transported over a single channel by addressed messages.

common channel signaling network: network comprising the combination of signaling data links between switching centers and signal transfer points, and the data links between STPs.

common information space: communication mode(s) common to the downstream and upstream directions.

common management information service element: in network management, provides management notification services that can be used to report any event about a managed object, and management oper-

ation services that define the operations required to perform actions on a managed object. Also, provides a framework for common management procedures that can be invoked from remote locations.

Common Management Information Services and Protocol over TCP/IP: management support software based on OSI network management standards, it incorporates ACSE and ROSE. CMOT runs with TCP or UDP.

communication: activity associated with distributing or exchanging information.

communication channel: one-way communication path.

communication circuit: communication path consisting of two channels, one in each direction between communicating entities.

communication controller: a type 4.0 SNA node that manages communication lines, cluster controllers, workstations, and the routing of data through a network.

communication network: consists of a finite set of facilities (stations, links, and nodes) that can be interconnected to support communication among several subscribers. Resources are allocated to users when they request the opportunity to communicate, and are sufficient to support many users at the same time.

communication protocol: rule or procedure that organizes and disciplines the process of sending messages.

communications quality voice: contains a limited range of frequencies and may have discernible noise and distortion. The speech quality is adequate for easy telephone communication including speaker recognition.

companding: action of reducing the dynamic range of a signal. When converting analog voice to digital voice, to improve signal-to-noise ratio, processing is included that distributes the signal samples evenly among the levels of a linear quantizer. The samples are compressed so the higher value amplitudes are reduced with respect to the lower-value amplitudes. With a fixed number of quantizing levels, this makes more levels available to lower-level signals and improves the signal-to-noise ratio at these levels. To convert digital voice samples back to analog voice, the amplitudes of the samples are expanded to restore the original values.

computer-based message system: electronic mail system that is supported by one, or more, mainframe computers.

computer conference: interactive store-and-forward service that allows participants to share a common data space and message community.

concentration and expansion processes: the number of ports on an end-office switch is much less than the total number of loops installed in the exchange area, but larger than the number of active circuits (under normal circumstances). A concentration stage is employed to connect the active loops on the distributing frame to the input ports of the switching stages. In reverse, the concentration stage acts as an expansion stage to connect active lines from the switching stages to the loops that complete the calls.

configuration management: in network management, actions of maintaining, adding, and updating records of hardware and software components, defining their functions and usability, initializing the network, and ensuring that the network is described correctly and completely at all times. Also, assigning, maintaining, and updating records of users, including billing responsibilities and communication privileges.

connectionless service: commonly provided over a packet network for short data messages. Packets carry originating and terminating addresses. At each network node, they wait their turn to be sent over a link that will get them closer to their destination. Subsequent packets are unlikely to follow the same path, so that the times

they take to reach their destinations will vary and they may arrive out of sequence. The receiver sends no acknowledgments.

connection-oriented service: a circuit is established between the sender and receiver (physical—as over a telephone network—or virtual or permanent virtual—as over a packet data network). The receiver acknowledges receipt of message segments (packets) and is likely to report the detection of errors and exercise flow control. The delay between packets may vary, but they are delivered in the sequence in which they were sent.

connection time: time between the moment the last digit is entered and the calling circuit is completed.

container: in synchronous digital hierarchy, part of the payload. Each container contains one or more tributary signals.

contention access: when using a contention technique, each station competes for transmission resources. They send when ready, send when ready *and* the bus is not occupied, or send when ready *and* the bus is not occupied *and* listen for a collision.

continuous phase FSK: digital modulation technique that avoids the phase discontinuities of FSK. Continuity of phase across the boundary between symbols is achieved by adjusting the phase of the frequency representing the next symbol to compensate for a phase shift introduced by the frequency keying process. In binary-CPFSK, the phase shift between symbols is 90°.

continuous-wave modulation: brings together a higher-frequency sinusoidal carrier signal and a lower-frequency message signal to create a modulated wave in which message signal energy is distributed at frequencies around the carrier signal frequency.

control point: in SNA APPN networks, supports communication between network management software entities to update routing tables, provide directory services, etc.

convergence sublayer: component of ATM adaptation layer. Performs a variety of functions depending on the service, such as separating fill (dummy octets) from user octets; correcting cell delay variations; recovering clock; monitoring lost cells; etc.

convolutional code: code generated by continuously performing logic operations on a moving, limited sequence of n bits contained in the message stream. For each bit in the message stream, the encoder produces a fixed number (two or more) of bits; their values depend on the value of the present message bit and the values of the preceding bits (i.e., the previous $n - 1$ bits).

cordless telephone: telephone instrument in which the path between handset and base station is completed by radio or infrared radiation.

core gateway: in Internet, a gateway administered by the Internet Network Operation Center. Principally, these gateways route traffic into, along, and out of the backbone network.

corporate 800 service: completes 1-800 calls over an enterprise network.

corporate credit card service: completes off-net calls from corporate credit card holders over the enterprise network.

corrupted protocol data unit: in communication security, authentic PDU that has been altered by accidental means.

country code: component of ITU's universal numbering plan that can have up to three-digits.

CRC-6 error event: in ESF error checking, when the received FCS does not match the FCS calculated at the receiver.

credit-based flow control: on the basis of the traffic it is capable of handling, the receiver monitors traffic on each incoming line and assigns credits to the upstream nodes. They may send message units up to their credit limit, but no more until they receive a new credit allocation. In effect, credit-based flow control stores traffic elements at upstream nodes in the network until they can be received without interference.

crosstalk: electromagnetic coupling between twisted pair circuits causes a signal in one to produce a disturbance in the other. Called crosstalk, it contributes to the noise background in the receiving circuit

Customer Information Control System: in SNA, teleprocessing monitor that supports multiple, on-line applications in a common network. Provides interfaces to files and database systems. Handles errors, security, and priority transactions.

customer loop: pair of wires (twisted pair) that connects the customer's station to the distributing frame of the serving central office.

custom local-area signaling services: in intelligent networks, based on the ability to know the calling party's number from messages on the signaling channel, a group of switch-controlled, common-channel signaling features that use existing customer lines to provide end-users with call-management capabilities.

cyclic redundancy check: error detecting procedure. The sender divides the message by a prime number to produce a remainder known as the frame check sequence. It is placed in the trailer of the frame. The receiver divides the message by the same prime number. If agreement is reached as to FCS, the message has been received without error.

D4–channel bank: digital loop carrier that multiplexes 48 or 96 individual voice or data channels over DS–1, DS–1C, or DS–2 facilities.

D5–channel bank: a digital loop carrier that multiplexes 96 individual voice or data channels over four DS–1 facilities or a DS–2 facility. Includes a common control subsystem that manages up to 20 units and supports fiber connections to the users' premises.

daily busy hour: on any particular day, the 60-minute period in which the traffic intensity on a group of telephone facilities is the highest.

data carrier: value-added carrier that provides data communication services.

data circuit terminating equipment: equipment that conditions data signals received from an associated DTE for transmission over communication connections, and restores received data signals to formats compatible with the DTE.

data compression: technique that reduces the number of bits transmitted during a communication session so that more information can be transmitted over a given facility.

data concentrator: frame buffering and multiplexing device that buffers frames received from several lower-speed lines until they can be sent to a central site on a single higher-speed line.

data flow control layer: level 5 in SNA, provides orderly data flow in a logical unit to logical unit (LU–LU) session. Enforces communication procedures between LUs and coordinates which LU can send and which LU can receive at any given time.

datagram: data unit that is routed across a packet network by nodal decisions (distributed routing) without establishing a connection or call record.

data integrity: proving that the data are complete, uncorrupted, and genuine.

data link control layer: level 2 in SNA, responsible for scheduling and error recovery of data across a link connecting two nodes. The data link control layer supports the link-level flow of data for both SDLC and System 370/390 data channel protocols, including ESCON.

data link control protocol: set of rules that governs the exchange of messages over a data link.

data link header: includes the flag, class of frame identifier, sequence number for use in the control byte, and other information.

data link PDU: protocol data unit that consists of data link header, network PDU, and data link trailer.

data link trailer: includes an FCS to check the frame, and a flag.

data services: services that collect data for processing in equipment that executes applications (programs), and that distribute the resulting information to persons and machines that must act on it.

data terminal equipment: data processing device that sends and receives data messages.

D-channel: ISDN channel that supports a packet-mode, ISDN protocol for call-control signaling, user-to-user signaling, and packet data. In addition, it is used for operations, administration, and maintenance messages. In the BRI, the speed of the D-channel is 16 kbits/s; in the PRI the speed is 64 kbits/s.

dc-term: in a Fourier series, the zero frequency component.

decapsulation: process of stripping protocol control information from the protocol data unit as it ascends the protocol stack of the receiving machine.

decoding noise: in digital voice, the difference between the reconstructed sampled signal at the output of the receiver's pulse generator, and the signal that would be produced if the sampled signal at the output of the sender's quantizer is transmitted without error.

delayed traffic: some (or all) blocked calls may be delayed in the network until facilities are available to handle them.

delta modulation: in digital voice, coding technique in which the sampling rate is many times the Nyquist rate for the input signal so that the correlation between consecutive samples is very high, and the difference between them is very small.

demodulation: signal processing technique in which the modulating signal is recovered from the modulated wave.

demultiplexing: action of recovering individual signals from a multiplexed signal stream.

deterministic function: at every instant, a deterministic function exhibits a value (including zero) that is related to values at neighboring times in a way that can be expressed exactly. Thus, it is possible to know the value of the signal at a future time, and to know the time in the future when the signal will have a specific value.

differential coding: in digital voice, coding that exploits the correlation between adjacent signal samples.

differential pulse code modulation: in digital voice, coding technique in which the differences between successive signal samples are sent to the receiver. At the receiver, they are integrated to predict the value of the signal. At the same time, the transmitter integrates the samples. When a major difference occurs between the transmitter's prediction and the actual value of the signal, a correction is sent to the receiver to restore its prediction to the real signal value.

differential phase-shift keying: digital phase modulation technique that eliminates the need for the synchronous carrier in the demodulation process. At the transmitter, the datastream is processed to produce a modulated wave in which the phase changes by 180° when one symbol (0 or 1) is transmitted, and does not change when the other symbol (1 or 0) is transmitted.

digital-added main line: 2B1Q digital subscriber equipment that implements an additional voice circuit over a single pair of wires for circuit expansion purposes.

Digital Advanced Mobile Phone System: defined by TIA Interim Standard 54, D–AMPS employs AMPS frequencies. It provides three digital channels (TDMA) in a 30-kHz channel, and employs a combination of analog and digital control channels. Dual-mode terminals operate with AMPS or D–AMPS to facilitate the evolution of AMPS to a higher-capacity digital system.

digital channel bank: *see* digital loop carrier.

digital cross-connect system: computer-based facility that reroutes channels contained in synchronous carrier transmission systems without demultiplexing. In SONET, redistributes (and adds or drops) individual channels (or virtual tributaries) among several STS-n links. Consolidates and segregates STS-1s, and can be used to separate high-speed traffic from low-speed traffic, so as to feed one to an ATM switch and the other to a TDM switch.

digital data communications message protocol: data communications procedure employing a header that contains exactly 10 octets, a text portion that contains a stated number of octets, and a trailer that consists of a 2-octet FCS.

digital data service: 56 kbits/s transport service for private-line applications. The DDS network is implemented as a set of interconnected hubs that perform as routing, maintenance, and administration centers. In turn, the hubs are connected to digital lines that collect customer channels from serving offices.

Digital European Cordless Telecommunications: standardized by ETSI in 1991, DECT is designed to deliver a wide range of communication services in the PCS band using FD/TDMA with time-division duplex.

digital function: single-valued, discrete state function that assumes a limited set of values. The most popular class of digital functions is two-valued (binary)—i.e., the members exist in two states only. Changes of state occur instantly, and the rate of change at that instant is infinite—otherwise, it is zero.

digital loop carrier: carrier system that time-division multiplexes subscriber loops and operates between the serving area interface and the serving central office. DLCs provide standard support functions for the user's terminal.

digital modulation: the amplitude, frequency, and/or phase of a sinusoidal carrier are caused to change instantaneously, according to the individual bits, or groups of bits, in the binary message signal.

digital signature standard: using the principle of public key cryptography, the DSS algorithm encrypts and verifies a short code that is attached to a data-file. This digital signature is created from a reduced representation of the data-file. The code provides electronic verification of the data and its source and leaves the entire file in cleartext.

digital service unit : converts unipolar digital signals received from DTE to bipolar signals and performs any special coding or other process required to make the signal compatible with the operation of the transport system—such as zeros-substitution codes.

digital subscriber line: *see* ISDN digital subscriber line.

digital termination service: local loop bypass microwave radio system that operates in the band 10.55 to 10.68 GHz, and provides digital channels at speeds up to 1.5 Mbits/s and distances up to 6 miles.

directory service: in SNA, provided by a node in response to a request from an originating LU to find the location of a terminating LU prior to starting an LU–LU session.

direct-sequence spread spectrum signal: the information signal modulates a *pn*-sequence to produce a signal that extends over the wideband spectrum of the digital sequence. In turn, this signal is used to modulate a sinusoidal carrier.

disabled: in network management, an operational state in which the managed object is not available.

disassociated signaling: signaling messages are sent over a route different from the route taken by the information that is exchanged between the parties.

discrete memoryless source: information source that emits a steady stream of mutually independent symbols.

distance learning: *see* education at a distance.

distributed broadband service: broadband ISDN service in which information transfer is one way, from distributor to customer.

distributed-queue dual-bus: metropolitan area network access arrangement that provides connectionless, connection-oriented, and isochronous transfer services. Each node gains access to a bus by joining a distributed queue of nodes waiting to transmit on the bus.

distributed routing: in a packet network, the nodes decide which paths the packet shall take to its destination on the basis of information about traffic conditions and equipment status. Each packet is routed without regard to packets that may have preceded it.

distribution cables: in the local loop, they provide connections from serving area interfaces to service access points close to the user's premises.

double-sideband suppressed-carrier AM: modulation produced by multiplying the carrier signal by the message signal. Energy associated with the message signal is distributed on both sides of the carrier frequency, but there is no separate carrier component. The transmission bandwidth is twice the bandwidth of the message signal.

downstream direction: direction in which richer messages flow.

drop and repeat node: SONET device configured to split SONET traffic and copy (repeat) individual channels on two, or more, output links. Applications include the distribution of residential video and alternate routing.

drop cables: connect from service access points (such as a pedestal or other distribution point) to the customer's premises.

DS–0: digital signal level 0, 64 kbits/s.

DS–0A: DS–0 signal containing data from a single subrate station. The bytes are repeated as necessary to match the sender's speed to the speed of the DS–0 line.

DS–0B: DS–0 signal containing data from several subrate stations. The octets contain a subrate synchronizing bit (bit 1) and six data bits (bits 2 through 7). The eighth bit is set to 1 to ensure meeting the 1s density requirement.

DS–1: digital signal level 1, 1.544 Mbits/s.

DS–1C: digital signal level 1C, 3.152 Mbits/s.

DS–2: digital signal level 2, 6.312 Mbits/s.

DS–3: digital signal level 3, 44.736 Mbits/s.

DS–4: digital signal level 4, 274.176 Mbits/s.

DS–4A: digital signal level 4A, 139.264 Mbits/s.

dual-tone multifrequency: signals used by telephone to signal called number, and other information.

dumb terminal: device that is dependent on the host for all but the most routine functions.

duobinary encoding: applied to the baseband signal, coding technique that achieves a signaling rate of $2B_T$ symbols/s in a bandwidth of B_T Hz. This is the maximum rate predicted by Nyquist.

duplex: information can flow in two directions at the same time. A circuit is employed; it may be *symmetrical* (i.e., the channel capacity is the same in either direction) or *asymmetrical* (i.e., the capacity of one channel, usually from source to receiver, is greater than the capacity of the other channel).

E–1: first level of voice multiplexing in Europe. Standardized by ETSI, combines 30 voice channels plus 2 common signaling channels at 2.048 Mbits/s.

echo suppresser: at the sending end, they create a signal that cancels the echo signals received from impedance mismatch point(s) on a transmission circuit.

education at a distance: use of telecommunication techniques to instruct students in several locations from a central location. The sound and sight of the instructor, and the images employed by the instructor, are transmitted to all receiving sites, and, when authorized to do so, students at the sites may ask questions of the instructor.

EIA232: standard that describes a multiwire cable that terminates in 25-pin connectors. At a speed of 20,000 bits/s, the standard limits the cable to 50 feet.

EIA449: standard that describes a cable that operates at 100 kbits/s, at distances up to 4,000 feet.

electronic data interchange: common message formats and procedures established by ANSI for transactions between industrial organizations.

Electronic Industries Association: nonprofit organization that represents the interests of manufacturers. The EIA is well known for its work in standardizing connections between data terminals and other equipment.

electronic mail: half-duplex record service.

empty token: in an IEEE 802.5 LAN, frame that consists of a start flag, an access control octet, and an end flag.

enabled: in network management, operational state in which the managed object is operable, available, and not in use.

encapsulation: procedure of attaching PCI to the SDU as it descends down the protocol stack of the sending machine. The headers and trailer are said to encapsulate the user data (the SDU).

encryption: action of disguising information so that it can be recovered relatively easily by persons who have the key, but is highly resistant to recovery by persons who do not have the key. Encryption is accomplished by scrambling the bits, characters, words, or phrases in the original message.

end node: in SNA APPN, node that exists at the edge of an APPN network and provides network access to end-users.

end-office: facility that contains the lowest node in the hierarchy of switches that comprise the network.

end-to-end encryption: the transport layer PDU is encrypted before being transferred to the network and data link layers.

end-user or transaction services layer: layer 7 in SNA, provides application services to the end-user, including end-user access to the network, requests for services from the network addressable unit or presentation services layer, and network management services.

enterprise systems connection: optical fiber connection operating up to 40 kilometers at 17 Mbits/s. ESCON was introduced by IBM to extend the mainframe, high-speed I/O channel directly to major, remote peripherals.

entropy: the entropy of a memoryless source is the average self-information of the symbols it emits.

entropy coding: coding in which the length of a codeword is proportional to the self-information of the symbol it represents. The shortest codeword is employed for the symbol used most, and the longest codeword for the symbol used least.

equalization: technique in which received signals are passed through a filter that approximates the inverse of the distortion of the channel.

equipment identity register: in mobile telephone network, database that lists subscriber's authorized equipment and denies service if attempts are made to use unregistered equipment.

erbium-doped fiber amplifier: provides uniform gain across a wide spread of wave-

lengths centered on 1,550 nm, the wavelength of minimum attenuation. Spans of 120 km (75 miles) between amplifiers can be employed. However, beyond 40 km, intersymbol interference produced by dispersion becomes a factor, and dispersion compensation is required.

ergodic process: statistical measures on sample functions of the random process in the time domain are equal to corresponding measures made on the ensemble of sample functions (the time average is equal to the ensemble average).

Erlang: 1 Erlang is the traffic due to a call that lasts for 1 hour; it is also the traffic due to 60 calls that last 1 minute each, or any combination of calls for which the sum of the holding times (in seconds) is 3,600.

Erlang B formula: determines the probability of call blocking in a communication (trunking) system that has no queueing capability. The blocked call is *lost* and must be *retried* at a later time.

Erlang C formula: determines the probability of call delay in a communication system that includes a queue to hold all call requests that cannot be assigned immediately to an idle channel.

error control: cooperative activity between sender and receiver in which the sender adds information to the character or frame to assist the receiver to determine whether an error has occurred. If it has, the sender and receiver work together to correct it. Consists of two parts: error detection and error correction.

errored second: in ESF error checking, any second in which one, or more, CRC-6 events are detected.

ESF error event: in ESF error checking, an OOF event, CRC-6 event, or both.

eternal function: function that exists for all time.

Ethernet: LAN pioneered by Xerox Corporation, and developed further by a team consisting of Xerox, Digital Equipment Corporation, and Intel Corporation. Later, the concept was included in IEEE 802.3. A bus-based LAN that employs carrier sense, multiple access, collision detection. While there are minor differences between the original Ethernet and IEEE 802.3, they are compatible and most persons describe either of them as Ethernet.

European Telecommunications Standards Institute: organization created to produce technical standards and achieve a unified European telecommunications market.

even parity check: the parity bit makes (or leaves) the number of 1s even.

event forward discriminator: in network management, directs reports to the appropriate destination.

excess information rate: number of bits sent in a certain time (bits/s) minus the committed information rate.

exchange area: area that contains all the customers whose loops terminate at a given end-office or wire center.

exchange code: identifies specific switching facilities located in an end-office or wire center. Many offices handle more than one exchange code.

Extended Binary Coded Decimal Interchange Code: code employing a sequence of 8 bits that contains 256, 8-bit patterns.

extended superframe format: 24 DS-1 frames form an extended superframe that contains 4,632 bits and takes 3 milliseconds to transmit at the DS-1 rate. The 24 framing bits are used for synchronization and other purposes.

exterior gateway: in Internet, gateway that exchanges resource information with gateways in different autonomous systems

extranet: private corporate network based on Internet to which customers and suppliers have access.

eye diagram: to assess the performance of a transmission line, a display that superimposes all of the waveform combinations over adjacent signaling intervals on an oscilloscope is used. This display is char-

acterized by eyes that show the signal level differences between 0s and 1s. Poor signal-to-noise ratio is shown by smaller eyes; good signal-to-noise ratio corresponds to larger eyes.

facsimile: intercommunication, simplex real-time service. A communication service that captures images from paper, transmits them to a distant site, and reconstructs them on paper.

factory automation: use of information technologies to improve the effectiveness of factory processes.

fade: decrease in signal strength that occurs when a receiver moves in an environment in which it is illuminated by the direct ray and reflected rays (multipath). Fades occur at approximately half-wavelength intervals along the unit's forward progress.

failed seconds state: in ESF error checking, state achieved after 10 consecutive severely errored seconds have occurred. The state remains in effect until the facility transmits 10 consecutive seconds that are not severely errored.

far-end crosstalk: crosstalk whose energy flows in the same direction to that of the signal energy in the disturbing channel.

fast-associated control channel: in GSM, control channel used for hand-offs and other functions.

fast packet switching: technique for transporting data, voice, or video messages in a fixed-length packet. FPS supports the transport of fixed-length cells that can contain data, voice, or video information. Asynchronous transfer mode is the preferred FPS technique.

fast select service: in a packet network, the call connect and disconnect procedures are circumvented by using a call request packet in which up to 128 bytes of user's data can be included. The network delivers the packet as an incoming call request. The receiver responds to the data with a clear indication packet that contains up to 128 bytes of information in reply.

fault management: in network management, provision, monitoring, and response to facilities that detect and isolate abnormal conditions in the network so as to restore the network to normal functioning and correct the aberrant condition(s).

fax-modem: combining the facsimile function and the modem function, fax-modems communicate with stand-alone machines or with a modem-equipped personal computer.

fax-modem–Class 1: extension of the Group 3 standard for facsimile machines. The computer handles call set-up and segmentation of the facsimile signal into data packets.

fax-modem–Class 2: the fax-modem is responsible for call set-up and segmentation into data packets.

fax-modem–Class 2.0: in addition to call set-up and segmentation of the facsimile signal into data packets, the fax-modem can also send and receive data files.

fax-relay: on circuits with voice compressors, a voice/fax activity detector is installed. When voice is detected, the signal is encoded, compressed, framed, and dispatched to a digital transmission facility. When fax is detected, the signal is demodulated to create the original (14.4 kbits/s) datastream, framed, and forwarded to the digital transmission facility. Fax data received from the digital transmission facility are remodulated and sent to the analog facility.

feeder cables: the backbone of the local loop network, feeder cables connect the distributing frame in the serving office to serving area interfaces.

fiber-and-coax network: hybrid network that provides duplex voice and data circuits and simplex broadband (television) channels to the public. Optical fibers are used for signal transport to local distribution points where the digital TDM signal streams are converted to an FDM stream and distributed to individual homes over shared coaxial cable. In each home, an interface device separates television from voice and data. Two-way operation over cable is achieved by allocating a portion of the spectrum for upstream signals.

fiber channel: designed to operate up to 10 kilometers at speeds up to 800 Mbits/s on optical fiber, FC operates over electrical media up to 100 meters (shielded twisted pair or coaxial cable). Intended to replace HIPPI, FC targets the interconnection of clusters of workstations, and the connection of workstations to supercomputers.

fiber-distributed data interface: FDDI is a dual-ring, optical fiber, data-only network that connects up to 500 stations in a ring of ≤ 100 kilometers. Data are passed at a speed of 100 Mbits/s using a token-passing protocol and 4B5B coding.

fiber-distributed data interface–II: development of FDDI that handles asynchronous packet data and isochronous voice and video.

fiber-distributed data interface follow-on LAN: gigabit version of FDDI that will be compatible with higher-speed SONET levels (STS–48, 2.488 Gbits/s, for instance) and BISDN.

fiber to the curb: extension of optical fiber loops to pedestals.

fiber to the home: extension of optical fiber loops to residences.

filter: in network management, uses Boolean operators to chain test conditions that determine which events in the network are to be reported.

firewall: software/hardware device that denies access to unauthorized callers. Situated between intranet and Internet, a firewall consists of a screening router and logic that implements a set of rules to determine which connections are allowed, and which are not. Called proxies, the rules restrict the number of services available to outside connections and prevent the manipulation of services to provide unauthorized levels of access. In addition, a firewall can be used to limit the flow of specific information to callers from within the intranet.

firewire: fast electronic bus designed to interconnect computer and multimedia devices at speeds up to 400 Mbits/s. A serial bus that operates over a cable containing two shielded twisted pairs, and two power wires, it is used to connect up to 63 devices in daisy-chain fashion.

foreign exchange line: tie line that connects a private network facility (e.g., a PBX) to a public exchange in a distant city so that callers may have the benefit of local service in the distant calling area.

forward error correction: technique that employs special codes that allow the receiver to detect and correct a limited number of errors without referring to the transmitter.

four-wire connection: duplex circuit that employs two pairs of wires, one for each direction of transmission. Four-wire connections must be employed when the distance is great enough to require repeaters or when the existence of two signals on the same bearer will cause them to interfere.

fractional T–1: fraction of a T-1 facility. FT-1 provides n DS–0 channels, where $n = 2, 4, 6, 8,$ or 12, as required.

fractional T–3: fraction of a T–3 facility—i.e., FT–3 = n T–1, where $1 < n < 28$.

frame check sequence: remainder after dividing the bitstream between the start flag and the end of the text field by a binary prime number.

frame-level multiplexing: frames from different input lines are interleaved on a common bearer.

frame relay: fast packet transport technique that transfers data in variable length frames over PVCs. Frame relay is a net-

work access technique. Within the network, the procedures employed may be frame relay, cell relay, X.25, or ISDN.

Frame Relay Forum: association of producers and users committed to promoting the application of frame relay in accordance with national and international standards.

frame relay PAD: creates frame-relay frames for equipment that is not able to do so and converts frame-relay frames to formats that are compatible with the end-users.

frame relay user–network interface: broadband network access interface that transfers information in variable-length frames (262 to 8,189 octets). Error detection (but not correction) is performed, and corrupted frames are destroyed (without requesting retransmission). Normally, the interface is limited to DS–1/E–1 speeds.

frame synchronization: state of transmission facility when sender and receiver agree on beginnings of frames.

free space: volume that surrounds the Earth; supports the propagation of radio waves for communication purposes.

frequency allocation: set of frequencies allocated to all stations offering a specific service.

frequency- and time-division multiplexing: the bandwidth of the bearer is divided up to produce frequency-divided channels. Then each channel is time-divided to produce time slots. A channel is made up of a sequence of time slots in a frequency-divided channel.

frequency assignment: specific frequency on which a particular station must operate to provide the service to which the frequency is allocated.

frequency diversity: in satellite communications, to avoid signal interference, satellites assigned to the same orbit positions operate in different frequency bands.

frequency-division multiple access: technique that assigns specific frequencies and bandwidths from the band of frequencies assigned to particular radio systems (communication satellites, mobile radios, etc.) to create individual message channels.

frequency-division multiplexing: to each signal a fraction of the bandwidth of the bearer is assigned so that the multiplexed signals exist at the same time, at different frequencies, on the same bearer.

frequency-domain coding: in digital voice, coding that divides the signal into a number of frequency bands and codes each separately. More bits may be employed in those bands in which activity is significant at the expense of those bands that are quiet.

frequency-hopping spread spectrum signal: the information signal modulates a sinusoidal carrier to produce a narrowband-modulated wave. Subsequently, the center frequency is changed by mixing the modulated wave with individual frequencies selected by a *pn*-sequence (the spreading signal). As a result, the transmitter *hops* from one frequency to another over a wide frequency band.

frequency modulation: modulation produced by varying the frequency of the carrier in sympathy with the amplitude of the message signal.

frequency reuse: principle of using a set of frequencies in one place and reusing them in another. The locations must be separated so that entities that use the same frequencies do not receive detectable signals from the wrong sources.

frequency reuse factor: in wireless networks, ratio of the number of cells using the same set of frequencies to the total number of cells in the system.

frequency shift keying: technique in which a carrier at one frequency represents a mark and a carrier at another frequency represents a space.

front-end processor: equipment that performs all the communication-related functions at a host site. The FEP relieves the much faster host of activities associated with the relatively slow processes required for data communication.

full-duplex: information can flow in two directions at the same time. A circuit is employed.

full-load signal: in digital voice, signal that just sweeps through all levels of a quantizer.

full-rate traffic channel: in GSM, occupies 24 slots in each 26-frame multiframe to provide a bit-rate of 22,800 bits/s.

functional component: elementary call-processing command that directs internal resources to implement specific intelligent network services.

fundamental component: in a Fourier series, component with the lowest frequency (but not zero).

fundamental frequency: in a Fourier series, frequency of the fundamental component.

Future Public Land Mobile Telephone System: ITU system that will provide global coverage for speech and low- to medium-bit-rate data ($\leq$ 64 kbits/s), with high-bit-rate data ($\leq$ 2.048 Mbits/s) available in limited areas. FPLMTS will be compatible with ISDN and SS7.

gateway: device used to connect networks when their differences involve layers above the network layer.

general packet radio service: in GSM, provides data distribution, messaging, data retrieval, and conferencing services at speeds to around 170 kbits/s. GPRS supports connectionless and connection-oriented (point-to-point and point-to-multipoint) packet transport services. The length of the messages can vary from less than 100 bytes to several thousand bytes.

generic program: in a switching machine, general-purpose software package installed in all machines of the same class that includes subprograms that perform call processing, system maintenance, and system administration.

geostationary orbit: orbit 22,300 miles above Earth's surface in which a satellite travels at approximately 6.900 miles/hr, appears stationary from Earth, and can illuminate around one-third of Earth's surface.

global functional plane: describes IN services from a service design perspective.

Global Positioning System: space segment consists of 24 satellites (21 in use plus 3 spares) distributed in 6 planes inclined at 55°. They orbit Earth every 12-hours at 12,554 miles (20,200 kms) above Earth. The primary function is to assist in the navigation of mobile stations by providing position, velocity, and exact time. For civilian use, navigation information is transmitted at 1575.42 MHz. A pseudo-noise code with length 1,023 bits and period 1 millisecond is used to produce a spread spectrum signal with bandwidth 1 MHz. Each satellite employs a different *pn*-code. At the receivers, code synchronization is provided by time signals from the satellites.

global system: in Internet, administered by a global authority, a group of autonomous networks interconnected by gateways. Within the global system, authorities responsible for individual autonomous systems agree on a common way to provide routing update messages to each other.

Global System for Mobile Communication: a pan-European system, GSM is a second-generation digital, cellular, land mobile telephone system. Operating in the 900-MHz region (890 to 915 MHz, and 935 to 960 MHz), GSM–900 employs 200-kHz channels separated by 45 MHz. Each channel is time-divided to create eight subchannels (FD/TDMA). GSM–1800 and GSM–1900 employ the same technology and operate in the PCS bands (around 1.8 GHz in Europe and 1.9 GHz in North America). GSM pioneered automatic roaming techniques, mobile-assisted handover, subscriber authentication, and user privacy (across the air interface)—features that have been adopted by North American systems.

go-back-n: ARQ technique in which the sender sends frames in a sequence and receives acknowledgments from the receiver. On detecting an error, the receiver discards the corrupted frame and ignores any further frames. The receiver notifies the sender of the number of the frame it expects to receive (i.e., the number of the first frame in error). On receipt of this information, the sender begins resending the data sequence starting from that frame.

graphics mail: intercommunication record service with point-to-point (sometimes point-to-many) simplex topology.

Group 1 Facsimile: scans documents at a resolution of approximately 100 lines per inch. Uses frequency modulation to transmit a page in 4 to 6 minutes.

Group 2 Facsimile: essentially the same as Group 1, but includes signal compression techniques to speed up delivery to 2 to 3 minutes.

Group 3 Facsimile: scans black-and-white documents at 200 x 100, or 200 x 200 dpi. Scanning generates digital signals that are subjected to variable-length coding to reduce redundant information. Transmits a page in less than 1 minute over modem-terminated, telephone line connections operating at speeds from 2,400 to 9,600 bits/s.

Group 4 Facsimile: black-and-white digital system that employs advanced, modified read coding to reduce redundant information. Transmits a page in less than 10 seconds over 64 kbits/s, digital connections. The group is divided into three classes according to combinations of horizontal and vertical resolutions that range from 200 dpi to 400 dpi.

half-duplex: information flows in two directions, but only in one direction at a time. If a single channel is used, it must be *turned around* to allow information flow in the reverse direction. If a circuit is used, only one channel is active at a time.

handoff: in wireless networks, action of changing the base station with which the mobile communicates.

hard to reach: protective routing scheme. When the answer to bid ratio of a switch drops below a given threshold, it is classified as hard to reach. On subsequent unsuccessful bids for connection, the originating switch is informed over SS7 of the HTR status of the destination switch. Individual switches reduce the number of times they request connection through a congested switch.

harmonic component: in a Fourier Series, component whose frequency is an integer multiple of the fundamental frequency.

harmonic frequency: in a Fourier series, frequency of a harmonic component.

H-channel: ISDN information channel (H for higher) that operates at 384 kbits/s (H0), 1,536 kbits/s (H11), or 1,920 kbits/s (H12).

hierarchical network: sometimes called a tree structure, lower-level nodes are connected to higher-level nodes. The highest node is the root node.

hierarchy: concerned with the ways in which nodes (switching centers, multiplexers, etc.) are arranged to achieve efficient division and routing of different classes of communications traffic.

high data rate digital subscriber line: provides duplex service over two unshielded twisted pairs at 1.544 Mbits/s (DS–1). Each pair operates at 784 kbits/s and employs echo cancellers to achieve duplex transmission. With 24-gauge copper pairs, service is limited to 12,000 feet—the nominal limit of the CSA. HDSL provides customers with repeaterless T–1 service from the CSA interface over pairs that do not require special engineering.

high-pass characteristic: performance such that energy is transmitted at frequencies greater than hf Hz, but not at frequencies below hf Hz.

high-performance parallel interface: provides a connection operating up to 120 feet at 800 Mbits/s on a single cable, and 1,600 Mbits/s on two cables. Over optical fiber, HIPPI operates up to 10 kilometers.

high-performance routing: in SNA, the entire message path is carried in the packet header. This eliminates table lookup at each intermediate node, and makes delivery much faster than ISR over low-error rate links.

high-speed serial interface: supports data transfer at speeds up to 52 Mbits/s.

highway interchange process: action of a space switching stage that, on instructions from the stored program control, places octets from incoming frames in the same time slots on different highways.

holding time: time between the moment when the circuit is first seized (the user goes off hook) and the moment when the circuit is released (the user goes on hook) and equipment becomes available for use by others.

home location register: in mobile telephone network, database that maintains and updates subscriber's location and service profile.

hookswitch: switch on telephone instrument that closes when the handset is removed from its cradle.

host: mainframe computer that performs processing and computing tasks in support of terminals. Also, a type 5 SNA node that is the main data processor in a subarea. It always contains an SSCP and supports one, or more, LUs.

Huffman code: variable-length code whose members are assigned to the symbols in some way in proportion to their use.

hybrid: in a telephone, balance transformer that separates the user's talking signal from the signal to which the user is listening.

hybrid fiber and coaxial cable network: network that provides duplex voice and data circuits and simplex broadband channels. It employs optical fibers for signal transport to local distribution points (analogous to serving area interface points) where the digital TDM signal streams are converted to an FDM stream and distributed to individual homes over shared coaxial cables. In each home, an interface device separates television from voice and data.

hybrid mode: in FDDI-II, asynchronous data and synchronous voice are carried in FDDI frames.

ideal transmission line: one without losses—losses due to the resistance of the conducting wire members and the leakance (conductance) between them.

IEEE 802.2: LAN standard for logical link control frame.

IEEE 802.3: LAN standard for CSMA/CD operation.

IEEE 802.4: LAN standard for token-bus operation.

IEEE 802.5: LAN standard for token-ring operation.

IEEE 802.6: standard for metropolitan area network.

illicit protocol data unit: in communication security, PDU introduced into a communication system for an illicit purpose. It may be an APDU that has been altered deliberately or a PDU generated by an illicit source.

illicit receiver: in communication security, receiver that is connected to the network without the knowledge and approval of the network owner.

illicit source: in communication security, sender that is connected to the network without the knowledge and approval of the network owner.

in-band signaling: the associated signals are analog tones that exist within the bandwidth used for the information exchange.

inclined Earth orbit: path on a plane that passes through the center of Earth situated between the polar and equatorial planes

information: that which is distributed or exchanged by communication. When associated with machines, it may be defined as organized or processed data. When associated with persons, it may be defined as the substance of messages—that which is known about, or may be inferred from, particular facts or circumstances. When associated with the exchange of symbols, it is the probability of the symbol (message) being sent.

Information Management System/Data Communications: in SNA, teleprocessing monitor that provides a comprehensive database management and data communications capability.

information system: software entity that organizes data to produce information for the benefit of an enterprise. The information may be evaluative, informative, or supportive

in-slot signaling: the associated digital signals use part, or all, of a bit position contained within a digital message octet (also known as bit robbing).

Institute of Electrical and Electronic Engineers: professional society that develops technical performance standards.

Integrated Services Digital Network: switched digital network that provides a range of voice, data, and image services through standard, multipurpose user-network interfaces based on 64 kbits/s clear channels.

integrated STP: in an SS7 network a signaling point that has an STP capability and also provides OSI model transport layer functions.

intellectual stage: with regard to using the telephone, managing the flow of information to create an understanding of the message.

intelligent multiplexer: multiplexer that has the ability to add, drop, and redistribute circuits among multiple trunks, on command. Supports large, complex mesh networks.

intelligent network: based on ISDN, a network that distributes call-processing capabilities across multiple network modules, employs reusable network capabilities to create a new feature or service, and employs standard protocols across network interfaces.

intelligent network capability set–1: CS–1 includes services that are of high commercial value and can be implemented without significant impact on existing installations. They are single-user, single-ended, single point of control, and single bearer services. Invoked by a single user, they carry no requirement for end-to-end messaging or control and are provided over the same medium as the communication.

intelligent network conceptual model: developed by ITU–T, the model consists of four planes: service plane, global functional plane, distributed functional plane, and physical plane.

intelligent network switch capabilities: identifies calls associated with intelligent network services, formulates requests for call-processing instructions from service logic elements, and implements instructions. Provides IN services to subtending users and lower-level switches equipped as network access points.

intelligent peripheral: in intelligent network, contains an adjunct service point that responds to requests for service processing. Provides and manages resources such as speech recognition, voice synthesis, announcements, and digit collection equipment as required by IN service logic.

intelligent terminal: device programmed to perform some of the tasks usually assigned to a host machine—such as screen formatting and data editing, or more complex assignments—so as to reduce dependence on the central processor.

interactive broadband service: B–ISDN service in which there is an exchange of video information and/or high-speed data.

interactive network: nonswitched network in which several (secondary) users obtain services from a specialized (primary) node within the network (e.g., data processing services provided by a central host). Also, one in which many equal status users exchange information (e.g., on a LAN, or over a VSAT network).

intercarrier interface: facilitates the interconnection of intra-LATA and inter-LATA SMDS networks.

intercommunication: telecommunication that takes place between two sites.

interexchange carrier: provides long-distance transport services between points of presence that are established for each LATA.

interior gateway: in Internet, gateway that exchanges resource information with gateways in the same autonomous system.

intermediate session routing: in SNA, the path control layer in each node assigns a unique identifier for the session. A 17-bit word, it is carried in the address field of the transmission header of the BIND command, and all subsequent messages for the session. At each intermediate node, a lookup table is required to identify the next leg of the route.

internal router: in SNA, all networks connected directly to this router belong to the same area.

International Electrotechnical Commission: promotes safety, compatibility, interchangeability, and acceptability of international electrical standards.

International Maritime Satellite Organization: established in 1979 to provide worldwide mobile satellite communications for the maritime community, INMARSAT has 79 member countries and operates global mobile satellite services on land, at sea, and in the air. To cover the globe, three Inmarsat-GEO satellites each employ a global beam and five spot beams. Stationed to cover the Pacific, Indian, and Atlantic oceans they support a wide range of services.

International Standards Organization: worldwide federation of national standards bodies concerned with the development of standards and related activities for the purpose of facilitating the international exchange of goods and services, and developing cooperation in the spheres of intellectual, scientific, technological, and economic activity

International Telecommunications Satellite Consortium: consortium of national communications satellite organizations that owns and operates over 20 spacecraft in geostationary orbit. Managed by COMSAT, a company created by act of Congress in 1962, INTELSAT provides repeater-in-the-sky services to most of the countries in the world for international (and some domestic) telecommunications. Terrestrial connections are provided by approximately 300 Earth stations that are owned and operated by local administrations and operating companies.

International Telecommunication Union: agency that recommends standards that facilitate international telecommunication. A specialized agency of the United Nations, ITU regulates radio communications, standardizes international telecommunications, and develops global networks.

International Telegraph Union: forerunner of International Telecommunication Union.

Internet: international network of packet networks operated by a government-university-industry consortium and serving data users around the world.

Internet2: private effort by the nation's universities and corporate partners (University Corporation for Advanced Internet Development) to facilitate the development and deployment of advanced network-based applications and network services to support the research and education objectives of the higher-education community. Internet2 will employ existing networks to connect members together.

Internet packet exchange: network layer in Novell's Netware protocol stack. IPX

makes a best-effort attempt to deliver packets to their destination. It uses a 12-byte address that includes the network address of an individual work group, the node address of the individual workstation in the workgroup, and a socket address that identifies a particular application on the station.

Internet protocol: connectionless Internet network layer procedure.

Internet Registry: administers Domain Name System. Associates domain names with IP addresses. Works through InterNIC in the United States, RIPE Network Coordination Center in Europe, and APNIC in the Asia-Pacific region.

Internet service provider: company that provides access to Internet facilities and supports data transport, information retrieval, electronic mail, and other services.

inter-office signaling: exchange of signaling messages between switches that serve the routes between calling and called parties.

interpersonal message service: with regard to X.400, service in which messages are delivered to a mailbox as they are created, and stored until delivery is requested by the owner of the mailbox.

interswitching system interface: in SMDS, facilitates the interconnection of switches from different manufacturers.

intranet: private network based on Internet protocols, but not connected to Internet, that distributes (proprietary) information to employees. Using a private network ensures the owner controls contents and uses.

invasion of company privacy: theft, interception, misuse, or aggregation of information concerning the operation of the business, particularly that information that may give a competitive advantage to the receiver.

invasion of individual privacy: theft, interception, misuse, or aggregation of information concerning an individual's thoughts, sentiments, emotions, lifestyle, or status.

inverse multiplexing: creating a higher-speed, point-to-point digital circuit from several independent, lower-speed digital channels that exist between the originating and terminating locations.

I/O channel: in an IBM mainframe, software-driven processor that executes commands and programs and provides high-speed input and output channels for the host.

ISDN B channel: 64 kbits/s switched digital service.

ISDN digital subscriber line: provides duplex service over a single UTP of 24-gauge copper wire. With a bit rate of 160 kbits/s, it operates up to 18,000 feet. To achieve this performance, 2B1Q (2-binary, 1-quaternary) coding is used.

ISDN H0: 384 kbits/s switched digital service.

ISDN H11: 1,536 kbits/s switched digital service.

ISDN service node: in an SS7 network, facility that supports provision of ISDN bearer services and supplementary services.

ISDN user part: SS7 protocol that represents the location of model layers invoked when the network employs the transport capabilities of the MTP and SCCP to provide call-related services to the user.

isochronous: synchronizing process in which timing information is embedded in the signals.

isochronous medium access controller: in FDDI–II, supports voice and video applications.

ISO/IEC Joint Technical Committee 1: formed in 1987 by the merger of information technology committees of ISO and IEC, JTC1 is responsible for consensus standards in the area of information processing systems.

jamming margin: in spread spectrum modulation, difference between the noise level at the receiver and the received signal level that will assure the desired value of $(S/N)_{out}$.

keying: process of modulating a continuous wave carrier with a binary, digital message signal. *See* ASK, FSK, and PSK.

knowledge worker: part business person, part computer scientist, and part engineer, performs information-based tasks and communicates knowledge and expertise directly to manufacturing, operations, and other segments of the enterprise.

LAP-B: component of ITU–T's X.25 standard.

LAP-D: for ISDN D-channel data transport.

LAP–D core: supports limited error control on a link-by-link basis. It recognizes flags (to define frame limits), executes bit stuffing (to achieve bit-transparency), generates or confirms frame-check sequences, destroys errored frames, and, using logical channel numbers, multiplexes frames over the links.

LAP–D remainder: performs error control on an end-to-end basis. Acknowledges receipt of frames, requests retransmission of destroyed frames, repeats unacknowledged frames, and performs flow control by limiting sending rate when requested by receiver.

LAP-F core: link access procedure that includes most of the capabilities of LAP–D including frame delimiting and alignment (using beginning and ending flags), inspection to ensure that the frame consists of an integer number of octets that is less than the maximum permitted, transparency (using zero-bit stuffing), frame multiplexing and demultiplexing (using the address field), and detection of transmission errors (using CRC). Flow control (using windowing) and error correction (using ARQ) are not included.

large-carrier amplitude modulation: modulation technique in which the carrier signal is multiplied by the factor $[1 + \mu m(t)]$. Known as the *modulation index*, $\mu \leq 1$. It controls the *depth* of modulation. $m(t)$ is the message signal. The information sidebands are distributed about the carrier signal. The modulated wave is double-side-band, and contains a large carrier signal component. The transmission bandwidth is twice the baseband or message signal bandwidth.

least-cost routing: automatic route advancement that is coupled with the requirement to start with the cheapest service and advance to the most expensive.

linear block code: a block of k information bits is presented to the encoder; the encoder responds with a unique code block of m bits ($m > k$). The set of code words is selected so that the logical distance between each code word and its neighbors is approximately the same.

linear prediction model: in digital voice, speech analysis and synthesis are performed using models in which successive samples of speech are compared with previous samples to obtain a set of LPCs that are used to reproduce the sound. In effect the transmitter listens to the input speech, calculates LPC strings, and transmits them to the receiver. It regenerates the sound using an artificial vocal tract. Reducing voice to a string of codes representing these sounds can produce a signal of the order of a few hundred bits per second.

line overhead: in SONET, part of the transport overhead that locates the first octet of the synchronous payload envelope (pointer function), and monitors line errors and communications between line-terminating equipment.

link-by-link encryption: the data link PDU is encrypted as it enters the physical layer.

local access and transport area: grouping of contiguous exchange areas served by a single local exchange carrier.

local-area network: bus- or ring-connected, limited-distance network that serves the data communication needs of users within a building or several buildings in proximity to one another.

local exchange carrier: company that terminates a collection of subscriber lines (the local loop) at a serving switch (the central office) or other routing device. Provides transport services directly between end-offices in a local access and transport area. Delivers traffic directed to an exchange area outside the LATA to interexchange carriers.

local network: star or hierarchically connected network provided by a local exchange carrier company to transport voice and data traffic within a defined area and to connect with networks of interexchange carrier companies.

locked: in network management, administrative state in which the managed object cannot be used.

logical channel: communication channel identified by a unique binary number. The channel is created when the identifier is assigned, and destroyed when the number is withdrawn.

logical channel number: in a packet network, defines the virtual connection from DTE1 to DCE1.

logical link control sublayer: defines the format and functions of the protocol data unit that is passed between service access points in the source and destination stations over an IEEE 802 LAN.

logical unit: in SNA, a software entity through which end-users communicate. In subarea networks, they are used only for end-user communication; in APPN, LUs are used for end-user communication and to support network management functions.

long-distance network: mesh or hierarchically connected network that spans states, countries, continents, or the world.

longitudinal redundancy check: parity bits are assigned to bit sequences formed from bits in specific positions in a block of characters. They form a block check character.

lossy transmission line: transmission line that takes account of the resistance of the conducting members and the conductance between them.

lost traffic: some blocked calls are likely to be abandoned by the callers and some may be transferred to the public network at the initiative of the caller. In either case, they are lost to the private network.

low Earth orbit: path a few hundred miles (400 to 900 miles) above Earth's surface. The orbit may be polar or inclined, and the footprint is around 1,000 to 3,500 miles in diameter.

low-entry networking node: in SNA APPN, supports simple point-to-point connections between adjacent LEN nodes. When connected to a network node, can communicate with other LEN nodes in the APPN network, and request directory searches and route calculations. LEN nodes contain LUs that support local end-users.

low-pass characteristic: performance such that energy is transmitted from 0 Hz (dc) to *gf* Hz, but not at frequencies above *gf* Hz.

M1: T–carrier multiplexer that combines 24 DS–0s into one DS–1, also known as M24, M24/56, and M24/64.

M1C: T–carrier multiplexer that combines two DS–1s into one DS–1C.

M5: subrate multiplexer with customer access = 5 9.6 kbits/s data channels. Output = DS–0B.

M10: subrate multiplexer with customer access = 10 4.8 kbits/s data channels. Output = DS–0B.

M12: T–carrier multiplexer that combines four DS–1s into one DS–2.

M13 T–carrier multiplexer that combines 28 DS–1s into 1 DS–3.

M20: subrate multiplexer with customer access = 20 2.4 kbits/s data channels. Output = DS–0B.

M23: T–carrier multiplexer that combines seven DS–2s into one DS–3.

M24: original T–1 multiplexer. Customer access = 24 DS–0 voice channels. Per-channel signaling uses eighth bit robbing in

every sixth frame. Framing uses 8,000 bits/s. Output = DS–1.

M24/56: multiplexer with customer access = 23 56 kbits/s data channels. One channel is used for common channel signaling. Output = DS–1.

M24/64: multiplexer with customer access = 23, 64 kbits/s data channels. Employs B8ZS, and one 64 kbits/s channel for data synchronization, alarms, and signaling. Output = DS–1.

M34: T–carrier multiplexer that combines six DS–3s into one DS–4.

M34NA: T–carrier multiplexer that combines three DS–3s into one DS–4NA.

M44: multiplexer with customer access = 44 DS–0 voice channels that are converted to 44, 32 kbits/s voice channels. Two DS–0 channels are used for common channel signaling. Output = DS–1.

M48: multiplexer with customer access = 48 DS–0 voice channels that are converted to 48, 32 kbits/s channels. Per-channel signaling uses eighth bit robbing in every sixth frame. Output = DS–1.

M88: multiplexer with customer access = 88 DS–0 voice channels that are converted to 88, 16 kbits/s voice channels. Two DS–0 channels are used for common channel signaling. Output = DS–1.

M96: multiplexer with customer access = 96, DS–0 voice channels that are converted to 96 16 kbits/s channels. Per-channel signaling uses eighth bit robbing in every sixth frame. Output = DS–1.

managed object: in network management, network entity that is included in the management activity. A managed object may be a switch, a PBX, a router, a multiplexer, a workstation, or other hardware device, or a buffer management routine, a routing algorithm, an access scheme, or other software package.

managed object class: in network management, objects with similar characteristics (i.e., attributes, operations, and notifications).

managing process: in network management, that part of the application process that is concerned with overall management tasks.

management information base: in network management, contains both long-term and short-term information concerning managed objects.

Manchester signal format: a 1 is a positive current pulse that changes to a negative current pulse of equal magnitude, and a 0 is a negative current pulse that changes to a positive current pulse. The changeover occurs exactly at the middle of the time-slot.

Manhattan street network: metropolitan area network that consists of orthogonal rings that pass information in alternating directions.

many to many: connects many locations to many locations.

m-**ary modulation**: digital modulation technique in which M distinct symbols are employed; each represents $\log_2 M$ bits. The symbols are generated by changing the amplitude, phase, or frequency (alone, or in combination) of a carrier (either a single carrier or a coherent quadrature-carrier arrangement) in M discrete ways. M-ary signaling schemes are used to transmit datastreams over bandpass channels when the desired data rate is more than can be achieved using binary schemes.

mass communication: telecommunication in which information flows simultaneously from a single (transmitting) site to a large number of (receiving) sites.

matched node: in SONET, pairs of MNs are used to interconnect SONET rings and provide alternate paths for recovery in case of link failure. SONET traffic is duplicated and sent over two paths between the rings. One set of MNs provides the active path; the other set is on standby in case of failure of the active connection. Pairs of DCCs or ADMs realize MNs.

maximum-length sequence: *pn*-sequence produced by an *m*-stage feedback shift-register whose period is $2^m - 1$ chips.

medium: entity that produces action at a distance.

medium access control sublayer: in LANs, defines the format and functions of headers and trailers that are added to PDUs so that entire frames can travel between source and destination addresses over IEEE 802 LANs.

medium Earth orbit: path about 6,000 miles above the surface of Earth. The orbit is may be polar or inclined and the footprint is around 5 to 6,000 miles in diameter.

mesh network: in a fully interconnected mesh, each node is connected to every other node.

message-framed data: characters are run together without start and stop bits to form a block of data. A start sequence and a stop sequence frame the block. The start sequence is called the header—it contains synchronizing, address, and control information. The stop sequence is called the trailer—it contains error-checking and terminating information. The entire data entity is called a frame. In response to a command from external equipment, the frame is transmitted as a continuous stream.

message-switched network: network in which complete data messages are received, stored, and, when facilities are available and the receiving station is ready, transmitted to their destinations.

message transfer agents: in X.400, agents that interact directly with user agents and other message transfer agents. They route messages, relay messages, and store messages as required.

message transfer part: SS7 protocol that provides reliable transfer of signaling information across the signaling network. This includes the capability to respond to network and system failures to ensure that reliable transfer is maintained.

message transfer service: in X.400, service (store, forward, and store) that delivers a message as soon as the receiver is ready to receive it. At the receiver, the message is stored until called for.

metropolitan area network: described by IEEE 802.6, a facility that spans an urban core, metropolitan area, or major industrial campus. Connecting the premises of many organizations, MANs provide the opportunity to employ public, switched, high-speed facilities to transport data, voice, and video.

microcell: in wireless networks, a small cell.

Microcom networking protocol: by assembling asynchronous data in blocks and performing cyclic redundancy checks, MNP permits the receiving modem to verify the integrity of the received stream. When an error is detected, the receiving modem requests retransmission of the frame. Versions MNP-1 through 4 provide increasingly complex procedures for error detection and correction. MNP-5 uses an adaptive algorithm to compress the data by approximately 2:1. MNP-7 uses entropy coding to achieve compressions of approximately 3:1. MNP-configured products exist in several versions (known as Classes) that focus on different operating environments.

minimum-shift keying: digital modulation technique in which advantage is taken of phase continuity and the waveforms representing the datastreams impressed on the quadrature carriers are sinusoidal (rather than NRZ pulses). Employing one-half of the frequency spacing used in (Sunde's) FSK, the main lobe of the PSD diagram spans a frequency range that is 25% less than ASK and 50% less than FSK. The power in the sidelobes drops off very quickly; more quickly than FSK and much more quickly than ASK.

mobile: mobile station in a mobile telephone network.

mobile agents: also known as **mobile objects** or **mobile executing units**. Self-contained programs designed to perform specific tasks for users, other agents, and other entities. A mobile agent may be a continuously executing program whose execution is interrupted momentarily when transported from one location to another, or it may be a program that runs to completion before being moved. In addition, an external device may initiate movement, or it may be triggered from within the program. Mobile agents travel between mobile agent systems.

mobile agent system: combination client-server system situated on a host that runs, sends, and receives the programs that are mobile agents. It provides access to the environment in which the mobile agent runs.

mobile assisted handover: the mobile monitors conditions on active channels in different sectors of the same cell and in surrounding cells. It reports the findings to the network on a slow-associated control channel in an active radio channel. The base controller determines which channel the mobile shall use and issues a handover command on the fast-associated control channel.

mobile carrier: company that operates terrestrial cellular radio facility to provide communication connections to (mostly) mobile subscribers in developed areas. Also, a company that operates a communication satellite system to provide communication connections to (mostly) mobile subscribers.

mobile telephone: telephone instrument in which part of the path between instrument and mobile telephone switching office is completed by radio. The radio link replaces the local loop so that the user (mobile) can roam freely.

mobile telephone switching office: in wireless networks, electronic switching system programmed to provide call-processing and system fault detection and diagnostics. The MTSO is connected directly by wireline, or microwave radio, to a set of base stations that communicate by radio with the mobiles in their cells. Also called mobile telephone serving office.

modem: data circuit-terminating equipment that converts digital signals to analog signals that match the bandwidth available on voice-grade lines.

modified Huffman code: combination of run-length coding and variable-length coding designed for use with facsimile signals.

modified READ code: run lengths are encoded by noting the position of change elements (i.e., where white changes to black, or black changes to white) in relation to a previous change element. Designed for use with facsimile machines.

modulated wave (signal): *see* modulation.

modulating signal: *see* modulation.

modulation: signal-processing technique in which, at the transmitter, one signal (the modulating signal) modifies a property of another signal (the carrier signal) so that a composite wave (the modulated wave) is formed. At the receiver, the modulating signal is recovered from the modulated wave (this action is known as demodulation). The bandwidth of the modulated wave is equal to, or greater than, the bandwidth of the modulating signal.

monitor bit: in an IEEE 802.5 LAN, bit used to detect unclaimed frames.

multiple path switching fabric: in ATM, several paths are available between each input/output pair.

multiplexer: equipment that combines digital signals (voice or data) into a TDM signal.

multiplexing: action of placing several signal streams on a single bearer.

multipoint to multipoint: connects many locations to many locations.

multiprotocol router: processes SNA, TCP/IP, Novell's IPX, and other frames over common transmission facilities. In effect, the router consists of separate parts that contain routing tables and logic peculiar to each protocol.

narrowband signal: an analog signal whose frequency components are contained within the nominal passband (4 kHz) of a voice channel.

National Institute for Standards and Technology (NIST): government unit (formerly the National Bureau of Standards) charged with providing an orderly basis for the conduct of business by maintaining measurement standards. Coordinates voluntary product standards. Develops computer software and data communication standards.

near-end crosstalk: crosstalk whose energy flows in the opposite direction to that of the signal energy in the disturbing channel.

network: set of links attached to nodes. The nodes serve to receive traffic from one link and send it on another.

network accessible unit: in SNA, software implementation of end-to-end protocols that implement the transmission control layer, the data flow control layer, the presentation services layer, and the transaction services layer. (NAU is not a physical unit; several NAUs may run in a single piece of hardware.)

network accessible unit or presentation services layer: layer 6 in SNA, defines and maintains communication procedures for transaction services programs. Implements functions for loading and invoking other transaction services programs, enforces correct syntax use and sequencing restrictions, and, by processing transaction services program commands, controls the interaction between these programs.

network access point: switches that are not equipped with intelligent network switching capabilities may be equipped as network access points. NAP capabilities permit access to a limited set of IN services. The NAP routes the call to an IN switching system, identifies the user, and requests IN treatment.

network ACD: in intelligent networks, enables subscribers to control the destination to which 1-800 calls are routed in order to match peak loads with availability of attendants and provide relief for overflow from congested ACDs.

network address translation: the firewall router is equipped with a table that lists the local IP addresses and the globally unique addresses of intranet clients and servers. Within intranet, the local IP addresses are used; outside intranet, the corresponding globally unique address is used. In this way the firewall router prevents outsiders from discovering the internal addresses and topology of the intranet.

network attendant service: in intelligent networks, routes calls requiring attendant assistance to a central location.

network control point: element in common channel-signaling network that contains databases that are used to implement special services.

Network Control Program: in SNA, provides network information to VTAM, controls cluster controllers and attached terminal nodes, routes data to communications controllers in other subareas, performs as a primary station for communication terminations, polls secondary stations, dials and answers switched lines, detects and corrects errors, maintains operating records, and allocates buffers for data coming from and going to the host, and to other stations.

network header: in OSI reference model, provides destination address (in a format understood by the chosen network) and sequence number.

network hierarchy: ways in which nodes (switching centers, multiplexers, etc.) are arranged to achieve efficient division and routing of different classes of communications traffic.

network information database: in intelligent networks, contains customer and network information.

network interface device: device installed by carrier at network extremity that performs testing, loopback, and powering functions.

network layer: layer 3 in OSI model. Provides end-to-end communications services to the transport layer. Makes use of the underlying data link layer services to establish, maintain, and terminate transmission across the network.

networks links advertisements: list routers connected to a network.

network management: planning, organizing, monitoring, and controlling of activities and resources within a network to provide a level of service that is acceptable to the users, at a cost that is acceptable to the owner.

network node: in SNA APPN, backbone node connected to others by transmission groups. Each network node contains CP software that supports communication between network management software entities to update routing tables, provide directory services, etc.

network protocol data unit: combination of network header and transport PDU.

network quality voice: contains the full range of frequencies without perceptible noise or distortion.

network resource manager: in IN, determines which module provides resources to continue call processing. May give routing instructions to establish connection to module.

network termination 1: digital interface point between ISDN and the customer's equipment. Facilitates functions principally related to the physical layer (i.e., physical and electrical termination of the network at the customer's premises, performance monitoring, power transfer, contention resolution, overvoltage protection, and testing).

network termination 2: digital interface point between ISDN and customer's equipment. Includes enhanced functions associated with the data link layer and the network layer. Facilitates protocol handling, switching, concentrating, and multiplexing functions that are associated with PBXs, cluster controllers, local networks, and multiplexers.

network topology: ways in which nodes and links are interconnected.

Next-Generation Internet: announced in 1997, multiagency U.S. government initiative to build a network 100 times as fast as the present Internet so as to upgrade connections between research universities and national laboratories. vBNS is an important part of this effort.

(N)-layer managed object: in network management, object specifically contained within an individual layer (e.g., data link protocol).

node-network interface standard: defines the forms, formats, and sequences of messages that flow between nodes and communication networks across node-network interfaces.

noise: sum of all the unwanted signals that are added to the message signal in the generation, transmission, reception, and detection processes employed in effecting telecommunication. The difference between the received (corrupted) signal and the intended (ideal) signal.

noncore gateway: in Internet, gateway not administered by Internet network operations center.

nonrepudiation: ensuring that agents cannot reject an action they authorized, or declare it to have no binding force.

nonreturn to zero signal format: currents are maintained for the entire bit period (time slot).

nonswitched network: provides communication services to users through a common facility without resorting to switching.

normalization: process in which a signal is scaled so that its maximum magnitude is 1.

normalized throughput: ratio of the time to deliver only the user's data bits to the time to deliver all bits.

North American Digital Cellular Standard: defined by TIA Interim Standard–136, NA–TDMA operates in two frequency bands. At AMPS frequencies, it provides six digital channels in a 30-kHz channel and employs a combination of analog and digital control channels. Dual-mode terminals operate in analog or digital modes. At PCS frequencies (around 1.9 GHz), NA–TDMA provides six digital channels in a 30-kHz channel and employs digital control channels exclusively. In the digital mode, automatic roaming, mobile-assisted handoff, terminal authentication, and user privacy are provided.

North American numbering plan: numbering plan for North America and certain countries in Caribbean Basin. Within the NANP area, geographic segmentation is accomplished by three-digit area codes.

null compression: data compression technique that reduces the number of zeros transmitted. A null-compression-indicating character is used followed by the number of all-zeros characters in the string.

null modem: cable that routes the transmit data circuit of one DTE to the receive data circuit of the other DTE, and vice versa, to achieve direct connection of DTEs.

Nyquist rate: sampling rate (f_s, in samples/s) exactly twice the highest frequency (W, in Hertz) in a band-limited signal ($0 < f \le W$).

Nyquist's limit: maximum signaling rate over a channel with a passband B Hz is $2B$ baud.

Nyquist's theorem: the signaling rate r in bits/s is related to the passband (B) in Hertz by $r \le 2B$.

octet-level multiplexing: octets from each input line are interleaved on the common bearer in time slots reserved for them. After each sequence of octets (one from each input line) has been serviced, the multiplexer adds a bit to define the frame.

odd parity check: the parity bit makes (or leaves) the number of 1s odd.

offered traffic: number of calls attempted multiplied by the average holding time of the carried traffic.

office automation: use of information technologies to improve the effectiveness of persons performing office-based tasks.

office database: in a switching machine, collection of data unique to a specific switching location that describes the office configuration and customer connections.

off-net call: call for which either the originating or the terminating end of the connection is within a private network, and the other end is in another network.

offset-QPSK: quaternary phase-shift keying in which one of the datastreams is delayed by one *bit* period. Phase changes are limited to 90°; this makes performance in a nonlinear environment significantly better than QPSK.

ones density: number of 1s in an octet, or other sequence. In T–1, the average ones density must be at least 12.5% (i.e., one 1 per octet).

one to all: connects one location to all possible locations.

one to many: connects one location to many other locations.

one to one: connects one location to one other.

on-net call: call for which the originating and terminating stations are within the same private network, and the traffic is carried over the facilities of the network.

open-shortest-path-first-protocol: participating routers are equipped with a database that includes information about system topology and the status of neighboring nodes. Routing is decided on the basis of the shortest route that is compatible with delay, throughput, security, and other requirements. Periodically, an updated database is broadcast to all routers in the system.

open systems architecture: permits transparent communication between cooperating machines.

operating time: time to establish and release a connection. It is the sum of call setup time and time to disconnect.

operations, maintenance, and administration part: SS7 protocol that provides the procedures required to monitor, coordinate, and control all the network resources needed to make communication based on SS7 possible.

operations system: in IN, performs network operations functions such as resource administration, surveillance, testing, traffic management, and data collection.

operator services system: in an SS7 network, facility that provides operator assistance, particularly directory assistance services.

optical carrier level 1: SONET optical carrier equivalent to STS–1.

optical carrier level N: SONET optical carrier equivalent to STS–N.

optical fiber: strand of exceptionally pure glass with a diameter about that of a human hair (125 micron = 0.005 inch). The refractive index varies from the center to the outside in such a way as to guide optical energy along its length.

OSI reference model: model of communication between cooperating machines that incorporates OSA. Consists of seven layers defined by a service definition (i.e., what functions does the layer perform) and a protocol specification (i.e., what procedures are employed to execute the functions).

other common carrier: independent carrier offering network services in competition with AT&T.

out-of-band signaling: the associated signals are analog tones that do not occupy the same bandwidth as the information exchange.

out-of-frame event: in ESF error checking, when two out of four consecutive framing bits are incorrect.

out-of-slot signaling: the associated digital signals use a time slot assigned exclusively for signaling.

outside plant: all facilities situated between the distribution frame of the serving office and the customers' premises.

packet: sequence of bits that is divided into two parts—one contains the user's information, and the other contains control information.

packet assembler and disassembler: defined by ITU–T Recommendation X.3, device that assembles packets from characters received from asynchronous DTEs and disassembles packets into characters for transmission to asynchronous DTEs. Usually, a PAD is implemented in software resident at the boundary of the network.

packet data channel: in GSM, PDCHs are organized in 51-frame multiframes that carry GSM voice-oriented information as well as GPRS information.

packet error rate: in a given time period, number of errored packets received divided by the number of packets sent.

packet filtering: after checking the header and contents in accordance with a set of rules (proxies), a firewall router permits or denies the passage of a data packet.

packet insertion rate: in a given time period, for a specific sender-receiver pair, number of packets received that were not sent by the sender divided by the number of packets sent to the receiver in the same time period.

packetized-ensemble-protocol: modem that employs a comb of 512 carrier tones within the voice band to send bits in parallel as QAM symbols.

packet loss rate: in a given time period, for a specific sender-receiver pair, number of packets sent, but not received, divided by the number of packets sent.

packet medium access controller: in FDDI–II, supports data access.

packet switch: network node that receives packets, performs error and congestion control, may store the packets briefly, and sends them on toward their destination over routes determined by information carried in the packet, by information stored at the node, and by the state of the network.

packet-switched network: network in which data messages are divided into segments that are inserted into packets. The packets are statistically multiplexed across links between store-and-forward nodes. Users employ X.25 protocols to connect synchronous DTE to the network. Within the network, the nodes route packets in accordance with information carried in individual packets, information stored in the node, and the state of the network.

pair gain: in the local loop, concentration and multiplexing is added at appropriate points to replace a larger number of underutilized loops by a smaller number of loops that are in use more often.

parity checking: error detection activity that adds a bit (0 or 1) to a sequence of bits in the user's data to make the total number of 1s in the sequence even or odd. Parity bits may be added to individual characters, in which case the process is known as vertical redundancy checking, or to bit sequences formed from bits assigned to specific positions in a block of characters, in which case the process is known as longitudinal redundancy checking.

passband: range of frequencies transmitted without distortion by a bearer and associated equipment.

path control layer: level 3 in SNA, responsible for routing data units through a network to the desired destination. The function is performed in subarea nodes and consists of several steps. Selecting the transmission group to be used between subarea nodes for a specific session. Determining which of the forward and reverse links available in the transmission group are to be used to make the explicit connection between the subarea nodes. Controlling the flow of data over the virtual route formed from all of the explicit links assigned to the session.

path control network: in SNA, layers 1, 2, and 3 of the SNA model implement the path control network. Normally, connections between nodes are implemented on parallel transmission links (transmission groups). To the rest of the network, the group appears to be a single, logical link. To the path control layer, however, each transmission group consists of a set of individual links, some in the forward direction and some in the reverse direction.

path overhead: in SONET, part of synchronous payload envelope that verifies connections, and monitors path errors, receiver status, and communication between path-terminating equipment.

payload: in a frame, data being transported between users.

peak busy hour: selected from a series of daily busy hours, the one that exhibits the highest traffic intensity.

peaked traffic: traffic in which many users demand service at the same time. The variance to mean ratio (α) is greater than 1.

peer polling: stations poll one another, in sequence, continually. If the polled station has something to send, it sends it; if it has nothing ready, it polls the next station in the sequence.

performance management: in network management, evaluate the performance of components, and the network as a whole, with regard to availability, response time, accuracy, throughput, and utilization.

periodic function: eternal function that repeats at regular intervals.

peripheral node: in SNA, terminal or cluster controller assigned to a specific subarea.

permanent virtual circuit: virtual circuit that is permanently assigned between two sta-

tions. Fixed logical channel numbers are used by the sending DTE and the receiving DCE. Because it is a permanent (virtual) connection, no connect or disconnect packets are required to set up the circuit, and information packets may be transmitted at will by the originating DTE. Commonly, PVCs are provided over data networks for longer messages that are not segmented, or segmented messages that must arrive in sequence and are sensitive to variations in intersegment delays.

Personal Access Communications System: serves pedestrians with inexpensive, lightweight terminals, and vehicles moving at moderate speeds, in urban and suburban areas. Operating in the 1.9-GHz PCS band, PACS employs FD/TDMA/Frequency Division Duplex channels. Each carrier is time divided into 2.5-millisecond frames that contain eight time slots. A forward channel consists of one time slot in each frame on a carrier in one band of carriers; the reverse channel consists of one time slot in each frame on a carrier in the other band of carriers. Slots are synchronized on all carriers. Forward channel slots precede reverse channel slots by 375 μseconds (1 time slot = 312.5 μs) so that a terminal does not transmit and receive simultaneously.

personal communication networks: first used in the United Kingdom, and then adopted in Europe, term employed to identify forms of metropolitan area portable radio telephony networks.

personal communication services: term employed by the FCC to mean a family of services provided over mobile or portable radio communications facilities for individuals and businesses on the move. Divided in three categories (narrowband, broadband, and unlicensed), the services will be provided in an integrated manner over a variety of competing networks

Personal Handyphone Service: PCS system developed by NTT. PHS provides wireless access to public and private communication networks for pedestrians with portable terminals, and, unique among PCS systems, it provides direct wireless communication between portable terminals.

personal services: voice, video, and document services used for the exchange of information among persons, and between persons and machines.

personal telecommunication: communication at a distance that involves the distribution or exchange of messages among persons, or between persons and machines.

phase modulation: modulation produced by varying the phase angle of the carrier in sympathy with the amplitude of the message signal.

phase-reversal keying: phase-shift keying that employs two states. The presence of the carrier signal in one phase condition corresponds to a 1-state, and the presence of the carrier signal 180° out of phase with it corresponds to a 0-state. Also known as binary-PSK.

phase-shift keying: digital modulation technique in which the transmitted signal is a sinusoid of constant amplitude whose phase is varied by the modulating datastream. When two states are employed, the presence of the carrier signal in one phase condition corresponds to a 1-state, and the presence of the carrier signal 180° out of phase with it corresponds to a 0-state. This is known as *binary*-PSK and sometimes as *phase-reversal* keying.

phoneme: elementary human sound uttered at rates up to 10 per second.

physical control layer: layer 1 in SNA, defines the electrical and signal characteristics necessary to establish, maintain, and terminate all physical connections of a link.

physical layer convergence protocol: defined by IEEE 802.6 and the ATM Forum, PLCP governs how ATM cells are carried in a DS–3 stream.

physical medium dependent sublayer: implements physical layer functions that depend on the medium selected (e.g., timing and coding).

physical stage: with regard to using the telephone, establishing a path along which signals can flow between users.

physical unit: in SNA subarea networks, software that enables software in managed entities to communicate with network management software.

plain old telephone service: transport and switching service for analog voice signals.

plesiochronous multiplexing: multiplexing process in which two or more signals are generated at nominally the same digital rate and their significant instants occur at nominally the same time.

point of presence: facility interface between LEC and IXC. A POP must be established on each trunk path that connects switches that belong to the local and long-distance carriers. On one side of the POP, the LEC is responsible for service; on the other side, the IXC is responsible.

point to all: connects one location to all possible locations.

point to multipoint: connects one location to many other locations.

point to point: connects one location to one other location.

polar earth orbit: orbit in which satellites travel around Earth, passing over the poles.

power spectral density: plot of the variation of signal power (in watts per Hertz) with frequency.

presentation header: in OSIRM, identifies the specific encoding, compression and encryption employed.

presentation layer: layer 6 of OSI model. Translates the information to be exchanged into terms that are understood by the end-systems.

presentation protocol data unit: combination of presentation header and application PDU.

pretty good privacy: combination of SKC and PKC for protecting electronic mail. The sender encrypts the message using a randomly generated key, then encrypts the key using the recipient's public key. The recipient uses his or her private key to decipher the random key, then uses the random key to decipher the message.

preventive cyclic retransmission: SS7 error control technique that uses forward error correction. It is employed on satellite links.

primary polling: in turn, the primary station asks each of the secondary stations if they have anything to send. If a secondary station has a frame(s) to send, it transmits it (them) to the primary, when polled; if the secondary station has nothing ready to send, it indicates nothing to send and the primary station polls the next station.

primary rate interface: ISDN channel consisting of 23 B-channels + 1 D-channel. The PRI is supplied over a single connection that operates at 1.544 Mbits/s. This is the T-1 rate; it contains 1.536 Mbits/s of user and signaling data, and 8 kbits/s for framing the bitstream.

primary reference clock: in network synchronization, timing source that establishes exact time. In AT&T's network, PRCs consists of an ensemble of primary or secondary atomic standards that achieve a frequency accuracy of better than 1×10^{-11}, and permit no more than two slips per year.

primary station: station that determines when secondary stations shall send, and when they shall receive.

priority bits: in an IEEE 802.5 LAN, they identify the level of priority a station must have to seize the token.

privacy: maintaining the confidentiality of data from agents not authorized to receive them.

private branch exchange: circuit switch that serves a community of stations. Usually, the community is located in a limited area that includes the PBX.

private network: created by a corporate entity. It is used to complete some fraction of the voice, data, image, and video communications employed in running the business. Regardless of who owns them, the facilities employed in private networks provide communication services for the benefit of the corporate community.

probabilistic function: function whose past and future values are described in statistical terms.

procedural stage: with regard to using the telephone, ensuring that the information is in a common language and is exchanged between the right parties.

process gain: in spread spectrum modulation, the ratio of the output signal-to-noise power ratio to the input signal-to-noise power ratio.

protocol control information: information exchanged by peer entities to perform certain tasks or settle on a format.

protocol data unit: defined by IEEE 802.2, frames exchanged between data link layers in a LAN. Also, the frames exchanged between peer entities in any protocol stack. The combination of SDU and PCI.

protocol specification: part of the layer standard, describes the procedures used within the layer, and between peer entities, to execute the functions defined by the service definition.

proxy: rule used by a firewall router to permit or deny the passage of a packet across the firewall.

pseudo-noise sequence: periodic binary sequence with a noiselike distribution of 1s and 0s in which the number of 1s is equal to the number of 0s plus or minus 1.

public-key cryptography: uses a secret key known only to the sender and a public key that is available to anyone who wants it.

public network: provides voice, data, image, or video communications services to any person with the ability to pay the fee published in the tariff describing the service. The person paying the fee controls the purpose and content of the communication, and benefits from the exchange.

pulse amplitude modulation: stream of pulses whose amplitudes are varied according to some rule.

pulse code modulation: form of digital voice that provides communications-quality voice signals. Based on sampling at 8,000 times per second, the 64 kbits/s analog signal reconstructed from the PCM stream cannot contain frequencies above 4,000 Hz.

quadrature binary modulation: using the quadrature property of orthogonal functions, the symbol rate can be doubled. By dividing the data into two streams (even-numbered bits and odd-numbered bits) of symbols that are 2 bit periods long, and using them to produce binary-shift keying of sine and cosine carriers (derived from the same source), a composite wave containing $2r$ bits/s is created. It occupies r Hz, the same bandwidth as a single wave carrying r bits/s.

quadrature-carrier amplitude modulation: digital modulation technique that employs orthogonal carriers derived from the same source. By dividing the datastream in two streams (even-numbered bits and odd-numbered bits) that amplitude modulate orthogonal carriers symbols are transmitted at rate $2r$ in a bandwidth of r Hz.

quadrature phase-shift keying: digital modulation technique that employs orthogonal sinusoidal carriers derived from a common source and phase-shift keying. The datastream is divided in two streams (of rate r symbols/s) that phase modulate the orthogonal carriers. By transmitting symbols at rate $2r$ symbols/s in a bandwidth of r Hz a spectral efficiency of 2 bits/s/Hz is achieved.

quadriphase phase-shift keying: *see* quadrature phase-shift keying.

quantizing: process that segregates sample values into ranges and assigns a unique,

discrete identifier to each range. Whenever a sample value falls within a range, the output is the discrete identifier assigned to the range.

quantizing noise signal: in digital voice, difference between the signal at the output of the reconstruction filter, and the original band-limited signal, assuming no errors are introduced in transmission.

quaternary-PSK: *see* quadrature phase-shift keying.

Radiocommunications Sector: sector of the ITU concerned with radio issues and the allocation of the electromagnetic spectrum.

raised-cosine pulse: signal consisting of one period of a cosine function ($-\pi$ to $+\pi$ radians) offset by its amplitude so that the values are always ≥ 0.

random function: probabilistic function whose values are limited to a given range. Over a long time, each value within the range will occur as frequently as any other value.

random traffic: traffic in which the calling pattern is random and the variance-to-mean ratio (α) is 1.

rate-based flow control: on the basis of the amount of traffic it is receiving, the receiver controls the rate at which input nodes accept traffic intended for it. As a result, messages may be queued at the edges of the network, and at the receiving node, but not in intervening nodes. In effect, rate-based control stores traffic at source nodes until the addressed node is ready to receive them.

reachability: routing information concerning how to reach individual hosts in an autonomous system.

real-time medium: entity that makes it possible to communicate at a distance at the same time.

reconstruction filter: in digital voice, low-pass filter that passes frequencies up to W Hz.

The input is a train of pulses whose amplitudes correspond to sample values of a signal limited to W Hz, and whose repetition rate corresponds to the sampling rate ($\geq 2W$ samples/s). The output is a close likeness of the original band-limited signal ($0 < f \leq W$ Hz).

record medium: by storing a limited amount of information, entity that supports communication at a later time. If the medium can be transported readily, communication can occur at another place, at a later time. In addition, if the record medium can be duplicated in quantity, independent communication can occur with a large number of persons, at a distance, at their convenience.

regeneration: strategy to overcome pulse distortion and achieve higher signaling rates on transmission lines carrying digital signals. The distance over which the pulse stream is transmitted before processing is reduced. This is done by introducing regenerators (also known as repeaters) at regular intervals along the transmission path. Their purpose is to read the pulse stream before it degrades substantially, generate a new stream, and pass it on to the next unit.

regenerator: *see* regeneration.

Reed-Solomon code: block code that employs groups of bits, not single bits. Known as symbols, the RS code has k' information symbols, r' parity symbols, and codewords of length $n' = k' + r'$ symbols. The number of symbols in a codeword is arranged to be $n' = 2^m - 1$ where m is the symbol length. The code is able to correct errors in $r'/2$ symbols. Reed-Solomon codes are not efficient codes for correcting random errors.

Regional Bell Operating Company: independent carrier that provides local exchange services within specific areas.

remote operations service element: in network management, general software support system that supports distributed, interactive applications processing between remote objects. In synchronous mode, ROSE requires the receiver to reply before the originator can invoke another operation. In asynchronous mode, ROSE permits the originator to continue invoking operations without waiting for a response.

repeater: device that links two networks together when their differences are limited to the implementation of their physical layers. Also *see* regeneration.

reservation-Aloha: in satellite communication, the Aloha process is modified to allow stations to reserve time slots. The technique guarantees that messages will arrive one at a time at the transponder and will be relayed without collisions.

reservation bits: in an IEEE 802.5 LAN, provide a mechanism for devices to request the opportunity to transmit.

residual error rate: total number of frames sent minus the number of good frames received divided by the total number of frames sent.

resistance design: in the local loop, a combination of wire sizes is chosen for each customer loop so that the total resistance does not exceed 1,300 ohms, the value that allows sufficient current to flow from the serving office battery for ringing.

response: in SDLC, frame that passes from a secondary station to the primary station.

resource management cell: in available bit-rate service (cell relay), resource management cells are sent from sender to receiver (*forward* RM cells), and then turned around to return to the sender (*backward* RM cells). Along the way, they provide rate information to the nodal processors and receive congestion notification(s).

retried traffic: blocked calls that are placed again at later times at the discretion of the callers.

return to zero: signal format in which bits are represented by a positive or a negative current pulse that returns to zero before the next bit in the sequence is handled.

richness: a message is said to be richer than another when it includes more message dimensions. Thus, video messages are richer than audio messages, audio messages are richer than graphics messages, and graphics messages are richer than text messages.

ring network: each station is connected to two other stations so that there is a continuous, single-thread connection among all units.

roaming: in mobile radio, condition that exists when the SID number of a mobile station does not match the SID number of the cell site receiving requests for service.

route protection: arrangement to recover transmission capacity should a route be interrupted.

routing information protocol: in Internet, the gateway selects a route that includes a minimum number of intermediate hops.

router: if their differences reach no higher than the network layers, a device that connects networks at the network level and operates with network layer addresses.

router links advertisements: distributed to internal routers, they contain information on the links in the area.

route selection service: in SNA APPN, provided by the CP in the network node in response to a request from an LU. The CP calculates an optimal route on the basis of the locations of LU1 and LU2 and information obtained from directory services.

run-length compression: superset of null compression that is used to compress any string of repeating characters.

sampling: process of determining the value of the amplitude of an analog signal at an instant in time. Usually, sampling is done in a regular repetitive manner at equal time intervals.

scoping: in network management, identifies the subtree of managed objects to which a particular filter is applied.

scrambling: action that breaks up long strings of the same symbol, or repeated patterns of symbols, and makes the signal train more random. The datastream is combined with a semirandom sequence to produce an acceptable ratio of 1s and 0s. After the scrambled stream is transmitted from sender to receiver, the receiver reconstructs the original datastream.

secret-key cryptography: uses a single secret key known only to the sender and receiver.

section overhead: in SONET, part of the transport overhead that is concerned with framing, frame identification, section error monitoring and communication between section-terminating equipment.

security management: in network management, control access to, and use of, the network according to rules established by the network owner.

segmentation and reassembly sublayer: component of ATM adaptation layer. At the sending end, receives 47-octet blocks from convergence sublayer, adds 1-octet headers to form SAR-PDUs, and delivers them to ATM layer. At the receiving end, receives 48-octet SAR-PDUs from ATM layer, strips 1-octet headers, and delivers 47-octet blocks to convergence sublayer.

selective dynamic overload: protective routing scheme that relieves congestion at an overloaded switch by sending an instruction to all switches connected to it to reduce the volume of their requests for service for a fixed time interval.

selective-repeat: an ARQ technique used on duplex connections, the sender only repeats those frames for which negative acknowledgments are received from the receiver, or no acknowledgment is received.

selective trunk reservation: protective routing scheme. When the number of idle trunks in a particular group drops below a specified threshold, the idle trunks remaining are used preferentially to complete single-link (first-routed) calls.

self-information: when a source emits a sequence of symbols, the self-information conveyed by each symbol is equal to the logarithm of the inverse of the probability it will be sent.

sequenced packet exchange: transport layer in Novell's Netware protocol stack. SPX provides virtual circuit service, uses a checksum for error detection, sends positive acknowledgments, on timeout retransmits packet, ensures sequencing, and provides flow control.

server: major data processing device that maintains databases and delivers data-files to clients, on demand.

service access point: point within the sending or receiving device that acts as the logical port for the exchange of data.

service control function: contains service logic that controls the implementation of IN services. Interfaces with service switching function, specialized resource function, and service data function.

service control point: in intelligent network, consists of processor-based service logic interpreter, network information data-bases, signaling network interface, and network resource manager. In an SS7 network, a facility that provides database and call processing procedures for special calls.

service data unit: consists of user data and control information created in the upper layers of the protocol stack.

service definition: part of the layer standard, describes the functions the layer performs and what services it provides (to the next upper layer).

service logic interpreter: in intelligent network, executes service logic programs and handles exchanges between IN modules.

service logic program: in IN, defines the operation of specific network features or service application. SLPs reside in service control points, adjunct processors, intelligent peripherals, and service nodes, or within the network switches themselves.

service node: in intelligent networks, provides service logic and implementing resources in a single network entity. Contains adjunct service point that responds to requests for service processing. Provides access to service logic programs that support a specific service, or group of services, or a full range of service functions (including attendant services).

service plane: describes IN services from a user's perspective.

services plane: plane containing orthogonal axes that divide real-time media (right-hand side) from record media (left-hand side), and intercommunication (upper half) from mass communication (lower half). The axes are marked off in richness in both directions from the crossing point, and the horizontal axis represents the richness in the upstream direction, while the vertical axis represents the richness in the downstream direction

service switching function: recognizes calls requiring intelligent network features. Interacts with call processing and service logic to provide the services requested.

service switching point: in an SS7 network, facility embedded in the switching point that processes calls that require remote database translations. The SSP recognizes special call types, communicates with the service control point, and handles the calls according to SCP instructions.

serving area: contains all of the premises connected to a serving area interface. An area served by a group of distribution cables.

serving area interface: cross connect point connecting a feeder cable to the distribution cables that serve a serving area.

session: in SNA, temporary logical connection between network-accessible units that exists for as long as the participating NAUs communicate. Sessions may be initiated for network management reasons, or for end-user reasons.

session header: in OSIRM, added to presentation PDU to define line discipline, direct establishment, maintenance, and termination of connection, and to identify any specific markers employed.

session protocol data unit: in OSIRM, combination of session header and presentation PDU.

setup channel: in mobile telephone networks, channel over which the signaling messages used to set up all calls are transmitted.

severely errored second: in ESF error checking, any second in which from 320 to 333 CRC-6 error events are detected, or one OOF event occurs.

Shannon-Fano code: code in which codeword lengths increase as the symbol self-information increases, and no member of the code is the prefix of any other member.

shutting down: in network management, administrative state in which the managed object will honor current users, but will not accept new users.

sights: static or dynamic relationships among objects arranged in space. Likely to present vistas, to witness events, or to assist with the visualization of ideas.

signal: manifestation of a message exchanged between points at a distance.

signal constellation: polar diagram in which points represent the phase and amplitude of QAM signals.

signaling: name given to the exchanges of messages that inform the users, and the telecommunications facilities, of what actions to take in order to set up, maintain, terminate, and bill a call.

signaling connection control part: SS7 protocol that enhances the services of the MTP to provide full OSI network layer capabilities. In particular, SCCP extends MTP's addressing and routing capabilities and provides connectionless (CL-SCCP) and connection-oriented (CO-SCCP) classes of services.

signaling end-point: SS7 node connected by high-speed data links to a network of signal transfer points. Can be a switching point, a service switching point, a service control point, an ISDN service node, or an operator services system.

signaling messages: signaling messages are of four kinds; alerting, supervising, controlling, and addressing.

signaling rate: number of independent symbols sent per second.

Signaling System 7: common-channel, network signaling system that supports normal message transfer activities as well as user-to-user signaling, closed-user groups, calling line identification, call forwarding, etc. The basic network architecture is a quad structure consisting of signaling end-points and signal transfer points. To provide added reliability, each SEP is connected to the SS7 network by duplicate STPs.

signaling tone: in mobile telephone networks, out-of-band tone burst (10 kHz). Among other things, the signaling tone is used to initiate ringing, to signal handoff to another cell site (tune the mobile to another set of frequencies), and to disconnect.

signal-to-noise ratio: ratio of the power in the message signal to the power in the noise signal. It is usual to express this ratio as a logarithm. The unit ratio is 1 decibel (db).

signal transfer point: facility that performs as a link concentrator and message switcher to interconnect signaling end-points. Routes signaling messages directly to the terminating switch, or directly to the STP that serves the terminating switch.

signal unit error rate monitor: device that monitors error performance of SS7 links.

Simple Network Management Protocol: support software that monitors and manages network elements (i.e., router/gateways, servers, multiplexers, etc.) in an internetwork environment. SNMP collects network statistics, changes network routing tables, and modifies the states of links and devices. SNMPv1 runs over a connectionless transport layer (UDP), describes a framework for formatting and storing management information, and defines management information elements. SNMPv2 adds a wider range of commands. SNMPv3 adds improved privacy and authentication features.

simplex: information flows in only one direction (a single channel is employed).

single-channel per carrier: demand assignment, multiple-access technique in which each carrier is used to provide a single channel.

single-line digital subscriber line: provides duplex service over one UTP at 1.544 Mbits/s and a 4-kHz telephone channel. With 24-gauge copper pairs, service is limited to 10,000 feet.

single-path switching fabric: in ATM, a unique path is available between each input/output pair.

single-sideband modulation: efficient, bandwidth-minimizing modulation technique in which all of the energy transmitted is devoted to a single copy of the message. Because the modulating technique produces sharp peak values when the message signal makes abrupt transitions, SSB cannot be used with digital signals. In addition, signals (such as television) that contain significant energy at low-frequencies cannot use SSB. When the higher-frequency sideband is employed, the modulated wave is described as upper-sideband SSB; when the lower-frequency sideband is employed, the modulated wave is described as lower-sideband SSB. The transmission bandwidth is the same as the bandwidth of the message signal.

skin effect: at low frequencies, the current flowing in a conductor is distributed uniformly over its cross-sectional area. As the operating frequency is raised the distribution is disturbed, the current density falls

at the center, and begins to increase in the outer layer of the conductor. At very high frequencies, the current flows along the *skin*.

sliding window flow control: process in which the transmitter can transmit up to W packets and must then wait for an acknowledgment from the receiving DTE. The value of W is negotiated between sending and receiving DTEs.

slip: in network synchronization, occurs when a timing error causes a time slot to be repeated or deleted.

slot: in an IEEE 802.6 MAN, consist of 53 octets (52 octet payload + 1 octet header). The number of slots depends on the data rate (13 slots at T–3, 45 slots at STS–3). When permitted, nodes write data to the slots and read and copy data from them.

slotted-Aloha: in satellite communications, when packets are sent to arrive at the transponder at the beginning of agreed-upon time slots. The technique requires that all stations be synchronized in some fashion so that their packets arrive at the transponder only at a slot start time. Under this circumstance, the period of vulnerability to a collision is reduced to one packet time so that the maximum throughput is $1/et_p$ which is twice the value for simple Aloha.

slow-associated control channel: in GSM, the signaling channel that handles call supervision and related activities.

smart terminal: device that contains memory and logic that can be set up to perform specific tasks.

SMDS interface protocol: three-level protocol that addresses, frames, and transmits PDUs using cell-relay techniques.

SMDS network interface: broadband network access arrangement that supports switched, high-speed, connectionless data services. Through the use of a three-level, SMDS interface protocol, SMDS transfers information in variable-length frames up to 9,188 octets over connections that employ cell-relay techniques. Each PDU is routed independently to its destination. Error detection (but not correction) is performed, and corrupted frames are discarded (without requesting retransmission).

smooth traffic: traffic in which calling is continuous. The variance-to-mean ratio (α) is zero.

Sockets: software entity that serves as the interface between the client and TCP/IP protocols.

soft handoff: initiated by the mobile, *soft handoff* is unique to NA–CDMA/IS–95 systems. The terminal makes measurements on the radio environment and informs the base station when handoff is required. The controlling switch selects a forward channel from a new base station and directs it to receive signals from the mobile. For a short while, the mobile receives signals from both cell sites until communication has been established with the new site.

sounds: as message components likely to discuss points, to describe situations, or to inspire feelings. The information is contained in the sequence and intensity of the sounds and in their pitch.

source coding: in digital voice, speech sounds are identified and encoded.

source entropy: average value of information per symbol emitted by a source.

space diversity: satellites are physically separated from each other. Either they occupy different orbits, or, if in the same orbit, they occupy different positions in the orbit.

space-division network: network in which input and output are connected by physical links that provide a single channel. They are neither time-divided nor frequency-divided.

space-time-space switch: combination of space-switching stages and time-switching stage to serve a large customer community. The time stage interconnects the space stage serving the calling party with the space stage serving the called party.

specialized common carrier: carrier authorized to carry data signals over private facilities between major cities.

specialized resource function: in intelligent networks, controls resources used by end-user to access network (e.g., DTMF receivers, protocol conversion, announcements, etc.).

specific management function area: in network management, area of specialized technical responsibility.

spectral efficiency: ratio of the bit rate (in bits/s) of the modulating signal to the transmission bandwidth (in Hz) of the modulated signal (i.e., r_b/B_T bits/s/Hz).

spillover: in digital modulation, name given to power that extends beyond the transmission bandwidth.

Spread Spectrum Cellular System: defined by TIA Interim Standard–95, NA–CDMA operates at AMPS frequencies with 1.3-MHz channels, and at PCS frequencies. Dual-mode terminals operate as AMPS terminals or CDMA terminals. Automatic roaming, mobile-assisted handoff, terminal authentication, and user privacy are provided.

spread spectrum signal: very wideband signal whose energy density is low enough so as not to produce noticeable interference in conventional AM or FM signals using frequencies in the same spectrum space. Produced by direct sequence or frequency-hopping spectrum-spreading techniques.

stand-alone STP: in an SS7 network, signaling point that provides only STP capability, or STP and SCCP capabilities.

standards: documented agreements containing precise criteria to be used as rules, guidelines, or definitions of characteristics. Their consistent application ensures that materials, products, processes, and services are fit for their purposes.

star network: all links radiate from a central node so that all user stations are connected to it, and all information exchanges must transit this node.

start-stop, data link control protocol: procedure that manages an asynchronous stream of characters framed by start and stop bits.

station signaling: exchange of signaling messages over local loops that connect the users' stations to the serving switch; station signaling is limited to an exchange area.

statistical multiplexing: *see* asynchronous multiplexing.

stop and wait: ARQ technique in which the sender sends a frame and waits for acknowledgment from the receiver.

store and forward: operation in which messages are generated, sent to an intermediate location, stored, and forwarded to the recipient on demand.

Stratum 1: in AT&T's network, group of 16 free-running primary reference clocks.

Stratum 2: in AT&T's network, a group of clocks located in toll offices. Disciplined by PRCs, they achieve a frequency accuracy of better than 1.6×10^{-8}, and permit no more than 10 slips per day.

Stratum 3: in AT&T's network, a group of clocks located in local networks. Disciplined by clocks in Stratum 2, they achieve a frequency accuracy of better than 4.6×10^{-6}, and permit no more than 130 slips per hour.

Stratum 4: group of clocks located in private networks. Disciplined by clocks in Stratum 3, they achieve a frequency accuracy of better than 3.2×10^{-5}, and permit no more than 15 slips per minute.

subarea: in SNA, network address associated with a node (subarea node) and its attached peripheral nodes.

subarea node: in SNA, communications controller or host.

subband coding: in digital voice, the input signal frequency band is divided into several subbands (typically four or eight). Each subband is encoded using an adaptive step size PCM (APCM) technique.

subnetwork access functions : sublayer of ISO's network layer protocol, concerned with data transfer within individual networks. The SNAF sublayer defines how network layer entities make use of the functions networks provide.

subnetwork-dependent convergence functions: sublayer of ISO's network layer protocol, concerned with correcting network anomalies and deficiencies. The SNDCF sublayer is included to accommodate situations in which some networks do not provide all the services assumed by SNICF.

subnetwork-independent convergence functions: sublayer of ISO's network layer protocol, concerned with internetwork routing and relay functions, and global network protocol. The SNICF sublayer contains functions that implement internetwork protocols; they do not depend on specific networks.

subrate data: data at 2.4, 4.8, 9.6, and 19.2 kbits/s.

subscriber code: identifies an individual station line that connects the subscriber to the serving switching facilities.

subscriber identity module: in GSM, the user has a subscriber identity module that is inserted at the mobile station. SIM carries the personal number assigned to the user. GSM associates it with the registered number of the mobile station and thus can route calls to the subscriber.

subscriber line interface circuit: in the local loop, circuit responsible for two- to four-wire conversion, signaling detection, and battery feed.

subscriber loop carrier: digital loop carrier.

subscriber terminal 1: terminal that is fully compliant with ISDN requirements. Connects directly to NT1.

subscriber terminal 2: terminal that does not fully comply with ISDN requirements. Must connect to NT1 through a terminal adapter.

subscription teletext: specialized information service for the dissemination of business and other commercially valuable information that can be used only by those who pay a fee.

substitution: in encryption, action in which new bit patterns, characters, words, or phrases are substituted for the originals without changing their order.

summary links advertisments: sent by border routers, they contain information on routes outside an area

superframe format: also called D4 superframe format, 12 DS-1 frames form a superframe that contains 2,316 bits and takes 1.5 milliseconds to transmit at the DS-1 rate. The 12 framing bits form a pattern that is used for synchronization and other purposes.

supervisory audio tone: in mobile telephone networks, out-of-band, continuous tone (either 5,700, 6,000 or 6,300 Hz). The SAT is present whenever a call is in progress. Its absence indicates that one of the parties has gone on hook, and the call is terminated.

SW56: 56 kbits/s switched digital service; this is switched DDS.

SW64: 64 kbits/s switched digital service; this is switched DS–0 or ISDN B channel.

SW384: 384 kbits/s switched digital service; this is ISDN H0.

SW1536: 1,536 kbits/s switched digital service; this is ISDN H11.

switched Ethernet: a switch replaces the Ethernet hub so that the two stations involved in a message transfer are connected directly over a high-speed channel. Thus, collisions are eliminated, stations do not have to wait for the bus to be quiet, and stations can operate at full bit rate.

Switched Multimegabit Data Service: high-speed, public, packet switched, connectionless data service. SMDS supports higher access speeds and greater throughputs than frame relay networks, and has no distance limitations.

switched network: consists of transmission equipment that can be interconnected on demand to provide individual communication channels between many independent sets of users.

switch fabric: distributed switching matrix in ATM switch.

switching matrix: makes connections so that a circuit is completed between the calling and called parties. The stored program control seeks out a suitable combination of time slots and multiplexed highways to perform this task, and instructs the hardware elements to employ them for the duration of the communication session.

switching point: in an SS7 network, end-office switch at which the customer gains access to the network. Using common-channel signaling, the switch provides normal call-switching functions—establish, maintain, and disconnect.

switching terminal exchange: DCE that implements X.75 procedures to connect to another packet network.

symbol rate: if symbols are generated at a rate of r per second to create a baseband signal with a bandwidth of W Hz, Nyquist showed that $r \leq 2W$. For a double-sideband modulated wave, whose transmission bandwidth is B_T Hz, $B_T = 2W$, so that, $r \leq B_T$.

symbols: numbers, letters, shapes, and logical relationships that specify facts, define entities, and seek to inform.

synchronous: describes activities that occur in synchrony with an external timing source.

synchronous data link control protocol: bit-oriented DLC protocol.

synchronous demodulation: process that recovers the message stream from a modulated wave by using the carrier as a reference. The unmodulated carrier signal may be transmitted separately to the receiver, or the receiver may reconstruct the carrier signal from the modulated wave.

synchronous digital hierarchy: designed to transport B–ISDN signals, employs a basic speed of 155.52 Mbits/s. Higher levels of SDH are multiples of basic speed—i.e., n 155.52 Mbits/s. Compatible with ISDN basic and primary rate signals [i.e., 64 kbits/s (DS-0) and 1.544 Mbits/s (DS–1) or 2.048 Mbits/s (E-1)], and SONET STS–N signals.

synchronous multiplexing: to combine n lower-speed, digital signals onto a higher-speed, single bearer, time slots are assigned to each signal in turn in a regular sequence paced by the system.

synchronous operation: data are sent or actions occur in response to a command from *external* equipment.

Synchronous Optical Network: network that employs optical fibers exclusively. Each facility conforms to standards that include electrical and optical speeds, and frame formats. It uses add and drop multiplexers, digital cross connects, and other terminals, to provide point-to-point, hub, and ring connections. They are used to expand interexchange, carrier serving area, and business customer routes.

synchronous payload envelope: in an STS–1 frame, 9 rows x 87 columns (783 octets) are called the STS–1 synchronous payload envelope (SPE). Of the 783, 9 octets (1 column) are reserved for path overhead information.

synchronous transport module level 1: in SDH, a frame of 19,440 bits that is transported at 155.52 Mbits/s.

synchronous transport module level N: in SDH, a frame of N x 19,440 bits that is transported at N x 155.52 Mbits/s. STM–N frames are created by interleaving N STM–1 frames, octet by octet.

synchronous transport signal level 1: in SONET, a frame of 6,480 bits that is transported at 51.84 Mbits/s. STS–1 signals are designed to carry DS–3 (synchronous or asynchronous) signals, or a combination

of DS–1, DS–1C, and DS–2 signals that is equivalent to DS–3.

synchronous transport signal level N: in SONET, a frame of N x 6,480 bits that is transported at N x 51.84 Mbits/s. STS–N signals are created by interleaving N STS–1 signals, octet by octet. For various reasons, the values N = 3, 12, 24, 48, and 96 are preferred.

synthetic quality voice: contains sounds generated by an electronic analog of the human speech production mechanism. May be more intelligible than communication quality voice, but speaker recognition is missing.

SYNTRAN DS–3: octets from 28, DS–1 channels of exactly 1.544 Mbits/s are multiplexed directly to DS–3. The format of the traditional DS–3 frame is restructured into a synchronous superframe so that DS–1 frames have fixed locations within the superframe. The purpose is to provide simple, reliable access to all DS–0 and DS–1 signals contained in the SYNTRAN DS–3 signal.

system-managed object: in network management, object pertaining to more than a single layer (e.g., router or gateway).

system management application service element: in network management, creates and uses the protocol data units that are exchanged during the management process. It employs communications services supplied by specific ASEs or by CMISE.

system management function: in network management, system function included in specific management function area.

Systems Network Architecture: IBM proprietary network in which the communication functions are tightly integrated with the computing resources.

system services control point: in SNA subarea networks, used by management software to communicate with software in managed entities.

T–1: T–carrier that employs a channel bank that processes 24 analog voice signals to become 24, 64 kbits/s, PCM signals, and octet multiplexes them to produce a bit-stream of 1.544 Mbits/s that is carried on conditioned, four wire circuits. The DS–1 frame contains 24 octets that represent samples from 24 voice signals and a single bit that is used for synchronizing (framing bit).

T–1C: T–carrier that operates at 3.152 Mbits/s. Interleaves two, DS–1 streams, bit by bit, and adds a control bit after every 52 bits (26 from each DS–1). To maintain the 1s density, one DS–1 stream is logically inverted before multiplexing.

T–2: T–carrier that operates at 6.312 Mbits/s. Interleaves four, DS–1 streams, bit by bit, and adds a control bit after every 48 payload bits. To maintain the 1s density, bits in DS–1 streams 2 and 4 are logically inverted before multiplexing.

T–3: T–carrier that operates at 44.736 Mbits/s. Bit interleaves groups of four DS–1s to create seven, DS–2 signals; in turn, the seven, DS–2 signals are bit-interleaved to create one DS 3 signal. In the conversion of seven, DS–2 to DS–3, a control bit is added for every 84 payload bits, and bits are stuffed in specific slots to equalize the speeds of the individual streams.

T–4: T–carrier that operates at 274.176 Mbits/s. Bit interleaves six DS–3 streams. Uses stuffing bits to achieve synchronization.

T–4A: T–carrier that operates at 139.264 Mbits/s. Bit interleaves three DS–3 streams Uses stuffing bits to achieve synchronization.

talkspurt: more or less connected period of vocal activity.

T-carrier hierarchy: multiplexing facilities that combine DS–0 signals into progressively higher speed streams. The four levels are T–1, T–2, T–3, and T–4, with an intermediate level, T–1C, between T–1 and T–2, and another intermediate level, T–4NA, between T–3 and T–4. The output signals are DS–1C, DS–2, DS–3, DS–4NA, and DS–4.

T–carrier multiplexers: produce composite signals that contain 24, 48, 96, 672, 2,016, and 4,032 digital voice signals. Also, provided certain restrictions are observed, they transport data signals. The family of T–carrier equipment extends from 1.544 Mbits/s to 274.176 Mbits/s in four levels.

telecommunication: communication from afar; the action of communicating at a distance.

Telecommunication Technology Committee: develops and disseminates Japanese domestic standards for deregulated technical items and protocols. TTC is composed of representatives of Japanese carriers, equipment manufacturers, and users.

telecommunications: technology of communication at a distance. An enabling technology, it makes it possible for information that is created anywhere to be used everywhere without delay.

telecommunications facility: combination of equipment services and associated support persons (if any) that implements a specific capability for communicating at a distance.

Telecommunications Information Networking Architecture Consortium: association of over 40 of the world's leading network operators, telecommunications equipment manufacturers, and computer manufacturers. They are committed to the definition, validation, and implementation of a common and open software architecture for the provision of telecommunication and information services.

telecommunications access method: in SNA, a queued access method that transfers messages from one remote terminal to another, and between terminals and application programs. Manages a communications network for the exclusive use of the applications it supports.

Telecommunications Development Sector: sector of ITU responsible for promoting the development of technical facilities and services likely to improve the operation of public networks.

Telecommunications Management Network: employs the OSI concepts of manager and agent, and application service elements, to manage telecommunications networks.

telecommunications network: array of facilities that provide custom routing and services for a large number of users so that they may distribute or exchange information simultaneously.

Telecommunications Standardization Sector: sector of ITU that develops and adopts recommendations (consensus standards) designed to facilitate global telecommunication.

telepoint: single base station providing one-way wireless, payphone access into the telephone network. It provides cordless pay phone service to customer-owned handsets within limited range of public base stations in malls and downtown areas.

teleprocessing access method: in SNA, a software interface between the host main memory and its peripherals that assists in moving information between the network and the host over the I/O channel. TP access methods reside in a host and run as independent jobs under the operating system.

teleprocessing monitor: in SNA, to implement real-time, on-line transaction processing, some applications are separated from the access method by a TP monitor that supervises multiple transaction programs running in a single partition.

teletex: simplex record service. A communication service that transports character-based messages to remote sites where they are printed out. Teletex provides direct electronic document distribution from electronic typewriters, word processors, or personal computers that are equipped with communications capabilities.

teletext: electronic magazine. A mass communication information retrieval service that supplies a hundred or more pages in the vertical blanking interval of a broadcast television signal without affecting the television service.

teletypewriter: device that accepts keyboard input, prints hard copy output, and stores and forwards messages.

telex: text message service provided by communicating teletypewriters and teleprinters.

terminal: device used to input and display data and information. It may have native computing and data processing capabilities

terminal adapter: provides the interface and protocol conversions needed for TE2 to connect to NT1.

terminal multiplexer: end-point or terminating device that connects originating or terminating electrical traffic to SONET. Converts to/from non-SONET channels to SONET channels. Has only one network connection.

text mail: intercommunication, half-duplex record service. The text message is composed at, and delivered by, data terminals; it may be stored in a central system of electronic mailboxes, or stored at one or other of the terminals serving the persons involved. When the intended recipients command, the messages are forwarded and displayed on a video terminal, or printed.

The Internet Society: international organization that promotes cooperation and coordination. Includes Internet Activities Board, Internet Engineering Task Force, and Internet Assigned Numbers Authority. They are concerned with network architecture, the evolution of TCP/IP protocols, and assigning IP addresses.

throughput: rate at which user's data bits are delivered—i.e., the number of user's data bits delivered in unit time.

tie line: nonswitched line directly connecting two network facilities (e.g., PBXs, data centers, etc.) for the exclusive, full-time use of those that lease it.

time-division duplex: over the same channel, messages are sent back and forth, a time-slot at a time.

time division multiple access: technique that assigns time slots to stations in the network to create individual digital channels.

time division multiplexing: to each signal the entire bandwidth of the bearer is assigned for very short periods of time so that multiplexed signals exist in series, at different times, on the same bearer.

time-slot interchange process: action of a time-switching stage that, on instructions from the stored program control, processes octets in incoming frames to produce outgoing frames on the same highway that contain the same octets in a different order.

time-slot synchronization: state of transmission facility when receiver can identify time slots within the frame and correctly allocate them to each channel.

time-space-time switch: combination of time-switching stages and a space-switching stage to serve a large customer community. The space stage interconnects the time stage serving the calling party with the time stage serving the called party.

time to dial-tone: time between the moment the user goes off-hook, and the moment dial tone is returned from the local switch.

time to disconnect: time between the moment the user goes on-hook, and the moment the circuit is available to others.

token: supervisory frame that grants access to the bearer so that a station may send.

token bit: in an IEEE 802.5 LAN, bit that gives the token status. If it is 0, a station that has sufficient priority may seize the token. If it is 1, the token has been seized by another station and the frame is in use.

token-ring local area network: LAN operating under IEEE 802.5 in which each station is connected to two others to form a single-thread loop that connects all of the stations. Data are transferred around the ring from station to station.

topology: concerned with the ways in which nodes and links are interconnected.

touchpad: number pad on telephone instrument that generates bursts of two frequencies when a button is pressed to enter a digit.

traditional DS–3: DS–3 signal is partitioned in frames containing 4,760 bits. Called M-frames, they consist of seven subframes of 680 bits each. The subframes are divided into eight blocks of 85 bits. In these blocks, the first bit is a control bit and the other 84 are payload bits. Each frame includes 54 overhead bits that are used to perform frame alignment, subframe alignment, performance monitoring, alarm and status reporting, and other functions.

traffic density: number of calls in progress at a given moment.

traffic intensity: traffic density averaged over a one-hour period.

transaction capabilities part: SS7 protocol that provides a set of tools in a connectionless environment that can be used by an application at one node to invoke execution of a procedure at another node, and exchange the results.

transaction terminal: device designed for capturing and processing specific on-line transactions. Examples are automatic teller machine and supermarket checkout terminal with bar-code reader.

transducer: device that converts messages from one medium to another.

transfer mode: manner of transfer of information and overhead bits between (among) users. Characterized by the switching/routing technique employed and the transmission technique that supports it. The transfer mode may be asynchronous (i.e., the clocks in the transmitter and receiver need not be synchronized) or synchronous (i.e., the clocks in the transmitter and receiver are synchronized).

transient function: function that exists for a limited time.

transmission control layer: level 4 in SNA, responsible for the synchronization and pacing of session-level data traffic. Performs enciphering and deciphering of data if required.

transmission control protocol: Internet transmission layer procedure.

transmission convergence sublayer: implements physical layer functions that are independent of the transmission medium (e.g., containers, sequence numbers).

transmission design: in the local loop, inductance is added to long loops to reduce the voice band attenuation of the connection so that the parties can hear each other.

transponder: amplifying and frequency changing device carried in a satellite that operates over a portion of the frequency band assigned to the satellite.

transport header: in OSIRM, contains the sequence number, and an FCS that checks the entire segment.

transport layer: layer 4 of OSI model. Maintains the integrity of end-to-end communication independent of the performance and number of networks involved in the connection between end systems. Serves as interface between the three lower layers that are concerned with providing transparent connections between users, and the three upper layers that ensure that data are delivered in correct and understandable form.

transport overhead: in SONET, in an STS–1 frame, 9 rows x 3 columns (27 octets) are occupied by transport overhead. It is made up of 9 octets (3 rows x 3 columns) of section overhead and 18 octets (6 rows x 3 columns) of line overhead.

transport protocol data unit: in OSIRM, combination of transport header and session PDU.

transposition: in encryption, action in which the order of the bit patterns, characters, words, or phrases is rearranged.

trellis code modulation: special form of QAM that incorporates twice as many signal points in the constellation as are needed to represent the data. This redundancy is used to reduce errors so that TCM modems are able to operate at higher speeds than their non-TCM counterparts.

tributary group: in SDH, several tributary units.

tributary signal: lower-speed signal consisting of a synchronized stream of octets that is incorporated in a higher-speed multiplexed stream.

tributary unit: in SDH, combination of virtual container and pointer information for a tributary signal.

tunnel: arrangement in which an outsider and the firewall use public key encryption to encipher the packets they exchange. Thus, the information is secure no matter what network the outsider chooses to use to connect to the firewall, and the firewall can authenticate the outsider. The process is known as tunneling.

two-wire connection: a pair of wires that constitutes a duplex circuit and carries information in both directions. Connects the user's terminal to the serving central office.

unacknowledged connectionless service: the receiving station does not acknowledge messages.

unchannelized T–1: for transmission over a DS–1 line, the bitstream is divided into segments of 192 bits to which framing bits are attached. Neither the sending nor the receiving equipment makes use of the boundaries that normally define channels.

unipolar signal format: a current (positive or negative) represents a 1 and a 0 is represented by the absence of a current.

universal digital channel: 2B1Q digital subscriber line equipment that implements six digital circuits on a single pair of wires for circuit expansion purposes.

universal numbering plan: worldwide numbering plan in which customer stations are identified by subscriber codes, exchange codes, area codes, and country codes.

universal personal telecommunication: ubiquitous telecommunication capability with which the users can communicate across town, or around the world, from anywhere. UPT services provide a subscriber with a unique personal number that can be used across any number of networks and with various access arrangements.

universal service requirement: within their operating areas, common carrier communication companies must serve all customers on an equal basis.

universal synchronous/ asynchronous receiver transmitter: integrated circuit chip that implements a start-stop DLC protocol.

unlocked: in network management, administrative state in which the managed object can be used.

upstream direction: direction in which the less rich messages flow.

user agents: in X.400, entities that interact directly with users and message transfer agents. They prepare messages, submit messages for routing and transmission to destinations, accept messages, deliver messages to users, and provide filing, retrieving, forwarding, and other message support functions.

user datagram protocol: connectionless Internet network layer procedure.

user-friendly: by most persons, the rules governing the operation of user-friendly services and equipment can be learned readily, and they cause little or no increase in anxiety level in users.

user-hostile: by most persons, the rules governing the operation of user-hostile services and equipment are difficult to learn, and they cause a significant increase in anxiety level in users.

user network interface standard: defines the forms, formats, and sequences of messages that flow between users and communication networks across user-network interfaces.

user-seductive: most persons are fascinated by the performance of user-seductive services and equipment, and are eager to use them. They dispel anxiety and produce pleasure in the user.

value-added carrier: besides transport, a value-added carrier provides additional services—e.g., error detection and correction.

vertical redundancy check: parity bits are added to individual characters.

very high data rate digital subscriber line: transmits downstream at 12.96, 25.82, or 51.84 Mbits/s. An upstream digital channel has speeds between 1.6 Mbits/s and 51.84 Mbits/s, and there is a 4-kHz duplex voice channel. Over a 24-gauge copper pair, service is limited to 1,000 feet at 51.84 Mbits/s.

very high performance Backbone Network Services: called vBNS, the backbone network consists of 12 nodes collocated with MCI's 622.08 Mbits/s national ATM network. Each node contains an ATM switch (operating at 622.08 Mbits/s), one or more IP (and some ATM) routers, ports that operate at DS–3, OC–3, and OC–12 for user-network access, and an OC–12 node-network connection. The nodes are fully connected by permanent virtual paths on SONET fibers operating at OC–12. Some 100 institutions, national laboratories, and supercomputer centers are connected.

very-small-aperture terminal network: comprised of a large number of geographically dispersed terrestrial microstations that, through a geostationary satellite, are connected to hub stations, or connected among themselves. Over a VSAT network, data communications can be provided among the scattered locations that comprise a retail chain, a major manufacturing firm, or a national distribution organization.

vestigial-sideband amplitude modulation: derived from LCAM, and approximating the bandwidth efficiency of SSB, in VSB one sideband is passed almost intact together with a carrier signal and a *vestige* of the other sideband. This can be achieved by a bandpass filter designed to have odd symmetry about the carrier frequency and a relative response of 50% at the carrier frequency. The power at the carrier frequency simplifies signal detection.

video conference: interactive, intercommunication, real-time service that may be augmented by graphics and text devices. Video, audio, and text space is used to connect remote conference rooms so that the sights, sounds, and symbols employed by their occupants are available for all to experience.

videoseminar: interactive, one-to-many, real-time service with unequal richness in the downstream and upstream directions.

videotex: electronic encyclopedia. A mass communication, information retrieval service that employs the telephone network in an interactive way to search through databanks containing thousands of pages of information.

viewdata: with the generic name wired-videotex, an information retrieval service that provided an electronic encyclopedia. The service has been discontinued.

virtual circuit: in a packet network, logical connection between DTEs that is defined by a VCI.

virtual circuit service: commonly provided over a packet network for segmented data messages (packets) that are not sensitive to variations in intersegment delays but must arrive at the destination in sequence. Delays between packets may vary. Each packet contains originating and terminating addresses that are used to send them toward the destination over the same link as the previous packets were sent. An initial packet may be sent to establish the virtual circuit, or the necessary informa-

tion may be stored with the routing information maintained at each node.

virtual container: in SDH, combination of container and path overhead for a tributary signal, or that part of the payload that contains several tributary groups.

virtual private network: network created out of carrier equipment that is operated as a private network with the special capabilities and performance required by the leasing corporation, but without the burden of investment, maintenance, and disaster recovery. It is configured to provide users with reports on calls, equipment usage, configuration status, and other information that is required to manage corporate telecommunications activities.

virtual path: consists of physical resources (links) that are shared among the virtual circuits it supports.

Virtual Telecommunications Access Method: in SNA, an access method that directly controls transmission of data to and from channel-attached devices. In an MVS environment, manages a communications network for the exclusive use of the applications it supports. To send data to, and receive data from, remote devices over SNA links, VTAM requires the use of Network Control Program (NCP).

virtual tributary: in SONET, a fixed number of octets that occupy 9 rows xn columns in the SPE.

visitor location register: in mobile telephone network, database that maintains and updates a visitor's location and service profile for local use.

Viterbi algorithm: implements maximum-likelihood decoding of convolutional codes. Employs an algorithm that limits searching the possible logical paths to a manageable number by creating a group of minimum (surviving) paths.

vocoding: source coding technique in which the speech signal is divided into several separate frequency bands (e.g., 16 bands of 200 Hz each). The magnitude of the energy in each band is transmitted to the receiver together with information on the presence or absence of pitch. At the receiver, voiced or unvoiced energy is applied to a set of receivers that produce recognizable speech.

voice mail: intercommunication, half-duplex record service. Recorded using telephone connections to voice storage devices (tapes or solid-state memories), voice messages are delivered on commands from the intended recipients. Delivery may be by the storage devices directly (answering machines at the recipients' telephones, for instance), or over telephone connections initiated by the recipients.

VT1.5: in SONET, virtual tributary for DS–1. It consists of 27 octets (9 rows x 3 columns, equivalent to 1.728 Mbits/s).

VT2: in SONET, virtual tributary for E–1. It consists of 36 octets (9 rows x 4 columns, equivalent to 2.304 Mbits/s).

VT3: in SONET, virtual tributary for DS–1C. It consists of 54 octets (9 rows x 6 columns).

VT6: in SONET, virtual tributary for DS–2. It consists of 108 octets (9 rows x 12 columns).

WATS line: connection that supports wide-area telephone service.

wavelength division multiplexing: to make full use of its bandwidth, several optical carriers are transmitted simultaneously in the same fiber.

white noise: random signal in which power is distributed uniformly at all frequencies. White noise has the property that different samples are uncorrelated, and that, if the probability density function of the amplitude spectrum is Gaussian, they are statistically independent. White, Gaussian noise is truly random.

wide-area network: star or hierarchically connected network that interconnects widely scattered facilities, usually for a private purpose.

wide-area telephone service: service that offers bulk rates for station-to-station calls that are dialed directly. Calling categories are intrastate and interstate, and incoming or outgoing.

wideband signal: analog signal with frequency components that exceed the nominal passband of a voice channel.

wide-sense stationary process: the expectation of the random process is constant, and the autocorrelation function depends only on the time interval over which it is evaluated.

wire center: physical location at which many cables come together to be served by a switching facility that spans several exchange codes.

wireless information network: network that supports communication between parties, at least one of which is not restricted to remain in close proximity to a fixed terminal. A network to which access is gained through wireless means. A network that supports communication between mobiles or between a mobile and a fixed facility.

workstation: device that contains significant processing power in its own right, but nevertheless needs to access other processors.

World System Teletext: ITU–T standard that includes most national teletext systems.

World Wide Web Consortium: industry consortium managed by MIT's Laboratory for Computer Science; develops common standards for the evolution of the World Wide Web.

X.3: ITU–T Recommendation that defines a PAD.

X.25: ITU–T Recommendation that defines a user-network interface procedure for a synchronous DTE to connect to a packet network. The protocol stack at the UNI has three layers: **packet layer**, or **X.25-3 layer** (network layer); **X.25-2 layer** (data link layer); **X.25-1 layer** (physical layer).

X.25 packet network: X.25 packet-switched networks multiplex packets of data across links between store-and-forward nodes. Users employ X.25 protocols to connect to the network. Within the network, the nodes route packets in accordance with information carried in individual packets, information stored in the node, and the state of the network.

X.28: ITU–T recommendation that defines the procedures whereby an asynchronous DTE is connected to a PAD.

X.29: ITU–T recommendation that defines the procedures that permit a packet-mode device to control the operation of the PAD.

X.75: ITU–T's recommendation that defines way in which packet networks are interconnected through STEs.

X.400: family of recommendations for message-handling systems prepared by ITU–T.

zero-byte time slot interchange: in a T–1 system carrying data, technique used to provide 64 kbits/s clear channels to the user yet maintain the 1s density requirements.

INDEX

About the Author

Born and educated in England, Bryan Carne received a Ph.D in Electrical Engineering from the University of London in 1952. Moving to Remington Rand's Advanced Development Laboratories in Norwalk, CT, he began his professional career working on problems associated with early Univac computers. A period of pilot production and manufacturing of proprietary devices followed. For ten years beginning in 1959, he worked in project engineering in manager, director, and general manager roles, for contractors associated with military communications and intelligence collection programs.

In 1969, Bryan Carne returned to civilian pursuits. He completed the Advanced Management Program at the Graduate School of Business at Harvard

University, Cambridge, MA and joined GTE Laboratories in Waltham, MA. There, he directed the Laboratories' telecommunications programs including proposals to create wired-cities, support of the AT&T–GTE Domestic Satellite System, development of the first optical fiber transmission system to carry network traffic, and software engineering and computer-aided design for telecommunications applications.

In 1986, Dr. Carne moved to academia as Visiting Professor of Electrical Engineering at Northeastern University in Boston, MA and later as BellSouth Visiting Professor of Telecommunications at Christian Brothers University in Memphis, TN. He is the author of thirty technical articles, four books on telecommunications, and an early book on artificial intelligence. He is a Life Senior Member of IEEE.

Today, Bryan Carne lives in Peterborough, NH. He divides his time between writing, hiking the mountain trails, and his grandchildren.